SCHOLAR Study Guide

Advanced Higher Mathe...
Course materials:
Part 1: Topics 1 to 4

Authored by:
Fiona Withey (Stirling High School)
Karen Withey (Stirling High School)

Reviewed by:
Margaret Ferguson

Previously authored by:
Jane S Paterson
Dorothy A Watson

Heriot-Watt University
Edinburgh EH14 4AS, United Kingdom.

First published 2019 by Heriot-Watt University.

This edition published in 2019 by Heriot-Watt University SCHOLAR.

Copyright © 2019 SCHOLAR Forum.

Members of the SCHOLAR Forum may reproduce this publication in whole or in part for educational purposes within their establishment providing that no profit accrues at any stage, Any other use of the materials is governed by the general copyright statement that follows.

All rights reserved. No part of this publication may be reproduced, stored in a retrieval system or transmitted in any form or by any means, without written permission from the publisher.

Heriot-Watt University accepts no responsibility or liability whatsoever with regard to the information contained in this study guide.

Distributed by the SCHOLAR Forum.

SCHOLAR Study Guide Advanced Higher Mathematics: Course Materials: Topics 1 to 4

Advanced Higher Mathematics Course Code: C847 77

ISBN 978-1-911057-76-5

Print Production and Fulfilment in UK by Print Trail www.printtrail.com

Acknowledgements

Thanks are due to the members of Heriot-Watt University's SCHOLAR team who planned and created these materials, and to the many colleagues who reviewed the content.

We would like to acknowledge the assistance of the education authorities, colleges, teachers and students who contributed to the SCHOLAR programme and who evaluated these materials.

Grateful acknowledgement is made for permission to use the following material in the SCHOLAR programme:

The Scottish Qualifications Authority for permission to use Past Papers assessments.

The Scottish Government for financial support.

The content of this Study Guide is aligned to the Scottish Qualifications Authority (SQA) curriculum.

All brand names, product names, logos and related devices are used for identification purposes only and are trademarks, registered trademarks or service marks of their respective holders.

Contents

1	Partial fractions	1
	1.1 Looking back	3
	1.2 Introduction to partial fractions	8
	1.3 Linear factors	11
	1.4 Repeated factors	14
	1.5 Irreducible quadratic factors	20
	1.6 Algebraic long division	25
	1.7 Reduce improper rational functions by division	28
	1.8 Learning points	33
	1.9 Extended information	34
	1.10 End of topic test	35
2	Differentiation	37
	2.1 Looking back	41
	2.2 Conditions for differentiability	74
	2.3 The product rule	88
	2.4 The quotient rule	92
	2.5 Differentiate cot(x), sec(x), cosec(x) and tan(x)	96
	2.6 Differentiate exp(x)	98
	2.7 Logarithmic differentiation	102
	2.8 Implicit differentiation	108
	2.9 Differentiating inverse functions	125
	2.10 Differentiation using the product, quotient and chain rules	138
	2.11 Parametric differentiation	141
	2.12 Related rate problems	155
	2.13 Displacement, velocity and acceleration	163
	2.14 Curve sketching in closed intervals	168
	2.15 Optimisation and other applications	169
	2.16 Learning points	175
	2.17 Proofs	179
	2.18 Extended information	180
	2.19 End of topic test	184
3	Integration	191
	3.1 Looking back	195
	3.2 Integrate using standard results	216
	3.3 Integration by substitution	221
	3.4 Integration involving inverse trigonometric functions and completing the square	235
	3.5 Partial fractions and integration	243
	3.6 Integration by parts	252

3.7	Area between a curve and an axis	260
3.8	Volume of revolution	269
3.9	Learning points	276
3.10	Extended information	279
3.11	End of topic test	281

4 Differential equations — 289

4.1	Looking back	291
4.2	First order linear differential equations introduction	301
4.3	First order linear differential equations with separable variables	308
4.4	Integrating factor	330
4.5	Second order linear differential equations with constant coefficients	344
4.6	Learning points	374
4.7	Proofs	377
4.8	Extended information	379
4.9	End of topic test	385

Glossary — 391

Hints for activities — 396

Answers to questions and activities — 398

Topic 1

Partial fractions

Contents

1.1 Looking back	3
1.1.1 Division by (x - a)	4
1.1.2 Factor theorem	6
1.1.3 Factorising polynomials	6
1.2 Introduction to partial fractions	8
1.3 Linear factors	11
1.4 Repeated factors	14
1.5 Irreducible quadratic factors	20
1.6 Algebraic long division	25
1.7 Reduce improper rational functions by division	28
1.8 Learning points	33
1.9 Extended information	34
1.10 End of topic test	35

TOPIC 1. PARTIAL FRACTIONS

Learning objective

By the end of this topic, you should be able to:

- use the method of partial fractions to express proper rational functions as a sum of partial fractions;
- use algebraic long division and factorise polynomials of up to degree 3.

1.1 Looking back

SummaryPolynomials

- A polynomial is an expression containing the sum or difference of algebraic terms with powers or the equivalent in factorised form.
  e.g. $2x^3 - 4x^2 - 3x + 10$ and $(x + 5)(x^2 - 5x - 3)$
- The degree of a polynomial is the value of the highest power.
  e.g. $2x^3 - 4x + 10$ has degree 3.
- Synthetic division is a method for factorising a polynomial.
  - $(x - a)$ is the divisor e.g. $(3x^3 + 5x + 4) \div (x - 1)$

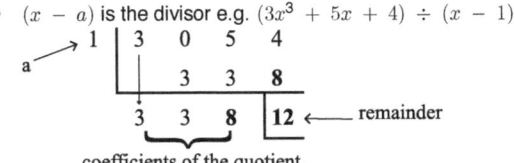

coefficients of the quotient

- If a polynomial divided by $(x - a)$ has remainder 0 then $(x - a)$ is a factor.
- If $(x - a)$ is a factor then the remainder under division by $(x - a)$ is 0.
- When trying a divisor of the form $(x - a)$ it is usually a good idea to start with $(x - 1)$.
  - If that does not work then think of the possible factors of the constant on the end of your polynomial.
  - Be systematic and don't rub out any attempts that do not work.
- Solving a polynomial is best done in factorised form and allows you to identify the roots (i.e. the places where the graph of the polynomial crosses the x-axis).
- To determine the equation of a polynomial from its graph:
  - use the roots to determine the factors e.g. roots a, b, c
    give factors $(x - a)(x - b)(x - c)$;
  - remember the polynomial may have a scalar k
    e.g. $y = k(x - a)(x - b)(x - c)$;
  - substitute the coordinates of the y-intercept to determine k;

1.1.1 Division by (x - a)

We can divide polynomials using the method of synthetic division.

> **Key point**
>
> **The Remainder Theorem**
>
> If a polynomial $f(x)$ is divided by $(x - a)$ the remainder is $f(a)$.

Example

Problem:

$f(x) = 3x^3 + 5x + 4$
Use synthetic division to find $(3x^3 + 5x + 4) \div (x - 1)$.

Solution:

$(x - a)$ is the divisor and in this example $a = 1$. Using synthetic division we will be able to find the quotient and the remainder.

Also, notice that we do not have a squared term so we must interpret this in a polynomial as $0x^2$. The polynomial should be interpreted as $3x^3 + 0x^2 + 5x + 4$.

```
 1 | 3   0   5   4
a  |     3   3   8
   | 3   3   8  |12|  ← remainder
     _____/
  coefficients of the quotient
```

So under division by $(x - 1)$ the quotient is $3x^2 + 3x + 8$ and the remainder is 12.

We can express the function as $f(x) = (x - 1)(3x^2 + 3x + 8) + 12$.

> **Key point**
>
> When $f(x)$ is divided by $(x - a)$ we can say
>
> $$f(x) = (x - a)Q(x) + R$$
>
> where $Q(x)$ is the quotient and R is the remainder.

Examples

1. Problem:

If $f(x) = 2x^4 + x^2 - x + 1$, divide $f(x)$ by $(x + 1)$.

Solution:

We must interpret the function as $2x^4 + 0x^3 + x^2 - x + 1$.
We can only divide by $(x - a)$ so we interpret the divisor as $(x - (-1))$.

TOPIC 1. PARTIAL FRACTIONS

This gives,

```
-1 | 2    0    1   -1    1
   |     -2    2   -3    4
     ‾2‾‾‾-2‾‾‾3‾‾‾-4‾|‾5
```

So the quotient is $2x^3 - 2x^2 + 3x - 4$ and the remainder is 5.

Hence $f(x) = (x + 1)(2x^3 - 2x^2 + 3x - 4) + 5$.

....

2. Problem:

If $f(x) = 2x^3 + 5x^2 - x - 1$, divide $f(x)$ by $(2x - 1)$.

Solution:

We can only divide by $(x - a)$ so we interpret the divisor as $2(x - 1/2)$.

This gives,

```
½ | 2    5   -1   -1
  |      1    3    1
    ‾2‾‾‾6‾‾‾2‾|‾0
```

So the quotient is $2x^2 + 6x + 2$ and the remainder is 0

So $f(x) = (x - 1/2)(2x^2 + 6x + 2)$ but we have to take the common factor of 2 out of the quotient and put it back into the divisor.

$$\left(x - \frac{1}{2}\right)(2x^2 + 6x + 2) = \left(x - \frac{1}{2}\right) \times 2\left(x^2 + 3x + 1\right)$$
$$= 2\left(x - \frac{1}{2}\right)\left(x^2 + 3x + 1\right)$$

giving $f(x) = (2x - 1)(x^2 + 3x + 1)$.

Division by (x - a) exercise   Go online

Q1: What is the remainder when $(4x^2 - 10x + 2) \div (x - 3)$?

....

Q2: What is the remainder when $(5x^3 - 7x^2 + 14) \div (x - 3)$?

....

Q3: What is the remainder when $(3x^2 - 8x + 4) \div (x + 3)$?

....

Q4: Express $(6x^3 + 7x^2 - 1) \div (3x - 1)$ in the form $(3x - 1)Q(x) + R$, where $Q(x)$ is the quotient and R is the remainder.

© HERIOT-WATT UNIVERSITY

1.1.2 Factor theorem

The Factor Theorem states that if $f(a) = 0$ then $(x - a)$ is a factor of $f(x)$ and if $(x - a)$ is a factor of $f(x)$ then $f(a) = 0$.

> **Key point**
>
> If a polynomial divided by $(x - a)$ has remainder 0 then $(x - a)$ is a factor and if $(x - a)$ is a factor then the remainder under division by $(x - a)$ is equal to 0.

Example

Problem:

Is $(x - 4)$ a factor of $f(x)$ where $f(x) = x^3 + 7x^2 - 26x - 72$?

Solution:

```
4 | 1   7   -26  -72
  |     4    44   72
    1  11    18 | 0
```

Since the remainder is 0, $(x - 4)$ is a factor of $f(x)$ and $f(x) = (x - 4)(x^2 + 11x + 18)$.

Notice that we can factorise the quotient and $f(x) = (x - 4)(x + 2)(x + 9)$ which gives the function in its fully factorised form.

Factor theorem exercise   Go online

Q5: Is $(x - 1)$ a factor of $f(x)$ where $f(x) = x^4 + 2x^3 - 7x^2 - 8x + 12$?

....

Q6: Is $(x - 2)$ a factor of $f(x)$ where $f(x) = x^3 + 3x^2 - 4x - 12$?

1.1.3 Factorising polynomials

> **Key point**
>
> Synthetic division can be used to help fully factorise a polynomial.

Example

Problem:

Factorise fully $x^4 + 2x^3 - 7x^2 - 8x + 12$.

Solution:

We are not given a divisor for this example so we have to use trial and error to find the first factor. The simplest divisor to try is $(x - 1)$ giving,

TOPIC 1. PARTIAL FRACTIONS

```
1 | 1   2   -7   -8   12
  |     1    3   -4  -12
  ---------------------
    1   3   -4  -12 | 0
```

Since the remainder is 0, $(x - 1)$ is a factor of
$x^4 + 2x^3 - 7x^2 - 8x + 12 = (x - 1)(x^3 + 3x^2 - 4x - 12)$.
Now we need to factorise the quotient $x^3 + 3x^2 - 4x - 12$. We could try $(x - 1)$ again but let's try $(x - 2)$ giving,

```
2 | 1   3   -4  -12
  |     2   10   12
  ------------------
    1   5    6 | 0
```

Since the remainder is 0, $(x - 2)$ is a factor of
$x^3 + 3x^2 - 4x - 12 = (x - 2)(x^2 + 5x + 6)$.

So it follows that $x^4 + 2x^3 - 7x^2 - 8x + 12 = (x - 1)(x - 2)(x^2 + 5x + 6)$.

Now all we have to do is try to factorise $x^2 + 5x + 6$... and so,
$$x^2 + 5x + 6 = (x + 2)(x + 3)$$
$$x^4 + 2x^3 - 7x^2 - 8x + 12 = (x - 1)(x - 2)(x + 2)(x + 3)$$

> **Key point**
>
> When trying a divisor of the form $(x - a)$ it is usually a good idea to start with $(x - 1)$. If that does not work then think of the possible factors of the constant on the end of your polynomial.
>
> For the constant -12 the factors and divisors are:
>
> | | |
> |---|---|
> | 1 and -12 | $(x - 1)$ and $(x + 12)$ |
> | 2 and -6 | $(x - 2)$ and $(x + 6)$ |
> | 3 and -4 | $(x - 3)$ and $(x + 4)$ |
> | 4 and -3 | $(x - 4)$ and $(x + 3)$ |
> | 6 and -2 | $(x - 6)$ and $(x + 2)$ |
> | 12 and -1 | $(x - 12)$ and $(x + 1)$ |
>
> Be systematic. Don't rub out any attempts that do not work just put a line through them. This will allow you to see what you have tried.

Example

Problem:

Fully factorise $x^3 + 2x^2 - 5x - 6$.

Solution:

For the constant -6 the factors and divisors are:

8   TOPIC 1. PARTIAL FRACTIONS

1 and -6	$(x - 1)$ and $(x + 6)$
2 and -3	$(x - 2)$ and $(x + 3)$
3 and -2	$(x - 3)$ and $(x + 2)$
6 and -1	$(x - 6)$ and $(x + 1)$

Try $(x - 1)$.

Evaluate $f(1)$:

```
1 | 1   2   -5   -6
  |     1    3   -2
  ─────────────────
    1   3   -2  |-8
```

Since the remainder $\neq 0$, $(x - 1)$ is not a factor.

Try $(x + 1)$.

Evaluate $f(-1)$:

```
-1 | 1   2   -5   -6
   |    -1   -1    6
   ──────────────────
     1   1   -6  | 0
```

Since the remainder = 0, $(x + 1)$ is a factor and
$$\begin{aligned} x^3 + 2x^2 - 5x - 6 &= (x + 1)\left(x^2 + x - 6\right) \\ &= (x + 1)(x + 3)(x - 2) \end{aligned}$$

Factorising polynomials exercise   Go online

Q7: Fully factorise $x^3 - 3x + 2$.

....

Q8: Fully factorise $2x^3 - 3x^2 - 11x + 6$.

1.2   Introduction to partial fractions

There are some very complex looking algebraic equations. To try to differentiate or integrate them as they stand would be very difficult. In this section methods for splitting them into manageable terms are investigated.

The following definitions will help to make this section clearer.

TOPIC 1. PARTIAL FRACTIONS

> **Key point**
>
> If $P(x) = a_n x^n + a_{n-1} x^{n-1} + a_{n-2} x^{n-2} + \ldots + a_2 x^2 + a_1 x^1 + a_0$
>
> where $a_0, \ldots, a_n \in \mathbb{R}$ then P is a *polynomial of degree n*.

Example

$x + 3$ has degree 1;

$x^2 - 2x + 3$ has degree 2;

$4x^3 + 2x^2 - 5$ has degree 3;

a constant such as 7 has degree 0

> **Key point**
>
> If $P(x)$ and $Q(x)$ are polynomials then $\frac{P(x)}{Q(x)}$ is called a *rational function*.

> **Key point**
>
> Let $P(x)$ be a polynomial of degree n and $Q(x)$ be a polynomial of degree m.
>
> If $n < m$ then $\frac{P(x)}{Q(x)}$ is a *proper rational function*.
>
> For example,
> $\frac{x^2 + 2x - 1}{x^3 - 3x + 4}$

> **Key point**
>
> Let $P(x)$ be a polynomial of degree n and $Q(x)$ be a polynomial of degree m.
>
> If $n \geq m$ then $\frac{P(x)}{Q(x)}$ is an *improper rational function*.
>
> For example,
> $\frac{x^3 + 2x - 1}{x^2 - 4}$ or $\frac{x^2 + 2x - 1}{x^2 - 4}$

An improper rational function can always be expressed as a polynomial plus a proper rational function (by using algebraic long division).

$$\underbrace{\frac{x^3 + 2x - 3}{x + 2}}_{\text{Improper Rational Function}} = \underbrace{x^2 - 2x + 6}_{\text{Polynomial}} - \underbrace{\frac{15}{x + 2}}_{\text{Proper Rational Function}}$$

© HERIOT-WATT UNIVERSITY

10 TOPIC 1. PARTIAL FRACTIONS

Note: The process of algebraic long division is covered in a later section in this topic.

We already know how to simplify algebraic fractions by finding a common denominator e.g.
$\frac{1}{1+x} + \frac{1}{x+2} = \frac{(x+2)+(1+x)}{(1+x)(x+2)} = \frac{2x+3}{(1+x)(x+2)}$

The opposite process, for example, expressing $\frac{2x+3}{(1+x)(x+2)}$ as $\frac{1}{1+x} + \frac{1}{x+2}$, is called putting proper rational functions into **partial fractions**.

> **Key point**
>
> The process of taking a proper rational function and splitting it into separate terms each with a factor of the original denominator as its denominator is called expressing the function as partial fractions.

The way in which the rational function splits up depends on whether the denominator is a quadratic equation or a cubic equation.

It also depends on whether this denominator has linear, repeated linear or quadratic factors (with no real roots).

The different ways in which a **proper rational function** with a denominator of degree at most three can be split into partial fractions are now explained.

Examples

1. Linear factors

Regardless of whether the denominator is a quadratic or a cubic, if it can be factorised into distinct factors that are linear (have a degree of 1) then its partial fractions take the following form:

This has a denominator of a quadratic with two distinct factors.
$\frac{\cdots}{(x+a)(x+b)} \equiv \frac{A}{(x+a)} + \frac{B}{(x+b)}$

This has a denominator of a cubic with three distinct factors.
$\frac{\cdots}{(x+a)(x+b)(x+c)} \equiv \frac{A}{(x+a)} + \frac{B}{(x+b)} + \frac{C}{(x+c)}$

....

2. Repeated linear factors

This has a denominator of a quadratic with two repeated factors.
$\frac{\cdots}{(x+a)^2} \equiv \frac{A}{(x+a)} + \frac{B}{(x+a)^2}$

This has a denominator of a cubic with three repeated factors.
$\frac{\cdots}{(x+a)^3} \equiv \frac{A}{(x+a)} + \frac{B}{(x+a)^2} + \frac{C}{(x+a)^3}$

....

3. Irreducible quadratic factors

This has an irreducible quadratic denominator (i.e. the denominator has no real roots).
$\frac{\cdots}{x^2+bx+c} \equiv \frac{Ax+B}{x^2+bx+c}$

© HERIOT-WATT UNIVERSITY

TOPIC 1. PARTIAL FRACTIONS

This has a cubic denominator. One factor is linear and the other is an irreducible quadratic.

$$\frac{\ldots}{(x+a)(x^2+bx+c)} \equiv \frac{A}{x+a} + \frac{Bx+C}{x^2+bx+c}$$

1.3 Linear factors

Before we begin the process of expressing a function as a sum of partial fractions we must ensure that we are working with a **proper rational function**, where the degree of the numerator is less than the degree of the denominator. If this is not the case, algebraic long division must be carried out before the process of partial fraction is applied.

Note: The process of algebraic long division is covered in a later section in this topic.

> **Key point**
>
> A linear polynomial is one which has at most a degree of 1, graphically this would be represented by a straight line.

The following examples will examine proper rational functions taking the general form shown below with linear factors and constants $A, B \ldots N$ to be determined. Each constant is written over one factor from the original denominator and added to the next term. Resulting in a sum of terms on the RHS.

$$\frac{\ldots}{(x+a)(x+b)\cdots(x+n)} \equiv \frac{A}{(x+a)} + \frac{B}{(x+b)} + \cdots + \frac{N}{(x+n)}$$

Note that you will be expected to factorise both quadratic and cubic polynomials.

Examples

1. Problem:

Express in partial fraction form $\frac{7x+1}{x^2+x-6}$

Solution:

Step 1: Factorise the denominator.

$x^2 + x - 6 = (x-2)(x+3)$

Step 2: Identify the type of partial fraction. In this case it has linear factors.

$$\frac{7x+1}{(x-2)(x+3)} = \frac{A}{(x-2)} + \frac{B}{(x+3)}$$

Step 3: Obtain the fractions with a common denominator.

$$\frac{7x+1}{(x-2)(x+3)} = \frac{A(x+3)}{(x-2)(x+3)} + \frac{B(x-2)}{(x-2)(x+3)}$$

Step 4: Equate numerators since the denominators are equal.

$7x + 1 \equiv A(x+3) + B(x-2)$

© HERIOT-WATT UNIVERSITY

Step 5: Choose two values of x to find the values of the two constants.

Whilst any two values for x will work, we can make the calculations easier by choosing $x = 2$ and $x = -3$ because either A or B will be multiplied by 0.

When $x = 2$:
$$7 \times 2 + 1 = A(2 + 3) + B(2 - 2)$$
$$15 = 5A$$
$$A = 3$$

When $x = -3$:
$$7 \times (-3) = A(-3 + 3) + B(-3 - 2)$$
$$-20 = -5B$$
$$B = 4$$

So our answer is,
$$\frac{7x + 1}{x^2 + x - 6} \equiv \frac{3}{(x - 2)} + \frac{4}{(x + 3)}$$

....

2.

This example will demonstrate an alternative method to selecting values of x called equating coefficients. This method of working out the constants A, B, C, ... has advantages in certain circumstances and this will be explored in further examples.

Problem:

Express $\frac{x + 4}{(x^2 - 7x + 10)}$ in partial fractions.

Solution:

Step 1: Factorise the denominator.

$$\frac{x + 4}{(x^2 - 7x + 10)} \equiv \frac{x + 4}{(x - 2)(x - 5)}$$

Step 2: Identify the type of partial fraction. In this case it has linear factors.

Let $\frac{x + 4}{(x^2 - 7x + 10)} \equiv \frac{A}{(x - 2)} + \frac{B}{(x - 5)}$

Step 3: Obtain the fractions with a common denominator.

$$\frac{x + 4}{(x^2 - 7x + 10)} \equiv \frac{A(x - 5)}{(x - 2)(x - 5)} + \frac{B(x - 2)}{(x - 2)(x - 5)}$$

Step 4: Equate numerators since the denominators are equal.

$$x + 4 = A(x - 5) + B(x - 2) \quad (*)$$

Step 5: Select values of x and substitute into (*)

Whilst we can choose any value of x we can make the calculations easier by choosing it such that $(x - 5) = 0$ and $(x - 2) = 0$ since then A and B will be multiplied by zero, respectively.

Hence the values taken are $x = 5$ and $x = 2$

Let $x = 5$:
$$x + 4 = A(x - 5) + B(x - 2)$$
$$5 + 4 = A(5 - 5) + B(5 - 2)$$
$$9 = 3B$$
$$B = 3$$
Let $x = 2$:
$$x + 4 = A(x - 5) + B(x - 2)$$
$$2 + 4 = A(2 - 5) + B(2 - 2)$$
$$6 = -3A$$
$$A = -2$$

By either method the result follows $\frac{x + 4}{(x^2 - 7x + 10)} \equiv \frac{-2}{(x - 2)} + \frac{3}{(x - 5)}$

....

3. Problem:

Express $\frac{x^2 - 13}{x^3 - 7x + 6}$ in partial fractions.

Solution:

Step 1: The denominator can be factorised using synthetic division.
$$\frac{x^2 - 13}{x^3 - 7x + 6} \equiv \frac{x^2 - 13}{(x - 1)(x - 2)(x + 3)}$$
Step 2: Identify the type of partial fraction. In this case again it has linear factors.

Let $\frac{x^2 - 13}{(x - 1)(x - 2)(x + 3)} \equiv \frac{A}{(x - 1)} + \frac{B}{(x - 2)} + \frac{C}{(x + 3)}$

Step 3: Obtain the fractions with a common denominator.
$$\frac{x^2 - 13}{(x - 1)(x - 2)(x + 3)} \equiv \frac{A(x - 2)(x + 3)}{(x - 1)(x - 2)(x + 3)} + \frac{B(x - 1)(x + 3)}{(x - 1)(x - 2)(x + 3)} + \frac{C(x - 1)(x - 2)}{(x - 1)(x - 2)(x + 3)}$$

Step 4: Equate numerators since the denominators are equal.

$x^2 - 13 = A(x - 2)(x + 3) + B(x - 1)(x + 3) + C(x - 1)(x - 2)$ (*)

Step 5: Select values of x and substitute into (*).

Whilst we can choose any value of x we can make the calculations easier by choosing it such that $(x - 2) = 0$, $(x + 3) = 0$ and $(x - 1) = 0$ since then A, B and C will be multiplied by zero respectively.

$x^2 - 13 = A(x - 2)(x + 3) + B(x - 1)(x + 3) + C(x - 1)(x - 2)$

Let $x = 2$ then,
$$-9 = 0 + 5B + 0$$
$$-9 = 5B$$
$$B = -\frac{9}{5}$$
Let $x = 1$ then,
$$-12 = -4A$$
$$A = 3$$
Let $x = -3$ then,
$$-4 = 20C$$
$$C = -\frac{1}{5}$$

Therefore, $\dfrac{x^2 - 13}{x^3 - 7x + 6} \equiv \dfrac{3}{(x-1)} - \dfrac{9}{5(x-2)} - \dfrac{1}{5(x+3)}$

Linear factors exercise Go online

Q9: Express $\dfrac{x+7}{x^2 - x - 2}$ into partial fractions.

Q10: Express $\dfrac{7x - 4}{2x^2 - 3x - 2}$ into partial fractions.

Q11: Express $\dfrac{8x}{x^3 + 3x^2 - x - 3}$ into partial fractions.

Q12: Express $\dfrac{-11x - 34}{x^3 + 5x^2 + 2x - 8}$ into partial fractions.

1.4 Repeated factors

When factorising quadratic and cubic functions factors may be repeated. This section will examine proper rational functions taking the form below with repeated factors on the denominator. Constants $A, B, \ldots, N$ are to be determined.

$$\dfrac{\ldots}{(x+a)^n} \equiv \dfrac{A}{(x+a)} + \dfrac{B}{(x+a)^2} + \cdots + \dfrac{N}{(x+a)^n}$$

Note that you will be expected to factorise both quadratic and cubic polynomials.

Examples

1. Problem:
Express in partial fraction form $\dfrac{5x + 18}{x^2 + 8x + 16}$

Solution:

Step 1: Factorise the denominator.
$x^2 + 8x + 16 = (x+4)(x+4)$

Step 2: Identify the type of partial fraction. In this case it has repeated linear factors.
$\dfrac{5x + 18}{(x+4)^2} = \dfrac{A}{x+4} + \dfrac{B}{(x+4)^2}$

Step 3: Obtain the fractions with a common denominator.
$\dfrac{5x + 18}{(x+4)^2} = \dfrac{A(x+4)}{(x+4)^2} + \dfrac{B}{(x+4)^2}$

Step 4: Equate numerators since the denominators are equal.
$5x + 18 = A(x+4) + B$ (*)

Step 5: Select values of x and substitute into (*).

© HERIOT-WATT UNIVERSITY

TOPIC 1. PARTIAL FRACTIONS

Whilst any value for x will work, we can make the calculations easier by choosing $x = -4$ as A will be multiplied by zero.

Let $x = -4$:
$$5x + 18 = A(x + 4) + B$$
$$5(-4) + 18 = A(-4 + 4) + B$$
$$-2 = B$$
$$B = -2$$

Now substitute $B = -2$ back into (*) and select a value of x to simplify the calculation.

Let $x = 0$:
$$5x + 18 = A(x + 4) + B$$
$$5(0) + 18 = A(0 + 4) - 2$$
$$18 + 2 = 4A$$
$$A = 5$$

We have $A = 5$ and $B = -2$.

Therefore: $\dfrac{5x + 18}{x^2 + 8x + 16} = \dfrac{5}{x + 4} - \dfrac{2}{(x + 4)^2}$

....

2. Problem:

Express in partial fraction form $\dfrac{2x - 10}{x^3 + 6x^2 + 12x + 8}$

Solution:

Step 1: Factorise the denominator. Synthetic division can be used to do this.

$$x^3 + 6x^2 + 12x + 8 = (x + 2)^3$$

Step 2: Identify the type of partial fraction. In this case it has repeated linear factors.

$$\dfrac{2x - 10}{(x + 2)^3} = \dfrac{A}{(x + 2)} + \dfrac{B}{(x + 2)^2} + \dfrac{C}{(x + 2)^3}$$

Step 3: Obtain the fractions with a common denominator.

$$\dfrac{2x - 10}{(x + 2)^3} = \dfrac{A(x + 2)^2}{(x + 2)^3} + \dfrac{B(x + 2)}{(x + 2)^3} + \dfrac{C}{(x + 2)^3}$$

Step 4: Equate numerators since the denominators are equal.

$$2x - 10 = A(x + 2)^2 + B(x + 2) + C \ (*)$$

Step 5: Select values of x and equate coefficients.

First we will select a value for x and substitute into (*).

Whilst any value for x will work, we can make the calculations easier by choosing $x = -2$ as A and B will be multiplied by zero.

Let $x = -2$:
$$2x - 10 = A(x + 2)^2 + B(x + 2) + C$$
$$2(-2) - 10 = A(-2 + 2)^2 + B(-2 + 2) + C$$
$$-14 = C$$
$$C = -14$$

In this case choosing another value of x would not eliminate A or B. Instead we use the method of equating coefficients. This is done in the following way.

© HERIOT-WATT UNIVERSITY

Substitute $C = -14$ back into (*), expand the brackets and collect like terms.

$2x - 10 = A(x + 2)^2 + B(x + 2) + C$

$2x - 10 = A(x^2 + 4x + 4) + Bx + 2B - 14$

$2x - 10 = Ax^2 + 4Ax + 4A + Bx + 2B - 14$

$2x - 10 = Ax^2 + (4A + B)x + 4A + 2B - 14$

Set the coefficient in front of x^2 on the LHS equal to the coefficient in front of x^2 on the RHS. In this case there is no x^2 on the LHS. Its coefficient is therefore zero. The corresponding term on the RHS is Ax^2. The coefficient of x^2 is A. We therefore equate zero and A.

$0 = A$

Set the coefficients in front of x on the LHS equal to the coefficients in front of x on the RHS.

$2 = 4A + B$

$2 = 4(0) + B$

$B = 2$

We have $A = 0$, $B = 2$ and $C = -14$.

Therefore: $\frac{2x - 10}{x^3 + 6x^2 + 12x + 8} \equiv \frac{2}{(x + 2)^2} - \frac{14}{(x + 2)^3}$

....

3. Problem:

Express in partial fraction form $\frac{5x + 2}{x^3 + x^2}$

Solution:

Step 1: Factorise the denominator.

$x^3 + x^2 = x^2(x + 1)$

Step 2: Identify the type of partial fraction. In this case it has distinct linear and repeated linear factors.

$\frac{5x + 2}{x^2(x + 1)} = \frac{A}{x} + \frac{B}{x^2} + \frac{C}{x + 1}$

Step 3: Obtain the fractions with a common denominator.

$\frac{5x + 2}{x^2(x + 1)} = \frac{Ax(x + 1)}{x^2(x+1)} + \frac{B(x + 1)}{x^2(x+1)} + \frac{Cx^2}{x^2(x+1)}$

Step 4: Equate numerators since the denominators are equal.

$5x + 2 = Ax(x + 1) + B(x + 1) + Cx^2$ (*)

Step 5: Select values of x and substitute into (*).

Whilst any values for x will work, we can make the calculations easier by choosing $x = -1$, $x = 0$ and $x = 1$.

Let $x = -1$:

$5x + 2 = Ax(x + 1) + B(x + 1) + Cx^2$

$5(-1) + 2 = A(-1)(-1 + 1) + B(-1 + 1) + C(-1)^2$

$-3 = C$

$C = -3$

Let $x = 0$:

$$5x + 2 = Ax(x+1) + B(x+1) + Cx^2$$
$$5(0) + 2 = A(0)(0+1) + B(0+1) + C(0)^2$$
$$2 = B$$
$$B = 2$$

Now substitute $B = 2$ and $C = -3$ back into (*).

Let $x = 1$:
$$5x + 2 = Ax(x+1) + B(x+1) + Cx^2$$
$$5(1) + 2 = A(1)(1+1) + 2(1+1) - 3(1)^2$$
$$7 = 2A + 1$$
$$A = 3$$

We have $A = 3$, $B = 2$ and $C = -3$.

Therefore: $\frac{5x+2}{x^3+x^2} = \frac{3}{x} + \frac{2}{x^2} - \frac{3}{x+1}$

....

4. Problem:

Express $\frac{x+3}{x^2 - 4x + 4}$ in partial fractions.

Solution:

Step 1: Factorise the denominator.

$$\frac{x+3}{x^2 - 4x + 4} \equiv \frac{x+3}{(x-2)^2}$$

Step 2: Identify the type of partial fraction. In this case it has repeated linear factors.

$$\frac{x+3}{(x-2)^2} \equiv \frac{A}{(x-2)} + \frac{B}{(x-2)^2}$$

Step 3: Obtain the fractions with a common denominator.

$$\frac{x+3}{(x-2)^2} \equiv \frac{A(x-2)}{(x-2)^2} + \frac{B}{(x-2)^2}$$

Step 4: Equate numerators since the denominators are equal.

$x + 3 = A(x-2) + B$ (*)

Step 5: Select values of x and substitute into (*).

Let $x = 2$:
$$x + 3 = A(x-2) + B$$
$$2 + 3 = A(2-2) + B$$
$$5 = B$$
$$B = 5$$

Since we know B and have only A left to work out we can choose any value of x and substitute it along with the value of B into (*). Choose x to make the calculation easy. Here we have chosen $x = 0$.

Let $x = 0$:
$$x + 3 = A(x-2) + B$$
$$0 + 3 = A(0-2) + 5$$
$$3 = -2A + 5$$
$$-2 = -2A$$
$$A = 1$$

Therefore: $\frac{x+3}{(x-2)^2} \equiv \frac{1}{(x-2)} + \frac{5}{(x-2)^2}$

....

5. Problem:

Express $\frac{2x^2 + 5x + 3}{x^3 + 6x^2 + 12x + 8}$ in partial fractions.

Solution:

Step 1: Use synthetic division to factorise the denominator.

$\frac{2x^2 + 5x + 3}{x^3 + 6x^2 + 12x + 8} = \frac{2x^2 + 5x + 3}{(x+2)^3}$

Step 2: Identify the type of partial fraction. In this case it has repeated linear factors.

$\frac{2x^2 + 5x + 3}{(x+2)^3} = \frac{A}{(x+2)} + \frac{B}{(x+2)^2} + \frac{C}{(x+2)^3}$

Step 3: Obtain the fractions with a common denominator.

$\frac{2x^2 + 5x + 3}{(x+2)^3} = \frac{A(x+2)^2}{(x+2)^3} + \frac{B(x+2)}{(x+2)^3} + \frac{C}{(x+2)^3}$

Step 4: Equate numerators since the denominators are equal.

$2x^2 + 5x + 3 = A(x+2)^2 + B(x+2) + C$ (*)

Step 5: Select values of x and equate coefficients.

First we will select a value for x and substitute into (*).

Let $x = -2$:

$2x^2 + 5x + 3 = A(x+2)^2 + B(x+2) + C$
$2(-2)^2 + 5(-2) + 3 = A(-2+2)^2 + B(-2+2) + C$
$8 - 10 + 3 = C$
$1 = C$
$C = 1$

Now substitute $C = 1$ back into (*), expand the brackets and collect like terms.

$2x^2 + 5x + 3 = A(x+2)^2 + B(x+2) + C$
$2x^2 + 5x + 3 = Ax^2 + 4Ax + 4A + Bx + 2B + 1$
$2x^2 + 5x + 3 = Ax^2 + (4A + B)x + (4A + 2B + 1)$

Set the coefficient in front of x^2 on the LHS equal to the coefficient in front of x^2 on the RHS.

$x^2: \quad 2 = A$
$\quad\quad A = 2$

Now, set the coefficient in front of x on the LHS equal to the coefficient in front of x on the RHS.

$x: \quad 5 = 4A + B$

Now, substitute $A = 2$ and evaluate for B.

$5 = 4(2) + B$
$5 = 8 + B$
$B = -3$

We have the values of A, B and C so we do not need to equate the constants.

We have $A = 2$, $B = -3$ and $C = 1$.

TOPIC 1. PARTIAL FRACTIONS

Therefore: $\dfrac{2x^2 + 5x + 3}{x^3 + 6x^2 + 12x + 8} = \dfrac{2}{(x+2)} - \dfrac{3}{(x+2)^2} + \dfrac{1}{(x+2)^3}$

....

6. Problem:

Express $\dfrac{4x^2 + 9}{x^3 + 4x^2 - 3x - 18}$ in partial fractions.

Solution:

Step 1: Use synthetic division to factorise the denominator.

$\dfrac{4x^2 + 9}{x^3 + 4x^2 - 3x - 18} = \dfrac{4x^2 + 9}{(x-2)(x+3)^2}$

Step 2: Identify the type of partial fraction. In this case it has a distinct linear factor and repeated linear factors.

$\dfrac{4x^2 + 9}{(x-2)(x+3)^2} = \dfrac{A}{(x-2)} + \dfrac{B}{(x+3)} + \dfrac{C}{(x+3)^2}$

Step 3: Obtain the fractions with a common denominator.

$\dfrac{4x^2 + 9}{(x-2)(x+3)^2} = \dfrac{A(x+3)^2}{(x-2)(x+3)^2} + \dfrac{B(x-2)(x+3)}{(x-2)(x+3)^2} + \dfrac{C(x-2)}{(x-2)(x+3)^2}$

Step 4: Equate numerators since the denominators are equal.

$4x^2 + 9 = A(x+3)^2 + B(x-2)(x+3) + C(x-2)$ (*)

Step 5: Select values of x and equate coefficients.

First we will select a value for x and substitute into (*).

Let $x = -3$:

$4x^2 + 9 = A(x+3)^2 + B(x-2)(x+3) + C(x-2)$

$4(-3)^2 + 9 = A(-3+3)^2 + B(-3-2)(-3+3) + C(-3-2)$

$36 + 9 = C(-5)$

$45 = -5C$

$C = -9$

Let $x = 2$:

$4x^2 + 9 = A(x+3)^2 + B(x-2)(x+3) + C(x-2)$

$4(2)^2 + 9 = A(2+3)^2 + B(2-2)(2+3) + C(2-2)$

$16 + 9 = A(5)^2$

$25 = 25A$

$A = 1$

Now substitute $A = 1$ and $C = -9$ back into (*), expand the brackets and collect like terms.

$4x^2 + 9 = A(x+3)^2 + B(x-2)(x+3) + C(x-2)$

$4x^2 + 9 = (x+3)^2 + B(x-2)(x+3) - 9(x-2)$

$4x^2 + 9 = x^2 + 6x + 9 + B(x^2 + x - 6) - 9x + 18$

$4x^2 + 9 = x^2 + 6x + 9 + Bx^2 + Bx - 6B - 9x + 18$

$4x^2 + 9 = x^2 + Bx^2 + Bx - 3x - 6B + 27$

$4x^2 + 9 = (1 + B)x^2 + (B - 3)x + (-6B + 27)$

Set the coefficient in front of x^2 on the LHS equal to the coefficient in front of x^2 on the RHS.

x^2: $4 = 1 + B$
$B = 3$

We have $A = 1$, $B = 3$ and $C = -9$.

Therefore: $\frac{4x^2 + 9}{(x - 2)(x + 3)^2} = \frac{1}{(x - 2)} + \frac{3}{(x + 3)} - \frac{9}{(x + 3)^2}$

Key point

When working out the values of A, B, C... always substitute values for x first. When this is exhausted equate coefficients.

Repeated factors exercise   Go online

Q13: Express $\frac{2x + 2}{(x + 3)(x + 3)}$ in partial fractions.

Q14: Express $\frac{x^2 - 7x + 2}{x^3 - 3x^2 + 3x - 1}$ in partial fractions.

Q15: Express $\frac{-3x^2 + 6x + 20}{x^3 - x^2 - 8x + 12}$ in partial fractions.

Q16: Express $\frac{x^2 + 11x + 15}{x^3 + 3x^2 - 4}$ in partial fractions.

1.5 Irreducible quadratic factors

Not all quadratic functions can be factorised into linear factors. When the resulting quadratic cannot be factorised to give real roots it is called irreducible. This section will examine proper rational functions taking the form below with irreducible quadratic factors on the denominator. Constants A and B are to be determined.

$$\frac{...}{ax^2 + bx + c} \equiv \frac{Ax + B}{ax^2 + bx + c}$$

Note that you will be expected to factorise both quadratic and cubic polynomials.

TOPIC 1. PARTIAL FRACTIONS

Examples

1. Problem:
Express in partial fraction form $\frac{x-12}{x^3+4x}$

Solution:

Step 1: Factorise the denominator.
$$\frac{x-12}{x^3+4x} = \frac{x-12}{x(x^2+4)}$$

Step 2: Identify the type of partial fraction. In this case it has a distinct linear factor and an irreducible quadratic factor. This can be checked by working out $b^2 - 4ac$.
$b^2 - 4ac = 0^2 - 4 \times 1 \times 4 = -16$ which is less than zero therefore $x^2 + 4$ is a irreducible quadratic factor.
The partial fraction form is therefore $\frac{x-12}{x(x^2+4)} = \frac{A}{x} + \frac{Bx+C}{x^2+4}$.

Step 3: Obtain the fractions with a common denominator.
$$\frac{x-12}{x(x^2+4)} = \frac{A(x^2+4)}{x} + \frac{(Bx+C)x}{x^2+4}$$

Step 4: Equate numerators since the denominators are equal.
$x - 12 = A(x^2+4) + (Bx+C)x$ (*)

Step 5: Select values of x and equate coefficients.

First we will select a value for x and substitute into (*).

Let $x = 0$:
$0 - 12 = A(0+4) + (0B+C)0$
$-12 = 4A$
$A = -3$

Now substitute $A = -3$ back into (*), expand the brackets and collect like terms.
$x - 12 = -3(x^2+4) + (Bx+C)x$
$x - 12 = -3x^2 - 12 + Bx^2 + Cx$
$x - 12 = (-3+B)x^2 + Cx - 12$

Set the coefficients in front of x^2 on the LHS equal to the coefficients in front of x^2 on the RHS.
$0 = -3 + B$
$B = 3$

Set the coefficients in front of x on the LHS equal to the coefficients in front of x on the RHS.
$1 = C$
$C = 1$

We have $A = -3, B = 3$ and $C = 1$.
Therefore: $\frac{x-12}{x^3+4x} = \frac{-3}{x} + \frac{3x+1}{x^2+4}$

....

2. Problem:

Express $\frac{4x+1}{x^3-x^2+x-6}$ in partial fractions.

© HERIOT-WATT UNIVERSITY

Solution:
Step 1: Factorise the denominator.
$$\frac{4x+1}{x^3-x^2+x-6} = \frac{4x+1}{(x-2)(x^2+x+3)}$$
Step 2: Identify the type of partial fraction. In this case it has a distinct linear factor and an irreducible quadratic factor.
$$\frac{4x+1}{(x-2)(x^2+x+3)} = \frac{A}{x-2} + \frac{Bx+C}{x^2+x+3}$$
Step 3: Obtain the fractions with a common denominator.
$$\frac{4x+1}{(x-2)(x^2+x+3)} = \frac{A(x^2+x+3)}{(x-2)(x^2+x+3)} + \frac{(Bx+C)(x-2)}{(x-2)(x^2+x+3)}$$
Step 4: Equate numerators since the denominators are equal.
$$4x+1 = A(x^2+x+3) + (Bx+C)(x-2) \quad (*)$$
Step 5: Select values of x and substitute into (*).
Let $x=2$:
$$4x+1 = A(x^2+x+3) + (Bx+C)(x-2)$$
$$4(2)+1 = A(2^2+2+3) + (2B+C)(2-2)$$
$$9 = 9A$$
$$A = 1$$
Now substitute $A=1$ back into (*) and select a value of x to simplify the calculation.
Let $x=0$:
$$4x+1 = A(x^2+x+3) + (Bx+C)(x-2)$$
$$4(0)+1 = 1(0^2+0+3) + (0B+C)(0-2)$$
$$1 = 3 - 2C$$
$$2C = 2$$
$$C = 1$$
Now substitute $A=1$ and $C=1$ back into (*) and select another value of x to simplify the calculation.
Let $x=1$:
$$4x+1 = A(x^2+x+3) + (Bx+C)(x-2)$$
$$4(1)+1 = 1(1^2+1+3) + (1B+1)(1-2)$$
$$5 = 5 - B - 1$$
$$B = -1$$
We have $A=1$, $B=-1$ and $C=1$.
Therefore: $\frac{4x+1}{(x-2)(x^2+x+3)} = \frac{1}{x-2} + \frac{-x+1}{x^2+x+3}$

....

3. Problem:

Express $\frac{-7x+5}{x^3-x^2+x+3}$ in partial fractions.

Solution:

Step 1: Use synthetic division to factorise the denominator.
$$\frac{-7x+5}{x^3-x^2+x+3} = \frac{-7x+5}{(x+1)(x^2-2x+3)}$$

TOPIC 1. PARTIAL FRACTIONS

Step 2: Identify the type of partial fraction. In this case it has a distinct linear factor and an irreducible quadratic factor.

$$\frac{-7x + 5}{(x + 1)(x^2 - 2x + 3)} = \frac{A}{x + 1} + \frac{Bx + C}{x^2 - 2x + 3}$$

Step 3: Obtain the fractions with a common denominator.

$$\frac{-7x + 5}{(x + 1)(x^2 - 2x + 3)} = \frac{A(x^2 - 2x + 3)}{(x + 1)(x^2 - 2x + 3)} + \frac{(Bx + C)(x + 1)}{(x + 1)(x^2 - 2x + 3)}$$

Step 4: Equate numerators since the denominators are equal.

$$-7x + 5 = A(x^2 - 2x + 3) + (Bx + C)(x + 1) \quad (*)$$

Step 5: Select values of x and substitute into (*).

Let $x = -1$:
$$-7x + 5 = A(x^2 - 2x + 3) + (Bx + C)(x + 1)$$
$$-7(-1) + 5 = A\left((-1)^2 - 2(-1) + 3\right) + (B(-1) + C)(-1 + 1)$$
$$12 = 6A$$
$$A = 2$$

Now substitute $A = 2$ back into (*) and select another value of x to simplify the calculation.

Let $x = 0$:
$$-7x + 5 = A(x^2 - 2x + 3) + (Bx + C)(x + 1)$$
$$-7(0) + 5 = 2(0^2 - 2(0) + 3) + (B(0) + C)(0 + 1)$$
$$5 = 6 + C$$
$$C = -1$$

Now substitute $A = 2$ and $C = -1$ back into (*) and select another value of x to simplify the calculation.

Let $x = 1$:
$$-7x + 5 = A(x^2 - 2x + 3) + (Bx + C)(x + 1)$$
$$-7(1) + 5 = 2(1^2 - 2(1) + 3) + (B(1) - 1)(1 + 1)$$
$$-2 = 4 + 2B - 2$$
$$2B = -4$$
$$B = -2$$

We have $A = 2$, $B = -2$ and $C = -1$.

Therefore: $\frac{-7x + 5}{(x + 1)(x^2 - 2x + 3)} = \frac{2}{x + 1} + \frac{-2x - 1}{x^2 - 2x + 3}$

Irreducible quadratic factors exercise   Go online

Q17: Express $\frac{x^2 - 4x + 14}{x^3 - x^2 + 2x - 8}$ in partial fractions.

....

Q18: Express $\frac{10x^2 + 16x + 9}{2x^3 + 7x^2 + 5x + 6}$ in partial fractions.

Partial fraction formation

Given $\frac{5x-1}{x^2-x-2}$, this has a denominator of a quadratic with two distinct factors.

Factorise the denominator: $\frac{5x-1}{x^2-x-2} \rightarrow \frac{5x-1}{(x-2)(x+1)}$

Equate with the general form of this type: $\frac{5x-1}{(x-2)(x+1)} \rightarrow \frac{A}{(x-2)} + \frac{B}{(x+1)}$

Obtain a common denominator for both fractions: $\frac{A}{(x-2)} + \frac{B}{(x+1)} \rightarrow \frac{A(x+1)}{(x-2)(x+1)} + \frac{B(x-2)}{(x-2)(x+1)}$

Equate numerator: $5x - 1 = A(x+1) + B(x-2)$

Select values of x and substitute to solve for A and B:
Let $x = -1$ then $-6 = A(0) + B(-3)$
Let $x = 2$ then $9 = A(3) + B(0)$

Hence $A = 3$ and $B = 2$ which gives

$$\frac{5x-1}{x^2-x-2} = \frac{3}{x-2} + \frac{2}{x+1}$$

Given $\frac{2x-5}{x^2-6x+9}$, this has a denominator of a quadratic with two repeated factors.

Factorise the denominator: $\frac{2x-5}{x^2-6x+9} \rightarrow \frac{2x-5}{(x-3)^2}$

Equate with the general form of this type: $\frac{2x-5}{(x-3)^2} \rightarrow \frac{A}{(x-3)} + \frac{B}{(x-3)^2}$

Obtain a common denominator for both fractions: $\frac{A}{(x-3)} + \frac{B}{(x-3)^2} \rightarrow \frac{A(x-3)}{(x-3)^2} + \frac{B}{(x-3)^2}$

Equate numerator: $2x - 5 = A(x-3) + B$

Select values of x and substitute to solve for A and B:
Let $x = 3$ then $-6 = A(0) + B(-3)$
Let $x = 0$ and $B = 1$ then $-5 = A(-3) + 1$

Hence $A = 2$ and $B = 1$ which gives

$$\frac{2x-5}{x^2-6x+9} = \frac{2}{x-3} + \frac{1}{(x-3)^2}$$

Given $\frac{5x^2+6x+4}{x^3+2x^2-2x-1}$, this has a cubic denominator. One factor is linear and the other is an irreducible quadratic.

Factorise the denominator: $\frac{5x^2+6x+4}{x^3+2x^2-2x-1} \rightarrow \frac{5x^2+6x+4}{(x-1)(x^2+3x+1)}$

Equate with the general form of this type: $\frac{5x^2+6x+4}{(x-1)(x^2+3x+1)} \rightarrow \frac{A}{(x-1)} + \frac{Bx+C}{(x^2+3x+1)}$

TOPIC 1. PARTIAL FRACTIONS

Obtain a common denominator for both fractions:

$$\frac{A}{(x-1)} + \frac{Bx+C}{(x^2+3x+1)}$$

$$\downarrow$$

$$\frac{A(x^2+3x+1)}{(x-1)(x^2+3x+1)} + \frac{Bx+C(x-1)}{(x-1)(x^2+3x+1)}$$

Equate numerator:

$$5x^2+6x+4 = A(x^2+3x+1) + (Bx+C)(x-1)$$

Let $x = 1$ then $12 = A(5) + (B(1)+C)(0)$

Let $x = 0$ and

Select values of x and substitute to solve for A and B:

$$A = 3$$

then $4 = 3(1) + (B(0)+C)(-1)$

Let $x = -1$, $A = 3$ and

$$C = -1$$

then $3 = 3(-1) + (B(-1)+(-1))(-2)$

Hence $A = 3$, $B = 2$ and $C = -1$ which gives

$$\frac{5x^2+6x+4}{x^3+2x^2-2x-1} = \frac{3}{x-1} + \frac{2x-1}{x^2+3x+1}$$

1.6 Algebraic long division

Key point

The **dividend** in a long division calculation is the expression which is being divided. As a fraction it is the numerator.

Key point

The **divisor** is the expression which is doing the dividing.

It is the expression outside the division sign. As a fraction it is the denominator.

Key point

The **quotient** is the answer to the division *not including* the remainder.

Key point

The **remainder** is what is left over after dividing.

Before performing long division with polynomials a numerical example will be used to illustrate the method.

© HERIOT-WATT UNIVERSITY

Examples

1. $351 \div 8 = 43\,r\,7$
The dividend is 351.
The divisor is 8.
The quotient is 43 and the remainder is 7.
In long division style this is written as:

$$\begin{array}{r} 43 \\ 8\overline{)351} \\ \underline{32}\downarrow \\ 31 \\ \underline{24} \\ 7 \end{array}$$

This would be written in fraction terms as $43\tfrac{7}{8}$.
The same technique can be used for dividing polynomials.

....

2. Problem:
Divide $x^3 - 2x + 5$ by $x^2 + 2x - 3$
Solution:
Step 1:
Lay out the long division taking account of 'missing terms'.

$$x^2 + 2x - 3 \overline{)x^3 + 0x^2 - 2x + 5}$$

Step 2:
Divide the first term of the divisor (x^2) into the first of the dividend (x^3) and write the answer at the top (x).

$$x^2 + 2x - 3 \overline{)\overset{\displaystyle x}{x^3 + 0x^2 - 2x + 5}}$$

Step 3:
Multiply each of the terms in the divisor by the first term of the quotient and write underneath the dividend.

$$x^2 + 2x - 3 \overline{)\overset{\displaystyle x}{\begin{array}{r} x^3 + 0x^2 - 2x + 5 \\ x^3 + 2x^2 - 3x \end{array}}}$$

Step 4:
Subtract to give a new last line in the dividend.

$$x^2 + 2x - 3 \overline{)\overset{\displaystyle x}{\begin{array}{r} x^3 + 0x^2 - 2x + 5 \\ \underline{x^3 + 2x^2 - 3x}\downarrow \\ -2x^2 + x + 5 \end{array}}}$$

TOPIC 1. PARTIAL FRACTIONS

Step 5:
Divide the first term of the divisor (x^2) into the first term of the last line ($-2x^2$) and write the answer at the top.

$$
\begin{array}{r}
x^2 + 2x - 3 \overline{)\begin{array}{rrrr} x^3 & +0x^2 & -2x & +5 \end{array}} \\
\begin{array}{rrr} x^3 & +2x^2 & -3x \end{array} \quad \downarrow \\
\hline
\begin{array}{rrr} -2x^2 & +x & +5 \end{array}
\end{array}
$$

with quotient $x - 2$ at top.

Step 6:
Multiply each of the terms in the divisor by the 2nd term of the quotient (-2) and write underneath the divisor.

$$
\begin{array}{r}
x - 2 \\
x^2 + 2x - 3 \overline{)\begin{array}{rrrr} x^3 & +0x^2 & -2x & +5 \end{array}} \\
\begin{array}{rrr} x^3 & +2x^2 & -3x \end{array} \quad \downarrow \\
\hline
\begin{array}{rrr} -2x^2 & +x & +5 \end{array} \\
\begin{array}{rrr} -2x^2 & -4x & +6 \end{array}
\end{array}
$$

Step 7:
Subtract to give a new last line in the dividend ($5x - 1$).
The division stops here in this case as the degree of the divisor (2) is greater than the degree of the last line (1).

$$
\begin{array}{r}
x - 2 \\
x^2 + 2x - 3 \overline{)\begin{array}{rrrr} x^3 & +0x^2 & -2x & +5 \end{array}} \\
\begin{array}{rrr} x^3 & +2x^2 & -3x \end{array} \quad \downarrow \\
\hline
\begin{array}{rrr} -2x^2 & +x & +5 \end{array} \\
\begin{array}{rrr} -2x^2 & -4x & +6 \end{array} \\
\hline
\begin{array}{rr} 5x & -1 \end{array}
\end{array}
$$

Therefore $\frac{x^3 - 2x + 5}{x^2 + 2x - 3} = x - 2 + \frac{5x - 1}{x^2 + 2x - 3}$
($x - 2$) is the quotient and ($5x - 1$) is the remainder.

Algebraic long division exercise   Go online

Q19: Divide $3x^3 - 2x^2 + 6$ by $x^2 + 4$

....

Q20: Divide $x^3 + 8x^2 + 13x - 10$ by $x + 5$.

....

© HERIOT-WATT UNIVERSITY

28 TOPIC 1. PARTIAL FRACTIONS

Q21: Evaluate $\frac{x^4 + 7x^2 + 13x - 4}{x^2 + 4x}$.

....

Q22: What is $\frac{x^5 - x^3}{x^5 + 1}$?

1.7 Reduce improper rational functions by division

A variation on the previous problems occurs when the initial expression is an improper rational function. In these circumstances it is necessary to divide through and obtain a polynomial and a proper rational function first. This rational function can then be expressed as a sum of partial fractions using the appropriate method from the types given in the previous sections.

Summary of partial fraction forms

Before we looking at reducing improper rational functions by division. Let's look back at the types of partial fractions from the previous sections. This activity provides a summary of the different partial fraction forms and when they should used.

Example : Partial fraction forms

In the denominator **each linear factor** e.g. $2x + 1$, requires one partial fraction of the form: $\frac{A}{2x + 1}$

each **repeated linear factor** e.g. $(2x + 1)^2$ requires two partial fractions: $\frac{A}{2x + 1}$ and $\frac{B}{(2x + 1)^2}$

each **irreducible quadratic factor** e.g. $x^2 + 5$, requires a partial fraction of the form: $\frac{Ax + B}{x^2 + 5}$

For the following fractions where the denominator has already been factorised identify the correct form of partial fractions.

Q23:

1) $\frac{1}{(x + 1)(x - 2)}$

2) $\frac{1}{(x^2 + 3)}$

3) $\frac{1}{(x + 1)^2}$

4) $\frac{1}{(x + 3)(x - 1)}$

5) $\frac{1}{(x^2 - 6)(x + 1)}$

6) $\frac{1}{(x + 1)(x - 2)^2}$

7) $\frac{1}{(x + 1)^2(x - 2)}$

8) $\frac{1}{(x^2 + 1)(x - 2)^2}$

a) $\frac{A}{(x + 1)} + \frac{B}{(x - 2)} + \frac{C}{(x - 2)^2}$

b) $\frac{A}{(x + 1)} + \frac{B}{(x + 1)^2} + \frac{C}{(x - 2)}$

c) $\frac{Ax + B}{(x^2 - 6)} + \frac{C}{(x + 1)}$

d) $\frac{Ax + B}{x^2 + 3}$

e) $\frac{A}{(x + 1)} + \frac{B}{(x - 2)}$

f) $\frac{Ax + B}{(x^2 + 1)} + \frac{C}{(x - 2)} + \frac{D}{(x - 2)^2}$

g) $\frac{A}{(x + 1)} + \frac{B}{(x + 1)^2}$

h) $\frac{A}{(x + 3)} + \frac{B}{(x - 1)}$

© HERIOT-WATT UNIVERSITY

TOPIC 1. PARTIAL FRACTIONS

Examples

1. Improper rational functions

Problem:

Express in partial fraction form $\frac{x^2 + 8x - 5}{x^2 + x - 6}$

Solution:

It is clear that we are working with an improper rational function as the degree of the numerator = degree of the denominator. We must therefore use algebraic long division before writing it as the sum of partial fractions.

Step 1: Divide using algebraic long division.

$$
\begin{array}{r}
1 \\
x^2 + x - 6 \overline{\smash{\big)}\, x^2 + 8x - 5} \\
\underline{x^2 + x - 6} \\
7x + 1
\end{array}
$$

This gives $\frac{x^2 + 8x - 5}{x^2 + x - 6} = 1 + \frac{7x + 1}{x^2 + x - 6}$.

Step 2: Now express $\frac{7x + 1}{x^2 + x - 6}$ in partial fractions by first factorising the denominator.

$\frac{7x + 1}{x^2 + x - 6} = \frac{7x + 1}{(x + 3)(x - 2)}$

Step 3: Now express $\frac{7x + 1}{(x + 3)(x - 2)}$ in partial fractions. In this case it has distinct linear factors.

$\frac{7x + 1}{(x + 3)(x - 2)} = \frac{A}{x + 3} + \frac{B}{x - 2}$

Step 4: Obtain the fractions with a common denominator.

$\frac{7x + 1}{(x + 3)(x - 2)} = \frac{A(x - 2)}{(x + 3)(x - 2)} + \frac{B(x + 3)}{(x + 3)(x - 2)}$

Step 5: Equate numerators since the denominators are equal.

$7x + 1 = A(x - 2) + B(x + 3)$ (*)

Step 6: Select values of x and substitute into (*).

Let $x = -3$:

$7x + 1 = A(x - 2) + B(x + 3)$
$7(-3) + 1 = A(-3 - 2) + B(-3 + 3)$
$-20 = -5A$
$A = 4$

Let $x = 2$:

$7x + 1 = A(x - 2) + B(x + 3)$
$7(2) + 1 = A(2 - 2) + B(2 + 3)$
$15 = 5B$
$B = 3$

Therefore:

$\frac{7x + 1}{(x + 3)(x - 2)} = \frac{4}{x + 3} + \frac{3}{x - 2}$

Step 7: Substitute partial fractions back into the expression.

$$\frac{x^2 + 8x - 5}{x^2 + x - 6} = 1 + \frac{7x + 1}{x^2 + x - 6}$$
$$= 1 + \frac{4}{x + 3} + \frac{3}{x - 2}$$

2. Problem:

Express $\frac{2x^3 - 7x^2 - 11x + 10}{(x + 2)(x - 4)}$ in partial fractions.

Solution:

If we expand the denominator it becomes $x^2 - 2x - 8$. It is then clear that we are working with an improper rational function as the degree of the numerator > degree of the denominator.

Step 1: Divide using algebraic long division.

$$\begin{array}{r} 2x - 3 \\ x^2 - 2x - 8 \overline{\smash{)}\, 2x^3 - 7x^2 - 11x + 10} \\ \underline{2x^3 - 4x^2 - 16x } \\ -3x^2 + 5x + 10 \\ \underline{-3x^2 + 6x + 24} \\ -x - 14 \end{array}$$

This gives $\frac{2x^3 - 7x^2 - 11x + 10}{(x + 2)(x - 4)} = 2x - 3 - \frac{x + 14}{(x + 2)(x - 4)}$.

Step 2: Now express $\frac{x + 14}{(x + 2)(x - 4)}$ in partial fractions. In this case it has a distinct linear factors.

$$\frac{x + 14}{(x + 2)(x - 4)} = \frac{A}{(x + 2)} + \frac{B}{(x - 4)}$$

Step 3: Obtain the fractions with a common denominator.

$$\frac{x + 14}{(x + 2)(x - 4)} = \frac{A(x - 4)}{(x + 2)(x - 4)} + \frac{B(x + 2)}{(x + 2)(x - 4)}$$

Step 4: Equate numerators since the denominators are equal.

$x + 14 = A(x - 4) + B(x + 2)$ (*)

Step 5: Select values of x and substitute into (*).

Let $x = -2$:

$x + 14 = A(x - 4) + B(x + 2)$
$-2 + 14 = A(-2 - 4) + B(-2 + 2)$
$12 = -6A$
$A = -2$

Let $x = 4$:

$x + 14 = A(x - 4) + B(x + 2)$
$4 + 14 = A(4 - 4) + B(4 + 2)$
$18 = 6B$
$B = 3$

Therefore:

$$\frac{x + 14}{(x + 2)(x - 4)} = \frac{-2}{x + 2} + \frac{3}{x - 4}$$

Step 6: Substitute partial fractions back into the expression.

$$\frac{2x^3 - 7x^2 - 11x + 10}{(x+2)(x-4)} = 2x - 3 - \frac{x+14}{(x+2)(x-4)}$$
$$= 2x - 3 - \left[\frac{-2}{x+2} + \frac{3}{x-4}\right]$$
$$= 2x - 3 + \frac{2}{x+2} - \frac{3}{x-4}$$

Note that the sum of partial fraction is being subtracted and so the signs of the partial fractions changes.

....

3. Problem:

Express $\frac{x^3 - 5x^2 + 6x - 8}{x^2 - 2x - 3}$ in partial fractions.

Solution:

It is clear that we are working with an improper rational function as the degree of the numerator > degree of the denominator.

Step 1: Divide using algebraic long division.

```
                   x -  3
x² - 2x - 3 ) x³ - 5x² + 6x -  8
              x³ - 2x² - 3x
              ─────────────
                  - 3x² + 9x -  8
                  - 3x² + 6x +  9
                  ─────────────
                         3x - 17
```

This gives $\frac{x^3 - 5x^2 + 6x - 8}{x^2 - 2x - 3} = x - 3 + \frac{3x - 17}{x^2 - 2x - 3}$.

Step 2: Now express $\frac{3x - 17}{x^2 - 2x - 3}$ in partial fractions by first factorising the denominator.

$\frac{3x - 17}{x^2 - 2x - 3} = \frac{3x - 17}{(x+1)(x-3)}$

Step 3: Identify the type of partial fraction. In this case it has a distinct linear factors.

$\frac{3x - 17}{(x+1)(x-3)} = \frac{A}{(x+1)} + \frac{B}{(x-3)}$

Step 4: Obtain the fractions with a common denominator.

$\frac{3x - 17}{(x+1)(x-3)} = \frac{A(x-3)}{(x+1)(x-3)} + \frac{B(x+1)}{(x+1)(x-3)}$

Step 5: Equate numerators since the denominators are equal.

$3x - 17 = A(x - 3) + B(x + 1)$ (*)

Step 6: Select values of x and substitute into (*).

Let $x = -1$:
$$3x - 17 = A(x - 3) + B(x + 1)$$
$$3(-1) - 17 = A(-1 - 3) + B(-1 + 1)$$
$$-20 = -4A$$
$$A = 5$$

Let $x = 3$:
$$3x - 17 = A(x - 3) + B(x + 1)$$
$$3(3) - 17 = A(3 - 3) + B(3 + 1)$$
$$-8 = 4B$$
$$B = -2$$

Therefore:
$$\frac{3x - 17}{(x + 1)(x - 3)} = \frac{5}{(x + 1)} - \frac{2}{(x - 3)}$$

Step 7: Substitute partial fractions back into the expression.
$$\frac{x^3 - 5x^2 + 6x - 8}{x^2 - 2x - 3} = x - 3 + \frac{3x - 17}{x^2 - 2x - 3}$$
$$= x - 3 + \frac{5}{x + 1} - \frac{2}{x - 3}$$

Reduce improper rational functions by division exercise   Go online

Q24: Express $\frac{x^3}{(x + 1)(x + 2)}$ in partial fractions.

....

Q25: Express $\frac{x^3 + 3x^2 + 4x + 3}{(x + 1)(x + 2)}$ in partial fractions.

1.8 Learning points

Partial fractions
Rational functions

- A **proper rational function** of two polynomials is when the degree of the numerator < the degree of the denominator.
$$\frac{x^2 + 2x - 1}{x^3 - 3x + 4}$$

- An **improper rational function** of two polynomials is when the degree of the numerator ≥ the degree of the denominator.
$$\frac{x^3 + 2x - 1}{x^2 + 4} \quad or \quad \frac{x^2 + 2x - 1}{x^2 + 4}$$

Algebraic long division

- The method of partial fractions is applied to proper rational functions only. If the function is improper, algebraic long division must be carried out first.

$$\begin{array}{r}
x - 2 \\
x^2 + 2x - 3 \overline{\smash{)} x^3 + 0x^2 - 2x + 5} \\
\underline{x^3 + 2x^2 - 3x} \\
-2x^2 + x + 5 \\
\underline{-2x^2 - 4x + 6} \\
5x - 1
\end{array}$$

- When setting up algebraic long division each power of the variable must be accounted for in the dividend.
In the example above $0x^2$ has been written in the dividend to retain the position of the x^2 terms.

Linear factors

- Any rational functions that have distinct linear factors on the denominator take the following form:
$$\frac{\ldots}{(x + a)(x + b)(x + c)} \equiv \frac{A}{(x + a)} + \frac{B}{(x + b)} + \frac{C}{(x + c)}$$

- Note that this rule can be extended to any number of distinct linear factors on the denominator.

Repeated linear factors

- Any rational functions that have repeated linear factors on the denominator take the following form:
$$\frac{\ldots}{(x + a)^3} \equiv \frac{A}{(x + a)} + \frac{B}{(x + a)^2} + \frac{C}{(x + a)^3}$$

- Note that this rule can be extended to any number of distinct linear factors on the denominator.

Irreducible quadratic factors

- An **Irreducible Quadratic** is a quadratic polynomial that cannot be factorised to give real roots i.e. $b^2 - 4ac < 0$
- Any rational functions that have irreducible quadratic factors on the denominator take the following form:
$$\frac{\cdots}{x^2 + bx + c} \equiv \frac{Ax + B}{x^2 + bx + c}$$
- Note that this rule can be extended to any irreducible polynomial factor on the denominator.

1.9 Extended information

Learning objective

To encourage an interest in related topics

The following web links should serve as an insight to the wealth of information and encourage readers to explore the subject further.
The authors do not maintain these web links and no guarantee can be given as to their effectiveness at a particular date.

http://www.quickmath.com/webMathematica3/quickmath/page.jsp?s1=algebra&s2=partialfractions&s3=basic
An extremely useful site for checking answers and solving problems.

1.10 End of topic test

End of topic 1 test — Go online

Linear Factors

Q26: Express $\dfrac{5x-8}{x^2-3x+2}$ in partial fractions.

Q27: Express $\dfrac{-6x-16}{x^2+4x+3}$ in partial fractions.

Q28: Express $\dfrac{-4x-10}{x^2+2x-8}$ in partial fractions.

Q29: Express $\dfrac{4x^2+x+1}{x^3-7x-6}$ in partial fractions.

Q30: Express $\dfrac{5x^2+6x+7}{x^3-2x^2-x+2}$ in partial fractions.

Repeated Linear Factors

Q31: Express $\dfrac{2x-7}{x^2-4x+4}$ in partial fractions.

Q32: Express $\dfrac{2x+8}{x^2+6x+9}$ in partial fractions.

Q33: Express $\dfrac{x^2+7x+19}{x^3+3x^2-4}$ in partial fractions.

Q34: Express $\dfrac{2x^2-5x+6}{x^3-6x^2+12x-8}$ in partial fractions.

Irreducible Factors

Q35: Express $\dfrac{4x^2+15x+23}{x^3+5x^2+11x+7}$ in partial fractions.

Q36: Express $\dfrac{8x-15}{x^3-3x^2+5x-3}$ in partial fractions.

Improper Fractions

Q37: Use algebraic division to express $\dfrac{x^3}{x^2+3x+2}$ as the sum of a polynomial and a proper fraction.

Q38: Now express $\frac{x^3}{x^2 + 3x + 2}$ in partial fractions.

....

Q39: Use algebraic division to express $\frac{2x^4 - 2x^3 - 2x^2 - 15x + 20}{x^2 - 3x + 2}$ as the sum of a polynomial and a proper fraction.

....

Q40: Now express $\frac{2x^4 - 2x^3 - 2x^2 - 15x + 20}{x^2 - 3x + 2}$ in partial fractions.

....

Q41: Use algebraic division to express $\frac{4x^3 - 16x^2 + 20x - 11}{x^2 - 4x + 4}$ as the sum of a polynomial and a proper fraction.

....

Q42: Now express $\frac{4x^3 - 16x^2 + 20x - 11}{x^2 - 4x + 4}$ in partial fractions.

....

Q43: Use algebraic division to express $\frac{-3x^4 + 11x^3 - 17x^2 - 4x + 18}{x^3 - 2x^2 + 2x + 5}$ as the sum of a polynomial and a proper fraction.

....

Q44: Now express $\frac{-3x^4 + 11x^3 - 17x^2 - 4x + 18}{x^3 - 2x^2 + 2x + 5}$ in partial fractions.

Topic 2

Differentiation

Contents

- 2.1 Looking back 41
  - 2.1.1 Rules for differentiation 44
  - 2.1.2 Differentiating products and quotients 46
  - 2.1.3 Differentiating sin x and cos x 48
  - 2.1.4 Differentiating using the chain rule 52
  - 2.1.5 Determining the equation of a tangent to a curve 54
  - 2.1.6 Curve sketching 57
  - 2.1.7 Leibniz Notation 61
  - 2.1.8 Closed intervals 63
  - 2.1.9 Applying differential calculus 66
  - 2.1.10 The derivative 70
- 2.2 Conditions for differentiability 74
  - 2.2.1 Differentiability at a point 74
  - 2.2.2 Differentiability over an interval 81
  - 2.2.3 Higher derivatives 83
  - 2.2.4 Discontinuities 85
- 2.3 The product rule 88
- 2.4 The quotient rule 92
- 2.5 Differentiate cot(x), sec(x), cosec(x) and tan(x) 96
- 2.6 Differentiate exp(x) 98
- 2.7 Logarithmic differentiation 102
  - 2.7.1 Differentiating ln(x) 102
  - 2.7.2 Logarithmic differentiation 105
- 2.8 Implicit differentiation 108
  - 2.8.1 Implicit and explicit functions 108
  - 2.8.2 Implicit differentiation 110
  - 2.8.3 Second derivative 114
  - 2.8.4 Applying implicit differentiation to problems 120
- 2.9 Differentiating inverse functions 125
  - 2.9.1 Differentiating inverse functions 126
  - 2.9.2 Differentiating inverse trigonometric functions 130
  - 2.9.3 Differentiating inverse trigonometric functions implicitly 136
- 2.10 Differentiation using the product, quotient and chain rules 138

- 2.11 Parametric differentiation 141
  - 2.11.1 Parametric curves 141
  - 2.11.2 First order parametric differentiation 147
  - 2.11.3 Motions in a plane 149
  - 2.11.4 Tangents to parametric curves 152
  - 2.11.5 Second order parametric differentiation 153
- 2.12 Related rate problems 155
  - 2.12.1 Explicitly defined rate related problems 155
  - 2.12.2 Implicitly defined rate related problems 156
  - 2.12.3 Related rate problems in practice 159
- 2.13 Displacement, velocity and acceleration 163
- 2.14 Curve sketching in closed intervals 168
- 2.15 Optimisation and other applications 169
- 2.16 Learning points 175
- 2.17 Proofs 179
- 2.18 Extended information 180
- 2.19 End of topic test 184

TOPIC 2. DIFFERENTIATION

Learning objective

By the end of this topic, you should be able to:

- identify continuous and discontinuous functions;
- understand when a function is not differentiable at a point or over an interval;
- differentiate beyond the first derivative;
- identify when to apply:
  - the product rule and apply it to a product of functions;
  - the quotient rule and apply it to simple rational functions;
- know the:
  - definitions of $\cot(x)$, $\sec(x)$, $\mathrm{cosec}(x)$ and $\tan(x)$ in terms of $\cos(x)$, $\sin(x)$ and $\tan(x)$;
  - graphs of the functions $\cot(x)$, $\sec(x)$ and $\mathrm{cosec}(x)$;
  - standard derivatives of $\cot(x)$, $\sec(x)$, $\mathrm{cosec}(x)$ and $\tan(x)$ and know how to derive them from their definitions;
- differentiate the exponential function and apply rules of differentiation to functions involving the exponential function;
- distinguish between implicit and explicit functions;
- apply:
  - differentiation to implicit functions;
  - the second derivative to implicit functions;
  - implicit differentiation within a contextual problem;
- understand how to find the derivative of the function $\ln(x)$;
- apply logarithmic differentiation to functions which are made up products, quotients and have variables as indices;
- know the conditions for an inverse function to exist;
- find the derivative of an inverse function;
- differentiate $\sin^{-1}(x)$, $\cos^{-1}(x)$ and $\tan^{-1}(x)$;
- sketch a curve from parametric equations;
- use parametric differentiation:
  - to find the first derivative;
  - to find the second derivative

© HERIOT-WATT UNIVERSITY

Learning objective continued

- calculate:
  - velocity components, speed and the instantaneous direction of motion;
  - the tangent to a curve from parametric equations;
- apply the chain rule to related rate problems where:
  - the functional relationship is given explicitly;
  - the functional relationship is given implicitly;
- solve practical related rate problems by first establishing a functional relationship between appropriate variables;
- define distance, displacement, speed, velocity and acceleration using calculus;
- calculate distance, displacement, speed, velocity and acceleration;
- define a closed interval;
- identify the maximum and minimum over a closed interval;
- use the skills in the differentiation topic to maximise or minimise a situation.

TOPIC 2. DIFFERENTIATION

2.1 Looking back

Summary of prior knowledge
The derivative: Higher

- The derived function represents:
    - the gradient of the tangent to the function;
    - the rate of change of the function.
- The gradient of a curve at a point is defined to be the gradient of the tangent to the curve at that point.
- Function notation
    - $f(x) = x^n \Rightarrow f'(x) = nx^{n-1}$ (dash notation is also known as Lagrange notation)
    - $f(x) = c \Rightarrow f'(x) = 0$ where c is a constant.
- Products and quotients must be simplified before differentiating.
- Simplify your derived expression if you can.
- Leibniz's notation
    - $y = x^n \Rightarrow \frac{dy}{dx} = nx^{n-1}$

Equation of a tangent: Higher

- To find the equation of a tangent at a point on a curve.
    - Identify or find the coordinates of the point of contact of the tangent
    - Find the derivative $\frac{dy}{dx}$.
    - Substitute the value of x into the derivative to get the gradient m.
    - Use $y - b = m(x - a)$ to get the equation.

Stationary points: Higher

- A stationary point could be one of the following:
    - A maximum turning point.
    - A minimum turning point.
    - A rising point of inflection.
    - A falling point of inflection.
- To determine stationary points and their nature:
    - find the derivative;
    - state that stationary points occur when $\frac{dy}{dx} = 0$;
    - solve for x when $\frac{dy}{dx} = 0$;
    - find the coordinates of the stationary points;

© HERIOT-WATT UNIVERSITY

- to determine the nature of the stationary points make a nature table;

x	$\rightarrow$	-1	$\rightarrow$	1	$\rightarrow$
$\frac{dy}{dx}$	+	0	-	0	+
shape	↗	→	↘	→	↗

- state the nature of the stationary points.

Curve sketching: Higher

- In order to make a good sketch of a curve some information about the curve is required:
    - The coordinates of the y-intercept.
    - The coordinates of the roots.
    - The coordinates of the stationary points and their nature.
    - The behaviour of the curve for large positive and negative values of x.

Sketching the graph of the derived function: Higher

- In order to sketch the graph of the derivative:
    - identify the stationary points;
    - make a nature table to identify where a function is increasing or decreasing;

x	$\rightarrow$	2	$\rightarrow$
shape	↘	→	↗
$f'(x)$	-	0	+
Derived function	below	on	above

   - sketch the derived function;
       - Stationary points on $f(x)$ become x-intercepts for $f'(x)$.
       - When $f(x)$ is increasing the graph of $f'(x)$ is above the x-axis.
       - When $f(x)$ is decreasing the graph of $f'(x)$ is below the x-axis.
- If $f(x)$ is a quadratic then the derived function $f'(x)$ will be a straight line.
- If $f(x)$ is a cubic then the derived function $f'(x)$ will be a quadratic function.

Closed intervals: Higher

- To determine the maximum and minimum values in a closed interval:
    - identify the stationary points and their coordinates;
    - make a nature table to determine the shape of the function;
    - determine the coordinates of the end points;
    - make a sketch;
    - identify the maximum and minimum values taking into consideration the end points as well.

TOPIC 2. DIFFERENTIATION

Rates of change: Higher

- Velocity, v, is the rate of change of displacement, s at time t, so $v = \frac{ds}{dt}$ or $s'(t)$.
- Acceleration, a, is a rate of change of the velocity, v at time t, so $a = \frac{dv}{dt}$ or $s''(t)$.

Optimisation: Higher

- Problems which involve finding a maximum or minimum value.
- The context is modelled by a function.
- The solution can be found by using differentiation to find stationary points.

Trigonometric derivatives: Higher

- $y = \sin x \Rightarrow \frac{dy}{dx} = \cos x$
- $y = \cos x \Rightarrow \frac{dy}{dx} = -\sin x$
- $f(x) = \sin ax \Rightarrow f'(x) = a \cos ax$
- $f(x) = \cos ax \Rightarrow f'(x) = -a \sin ax$

Derivatives of composite functions: Higher

- If $f(x) = (x + a)^n \Rightarrow f'(x) = n(x + a)^{n-1}$
- If $f(x) = (ax + b)^n \Rightarrow f'(x) = an(ax + b)^{n-1}$
- If $h(x) = g(f(x)) \Rightarrow h'(x) = g'(f(x)) \times f'(x)$
- If $y = y(u)$ and $u = u(x)$ then $\frac{dy}{dx} = \frac{dy}{du} \times \frac{du}{dx}$

Logarithmic rules: Higher

- $\log_a xy = \log_a x + \log_a y$
- $\log_a \frac{x}{y} = \log_a x - \log_a y$
- $\log_a x^n = n \log_a x$

The English mathematician *Sir Isaac Newton* and the German mathematician *Gottfried Leibniz* each independently developed theories regarding differentiation. There is more information about these famous mathematicians in the Extended Information section near the end of this topic.

2.1.1 Rules for differentiation

Let's look at how to differentiate four different functions.

1. $f(x) = x^2$
2. $f(x) = x^{-5}$
3. $f(x) = x^{\frac{1}{2}}$
4. $f(x) = 4x^3 - 2x$

1.

$f(x) = x^2$ $f'(x) = 2 \times x^2$ Bring down the power
                $f'(x) = 2\, x^{2-1}$ Reduce the power by 1
                $f'(x) = 2\, x^1$ Simplify
                $f'(x) = 2\, x$

2.

$f(x) = x^{-5}$ $f'(x) = -5 \times x^{-5}$ Bring down the power
                $f'(x) = -5\, x^{-5-1}$ Reduce the power by 1
                $f'(x) = -5\, x^{-6}$ Simplify
                $f'(x) = \frac{-5}{x^6}$

3.

$f(x) = x^{\frac{1}{2}}$ $f'(x) = \frac{1}{2} \times x^{\frac{1}{2}}$ Bring down the power
                $f'(x) = \frac{1}{2} x^{\frac{1}{2}-1}$ Reduce the power by 1
                $f'(x) = \frac{1}{2} x^{-\frac{1}{2}}$ Simplify
                $f'(x) = \frac{1}{2x^{\frac{1}{2}}}$ Remember the power of a half is the square root
                $f'(x) = \frac{1}{2\sqrt{x}}$

4.

$f(x) = 4x^3 - 2x$ $f'(x) = 3 \times 4x^3 - 1 \times 2x^1$ Bring down the power
                $f'(x) = 12x^{3-1} - 2x^{1-1}$ Reduce the power by 1
                $f'(x) = 12x^2 - 2x^0$ Simplify
                $f'(x) = 12x^2 - 2$ Remember $x^0 = 1$

TOPIC 2. DIFFERENTIATION

> **Key point**
>
> $f(x) = x^n \Rightarrow f'(x) = nx^{n-1}$
> $f(x) = c \Rightarrow f'(x) = 0$ where c is a constant.

Examples

1. Problem:

If $f(x) = 5x^2 - 4x + 7$ what is $f'(x)$?

Solution:

$f(x) = 5x^2 - 4x + 7$
$f'(x) = 2 \times 5x^1 - 4x^0 + 0$
$ = 10x - 4$

2. Problem:

Differentiate $g(x) = x^{\frac{4}{3}}$

Solution:

$\begin{aligned} g'(x) &= \frac{4}{3} x^{\frac{4}{3} - 1} \\ &= \frac{4}{3} x^{\frac{4}{3} - \frac{3}{3}} \\ &= \frac{4}{3} x^{\frac{1}{3}} \end{aligned}$

3. Problem:

Differentiate $f(x) = \sqrt{x}$

Solution:

First rewrite $f(x)$ in terms of x^n.

$f(x) = \sqrt{x} = x^{\frac{1}{2}}$

Therefore,

$\begin{aligned} f'(x) &= \frac{1}{2} x^{-\frac{1}{2}} \\ &= \frac{1}{2x^{\frac{1}{2}}} \\ &= \frac{1}{2\sqrt{x}} \end{aligned}$

4. Problem:

Differentiate $f(x) = \frac{1}{x}$

Solution:

Again rewrite $f(x)$ in terms of x^n

$f(x) = \frac{1}{x} = x^{-1}$

© HERIOT-WATT UNIVERSITY

Therefore,

$f'(x) = -x^{-2} = -\frac{1}{x^2}$

5. Problem:

If $f(x) = 2x^4 - \frac{2}{\sqrt[3]{x}}$ what is $f'(x)$?

Solution:

We can't differentiate until we have written the second term in index form.

So, $f(x) = 2x^4 - 2x^{-\frac{1}{3}}$

$f'(x) = 4 \times 2x^3 - \left(-\frac{1}{3}\right) \times 2x^{-\frac{4}{3}}$

$\quad = 8x^3 + \frac{2}{3}x^{-\frac{4}{3}} \qquad Simplify$

$\quad = 8x^3 + \frac{2}{3\sqrt[3]{x^4}} \qquad Note: \ x^{-\frac{4}{3}} = \frac{1}{\sqrt[3]{x^4}}$

6. Problem:

$f(x) = \frac{3}{\sqrt{x}} + \frac{\sqrt{x}}{5}$

Solution:

We can't differentiate until we have written both terms in index form.

$f(x) = \frac{3}{\sqrt{x}} + \frac{\sqrt{x}}{5} = 3x^{-\frac{1}{2}} + \frac{1}{5}x^{\frac{1}{2}}$

Now differentiate as follows,

$f'(x) = \left(-\frac{1}{2}\right) \times 3x^{-\frac{3}{2}} + \frac{1}{2} \times \frac{1}{5}x^{-\frac{1}{2}}$

$f'(x) = -\frac{3}{2}x^{-\frac{3}{2}} + \frac{1}{10}x^{-\frac{1}{2}} = -\frac{3}{2\sqrt{x^3}} + \frac{1}{10\sqrt{x}}$

Key point

Make sure that each term in $f(x)$ takes the form ax^n before you start to differentiate.

If $f(x) = ax^n$ then $f'(x) = n \times ax^{n-1}$

2.1.2 Differentiating products and quotients

Differentiating products

Differentiate the function: $f(x) = (x + 3)(2x - 5)$

This function is written as a product. First you should expand the expression for $f(x)$.

$f(x) = (x + 3)(2x - 5) = 2x^2 - 5x + 6x - 15 = 2x^2 + x - 15$

Now you are ready to differentiate.

$f'(x) = 4x + 1$

TOPIC 2. DIFFERENTIATION

Examples

1. Problem:

If $f(x) = (x + 7)(x - 3)$ what is $f'(x)$?

Solution:

We can't differentiate until we have expanded the brackets.

$f(x) = (x + 7)(x - 3) = x^2 + 4x - 21$

$f'(x) = 2x + 4$

....

2. Problem:

Differentiate the function $f(x) = \left(\sqrt{x} + \frac{1}{\sqrt{x}}\right)^2$

Solution:

Multiply out the brackets in the function so that it can be expressed as a sum of individual terms.

$$\begin{aligned} f(x) &= \left(\sqrt{x} + \frac{1}{\sqrt{x}}\right)^2 \\ &= \left(\sqrt{x} + \frac{1}{\sqrt{x}}\right)\left(\sqrt{x} + \frac{1}{\sqrt{x}}\right) \\ &= \sqrt{x}^2 + 2\frac{\sqrt{x}}{\sqrt{x}} + \frac{1}{(\sqrt{x})^2} \\ &= x + 2 + \frac{1}{x} \\ &= x + 2 + x^{-1} \end{aligned}$$

Now differentiate term by term.

$$\begin{aligned} f'(x) &= 1 - x^{-2} \\ &= 1 - \frac{1}{x^2} \end{aligned}$$

Differentiating quotients

Differentiate the function $f(x) = \frac{x^5 + 3x^2 - x}{x^2}$

This function is written as a quotient. We need to simplify the expression for $f(x)$ before we are able to differentiate.

$$\begin{aligned} f(x) &= \frac{x^5 + 3x^2 - x}{x^2} \\ &= \frac{x^5}{x^2} + \frac{3x^2}{x^2} - \frac{x}{x^2} \end{aligned}$$

Notice that we break the quotient apart into separate functions.

$$\begin{aligned} f(x) &= \frac{x^5}{x^2} + \frac{3x^2}{x^2} - \frac{x}{x^2} \\ &= x^3 + 3 - x^{-1} \end{aligned}$$

© HERIOT-WATT UNIVERSITY

48 TOPIC 2. DIFFERENTIATION

We are now able to differentiate.
$$f'(x) = 3x^2 + x^{-2}$$
$$= 3x^2 + \frac{1}{x^2}$$

Examples

1. Problem:
If $f(x) = \frac{(x+2)(x-3)}{x^2}$ what is $f'(x)$?

Solution:
We cannot differentiate until we have expanded the brackets and the quotient is broken up.
$$f(x) = \frac{(x+2)(x-3)}{x^2}$$
$$= \frac{x^2 - x - 6}{x^2}$$
$$= \frac{x^2}{x^2} - \frac{x}{x^2} - \frac{6}{x^2}$$
$$= 1 - x^{-1} - 6x^{-2}$$

We are now able to differentiate.
$$f'(x) = x^{-2} + 12x^{-3}$$
$$= \frac{1}{x^2} + \frac{12}{x^3}$$

....

2. Problem:
If $f(x) = \frac{x-1}{\sqrt{x}}$ what is $f'(x)$?

Solution:
In this case, the denominator will be written with a negative power. The following expression can then be multiplied out into a form that can then be differentiated.
$$f(x) = \frac{x-1}{\sqrt{x}}$$
$$= x^{-\frac{1}{2}}(x-1) \quad \text{Note that: } x^{-\frac{1}{2}} \times x = x^{-\frac{1}{2}+1} = x^{\frac{1}{2}}$$
$$= x^{\frac{1}{2}} - x^{-\frac{1}{2}}$$

We are now able to differentiate.
$$f'(x) = \frac{1}{2}x^{-\frac{1}{2}} + \frac{1}{2}x^{-\frac{3}{2}}$$
$$= \frac{1}{2\sqrt{x}} + \frac{1}{2\sqrt{x^3}} \quad \text{Note that: } x^{-\frac{1}{2}} = \frac{1}{\sqrt{x}} \quad \text{and} \quad x^{-\frac{3}{2}} = \frac{1}{\sqrt{x^3}}$$

2.1.3 Differentiating sin x and cos x

A ball, attached to the end of a stretched spring, is released at time $t = 0$. The displacement y (cm) of the ball from the x-axis at time t (seconds) is given by the formula $y = 10 \sin t$.

TOPIC 2. DIFFERENTIATION   49

Differentiation of the sine function   Go online

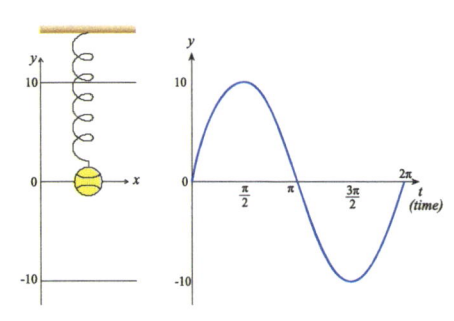

The following questions might be posed:

- What is the speed of the ball after 2 seconds?
- When is the ball first stationary?

Since speed is the rate of change of distance with respect to time, the speed of the ball is given by the differential equation.

$$\frac{dy}{dt} = \frac{d}{dt}(10 \sin t)$$

Derivative of sin x   Go online

The graph for $y = \sin x$ is shown.

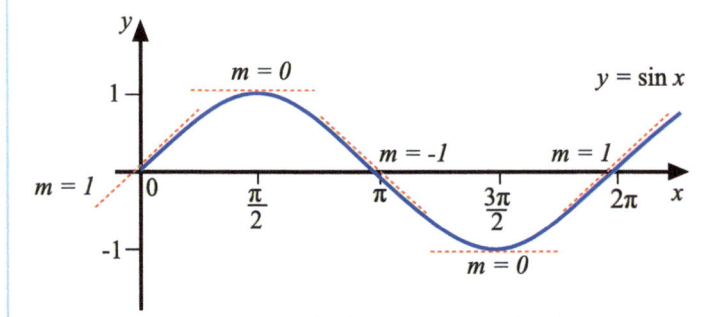

Notice that the tangent to the curve is drawn at various points. The value for the gradient of the tangent at these points is recorded in the following table.

© HERIOT-WATT UNIVERSITY

50   TOPIC 2. DIFFERENTIATION

x	0	$\frac{\pi}{2}$	π	$\frac{3\pi}{2}$	2π
m_T	1	0	-1	0	1

When these points are plotted and joined with a smooth curve the result is as follows.

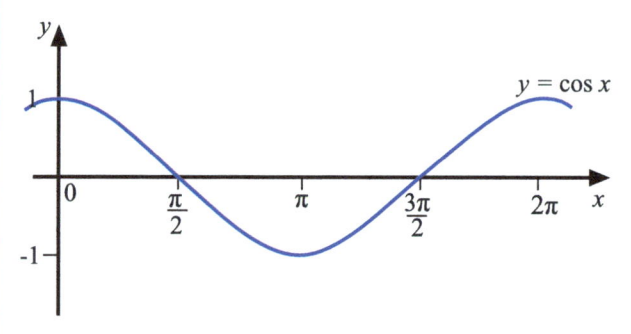

Since the gradient of the tangent at the point $(a, f(a))$ on the curve $y = f(x)$ is $f\,'(a)$, then the above graph represents the graph of the derivative of sin x.

Then for $y = \sin x$ it appears that $\frac{dy}{dx} = \cos x$

(This can be checked by calculating gradients at intermediate points.)

Differentiating sin x and cos x exercise   Go online

The derivative of the sine function is required to answer these questions.

Q1:

Study the graph for $y = \cos x$ as shown here.

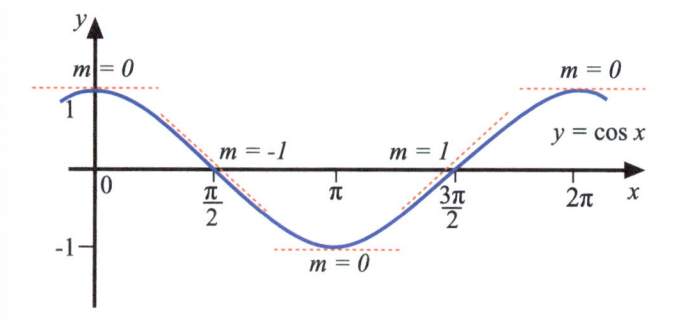

TOPIC 2. DIFFERENTIATION

Complete the following table for the gradient of the tangents.

x	0	$\frac{\pi}{2}$	π	$\frac{3\pi}{2}$	2π
m_T					

Plot the points and join with a smooth curve.

....

Q2: When $y = \cos x$ it appears that $\frac{dy}{dx} =$?

a) $\sin x$
b) $\cos x$
c) $-\sin x$
d) $-\cos x$

Key point

When $f(x) = \cos x$ then $f'(x) = -\sin x$
When $f(x) = \sin x$ then $f'(x) = \cos x$

Note: x must be measured in *radians*

Examples

1. Problem:

Find $f'(x)$ when $f(x) = 3 \cos x$.

Solution:

When $f(x) = 3 \cos x$
then $f'(x) = -3 \sin x$

....

2. Problem:

Find $f'(x)$ when $f(x) = 2 \sin x - \cos x + 4x^2$.

Solution:

$f'(x) = 2 \cos x - (-\sin x) + 8x$
$ = 2 \cos x + \sin x + 8x$

2.1.4 Differentiating using the chain rule

The Theory
Function notation

$h(x) = (ax + b)^n$ is an example of a composite function.
Let $f(x) = ax + b$ and $g(x) = x^n$ then $h(x) = (ax + b)^n = (f(x))^n = g(f(x))$
When $f(x) = ax + b$ then $f'(x) = a$
When $g(x) = x^n$ then $g'(x) = nx^{n-1}$
Also when $g(f) = (f)^n$ then $g'(f) = n(f)^{n-1} = n(ax + b)^{n-1}$
However, when $y = (ax + b)^n$ then $\frac{dy}{dx} = an(ax + b)^{n-1}$
Writing this in function notation gives,
When $h(x) = g(f(x))$ then

$$h'(x) = an(ax + b)^{n-1} = n(ax + b)^{n-1} \times a = g'(f(x)) \times f'(x)$$

This result is known as **the chain rule**.

Leibniz notation

The chain rule can also be written in Leibniz notation.
Let $y = (ax + b)^n$ but this time let $u = ax + b$
so $y = u^n$
Since $u = ax + b$ then $\frac{du}{dx} = a$
Since $y = u^n$ then $\frac{dy}{du} = nu^{n-1} = n(ax + b)^{n-1}$
As before, when $y = (ax + b)^n$ then $\frac{dy}{dx} = an(ax + b)^{n-1}$
In Leibniz notation this is as follows
When $y = u^n$ then

$$\frac{dy}{dx} = an(x + a)^{n-1} = n(x + a)^{n-1} \times a = \frac{dy}{du} \times \frac{du}{dx}$$

It will be useful to remember both forms of the chain rule. (However, you are not required to be able to prove either of them.)

Key point

Function notation: $h'(x) = g'(f(x)) \times f'(x)$

Leibniz notation: $\frac{dy}{dx} = \frac{dy}{du} \times \frac{du}{dx}$

Many types of composite functions can be differentiated using the chain rule. Note that either function or Leibniz notation can be used.

Examples

1. Problem:
Find $\frac{dy}{dx}$ when $y = \sqrt{x^3 - 4x^2 - 5}$.
Solution:
It is often easier to follow these steps.

TOPIC 2. DIFFERENTIATION

Step 1:	Turn the square root into a power.	$y = (x^3 - 4x^2 - 5)^{\frac{1}{2}}$
Step 2:	Bring down the power.	$\frac{1}{2}$
Step 3:	Write down the bracket.	$\frac{1}{2}(x^3 - 4x^2 - 5)$
Step 4:	Reduce the power by 1.	$\frac{1}{2}(x^3 - 4x^2 - 5)^{-\frac{1}{2}}$
Step 5:	Differentiate the bracket.	$\frac{1}{2}(x^3 - 4x^2 - 5)^{-\frac{1}{2}} \times (3x^2 - 8x)$
Step 6:	Simplify the answer.	$\frac{dy}{dx} = \frac{3x^2 - 8x}{2\sqrt{x^3 - 4x^2 - 5}}$

....

2. Problem:

Find $f\,'(x)$ when $f(x) = \cos(2x + 1)$.

Solution:

Step 1:	Differentiate cos.	$-sin$
Step 2:	Write down the bracket.	$-sin\,(2x + 1)$
Step 3:	Differentiate the bracket.	$-sin\,(2x + 1) \times 2$
Step 4:	Simplify the answer.	$-2sin\,(2x + 1)$

....

3. Problem:

Find the derivative of $\sin^2(3 - 2x)$.

Solution:

Step 1:	Remember:	$\sin^2(3 - 2x) = (sin(3 - 2x))^2$
Step 2:	Bring down the power.	2
Step 3:	Write down the bracket.	$2sin(3 - 2x)$
Step 4:	Reduce the power by 1.	$2sin(3 - 2x)^1$
Step 5:	Differentiate the bracket.	
	Now the bracket is:	$sin(3 - 2x)$
	We have to use the chain rule here too.	$cos(3 - 2x) \times -2 = -2cos(3 - 2x)$
	Giving:	$2sin(3 - 2x) \times -2cos(3 - 2x)$
Step 6:	Simplify the answer.	$\frac{d}{dx} = -4\sin(3 - 2x)\cos(3 - 2x)$

The following derivatives are given as a reminder. However, they should be learned.

$f(x)$	$f\,'(x)$
$\sin(ax)$	$a \times \cos(ax)$
$\cos(ax)$	$-a \times \sin(ax)$

© HERIOT-WATT UNIVERSITY

Differentiating using the chain rule exercise

Go online

Try these questions on paper and then check your answers.

Q3: Find $h'(x)$ when $h(x) = \sqrt{x^2 + 6x}$

Q4: If $f(x) = \cos^3 x$, find $f'(x)$.

Q5: Differentiate $f(x) = (x^3 + 2x^2 - 3)^5$

Q6: Find $k'(x)$ when $k(x) = \frac{4}{(1-3x)^4}$

Q7: If $h(x) = \sin\left(5x - \frac{\pi}{3}\right)$, find $h'(x)$.

2.1.5 Determining the equation of a tangent to a curve

Key point

The gradient of a function $f(x)$ at the point P(a,b) is called the derivative of f at a and is written as $f'(a)$.

Examples

1. Problem:

If $f(x) = 3x^2 - 4x - 2$, what is the gradient of the tangent at the point (0,-2)?

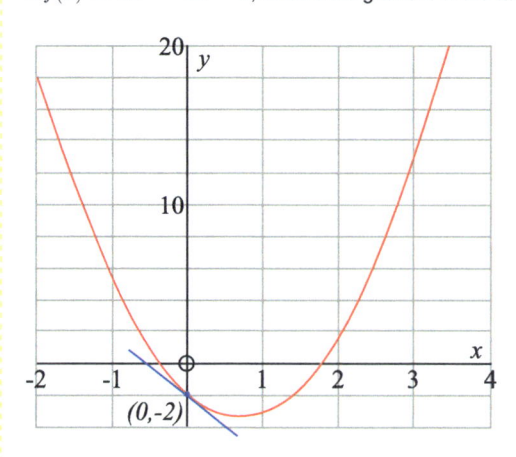

Solution:

To find the gradient of a tangent to a curve we need the derivative.

$f'(x) = 6x - 4$

We want the gradient at the point (0,-2), i.e. when $x = 0$

$f'(0) = 6 \times 0 - 4 = -4$

$m_{tangent}$ = -4

....

2. Problem:

What are the coordinates of the points on the curve with equation $y = x^3 - 9x^2 + 14x$ where the gradient of the tangent is -1?

Solution:

The gradient of a tangent is the derivative $\frac{dy}{dx} = 3x^2 - 18x + 14$

but we know that the gradient is -1 so $3x^2 - 18x + 14 = -1$

Solve to find the x-coordinates of the points of contact of the tangents.

$3x^2 - 18x + 15 = 0$

$3(x^2 - 6x + 5) = 0$

$3(x - 1)(x - 5) = 0$

$x - 1 = 0 \quad and \quad x - 5 = 0$

$\quad\quad x = 1 \quad and \quad x = 5$

Now we need the y-coordinates so substitute back into the equation of the curve $y = x^3 - 9x^2 + 14x$

When $x = 1, y = 1^3 - 9 \times 1^2 + 14 \times 1 = 6$ giving (1,6)

and when $x = 5, y = 5^3 - 9 \times 5^2 + 14 \times 5 = -30$ giving (5,-30)

> **Key point**
>
> The tangent to a curve is a straight line so the equation for the tangent can be written in the form $y - b = m(x - a)$

The following strategy is useful when finding the equation of the tangent.

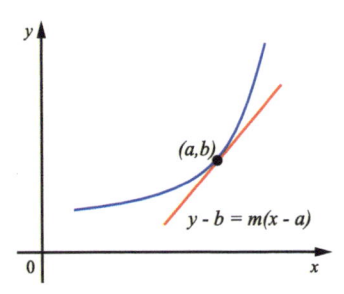

1. Find the coordinates of (a,b), the point of contact of the tangent with the curve.
2. Calculate the gradient, m, which is equal to the value of $\frac{dy}{dx}$ at $x = a$.

Examples

1. Problem:

Find the equation of the tangent to the curve $y = 3\sqrt{x}$ at $x = 4$

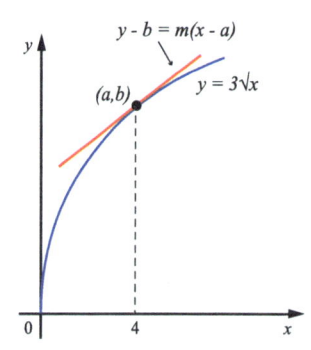

Solution:

Step 1: Find the coordinates of the point of contact.
Substitute $x = 4$ into $y = 3\sqrt{x}$
$y = 3\sqrt{x} = 3\sqrt{4} = 6$
So the point of contact is (4,6).

Step 2: Differentiate to find the gradient of the tangent.
$y = 3\sqrt{x} = 3x^{\frac{1}{2}}$
$\frac{dy}{dx} = \frac{1}{2} \times 3x^{-\frac{1}{2}} = \frac{3}{2\sqrt{x}}$
We want the gradient when $x = 4$ so $m = \frac{3}{2\sqrt{4}} = \frac{3}{4}$

Step 3: Find the equation of the tangent.
Using $y - b = m(x - a)$ with $m = \frac{3}{4}$ and the point (4,6) we get

$$y - b = m(x - a)$$
$$y - 6 = \frac{3}{4}(x - 4)$$
$$4(y - 6) = 3(x - 4)$$
$$4y - 24 = 3x - 12$$
$$4y - 3x - 12 = 0$$

2. Problem:

The parabola with equation $y = x^2 - 12x + 40$ has a tangent at the point (7,5).

a) Find the equation of this tangent.
   This tangent is also a tangent to the parabola with equation $y = -x^2 - 10$
b) Find the coordinates of the point of contact.

Solution:

a) $y = x^2 - 12x + 40$ so $\frac{dy}{dx} = 2x - 12$
   When $x = 7, m = 2 \times 7 - 12 = 2$
   When we have $m = 2$ and the point (7,5) we get $y - 5 = 2(x - 7)$
   So the equation of the tangent is $y = 2x - 9$

b) The tangent occurs when $-x^2 - 10 = 2x - 9$
   $-x^2 - 2x - 1 = 0$
   $-(x^2 + 2x + 1) = 0$
   $-(x + 1)(x + 1) = 0$
   Note that equal roots confirms tangency.
   $x + 1 = 0$ so $x = -1$ and using the equation of the tangent $y = 2x - 9$
   $y = 2 \times -1 - 9 = -11$
   So the point of contact is (-1,-11).

Determining the equation of a tangent to a curve exercise   Go online

Q8: What is the equation of the tangent to the curve $f(x) = (x^2 + 3)(x - 2)$ when $x = -1$?
Give your answer in the form $y = mx + c$.

2.1.6 Curve sketching

Key point

In order to make a good sketch of a curve some information about the curve is required.

1. The coordinates of the y-intercept
2. The coordinates of the roots.
3. The coordinates of the stationary points and their nature.
4. The behaviour of the curve for large positive and negative values of x.

Examples

1. Problem:
Sketch the curve $y = 3x^2 - x^3$

Solution:
Step 1: Write down the y-intercept.
The y-intercept occurs when $x = 0$ so $y = 3 \times 0^2 - 0^3 = 0$
Hence the y-intercept is (0, 0).
Step 2: Find the roots.
The roots occur when $y = 0$
Factorise $3x^2 - x^3 = 0$ gives,
$$x^2(3 - x) = 0$$
$$x^2 = 0 \quad or \quad (3 - x) = 0$$
$$x = 0 \quad or \quad x = 3$$
Hence the roots are (0,0) and (3,0).
Step 3: Find the stationary points and their nature.
$\frac{dy}{dx} = 6x - 3x^2$
Stationary points occur when $\frac{dy}{dx} = 0$
$$6x - 3x^2 = 0$$
$$3x(2 - x) = 0$$
$$3x = 0 \quad or \quad (2 - x) = 0$$
$$x = 0 \quad or \quad x = 2$$
When $x = 0$ then $y = 3 \times 0^2 - 0^3 = 0$
When $x = 2$ then $y = 3 \times 2^2 - 2^3 = 4$
Thus the stationary points are (0, 0) and (2, 4).

x	-1 $\to$	0	1 $\to$	2	3 $\to$
$3x$	-	0	+	+	+
$(2 - x)$	+	+	+	0	-
$3x(2 - x)$	-	0	+	0	-
shape	↘	→	↗	→	↘

This gives us a minimum turning point at (0,0) and a maximum turning point at (2,4).
Step 4: Check the behaviour of the curve when x takes large positive and negative values.
For large positive and negative values of x, y behaves like $-x^3$, this is the term with the highest degree.
When x is very large and negative $-x^3$ is very large and positive, and we write this as,
$x \to -\infty$ then $y \to +\infty$
When x is very large and positive $-x^3$ is very large and negative, and we write this as,
$x \to +\infty$ then $y \to -\infty$
This matches what is happening at the start and end of our nature table.

Step 5: Plot the points that you have found to make a good sketch of the curve.

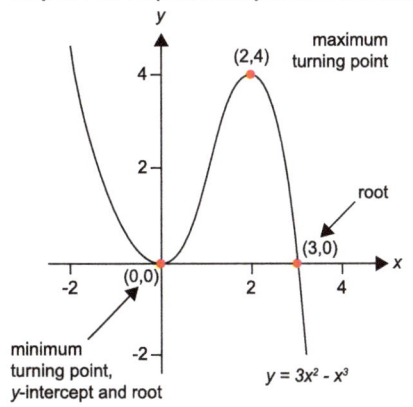

(graph showing y = 3x² - x³ with maximum turning point (2,4), root (3,0), minimum turning point, y-intercept and root (0,0))

Notice that the shape matches our nature table and the important points on the curve are clearly labelled.

....

2. Problem:

Sketch the curve $f(x) = x^3 - 3x + 2$

Solution:

Step 1: Write down the y-intercept.

The y-intercept occurs when $x = 0$ so $y = 0^3 - 3 \times 0 + 2 = 2$

Hence the y-intercept is (0,2).

Step 2: Find the roots.

The roots occur when $f(x) = 0$

$x^3 - 3x + 2$ is a polynomial so to factorise we need synthetic division.

Firstly we could try dividing by $(x - 1)$. We place 1 into the synthetic division. If the remainder is equal to 0 then $(x - 1)$ is a factor.

```
1 | 1   0   -3   2
  |     1    1  -2
  ----------------
    1   1   -2 | 0
```

$x^3 - 3x + 2 = (x - 1)(x^2 + x - 2)$
$ = (x - 1)(x + 2)(x - 1)$

$(x - 1)(x + 2)(x - 1) = 0$
$x - 1 = 0 \quad or \quad x + 2 = 0$
$ x = 1 \quad or \quad x = -2$

Hence the roots are (1,0) and (-2,0).

Step 3: Find the stationary points and their nature.

$\frac{dy}{dx} = 3x^2 - 3$

Stationary points occur when $\frac{dy}{dx} = 0$

$3x^2 - 3 = 0$
$3(x^2 - 1) = 0$
$3(x - 1)(x + 1) = 0$
$x - 1 = 0 \;\; or \;\; x + 1 = 0$
$x = 1 \;\; or \;\; x = -1$

When $x = 1$ then $y = 1^3 - 3 \times 1 + 2 = 0$
When $x = -1$ then $y = (-1)^3 - 3 \times (-1) + 2 = 4$

Thus the stationary points are (1, 0) and (-1, 4).

x	-2 $\to$	-1	0 $\to$	1	2 $\to$
$(x - 1)$	-	-	-	0	+
$(x + 1)$	-	0	+	+	+
$3(x - 1)(x + 1)$	+	0	-	0	+
shape	↗	→	↘	→	↗

This gives us a maximum turning point at (-1,4) and a minimum turning point at (1,0).

Step 4: Check the behaviour of the curve when x takes large positive and negative values.

For large positive and negative values of x, y behaves like x^3, this is the term with the highest degree.

As $x \to -\infty$ then $y \to -\infty$
As $x \to +\infty$ then $y \to +\infty$

This matches what is happening at the start and end of our nature table.

Step 5: Plot the points that you have found to make a good sketch of the curve.

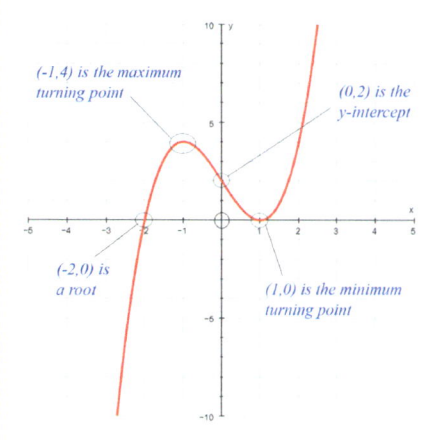

(-1,4) is the maximum turning point
(0,2) is the y-intercept
(-2,0) is a root
(1,0) is the minimum turning point

TOPIC 2. DIFFERENTIATION

Notice that the shape matches our nature table and the important points on the curve are clearly labelled.

2.1.7 Leibniz Notation

In calculus, Leibniz's notation, named in honour of the 17th Century German philosopher and mathematician Gottfried Wilhelm Leibniz, uses $\frac{dy}{dx}$ instead of $f'(x)$ for the derivative.

> **Key point**
>
> Given a function in the form $f(x) = x^n$, we would express the derivative in the form $f'(x) = nx^{n-1}$.
>
> Given a function in the form $y = x^n$, we would use Leibniz notation to express the derivative as $\frac{dy}{dx} = nx^{n-1}$.

For $\frac{dy}{dx}$ we say "dy by dx".

Examples

1. Problem:

If $y = 2x^3 - 5x$, find the value of the derivative at $x = 1$

Solution:

Find the derivative $\frac{dy}{dx} = 6x^2 - 5$

Substitute $x = 1$ into the derivative: $\frac{dy}{dx} = 6 \times 1^2 - 5 = 1$

....

2. Problem:

If $y = \frac{(x-1)(x+3)}{x}$, find $f'(-2)$

Solution:

Simplify the function:

$$y = \frac{(x-1)(x+3)}{x}$$
$$= \frac{x^2}{x} + \frac{2x}{x} - \frac{3}{x}$$
$$= x + 2 - 3x^{-1}$$

Differentiate:

$$\frac{dy}{dx} = 1 + 3x^{-2}$$
$$= 1 + \frac{3}{x^2}$$

Substitute $x = -2$ into $f'(x)$:

© HERIOT-WATT UNIVERSITY

$$\frac{dy}{dx} = 1 + \frac{3}{(-2)^2}$$
$$= 1 + \frac{3}{4}$$
$$= 1\frac{3}{4}$$

3. Problem:
What is the gradient of the tangent to the function $y = x^3 - 3x^2 + x + 6$ at the point (2,1).

Solution:
$\frac{dy}{dx} = 3x^2 - 6x + 1$

To find the gradient of the tangent at (2,1), substitute $x = 2$ into $\frac{dy}{dx}$ giving:

$m_{\text{tangent}} = 3(2)^2 - 6(2) + 1$
$= 12 - 12 + 1$
$= 1$

Key point

Remember that the derived function, $f'(x)$ or $\frac{dy}{dx}$, represents:

1. the gradient of the tangent to the function;
2. the rate of change of the function.

Examples

1. Problem:
Find the rate of change of $8\sqrt[4]{x^3}$ at $x = 16$

Solution:
$8\sqrt[4]{x^3} = 8x^{\frac{3}{4}}$

Since the expression was not given as $y = 8\sqrt[4]{x^3}$, we express the derivative as

$\frac{d}{dx} = 6x^{-\frac{1}{4}} = \frac{6}{\sqrt[4]{x}}$

Substituting $x = 16$ into the derivative gives $\frac{6}{\sqrt[4]{16}} = \frac{6}{2} = 3$

TOPIC 2. DIFFERENTIATION

2. Problem:

Water flowing in a stream travels $\sqrt{t^3}$ metres from its source in t seconds.

Calculate the speed of the water or rate of change after 16 seconds.

Solution:

The distance travelled by the water is $d = \sqrt{t^3} = t^{\frac{3}{2}}$

Thus the speed of the water is given by:

$$d'(t) = \frac{3}{2}t^{\frac{1}{2}}$$
$$= \frac{3}{2}\sqrt{t}$$

After 16 seconds, the speed of the water is:

$$d'(16) = \frac{3}{2}\sqrt{16}$$
$$= 6 \ ms^{-1}$$

Leibniz notation exercise Go online

Q9: If $y = 3x^5 + 3x^2$, find the value of the derivative $x = 2$.

....

Q10: If $y = \frac{4x^3 + 2x^2 - 3}{x}$, find $\frac{dy}{dx}$ when $x = 1$.

....

Q11: What is the gradient of the tangent to the function $y = \sin(3\theta)$ at $\theta = \frac{\pi}{3}$?

....

Q12: Find the rate of change of $S = \sqrt{(3t - 4)}$

2.1.8 Closed intervals

Key point

The maximum or minimum value in a closed interval occurs at either the stationary point or an endpoint.

Maximums and minimums on a closed interval   Go online

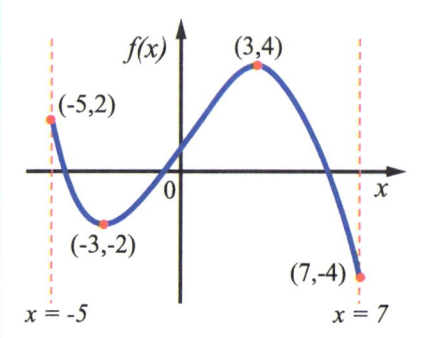

- The graph above has been drawn with a closed interval of $-5 \leq x \leq 7$.
- The maximum value of the function is 4 which occur at the stationary point (3,4).
- The minimum value of the function is -4 which occur at the end point (7,-4).

Notice that the minimum turning point is at (-3,-2), but in this case the end point is at (7,-4) which gives a lower value for the function.

Whereas the maximum turning point is at (3,4) which gives a higher value than the end point which is at (-5,2).

Example

Problem:

Find the maximum and minimum values of $f(x) = 3x - x^3$ in the closed interval $-1 \cdot 5 \leq x \leq 3$.

Solution:

It is much easier to spot the maximum and minimum values if we make a sketch.

Step 1: Find the stationary points and their nature.

$f'(x) = 3 - 3x^2$

Stationary points occur when $f'(x) = 0$

$$3 - 3x^2 = 0$$
$$3(1 - x^2) = 0$$
$$3(1 - x)(1 + x) = 0$$
$$1 - x = 0 \text{ or } 1 + x = 0$$
$$x = 1 \text{ or } x = -1$$

When $x = 1$ then $y = 3 \times 1 - 1^3 = 2$

When $x = -1$ then $y = 3 \times (-1) - (-1)^3 = -2$

Thus the stationary points are (1,2) and (-1,-2).

x	-2 →	-1	0 →	1	2 →
$(1 - x)$	+	+	+	0	-
$(1 + x)$	-	0	+	+	+
$3(1 - x)(1 + x)$	-	0	+	0	-
shape	↘	→	↗	→	↘

This gives us a maximum turning point at (1,2) and a minimum turning point at (-1,-2).

Step 2: Find the end points.

When $x = -1 \cdot 5$ then $f(-1 \cdot 5) = -1 \cdot 125$

When $x = 3$ then $f(3) = -18$

Step 3: Make a sketch.

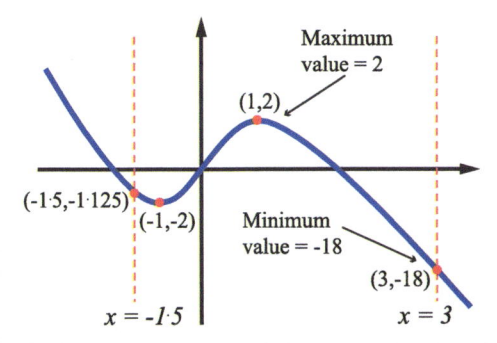

Hence the maximum value is 2 at the maximum turning point (1,2) and the minimum value is -18 at the end point (3,-18).

Closed intervals exercise

Q13:

Find the maximum and minimum values of $y = x^3 - 12x$ in the closed interval $-5 \leq x \leq 3$.

2.1.9 Applying differential calculus

Differential calculus can also be used to solve problems in context.

Rates of Change

Problems can often be solved by forming some kind of mathematical model such as a formula, an equation or a graph. The derivative of a function at a particular value describes the rate of change of the function at that value.

Example : Rate of change of volume

Problem:

A sphere with radius r and volume $V = \frac{4}{3}\pi r^3$ is inflated.

Find:

a) the rate of change of the volume of the sphere with respect to the radius;

b) the rate of change of the volume when the radius is 2 cm.

Solution:

a) The rate of change of the volume of the sphere with respect to the radius is $\frac{dV}{dr}$.

$$V = \frac{4}{3}\pi r^3$$
$$\frac{dV}{dr} = 3 \times \frac{4}{3}\pi r^2 \text{ (Note that } \frac{4}{3}\pi \text{ is a constant.)}$$
$$= 4\pi r^2$$

b) The rate of change of the volume when the radius is 2 cm is:

$$4\pi \times 2^2 = 16\pi$$

> **Key point**
>
> Velocity or speed is a rate of change of distance (or displacement) s at time t so the velocity, $v = \frac{ds}{dt}$ or $s'(t)$.

> **Key point**
>
> Acceleration, a, is a rate of change for velocity (or speed), v at time t, so $a = \frac{dv}{dt}$ or $s''(t)$.

Example : Distance (s), velocity (v) and acceleration (a)

Problem:

The displacement of a particle at time t, relative to its starting position, is given by the formula
$s(t) = 2t^3 - 6t^2 + 10t + 5$

$s(t)$ is the distance in centimetres and t is the time in seconds.

TOPIC 2. DIFFERENTIATION

Calculate:

a) the velocity of the particle when $t = 4$

b) the acceleration of the particle when $t = 4$

Solution:

a) Velocity, $\frac{ds}{dt} = 6t^2 - 12t + 10$
When $t = 4$ then velocity $= 6 \times 4^2 - 12 \times 4 + 10 = 58 cm\ s^{-1}$

b) To find an expression for acceleration we find the second derivative of displacement. This is the same as differentiating our formula for velocity.
Acceleration, $a = \frac{dv}{dt} = 12t - 12$
When $t = 4$ then acceleration $= 12 \times 4 - 12 = 36\ cms^{-2}$

Optimisation

Many problems involve finding a maximum or minimum value where the context is modelled on a function. These values can be found by using differentiation to find stationary points.

Examples

1. Maximum height

Problem:

The height, h, of a ball thrown upwards is given by the formula $h(t) = 20t - 5t^2$, where t is the time in seconds from when the ball is thrown.

a) When does the ball reach its maximum height?

b) Calculate the maximum height that the ball reaches?

Solution:

a) The ball reaches its maximum height at a stationary point i.e. when $h'(t) = 0$.

$h'(t) = 20 - 10t$
$20 - 10t = 0$
$20 = 10t$
$t = 2$

It is important to check that the stationary point is a maximum by making a nature table.

t	1 →	2	3 →
$h'(t)$	+	0	-
Shape	↗	→	↘

Thus there is a maximum at $t = 2$ and the ball reaches its maximum height after 2 seconds.

© HERIOT-WATT UNIVERSITY

b) The ball reaches its maximum height when $t = 2$
To find the height we must let $t = 2$ in $h(t)$: $h(2) = 20 \times 2 - 5 \times 2^2 = 20$
Hence the ball reaches a maximum height of 20 metres.

....

2. Maximum Area
Problem:
A farmer has 80 metres of fencing to make a rectangular sheep pen against an existing wall.

a) Show that the area of the sheep pen is given by $A = 80x - 2x^2$
b) Find the greatest area that the farmer can enclose.

Solution:

a) First assign variables for the length and breadth of the rectangle.
If x is the breadth of the rectangle then, since there is 80 metres of fencing in total, the length of the rectangle must be $80 - 2x$ metres.
Thus the area of the rectangle is: $A(x) = length \times breadth = (80 - 2x)x = 80x - 2x^2$

b) The pen will have its maximum area at a stationary point, i.e. $A'(x) = 0$
$$A'(x) = 80 - 4x$$
$$80 - 4x = 0$$
$$80 = 4x$$
$$x = 20$$

Check that the stationary point at $x = 20$ is a maximum by making a nature table.

x	19 →	20	21 →
$A'(x)$	+	0	-
shape	↗	→	↘

Thus there is a maximum area at $x = 20$

The dimensions of the pen are:
$length = 80 - 2x$ and $breadth = x$
$ = 40\ m$ $ = 20\ m$

Maximum area = length × breadth
$= 40 \times 20$
$= 800\ m^2$

....

3. Problem:

A family hope to build a room in their attic.

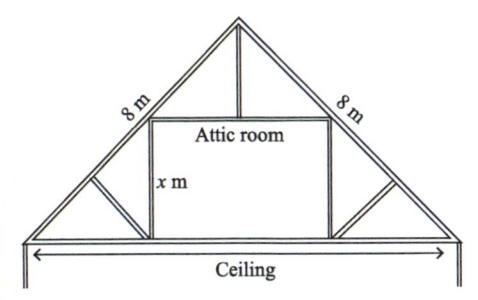

The roof space is in the shape of an isosceles right-angled triangle.

a) Find the length of the ceiling.
b) Given that the height of the attic room is x metres, show that the area of the room is given by $A = 8\sqrt{2}x - 2x^2$
c) Find the dimensions of the attic room which maximises the area of the space.

Solution:

a) The roof space is a right angled triangle.

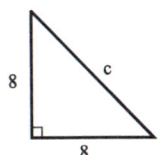

The length of the ceiling $= \sqrt{8^2 + 8^2} = 8\sqrt{2}$

b) The shaded triangles are all isosceles right-angled triangles.

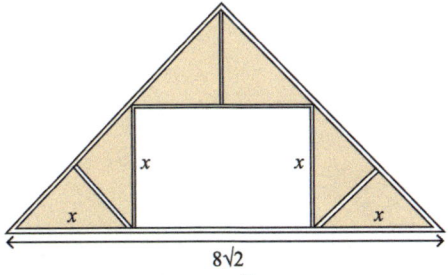

Length of attic room $= 8\sqrt{2} - 2x$
Area of the attic room
$= \left(8\sqrt{2} - 2x\right) \times x$ as required.
$= \left(8\sqrt{2}\right)x - 2x^2$

c) The attic room will have its maximum area at a stationary point, i.e. $\frac{dA}{dx} = 0$

$$\frac{dA}{dx} = 8\sqrt{2} - 4x$$
$$8\sqrt{2} - 4x = 0$$
$$8\sqrt{2} = 4x$$
$$x = 2\sqrt{2}$$

Check that the stationary point at $x = 2\sqrt{2}$ is a maximum by making a nature table.

x	$2 \rightarrow$	$2\sqrt{2}$	$3 \rightarrow$
$\frac{dA}{dx}$	+	0	−
Shape	↗	→	↘

Thus there is a maximum area at $x = 2\sqrt{2}$

The dimensions of the attic room are:
$length = 8\sqrt{2} - 2x$ and $breadth = x$
$\qquad = 4\sqrt{2}\ m$ $\qquad = 2\sqrt{2}\ m$

Maximum area = length × breadth
$\qquad = 4\sqrt{2} \times 2\sqrt{2}$
$\qquad = 16\ m^2$

2.1.10 The derivative

The diagram that follows shows a distance-time graph for a car travelling along a road on the first 4 seconds of its journey.

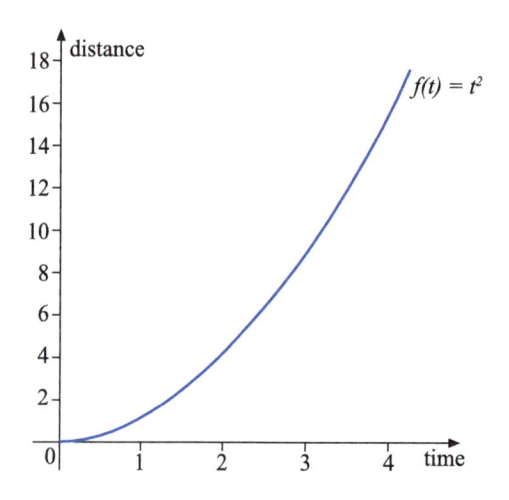

Remember, it is possible to calculate average speed as, average speed $= \frac{distance}{time}$

TOPIC 2. DIFFERENTIATION

Thus the average speed for the car in the first three seconds of its journey is,

$$\text{average speed} = \frac{\text{distance}}{\text{time}}$$
$$= \frac{9}{3}$$
$$= 3 \; ms^{-1}$$

Look again at the distance-time graph and answer the following questions.

Q14: When does it seem that the car is travelling faster?

a) $t = 1$
b) $t = 4$

....

Q15: How can you tell this from the graph?

....

Q16: Why can the *average* speed only be calculated for the first few seconds of the journey and not the exact speed?

Consider the question, what is the speed at exactly $t = 2$?

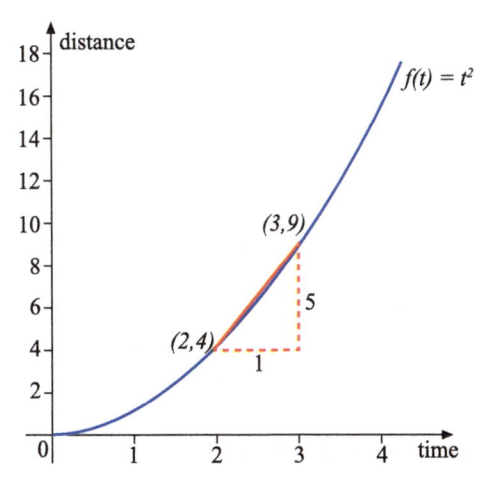

A first estimate can be found by calculating the average speed between $t = 2$ and $t = 3$ from the diagram above.

72   TOPIC 2. DIFFERENTIATION

$$\begin{aligned} average\ speed &= \frac{\text{distance}}{\text{time}} \\ &= \frac{f(3) - f(2)}{3 - 2} \\ &= \frac{9 - 4}{3 - 2} \\ &= \frac{5}{1} \\ &= 5\ ms^{-1} \end{aligned}$$

To obtain a better estimate for the instantaneous speed at $t = 2$, we could choose shorter time intervals. See the following table of results:

Time Interval	Distance	Time	Average Speed
2 - 3	5	1	5
2 - 2·5	2·25	0·5	4·5
2 - 2·2	0·84	0·2	4·2
2 - 2·1	0·41	0·1	4·1
2 - 2·01	0·0401	0·01	4·01
2 - 2·001	0·004001	0·001	4·001

Notice that, as the time tends to zero, the speed tends to a limit of 4 ms⁻¹. It seems reasonable to conclude that the instantaneous speed of the car at $t = 2$ is 4 ms⁻¹.

In a similar way, it is also possible to calculate the instantaneous speed at $t = 1, 2, 3, 4, 5\ldots$ The results are shown here.

Time	1	2	3	4	5
Instantaneous Speed	2	4	6	8	10

Perhaps you have already noticed for the function $f(t) = t^2$ that the instantaneous speed at time t can be calculated as $2t$ ms⁻¹.

The instantaneous speed, or rate of change of distance with respect to time, can be written as $f'(t)$; we say this as "f dashed t" and it is known as the derived function of $f(t)$.

In a similar way, it is also possible to consider other functions apart from $f(t) = t^2$ and the results that are obtained for the derived function are as listed here.

Distance $f(t)$	t	t^2	t^3	t^4	t^5	$\ldots$	t^n
Speed $f'(t)$	1	$2t$	$3t^2$	$4t^3$	$5t^4$	$\ldots$	nt^{n-1}

Notice that we have obtained a general rule for finding the **derived function**.

Key point

When $f(t) = t^n$ then $f'(t) = nt^{n-1}$

We differentiate $f(t)$ to obtain $f'(t)$

This process is called differentiation.

© HERIOT-WATT UNIVERSITY

Differentiation from first principles

The instantaneous speed of a car at time $t = 2$ seconds can be obtained by calculating the average speed over smaller and smaller time intervals. It is important to relate this to the features of the graph for $f(t)$.

The first estimate was the average speed between $t = 2$ and $t = 3$. Notice that:

$$\text{average speed} = \frac{\text{distance}}{\text{time}}$$
$$= \text{gradient of chord AB}$$

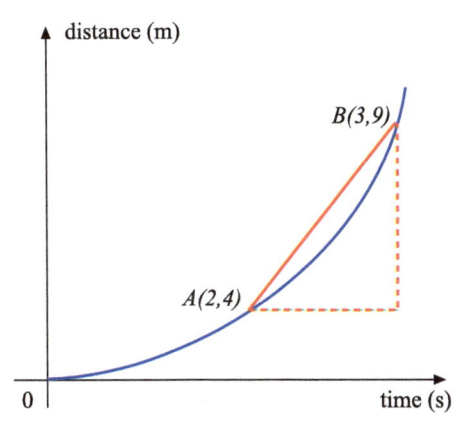

Better estimates for the instantaneous speed at $t = 2$ are obtained by taking progressively shorter time intervals. Notice that the gradients of the chords, AB_1, AB_2 and AB_3, move closer to the gradient of the tangent to the curve at $t = 2$. Indeed, the instantaneous speed at $t = 2$ is equal to the gradient of the tangent to the curve at that point.

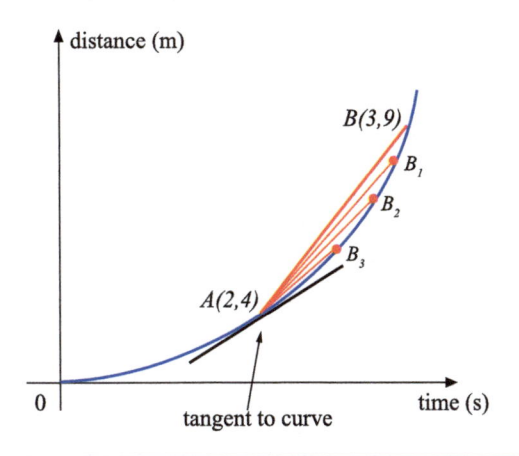

74   TOPIC 2. DIFFERENTIATION

In general, the instantaneous speed, $f\,'(t)$, is equal to the gradient of the tangent to the curve at that time.

The gradient of a tangent can now be used to calculate the instantaneous rate of change of any function.

$f\,'(x)$ is the derived function or the derivative of $f(x)$. It represents the gradient of the tangent to the graph of the function and the rate of change of the function.

The gradient of a curve at a point is defined to be the gradient of the tangent to the curve at that point.

When we use this method to obtain $f\,'(x)$, this is called differentiating from first principles. Differentiation is one area of calculus.

> **Key point**
>
> The derived function represents:
>
> 1. the gradient of the tangent to the function.
> 2. the rate of change of the function.

2.2   Conditions for differentiability

Not all functions can be differentiated. This section will look at what it means for a function to be differentiable as well as looking at circumstances in which a function is not differentiable.

2.2.1   Differentiability at a point

> **Key point**
>
> In general, a function $f\,(x)$ is differentiable at $x = a$ if the curve $y = f\,(x)$ has a tangent at the point $(a,\ f\,(a))$

This is because the derivative is defined as:

$$f'(x) = \lim_{h \to \infty} \left\{ \frac{f(x+h) - f(x)}{h} \right\} \text{ i.e. } \lim_{h \to \infty} \left\{ \frac{\text{change in } y}{\text{change in } x} \right\} = m$$

Essentially, if the gradient of the tangent at the point $x = a$ *exists* and *can be found*, and this occurs for all values of $f(x)$ then the function is differentiable.

The following diagram illustrates this:

a)

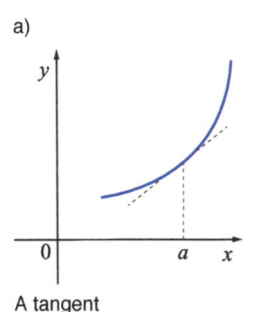

A tangent

Now compare this to two functions that are not differentiable:

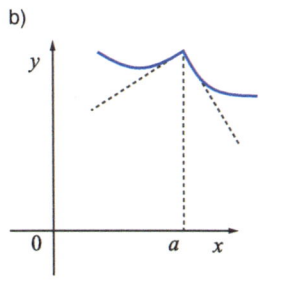

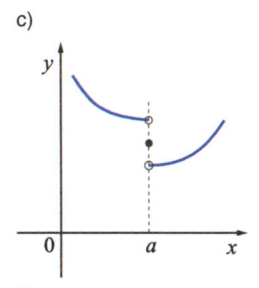

b)

c)

No tangent

No tangent

The tangent line incorporates two properties that allow a function to be differentiable:

- Smoothness
- Continuity

Diagram b) is not smooth at $x = a$ since the gradient of the tangent calculated approaching from the left does not equal the gradient of the tangent calculated approaching from the right: there is a corner. Therefore this function is not differentiable at $x = a$.

Diagram c) is not continuous (it is discontinuous) at $x = a$. There is a gap between the y coordinates at $x = a$. If a function is not continuous at $x = a$, then it is not differentiable at $x = a$

At this point we will define continuity at a as $f(a) = \lim_{x \to a} f(x)$. This means that as you approach $f(a)$ from the left you should get the same value as you do approaching $f(a)$ from the right. In diagram c), it is clear that the y-coordinates are different depending on which direction you come from along the x-axis.

There is a proof in the proofs section which justifies the connection between differentiability and continuity. This is not required to be reproduced for exam purposes, simply for your interest.

Proof 1: Justification of connection between differentiability and continuity

Prove that $\lim_{h \to 0} g(x+h) - \lim_{h \to 0} g(x) = 0 \Rightarrow \lim_{h \to 0} g(x+h) = g(x)$

$$\lim_{h \to 0} [g(x+h) - g(x)] = \lim_{h \to 0} \left[\frac{g(x+h) - g(x)}{h} \times h \right]$$
$$= \lim_{h \to 0} \frac{g(x+h) - g(x)}{h} \times \lim_{h \to 0} h$$
$$= g'(x) \times 0$$
$$= 0$$

Thus,
$\lim_{h \to 0} [g(x+h) - g(x)] = 0$
Hence,
$\lim_{h \to 0} g(x+h) - \lim_{h \to 0} g(x) = 0 \Rightarrow \lim_{h \to 0} g(x+h) = g(x)$

However, just because a function is smooth and continuous does not mean it is differentiable. Remember, to be differentiable the gradient of the tangent must exist and be calculable as well.

Examples

1.

Notice that the graph of $f(x) = x^{\frac{1}{3}}$ is smooth and continuous, but is not differentiable at $x = 0$

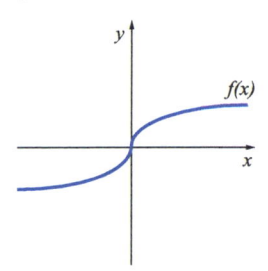

As x tends to 0^+, the gradient m_T is undefined or tends to infinity. If the gradient cannot be defined then it is not differentiable at this point.

....

2.
Consider the modulus function $f(x) = |x|$

This defines a function on $\mathbb{R}$ where:

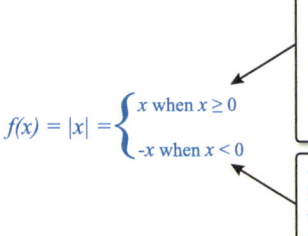

$$f(x) = |x| = \begin{cases} x \text{ when } x \geq 0 \\ -x \text{ when } x < 0 \end{cases}$$

x when *x* ≥ 0 means that when *x* is bigger than or equal to zero (a positive number) the function is represented by the line *y* = *x*, the *y* coordinates are positive.

-*x* when *x* < 0 means that when *x* is less than zero (a negative number) the function is represented by the line *y* = -(-*x*)
i.e. *y* = *x* , the *y* coordinates are positive.

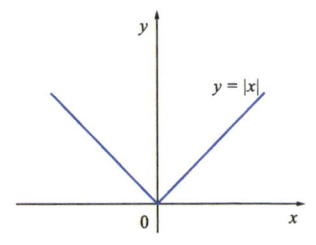

Notice that the graph has a sharp point at $x = 0$ (it is not smooth) and therefore does not have a tangent at this point. As a consequence, it is not differentiable at $x = 0$

....

3. This function is discontinuous at $x = -\frac{\pi}{2}$ and $x = \frac{\pi}{2}$.
It is therefore not differentiable at these points.

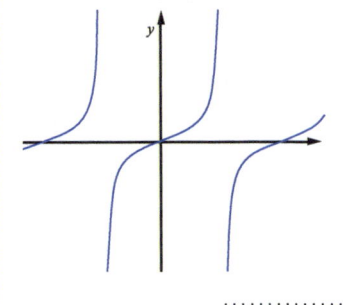

....

TOPIC 2. DIFFERENTIATION

4. Consider the function:

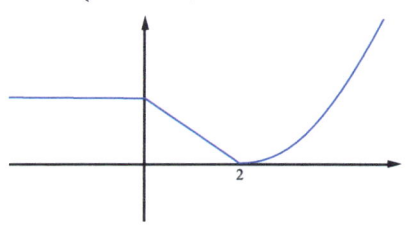

$$f(x) = \begin{cases} 4 & , \quad x \leq 0 \\ -2x + 4 & , \quad 0 < x \leq 2 \\ x^2 & , \quad x > 2 \end{cases}$$

This function is not smooth at $x = 0$ and $x = 2$. It is therefore not differentiable at these points.

Differentiability at a point: Continuous or discontinuous   Go online

Q17: Identify which graphs are continuous and which are discontinuous. You will also need to determine if they are smooth or not.

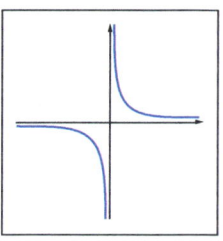

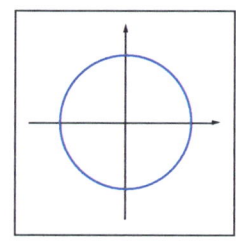

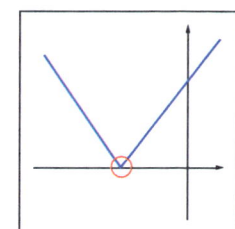

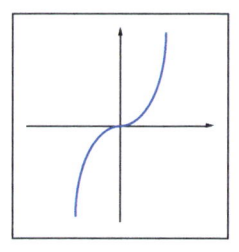

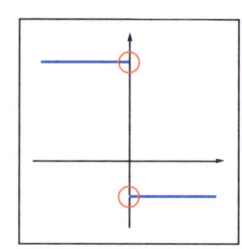

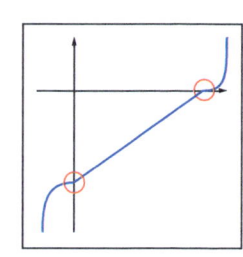

TOPIC 2. DIFFERENTIATION

> **Key point**
>
> An easy way to distinguish between continuous and discontinuous graphs is that:
>
> - continuous graphs can be drawn without lifting your pen from the paper;
> - discontinuous graphs your pen has to be lifted.

Differentiability at a point exercise   Go online

Q18: Is the graph continuous or discontinuous?

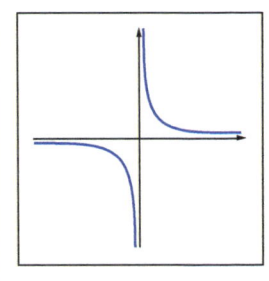

a) Continuous
b) Discontinuous

....

Q19: In the graph below, for what value of x is the graph undefined?

....

Q20: Is the graph differentiable for all values of x?

a) Yes
b) No

....

Q21: For what value of x is the function not differentiable?

....

Q22: Is this function differentiable everywhere?

$f(x) = (x^2 - 2)^2 (x^2 + 2)^2$

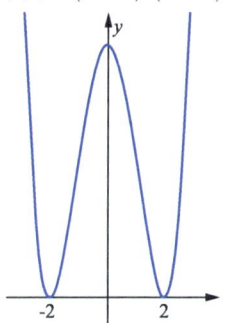

....

Q23: Is this function differentiable everywhere?

$f(x) = \log x$

....

Q24: Is this function differentiable everywhere?

$f(x) = \begin{cases} x^2, & x \leqslant 1 \\ x + 2, & -1 < x \leqslant 2 \\ 4 - (x-2)^2, & x > 2 \end{cases}$

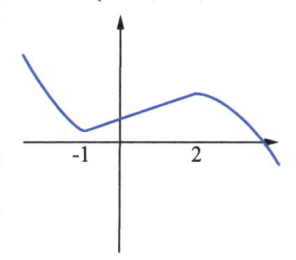

....

TOPIC 2. DIFFERENTIATION

Q25: Which of these graphs are not differentiable everywhere?

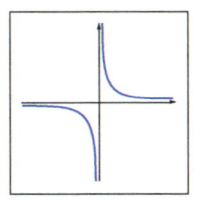

Graph A

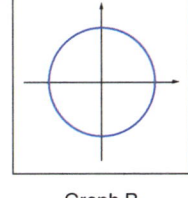

Graph B

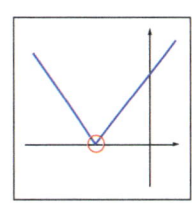

Graph C

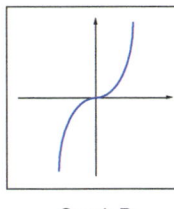

Graph D

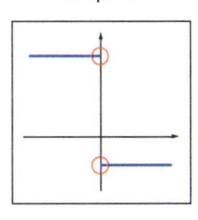

Graph E

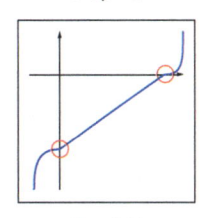

Graph F

a) B, C, D, F
b) A, C, D, E
c) B, C, E, F
d) A, C, E, F

....

Q26: Why are these graphs not differentiable?

2.2.2 Differentiability over an interval

Key point

Notation for the closed interval $[a, b]$ given as $[a, b] = \{x \in \mathbb{R} : a \leq x \leq b\}$ means the closed interval with end points a and b.

$f(x)$ is differentiable over the interval $[a, b]$ if $f'(x)$ exists for each $x \in [a, b]$

Differentiability over an interval

Example The graph shown is $f(x) = \sqrt{x}$ for $x \in [1, 10]$

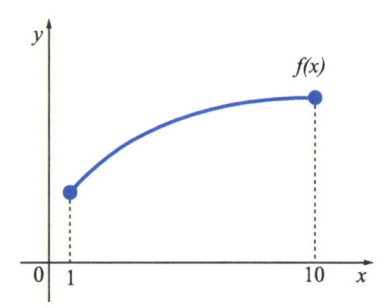

$f(x)$ is differentiable in this interval because $f'(x)$ exists for each $x \in [1, 10]$.

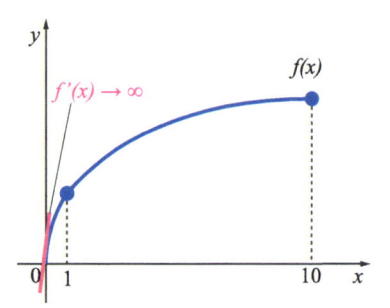

Whereas for the interval $[0, 10]$, $f(x)$ is not differentiable at $x = 0$ because as $x \to 0$, $f'(x) \to \infty$

Differentiability over an interval exercise

Q27: Is the function below differentiable over the interval [-2,3]?
Justify your answer.

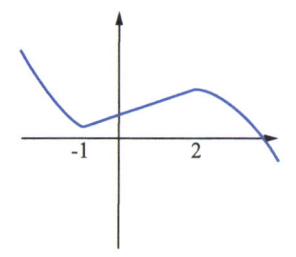

TOPIC 2. DIFFERENTIATION

Q28: Is the function below differentiable over the interval [-5,-1]?
Give a possible closed interval for which it is not differentiable.

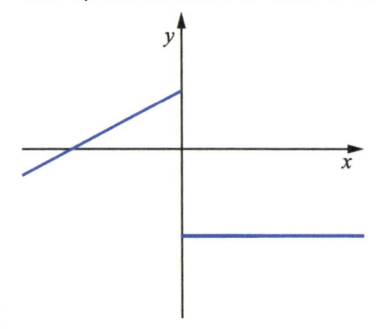

Q29: Give a possible closed interval in which this function is differentiable.

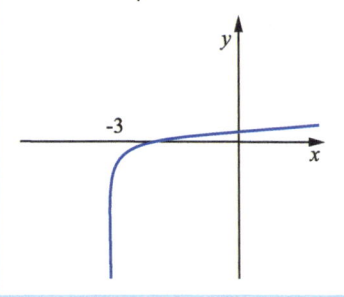

2.2.3 Higher derivatives

If $y = f(x)$, then its derivative $f'(x)$ is also a function of x and may itself have a derivative.

The derivative of $f'(x)$ is denoted by $f''(x)$ or $f^{(2)}(x)$

This is read as 'f double dash x' or 'f two of x'.

In Leibniz notation, the derivative of $\frac{dy}{dx}$ is denoted by $\frac{d^2y}{dx^2}$

This is read as 'd two y by dx squared'.

Key point

$f''(x)$ or $\frac{d^2y}{dx^2}$ is called the second derivative of the function.

When $f'(x)$ and $f''(x)$ exist and are differentiable, then the derivative of $f''(x)$ is denoted by $f'''(x)$ or $f^{(3)}(x)$

Similarly, in Leibniz notation when $\frac{dy}{dx}$ and $\frac{d^2y}{dx^2}$ exist and are differentiable, then $\frac{d}{dx}\left(\frac{d^2y}{dx^2}\right) = \frac{d^3y}{dx^3}$

© HERIOT-WATT UNIVERSITY

84   TOPIC 2. DIFFERENTIATION

More generally, for $n = 2, 3, 4, \ldots$ the nth derivative of $f(x)$, when it exists, is denoted by: $f^{(n)}$ or $\frac{d^n y}{dx^n}$

Examples

1. Problem:
Find the first, second, third and fourth derivative of the function $y = 3x^3 - 6x + 4$
Solution:
The first derivative is $\frac{dy}{dx} = 9x^2 - 6$

The second derivative is $\frac{d^2 y}{dx^2} = 18x$ which is the result of finding the derivative of $\frac{dy}{dx}$
The third derivative is $\frac{d^3 y}{dx^3} = 18$ which is the result of finding the derivative of $\frac{d^2 y}{dx^2}$
The fourth derivative is $\frac{d^4 y}{dx^4} = 0$ which is the result of finding the derivative of $\frac{d^3 y}{dx^3}$

....

2. Problem:
Find $f'(x)$, $f''(x)$, $f^{(3)}(x)$ and $f^{(4)}(x)$ for the function $f(x) = \sin 3x$
Solution:
$f'(x) = 3\cos 3x$
$f''(x) = -9\sin 3x$ by differentiating $f'(x)$
$f^{(3)}(x) = -27\cos 3x$ by differentiating $f''(x)$
$f^{(4)}(x) = 81\sin 3x$ by differentiating $f^{(3)}(x)$

....

3. Problem:
Find all the derivatives of $y = 5x^3 + 2x^2 + 3x - 1$
Solution:
$\frac{dy}{dx} = 15x^2 + 4x + 3$
$\frac{d^2 y}{dx^2} = 30x + 4$
$\frac{d^3 y}{dx^3} = 30$
$\frac{d^4 y}{dx^4} = 0$ and $\frac{d^n y}{dx^n} = 0$ for $n = 5, 6, 7, \ldots$

© HERIOT-WATT UNIVERSITY

Higher derivatives exercise

Q30: Find all the derivatives of $f(x) = 4x^4 + 5x^3 - 3$

Q31: Find $f''(x)$ when $f(x) = \frac{3}{(2-x)}$

Q32: What is the third derivative of $y = 4\cos\left(2x - \frac{\pi}{3}\right)$?

Q33: How many times can you differentiate the function $f(x) = 2x^{10} + 3x^7 - 2x + 9$ before it becomes zero?

2.2.4 Discontinuities

Derivatives can have discontinuities.

Study the following example.

Example

Consider the graph of $f(x) = x^{\frac{4}{3}}$

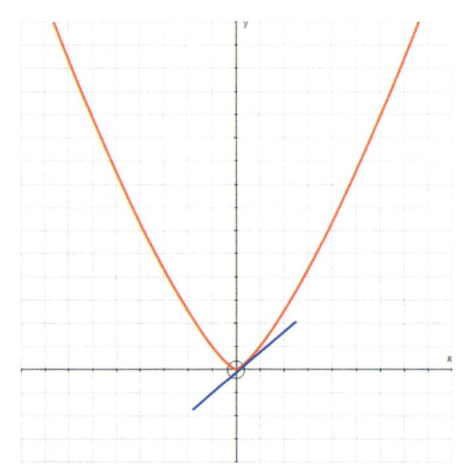

It is possible to find and graph $f'(x) = \frac{4}{3}x^{\frac{1}{3}}$

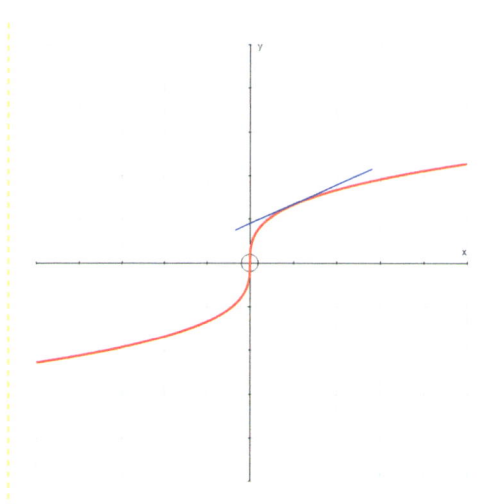

Notice that, as $x \to 0$, $m_T \to \infty$ and so $f''(x)$ is undefined at $x = 0$
Therefore, $f'(x)$ is not differentiable at $x = 0$

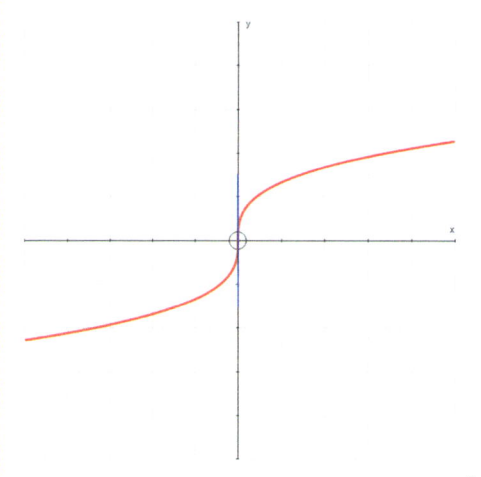

It is also possible to find and graph $f''(x) = \frac{4}{9}x^{-\frac{2}{3}}$

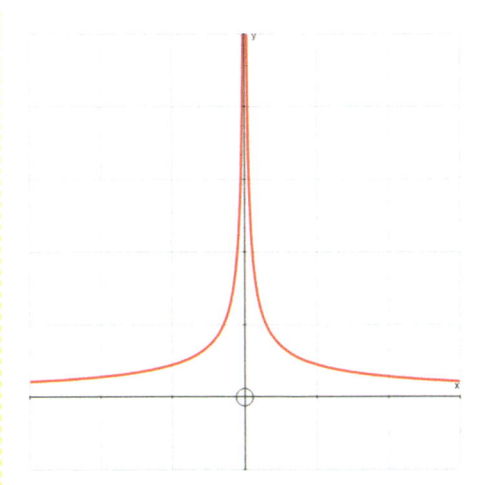

Notice that the limit does not exist for f'' as $(x) \to 0$

The graph is discontinuous at $x = 0$

So, in this example, $f(x)$ and $f'(x)$ are continuous but $f''(x)$ is discontinuous.

Key point

If $f(x)$ and $f'(x)$ is continuous it does not mean that $f''(x)$ is continuous or further derivatives are automatically continuous.

Discontinuity exercise   Go online

Let $f(x) = 5(x+1)^{\frac{7}{5}}$.

Q34: Calculate $f'(x)$ and $f''(x)$

....

Q35: At which value of x is $f'(x)$ not differentiable and $f''(x)$ discontinuous?

....

Q36: Let $f(x) = 3(2x+1)^{\frac{4}{3}}$. Calculate $f'(x)$ and $f''(x)$.

....

88 TOPIC 2. DIFFERENTIATION

Q37: The graphs for $f(x) = 3(2x+1)^{\frac{4}{3}}$ along with its first and second derivatives are shown. At which values of x is $f'(x)$ and $f''(x)$ not defined?

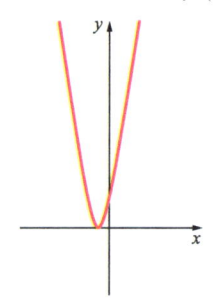

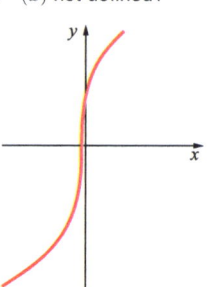

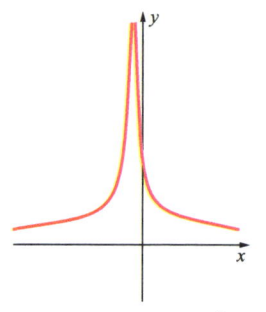

$f(x) = 3(2x+1)^{\frac{4}{3}}$ $f'(x) = 8(2x+1)^{\frac{1}{3}}$ $f''(x) = \frac{16}{3}(2x+1)^{-\frac{2}{3}}$

....

Q38: Consider $f(x) = \begin{cases} x^2 - 3, & \text{for } x < 0 \\ x - 3, & \text{for } x \geqslant 0 \end{cases}$

a) Make a sketch of this graph.
b) Calculate $f'(x)$ and sketch this graph.
c) For what values of x is $f'(x)$ not differentiable?

2.3 The product rule

> **Key point**
>
> The **product rule** is a method used to differentiate the product of two or more functions.
>
> It states that, when $k(x) = f(x)\,g(x)$, then $k'(x) = f'(x)\,g(x) + f(x)\,g'(x)$

There is a proof in the proofs section which justifies the product rule being used to differentiate the product of two or more functions. This is not required to be reproduced for exam purposes, simply for your interest.

TOPIC 2. DIFFERENTIATION

Proof 2: The product rule

This result can be proved using differentiation from 'first principles'.

Step 1: replace $k(x)$ with $f(x)g(x)$
Let $k(x) = f(x)\,g(x)$
then $k'(x) = \lim_{h \to 0} \left\{ \dfrac{k(x+h) - k(x)}{h} \right\}$
$= \lim_{h \to 0} \left\{ \dfrac{f(x+h)\,g(x+h) - f(x)\,g(x)}{h} \right\}$

Step 2: introduce a term $f(x)\,g(x+h)$
$= \lim_{h \to 0} \left\{ \dfrac{f(x+h)g(x+h) + f(x)g(x+h) - f(x)g(x)}{h} \right\}$

Step 3: since we introduced this term, we need to take $f(x)\,g(x+h)$ away again
$= \lim_{h \to 0} \left\{ \dfrac{f(x+h)g(x+h) - f(x)g(x+h) + f(x)g(x+h) - f(x)g(x)}{h} \right\}$

Step 4: now rearrange to take out a factor of $g(x+h)$ and $f(x)$
$= \lim_{h \to 0} \left\{ \dfrac{(f(x+h) - f(x))g(x+h) + f(x)(g(x+h) - g(x))}{h} \right\}$

Step 5: now separate into two expressions
$= \lim_{h \to 0} \left\{ \dfrac{f(x+h) - f(x)}{h} \times g(x+h) \right\} + \lim_{h \to 0} \left\{ \dfrac{g(x+h) - g(x)}{h} \times f(x) \right\}$

Step 6: by taking the limit of each expression, we now have the definition of differentiation from first principles
$= \lim_{h \to 0} \left\{ \dfrac{f(x+h) - f(x)}{h} \right\} \times \lim_{h \to 0} g(x+h) + \lim_{h \to 0} \left\{ \dfrac{g(x+h) - g(x)}{h} \right\} \times \lim_{h \to 0} f(x)$

Step 7: rewrite this with dashed notation
$= f'(x)g(x) + g'(x)f(x)$
$= f'(x)g(x) + f(x)g'(x)$

The proof in full looks like this:
Let $k(x) = f(x)g(x)$
then $k'(x) = \lim_{h \to 0} \left\{ \dfrac{k(x+h) - k(x)}{h} \right\}$
$= \lim_{h \to 0} \left\{ \dfrac{f(x+h)g(x+h) - f(x)g(x)}{h} \right\}$
$= \lim_{h \to 0} \left\{ \dfrac{f(x+h)g(x+h) - f(x)g(x+h) + f(x)g(x+h) - f(x)g(x)}{h} \right\}$
$= \lim_{h \to 0} \left\{ \dfrac{(f(x+h) - f(x))\,g(x+h) + f(x)\,(g(x+h) - g(x))}{h} \right\}$
$= \lim_{h \to 0} \left\{ \dfrac{f(x+h) - f(x)}{h} \times g(x+h) \right\} + \lim_{h \to 0} \left\{ \dfrac{g(x+h) - g(x)}{h} \times f(x) \right\}$
$= \lim_{h \to 0} \left\{ \dfrac{f(x+h) - f(x)}{h} \right\} \times \lim_{h \to 0} g(x+h) + \lim_{h \to 0} \left\{ \dfrac{g(x+h) - g(x)}{h} \right\} \times \lim_{h \to 0} f(x)$
$= f'(x)g(x) + g'(x)f(x)$
$= f'(x)g(x) + f(x)g'(x)$

© HERIOT-WATT UNIVERSITY

TOPIC 2. DIFFERENTIATION

> **Key point**
>
> In short Proof 2 can be summarised as: $\frac{d}{dx}(fg) = f'g + fg'$

Proof 1 in the proofs section justifies $\lim_{h \to 0} g(x+h) = g(x)$. This is not required to be reproduced for exam purposes, simply for your interest.

Proof 2 can be extended to more than two factors.

When $k(x) = f(x)\,g(x)\,h(x)$
then $k'(x) = f'(x)\,g(x)\,h(x) + f(x)\,g'(x)\,h(x) + f(x)\,g(x)\,h'(x)$

> **Key point**
>
> In short: $\frac{d}{dx}(f\,g\,h) = f'\,g\,h + f\,g'\,h + f\,g\,h'$

Leibniz Notation

If $y = uv$, where u and v are functions of x, then: $\frac{dy}{dx} = \frac{du}{dx}v + u\frac{dv}{dx}$

> **Key point**
>
> In short: $y' = u'v + uv'$

Examples

1.

Problem:

Calculate $k'(x) = x^2(x+2)^3$

Solution:

Let $f = x^2$ and $g = (x+2)^3$
then $f' = 2x$ and $g' = 3(x+2)^2$

Now $(fg)' = f'g + fg'$
$k'(x) = 2x(x+2)^3 + x^2 \times 3(x+2)^2$
$k'(x) = x(x+2)^2(2(x+2) + 3x)$ (Notice that a common factor is taken out here.)
$k'(x) = x(x+2)^2(5x+4)$

....

2.

Problem:

Using Leibniz notation, calculate $\frac{dy}{dx}$ when $y = \sqrt{x}\,(2x+3)^2$

Solution:

Let $u = \sqrt{x} = x^{\frac{1}{2}}$ and $v = (2x+3)^2$

then $\frac{du}{dx} = \frac{1}{2}x^{-\frac{1}{2}}$ and $\frac{dv}{dx} = 2(2x+3) \times 2 = 4(2x+3)$

Now:

$$\frac{d}{dx}(uv) = \frac{du}{dx}v + u\frac{dv}{dx}$$

$$\frac{dy}{dx} = \frac{1}{2}x^{-\frac{1}{2}}(2x+3)^2 + x^{\frac{1}{2}} \times 4(2x+3)$$

$$= \frac{1}{2}x^{-\frac{1}{2}}(2x+3)\left((2x+3) + 8\frac{x^{\frac{1}{2}}}{x^{-\frac{1}{2}}}\right)$$

$$= \frac{1}{2}x^{-\frac{1}{2}}(2x+3)(10x+3)$$

The product rule exercise   Go online

Q39: Show that $\frac{d}{dx}\{(x^2+7)(x^3-6x+9)\} = 5x^4 + 3x^2 + 18x - 42$

Q40: Show that $\frac{d}{dx}(\sin x \, \cos x) = \cos 2x$

Q41: Given $f(x) = x(2-3x^2)^5$, obtain $f'(x)$ and simplify your answer.

Q42: Given $f(x) = 5x\sin^2 x$, $-\pi < x < \pi$, obtain $f'(x)$ and evaluate $f'\left(\frac{\pi}{4}\right)$.

Q43: Given $f(x) = x^2\sqrt{5x^2-8}$, obtain $f'(x)$ and simplify your answer.

Q44: Given $f(x) = \sin 4x \cos(2x+3)$, obtain $f'(x)$ and simplify your answer.

Q45: Given $f(x) = x(x+3)^2 \sin x$, obtain $f'(x)$ and simplify your answer.

Q46: Given $f(x) = x^2 \sin 2x \cos 3x$, obtain $f'(x)$ and simplify your answer.

2.4 The quotient rule

> The **quotient rule** gives us a method that allows us to differentiate algebraic fractions.
>
> It states that when $k(x) = \frac{f(x)}{g(x)}$, then $k'(x) = \frac{f'(x)g(x) - f(x)g'(x)}{[g(x)]^2}$

There is a proof in the proofs section which justifies the quotient rule being used to differentiate algebraic fractions. This is not required to be reproduced for exam purposes, simply for your interest.

Proof 3: The quotient rule

Let $k(x) = \frac{f(x)}{g(x)} = f(x)[g(x)]^{-1}$

We can use the product rule to differentiate this:

Let $f = f(x)$ and $g = [g(x)]^{-1}$ then:

$f' = f'(x)$ and $g' = -[g(x)]^{-2} \times g'(x)$ (by the chain rule)

Step 1: from the product and chain rule
$k'(x) = f'(x)[g(x)]^{-1} + f(x)\left(-[g(x)]^{-2} \times g'(x)\right)$

Step 2: take out a factor of $[g(x)]^{-2}$
$k'(x) = [g(x)]^{-2} [f'(x)g(x) - f(x)g'(x)]$

Step 3: rewrite as a fraction
$k'(x) = \frac{f'(x)g(x) - f(x)g'(x)}{[g(x)]^2}$

Key point

In short:

$$\frac{f'}{g} = \frac{f'g - fg'}{g^2}$$

Leibniz Notation

If $y = \frac{u}{v}$, where u and v are functions of x, then:

Key point

$$\frac{dy}{dx} = \frac{\frac{du}{dx}v - u\frac{dv}{dx}}{v^2}$$

TOPIC 2. DIFFERENTIATION

Examples

1. Problem:

Differentiate $k(x) = \frac{(2x-3)^2}{(x+2)^2}$

Solution:

Using the quotient rule: $\frac{f'}{g} = \frac{f'g - fg'}{g^2}$

Step 1:

Let $f = (2x-3)^2$ and $g = (x+2)^2$

so, $f' = 2(2x-3) \times 2 = 4(2x-3)$ (using the chain rule) and $g' = 2(x+2)$

giving, $\frac{f'}{g} = \frac{4(2x-3)(x+2)^2 - (2x-3)^2 2(x+2)}{((x+2)^2)^2}$

Step 2: take out the common factor of $2(2x-3)(x+2)^2$

$$\frac{f'}{g} = \frac{4(2x-3)(x+2)^2 - (2x-3)^2 2(x+2)}{((x+2)^2)^2}$$

$$= \frac{2(2x-3)(x+2)(2(x+2) - (2x-3))}{(x+2)^4}$$

$$= \frac{2(2x-3)(7)}{(x+2)^3}$$

$$= \frac{14(2x-3)}{(x+2)^3}$$

....

2. Problem:

Differentiate $f(x) = \frac{2-3x}{\sqrt{2+3x}}$

Solution:

Step 1: Using the Quotient Rule: $\frac{f'}{g} = \frac{f'g - fg'}{[g]^2}$

Let $f = 2 - 3x$ and $g = (2+3x)^{\frac{1}{2}}$

Then $f' = -3$ and $g' = \frac{3}{2}(2+3x)^{-\frac{1}{2}}$

Giving,

$$\frac{f'}{g} = \frac{-3(2+3x)^{\frac{1}{2}} - (2-3x)\frac{3}{2}(2+3x)^{-\frac{1}{2}}}{\left((2+3x)^{\frac{1}{2}}\right)^2}$$

Step 2: Simplify the denominator

Taking out a common factor of $\frac{3}{2}(2+3x)^{-\frac{1}{2}}$, or if you prefer $\frac{3}{2(2+3x)^{\frac{1}{2}}}$

Before we do this, we will re-write the numerator in a way that shows clearly how this factor is taken out:

$$\frac{f'}{g} = \frac{\frac{-6(2+3x)}{2(2+3x)^{\frac{1}{2}}} - (2-3x)\frac{3}{2(2+3x)^{\frac{1}{2}}}}{2+3x}$$

© HERIOT-WATT UNIVERSITY

Now take out the common factor of $\frac{3}{2(2+3x)^{\frac{1}{2}}}$

$$\frac{f'}{g} = \frac{\frac{3}{2(2+3x)^{\frac{1}{2}}}[-2(2+3x)-(2-3x)]}{2+3x}$$

Step 3: Simplify by taking $2(2+3x)^{\frac{1}{2}}$ to the denominator and tidying the bracket

$$\frac{f'}{g} = \frac{3[-2(2+3x)-(2-3x)]}{2(2+3x)^{\frac{3}{2}}}$$

$$= \frac{3[-6-3x]}{2(2+3x)^{\frac{3}{2}}}$$

$$= \frac{-9[2+x]}{2(2+3x)^{\frac{3}{2}}}$$

$$= \frac{-9[x+2]}{2(2+3x)^{\frac{3}{2}}}$$

....

3. Problem:

Differentiate $f(x) = \frac{\sin^2(x+1)}{x^3}$

Solution:

Step 1: Using the Quotient Rule: $\frac{f'}{g} = \frac{f'g - fg'}{[g]^2}$

Let $f = \sin^2(x+1)$ and $g = x^3$

Then $f' = 2\sin(x+1)\cos(x+1)$ and $g' = 3x^2$

Which gives us,

$$\frac{f'}{g} = \frac{2\sin(x+1)\cos(x+1)x^3 - \sin^2(x+1)3x^2}{(x^3)^2}$$

Step 2: Simplify the denominator and take out a common factor

Take out the common factor of $x^2 \sin(x+1)$

$$\frac{f'}{g} = \frac{x^2\sin(x+1)[2x\cos(x+1) - 3\sin(x+1)]}{x^6}$$

Step 3: Cancel the common factor of x^2 from the numerator and denominator

$$\frac{f'}{g} = \frac{\sin(x+1)[2x\cos(x+1) - 3\sin(x+1)]}{x^4}$$

....

4. Problem:

Find $\frac{dy}{dx}\left(\frac{\sin^3(x)}{\cos(x)}\right)$

Solution:

Step 1: Using the Quotient Rule: $\frac{f'}{g} = \frac{f'g - fg'}{[g]^2}$

Let $f = \sin^3(x)$ and $g = \cos(x)$

Then $f' = 3\sin^2(x)\cos(x)$ and $g' = -\sin(x)$

Giving,

$$\frac{f'}{g} = \frac{3\sin^2(x)\cos(x)\cos(x) - \sin^3(x)(-\sin(x))}{(\cos(x))^2}$$

TOPIC 2. DIFFERENTIATION

Step 2: Simplifying the numerator and denominator:

$$\frac{f'}{g} = \frac{3\sin^2(x)\cos^2(x) + \sin^4(x)}{\cos^2(x)}$$

Step 3: Remembering that $\tan(x) = \frac{\sin(x)}{\cos(x)}$ we can re-write that as:

$$\frac{f'}{g} = 3\sin^2(x) + \frac{\sin^4(x)}{\cos^2(x)}$$

$$= 3\sin^2(x) + \sin^2(x) \times \frac{\sin^2(x)}{\cos^2(x)}$$

$$= 3\sin^2(x) + \sin^2(x) \times \frac{1}{\tan^2(x)}$$

$$= 3\sin^2(x) + \sin^2(x) \times \frac{1}{\tan^2(x)}$$

The quotient rule exercise Go online

Q47: Show that $\frac{d}{dx}\left(\frac{4x^2}{\sqrt{3x-1}}\right) = \frac{2x}{(3x-1)^{\frac{3}{2}}}(9x-4)$

Q48: Show that $\frac{d}{dx}\left(\frac{\cos^2 x}{\sin x}\right) = -2\cos x - \frac{\cos x}{\tan^2 x}$

Q49: Differentiate $\frac{5x}{2x-3}$

Q50: Differentiate $\frac{6x^3 + 2x - 3}{x^2 - 4x + 1}$

Q51: Differentiate $\frac{\sin^3 x}{x^2}$

Q52: Differentiate $f(x) = \frac{1+5x}{\sqrt{2x-5}}$

2.5 Differentiate cot(x), sec(x), cosec(x) and tan(x)

Some new trigonometric functions and their graphs are introduced in this section.

Secant function

$$\sec x = \frac{1}{\cos x}$$

Cosecant function

$$\operatorname{cosec} x = \frac{1}{\sin x}$$
or
$$\csc x = \frac{1}{\sin x}$$

Cotangent function

$$\cot x = \frac{1}{\tan x}$$

We can find the derivative of these functions using the quotient rule.

Examples

1. Differentiating tan x

Problem:

Find $k'(x)$ when $k(x) = \tan x$

Solution:

Step 1: rewrite $k(x) = \tan x$

$k(x) = \tan x = \frac{\sin x}{\cos x}$

Step 2: use the quotient rule

$\frac{d}{dx}\left(\frac{f}{g}\right) = \frac{f'g - fg'}{g^2}$

Let $f = \sin x$ and $g = \cos x$

Then $f' = \cos x$ and $g' = -\sin x$

giving,

$k'(x) = \frac{\cos x \cos x - \sin x (-\sin x)}{\cos^2 x}$

$= \frac{\cos^2 x + \sin^2 x}{\cos^2 x}$

Step 3: remember that $\cos^2 x + \sin^2 x = 1$ and $\sec x = \frac{1}{\cos x}$.

$k'(x) = \frac{1}{\cos^2 x} = \sec^2 x$

....

2. Differentiating sec x

Problem:

Find $k'(x)$ when $k(x) = \sec x$

Solution:

Step 1: rewrite $k(x) = \sec x$

$$k(x) = \sec x = \frac{1}{\cos x}$$

Step 2: use the quotient rule

$$\frac{d}{dx}\left(\frac{f}{g}\right) = \frac{f'g - fg'}{g^2}$$

Let $f = 1$ and $g = \cos x$

Then $f' = 0$ and $g' = -\sin x$

giving,

$$k'(x) = \frac{0 \cos x - 1(-\sin x)}{(\cos x)^2}$$

$$= \frac{\sin x}{(\cos x)^2}$$

Step 3: remember $\tan x = \frac{\sin x}{\cos x}$ and $\sec x = \frac{1}{\cos x}$.

$$k'(x) = \frac{1}{\cos x} \times \frac{\sin x}{\cos x}$$

Step 4: simplify

$$k'(x) = \sec x \tan x$$

Differentiate cot(x), sec(x), cosec(x) and tan(x) exercise   Go online

Q53: Show that $\frac{d}{dx}(\cot x) = -\csc^2 x$

....

Q54: Show that $\frac{d}{dx}(\csc x) = -\csc x \cot x$

Key point

$$\sec x = \frac{1}{\cos x} \qquad \frac{dy}{dx}(\sec x) = \sec x \tan x$$

$$\csc x = \frac{1}{\sin x} \qquad \frac{dy}{dx}(\csc x) = -\csc x \cot x$$

$$\cot x = \frac{1}{\tan x} \qquad \frac{dy}{dx}(\cot x) = -\csc^2 x$$

$$\tan x = \frac{\sin x}{\cos x} \qquad \frac{dy}{dx}(\tan x) = \sec^2 x$$

2.6 Differentiate exp(x)

Exponential graphs — Go online

Compare the graphs for $f(x) = 2^x$ and $f(x) = 3^x$

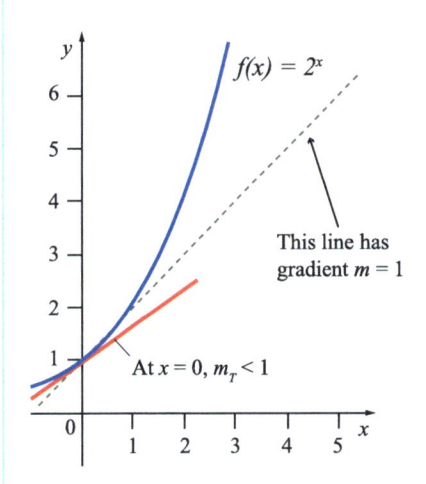

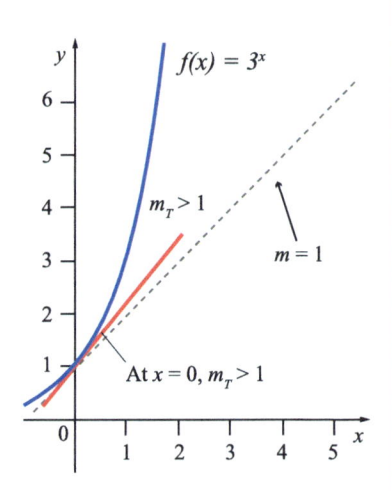

At $x = 0$, the gradient of the tangent to $f(x) = 2^x$ ($m_T = f'(x)$) is less than 1 whereas the gradient of the tangent to $f(x) = 3^x$ is greater than 1.

Thus, there must be a value, denoted as e, so that $2 < e < 3$ and at $x = 0$ the gradient of the tangent to $f(x) = e^x = 1$

From 'first principles', the gradient of a tangent to these curves is given by these formulae:

When $f(x) = 2^x$ then $f'(x) = \lim_{h \to 0} \frac{2^{x+h} - 2^x}{h}$

When $f(x) = 3^x$ then $f'(x) = \lim_{h \to 0} \frac{3^{x+h} - 3^x}{h}$

Thus, for $f(x) = 2^x$ the gradient of the tangent to the curve at $x = 0$ is given by
$f'(0) = \lim_{h \to 0} \frac{2^h - 1}{h}$

and for $f(x) = 3^x$ the gradient of the tangent to the curve at $x = 0$ is given by $f'(0) = \lim_{h \to 0} \frac{3^h - 1}{h}$

For very small h, $\lim_{h \to 0} \frac{2^h - 1}{h} < 1$ and $\lim_{h \to 0} \frac{3^h - 1}{h} > 1$

This means that, at $x = 0$, the gradient of the tangent to $f(x) = 2^x$ is less than 1, whereas at $x = 0$ the gradient of the tangent to $f(x) = 3^x$ is greater than 1 (as we have already observed from the graphs for $f(x) = 2^x$ and $f(x) = 3^x$).

Check this for some small values of h, e.g. $h = 0 \cdot 5, 0 \cdot 2$, etc.

There must be a value, which is denoted as e, so that the gradient of $2^x < e^x < 3^x$.

TOPIC 2. DIFFERENTIATION

In fact $\lim_{h \to 0} \frac{e^h - 1}{h} = 1$

Key point

e $\approx 2 \cdot 71828\ldots$ is a real number constant that occurs in many mathematical problems.

Now from 'first principles', when, $f(x) = e^x$ then:

$$\begin{aligned} f'(x) &= \lim_{h \to 0} \frac{e^{x+h} - e^x}{h} \\ &= \lim_{h \to 0} \frac{e^x(e^h - 1)}{h} \\ &= \lim_{h \to 0} e^x \lim_{h \to 0} \frac{e^h - 1}{h} \\ &= e^x \times 1 \\ &= e^x \end{aligned}$$

Key point

When $f(x) = e^x$, then $f'(x) = e^x$

e^x is also denoted by $\exp(x)$ and is called the exponential function.

An effective way to approximate the value of e is to use the following infinite sum:

$e = \frac{1}{0!} + \frac{1}{1!} + \frac{1}{2!} + \frac{1}{3!} + \frac{1}{4!} + \ldots = \sum_{r=0}^{\infty} \frac{1}{r!}$

Top tip

! notation means that something is a factorial.

4! (4 factorial) is the same as $4 \times 3 \times 2 \times 1 = 24$

$5! = 5 \times 4 \times 3 \times 2 \times 1 = 120$

The factorial button on a calculator is $n!$

In general $n! = (n)(n-1)(n-2)(n-3)\ldots(2)(1)$

(The concept of an infinite sum is studied in more detail in the topic *Sequences and series*)

Top tip

Note: On a calculator, the constant $e = \exp(1) = e^1$

Examples

1. Problem:

Differentiate $y = \exp(4x)$

100 TOPIC 2. DIFFERENTIATION

Solution:
Use the chain rule with $u = 4x$ then $y = \exp(4x) = \exp(u)$
Hence:

$$\begin{aligned}\frac{dy}{dx} &= \frac{dy}{du} \times \frac{du}{dx} \\ &= \exp(u) \times 4 \\ &= 4\exp(4x)\end{aligned}$$

....

2. Problem:

Differentiate $y = e^{x^2}$

Solution:
Use the chain rule with $u = x^2$ then $y = e^u$
Hence,

$$\begin{aligned}\frac{dy}{dx} &= \frac{dy}{du} \times \frac{du}{dx} \\ &= e^u \times 2x \\ &= 2xe^{x^2}\end{aligned}$$

....

3. Problem:

Differentiate $y = 2x\, e^{x^2}$

Solution:
Use the product rule with $f = 2x$ and $g = e^{x^2}$
$f' = 2$ and $g' = 2xe^{x^2}$ (using the chain rule)
Therefore:

$$\begin{aligned}\frac{dy}{dx} &= f'g + fg' \\ &= 2e^{x^2} + 2x \times 2xe^{x^2} \\ &= 2e^{x^2}(1 + 2x^2)\end{aligned}$$

....

4. Problem:

Differentiate $y = e^{\sec x} \sin x$

Solution:
Use the product rule with,
$f = e^{\sec x}$ and $g = \sin x$
$f' = \sec x \tan x e^{\sec x}$ and $g' = \cos x$
Therefore, $-\frac{\pi}{2} < x < \frac{\pi}{2}$

© HERIOT-WATT UNIVERSITY

$$y' = f'g + fg'$$
$$= \sec x \tan x e^{\sec x} \sin x + \cos x e^{\sec x}$$
$$= e^{\sec x} (\sec x \tan x \sin x + \cos x)$$
$$= e^{\sec x} (\tan^2 x + \cos x)$$

....

5. Problem:

Find the coordinates and nature of any stationary points on the $y = 3xe^x$

Solution:

Step 1: Differentiate $y = 3xe^x$

Use the product rule where,

$f = 3x$ and $g = e^x$

$f' = 3$ and $g' = e^x$

Therefore,

$$y' = f'g + fg'$$
$$= 3e^x + 3xe^x$$
$$= 3e^x(1+x)$$
$$\frac{dy}{dx} = 3e^x(1+x)$$

Step 2: For stationary points $\frac{dy}{dx} = 0$

$3e^x(1+x) = 0$

$(x+1) = 0$

$x = -1$

Step 3: Create a nature table to identify the stationary point

x	$\to$	-1	$\to$
$\frac{dy}{dx}$	-	0	+
slope	↗	→	↘

There is a maximum turning point at $\left(-1, -\frac{3}{e}\right)$

Differentiate exp(x) exercise   Go online

Q55: Differentiate $f(x) = e^{5x}$

....

Q56: Differentiate $f(x) = e^{\cot x}$, $0 < x < \pi$

....

Q57: Differentiate $f(x) = e^{(6x-1)}$

....

102 TOPIC 2. DIFFERENTIATION

Q58: Differentiate $f(x) = x^2 e^{\cos x}$, $0 < x < \pi$

....

Q59: Differentiate $f(x) = e^{\tan 3x}$, $0 < x < \frac{\pi}{6}$

....

Q60: Differentiate $f(x) = (3 - 4x)e^{4x^3}$

....

Q61: Differentiate $f(x) = \cos^2 x \, e^{\tan x}$ for $-\frac{\pi}{2} < x < \frac{\pi}{2}$.
Factorise and simplify your answer.

....

Q62: Find the coordinates and nature of any turning points on the curve $y = 5xe^x$

2.7 Logarithmic differentiation

Growth and decay can be described using exponential functions where the dependent variable is a power. For example, 2^x, 3^{x^2}, etc. If we wish to know the rate of change of this growth or decay we need to differentiate. We are unable to differentiate this power directly so we have to transform this model into one that we know how to differentiate.

This section will look at how applying logarithms allows us to differentiate functions like these. It will also look at an alternative, possibly simpler, way of differentiating complicated looking functions involving quotients and products.

2.7.1 Differentiating ln(x)

Remember that for $f(x) = e^x$, an inverse function $f^{-1}(x)$ exists.

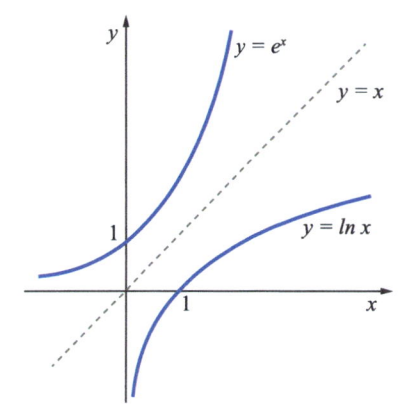

TOPIC 2. DIFFERENTIATION

$f^{-1}(x)$ is the logarithmic function with base e and is written as $y = \log_e x = \ln x$

As can be seen in the graph, $y = \ln x$ is the reflection of $y = e^x$ in the line $y = x$

Thus, if $y = \ln x$ then $x = e^y$.

Now, if we have $y = \ln x$ and $x = e^y$, then we can say that $\frac{dx}{dy} = e^y$ (differentiating with respect to y) and therefore:

$$\frac{dy}{dx} = \frac{1}{\left(\frac{dx}{dy}\right)}$$
$$= \frac{1}{e^y}$$
$$= \frac{1}{x}$$

> **Key point**
>
> When $y = \ln x$ then $\frac{dy}{dx} = \frac{1}{x}$

There is a proof in the proofs section which justifies $\frac{dy}{dx} = \frac{1}{\left(\frac{dx}{dy}\right)}$. This is not required to be reproduced for exam purposes, simply for your interest.

> **Proof 5: Rule connecting $^{dy}/_{dx}$ and $^{dx}/_{dy}$**
>
> Prove that $\frac{dy}{dx} = \frac{1}{\left(\frac{dx}{dy}\right)}$
>
> > By definition:
> >
> > $$\frac{dy}{dx} = \lim_{\Delta x \to 0}\left(\frac{\Delta y}{\Delta x}\right)$$
> > $$= \lim_{\Delta x \to 0}\left(\frac{1}{\frac{\Delta x}{\Delta y}}\right)$$
>
> But, as $\delta x \Rightarrow 0$, then so also $\delta y \Rightarrow 0$ so:
>
> $$\frac{dy}{dx} = \lim_{\delta x \to 0}\left(\frac{1}{\frac{\Delta x}{\Delta y}}\right)$$
> $$= \frac{1}{\left(\frac{dx}{dy}\right)}$$

> **Key point**
>
> In general, $\frac{d}{dx}\ln|f(x)| = f'(x) \times \frac{1}{f(x)}$

© HERIOT-WATT UNIVERSITY

> We cannot evaluate the logarithm of a negative number. To ensure that the number evaluated will always be positive we use modulus sign $|f(x)|$. If it is obvious that $f(x)$ is positive some texts use ($f(x)$) instead.

Examples

1. Problem:
Differentiate $y = \ln|4x + 1|$.
Solution:
Applying the chain rule to $\ln|f(x)|$:
$\frac{d}{dx}\ln|f(x)| = f'(x) \times \frac{1}{f(x)}$
Apply the chain rule:
Let $f(x) = 4x + 1$ and $f'(x) = 4$
$\frac{d}{dx} = 4 \times \frac{1}{4x+1}$

....

2. Problem:
Differentiate $f(x) = \ln|\sin x|$.
Solution:
Apply the chain rule:
$\frac{d}{dx}\ln|f(x)| = f'(x) \times \frac{1}{f(x)}$
Let $f(x) = \sin x$ then $f'(x) = \cos x$
$\frac{d}{dx} = \cos x \times \frac{1}{\sin x}$
$= \frac{1}{\tan x}$
$= \cot x$

....

3. Problem:
Differentiate $f(x) = \ln\left|\frac{x+2}{x+3}\right|$.
Solution:
Apply logarithmic rules: $\log_a \frac{x}{y} = \log_a x - \log_a y$
$f(x) = \ln(x+2) - \ln(x+3)$
So, $\frac{d}{dx}(\ln|x+2|)$
Let $g(x) = \ln|x+2|$ then $g'(x) = 1$

Then,

$$\frac{d}{dx}\ln|g(x)| = g'(x) \times \frac{1}{g(x)}$$
$$= 1 \times \frac{1}{x+2}$$
$$= \frac{1}{x+2}$$

And $\frac{d}{dx}(\ln|x+3|)$

Let $h(x) = \ln|x+3|$ then $h'(x) = 1$

Then,

$$\frac{d}{dx}\ln|h(x)| = h'(x) \times \frac{1}{h(x)}$$
$$= 1 \times \frac{1}{x+3}$$
$$= \frac{1}{x+3}$$

Now,

$$f'(x) = g'(x) - h'(x)$$
$$= \frac{1}{x+2} - \frac{1}{x+3}$$

Differentiating ln(x) exercise   Go online

Q63: Differentiate $y = \ln|5x - 2|$

....

Q64: Differentiate $y = \ln\left|\sqrt{x^2 + 3}\right|$

....

Q65: Differentiate $y = \frac{x^3}{\ln x}$

....

Q66: Differentiate $y = \ln(\tan x)$

2.7.2 Logarithmic differentiation

Sometimes, taking the natural logarithms of an expression makes it easier to differentiate. This can be the case for extended products and quotients or when the variable appears as a power.

The natural logarithm is the logarithm with base e.

The notation is $\log_e |x| = \ln|x|$.

The examples may make this clearer.

Examples

1. Problem:

Differentiate $y = 5^x$

Solution:

Apply ln(x) to both sides: $\ln|y| = \ln|5^x|$

Apply logarithmic rules:

$\log_a x^n = n\log_a x \Rightarrow \ln|y| = x\ln|5|$

Apply implicit differentiation to both sides:

$\dfrac{1}{y}\dfrac{dy}{dx} = \ln|5|$ ⠀⠀⠀ $\ln|5|$ is a constant

$\dfrac{dy}{dx} = y\ln|5|$ ⠀⠀⠀ rearrange for $\dfrac{dy}{dx}$

$\dfrac{dy}{dx} = 5^x \ln|5|$ ⠀⠀⠀ replace y with the original function $\ln|5|$

....

2. Problem:

Differentiate $y = x^{\sin x}$

Solution:

Apply ln(x) to both sides: $\ln|y| = \ln\left|x^{\sin x}\right|$

Apply logarithmic rules:

$\log_a x^n = n\log_a x \Rightarrow \ln|y| \sin x \ln|x|$

Apply implicit differentiation to both sides:

$\dfrac{1}{y}\dfrac{dy}{dx} = \cos x \ln|x| + \sin x \times \dfrac{1}{x}$ ⠀⠀⠀ Apply the product rule to the RHS

$\dfrac{dy}{dx} = y\left(\cos x \ln|x| + \dfrac{\sin x}{x}\right)$ ⠀⠀⠀ Rearrange for $\dfrac{dy}{dx}$

$\dfrac{dy}{dx} = x^{\sin x}\left(\cos x \ln|x| + \dfrac{\sin x}{x}\right)$ ⠀⠀⠀ Replace y with the original function $x^{\sin x}$

....

3. Problem:

Differentiate $y = \dfrac{x^3}{\sqrt{x+4}}$

Solution:

Apply ln(x) to both sides: $\ln|y| = \ln\left|\dfrac{x^3}{\sqrt{x+4}}\right|$

Apply logarithmic rules:

$\log_a \dfrac{x}{y} = \log_a x - \log_b y$

$\log_a x^n = n\log_a x$

$\ln|y| = 3\ln|x| - \dfrac{1}{2}\ln|x+4|$

Apply implicit differentiation to both sides:

TOPIC 2. DIFFERENTIATION

$$\frac{1}{y}\frac{dy}{dx} = 3 \times \frac{1}{x} - \frac{1}{2} \times \frac{1}{x+4}$$

$$\frac{dy}{dx} = y\left(\frac{3}{x} - \frac{1}{2(x+4)}\right)$$

$$\frac{dy}{dx} = \frac{x^3}{\sqrt{x+4}}\left(\frac{3}{x} - \frac{1}{2(x+4)}\right)$$

4. Problem:

Differentiate $y = \sqrt{\frac{1+x}{1-x}}$

Solution:

Apply $\ln(x)$ to both sides: $\ln|y| = \ln\left|\sqrt{\frac{1+x}{1-x}}\right|$

Apply logarithmic rules:

$$\ln|y| = \ln\left|\left(\frac{1+x}{1-x}\right)^{\frac{1}{2}}\right|$$

$$\ln|y| = \frac{1}{2}\{\ln|(1+x)| - \ln|(1-x)|\}$$

Care must be applied when differentiating $\ln|1-x|$ using the chain rule:

$$\frac{d}{dx}[\ln(f(x))] = f'(x) \times \frac{1}{f(x)}$$

$$\frac{d}{dx}[\ln|(1-x)|] = (-1) \times \frac{1}{1-x}$$

Apply implicit differentiation to both sides:

$$\frac{1}{y}\frac{dy}{dx} = \frac{1}{2}\left|\frac{1}{1+x} - (-1) \times \frac{1}{1-x}\right|$$

$$\frac{dy}{dx} = y\left|\frac{1}{2(1+x)} + \frac{1}{2(1-x)}\right|$$

$$\frac{dy}{dx} = \left|\sqrt{\frac{1+x}{1-x}}\right|\left|\frac{1}{2(1+x)} + \frac{1}{2(1-x)}\right|$$

Logarithmic differentiation exercise   Go online

Q67: Differentiate $y = (2x+5)^3(x^2+1)^{-2}$, where $x > 0$

Q68: Differentiate $y = \frac{\sqrt{5-2x}}{(x+1)^2}$, where $x > 0$

Q69: Differentiate $y = x^x$

Q70: Differentiate $y = \ln x^{\ln x}$

Q71: Differentiate $y = \dfrac{(2x+1)^{\frac{1}{2}}(3x-1)^{\frac{2}{3}}}{(4x+3)^{\frac{3}{4}}}$

Q72: $y = \dfrac{(2x+3)^2}{\sqrt{x+1}}$

Q73: $y = \dfrac{2^x}{2x+1}$

Q74: $y = \sqrt{\dfrac{3+x}{3-x}}$

Q75: $y = (1+x)(2+3x)(x-5)$

Q76: $y = \dfrac{x^2\sqrt{7x-3}}{1-x}$

2.8 Implicit differentiation

Sometimes with a function, it is impossible to separate the y on one side and the dependent variable, x, on the other. These are known as implicit functions. However, we may still need to differentiate these functions. This section will look at how to tackle these problems by differentiating implicitly.

2.8.1 Implicit and explicit functions

Example

Problem:

Re-arrange the equation $xy - y = 4x^3 - x^2$ to show that y can be expressed as a clearly defined function of x (make y the subject of the equation).

Solution:

$$xy - y = 4x^3 - x^2$$
$$y(x-1) = 4x^3 - x^2$$
$$y = \dfrac{4x^3 - x^2}{x-1}$$

y is an explicit function of x as it is a clearly defined function of x.

Key point

An **explicit function** is when y can be written as an expression in terms of x only, and for any value of x that is put into the function there is only one corresponding y value.

Example

These are examples of implicit functions:

$$y^2 + 2xy - 5x^2 = 0$$

y cannot be made the subject of the formula; it is therefore implicit.

The unit circle, centre the origin and radius 1.

$$x^2 + y^2 = 1$$
$$y^2 = 1 - x^2$$
$$y = \pm\sqrt{1 - x^2}$$

y is defined as two functions on the interval [-1, 1]. It is implicit because for every one value of x, there are two values for y.

Key point

If x is given any value, then corresponding value(s) of y can be found; it is therefore not explicit and is an **implicit function**.

Explicit and implicit functions exercise

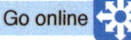

Go online

Identify the following functions as being either implicit or explicit.

Q77: $3x^2 + 7xy + 9y^2 = 6$

a) Explicit
b) Implicit

....

Q78: $xy = 2x - 1 + 2y$

a) Explicit
b) Implicit

....

Q79: $3x - 5 = e^{xy}$

a) Explicit
b) Implicit

....

Q80: $8y - 7x = 2x^2 + 5y^2 + 6y$

a) Explicit
b) Implicit

© HERIOT-WATT UNIVERSITY

Q81: $x - \cos y = x^2 y$

a) Explicit
b) Implicit

Q82: $3x^3 y = \tan 2x$

a) Explicit
b) Implicit

2.8.2 Implicit differentiation

Given an implicit equation in variables x and y it is possible to find $\frac{dy}{dx}$. Implicit equations can be differentiated using the chain rule. The following examples may help.

Basic implicit differentiation

Before differentiating implicitly we will need a few reminders in order to understand the examples.

The notation $\frac{d}{dx}$ means to differentiate an expression with respect to x. This expression would usually be given in brackets beside it.

For example, $\frac{d}{dx}\left(5x^3 - 2x^2 + 3\right) = 15x^2 - 4x$

The notation $\frac{dy}{dx}$ means that you are differentiating the function, represented by y, with respect to x. It is known that y is a function of x.

For example, if $y = 5x^3 - 2x^2 + 3$ then $\frac{dy}{dx} = 15x^2 - 4x$

If we wanted to show that we are differentiating the function y, but do not know what it is explicitly in terms of x, then we would just write $\frac{dy}{dx}$ to show that it is being differentiated.

Now let us consider differentiating the function $\left(5x^3 - 2x^2 + 3\right)^2$.

In order to differentiate this we need to apply the chain rule.

Let $y = 5x^3 - 2x^2 + 3$, so we have $z = (y)^2$.

The chain rule gives: $\frac{dz}{dx} = \frac{dz}{dy} \times \frac{dy}{dx}$

$\frac{dz}{dy} = 2y$ and $\frac{dy}{dx} = 15x^2 - 4x$

$\frac{dz}{dx} = 2y \times \left(15x^2 - 4x\right)$

Now if we did not know that $y = 5x^3 - 2x^2 + 3$ then $\frac{dz}{dx} = 2y\frac{dy}{dx}$.

This is essentially what implicit differentiation is. It is applying the chain rule to the variable y and representing the derivative of y with respect to x as $\frac{dy}{dx}$.

Examples

1. Implicit differentiation of y^2

Problem:

$\frac{d}{dx}\left(y^2\right) = ?$

Solution:

Let $z = y^2$

Apply the chain rule: $\frac{dz}{dx} = \frac{dz}{dy} \times \frac{dy}{dx}$

$\frac{dz}{dy} = 2y$

Since we do not know what the function y is explicitly then it is represented by $\frac{dy}{dx}$.

So,

$$\frac{dz}{dx} = \frac{dz}{dy} \times \frac{dy}{dx}$$

$$\frac{dz}{dx} = 2y \times \frac{dy}{dx}$$

We have $\frac{d}{dx}\left(y^2\right) = 2y\frac{dy}{dx}$.

....

2. Implicit differentiation of y^3

Problem:

$\frac{d}{dx}\left(y^3\right) = ?$

Solution:

Let $z = y^3$

Apply the chain rule: $\frac{dz}{dx} = \frac{dz}{dy} \times \frac{dy}{dx}$

$\frac{dz}{dy} = 3y^2$

Since we do not know what the function y is explicitly then it is represented by $\frac{dy}{dx}$.

So,

$$\frac{dz}{dx} = \frac{dz}{dy} \times \frac{dy}{dx}$$

$$\frac{dz}{dx} = 3y^2 \times \frac{dy}{dx}$$

We have $\frac{d}{dx}\left(y^3\right) = 3y^2\frac{dy}{dx}$

....

3. Implicit differentiation of xy

Problem:

$\frac{d}{dx}\left(xy\right) = ?$

Solution:

Let $z = xy$

Since this is a product to of two expressions both in terms of x we need to use the product rule and then apply the chain rule where necessary.

Apply the product rule:

Let $u = x$ and $v = y$.

$\frac{dz}{dx} = \frac{du}{dx}v + u\frac{dv}{dx}$

So, $\frac{du}{dx} = 1$

Therefore,

$\frac{dv}{dx} = \frac{dv}{dy} \times \frac{dy}{dx}$

$\frac{dv}{dx} = 1 \times \frac{dy}{dx}$

$\frac{dv}{dx} = \frac{dy}{dx}$

So,

$\frac{dz}{dx} = 1 \times y + x\frac{dy}{dx}$

$\frac{dz}{dx} = y + x\frac{dy}{dx}$

We have: $\frac{d}{dx}(xy) = x\frac{dy}{dx} + y$

Key point

When y is multiplied by a function of x the product rule needs to be used.

$$\frac{d}{dx}(xy) = \frac{d}{dx}(x)y + x\frac{d}{dx}(y)$$
$$= 1y + x\frac{dy}{dx}$$
$$= y + x\frac{dy}{dx}$$

This result is used in the next example.

Example

Problem:

Find $\frac{dy}{dx}$ in terms of x and y when $3xy + 5y^2 - x^2 = 2y$

Solution:

In this case we are not asked to just differentiate, but to differentiate and then find an expression for $\frac{dy}{dx}$.

Differentiate this implicit function term by term.

TOPIC 2. DIFFERENTIATION

$$\frac{d}{dx}(3xy + 5y^2 - x^2) = \frac{d}{dx}(2y)$$

$$\frac{d}{dx}(3xy) + \frac{d}{dx}(5y^2) - \frac{d}{dx}(x^2) = \frac{d}{dx}(2y)$$

$$\frac{d}{dx}(3x)y + 3x\frac{dy}{dx} + \frac{d}{dx}(5y^2)\frac{dy}{dx} - \frac{d}{dx}(x^2) = 2\frac{dy}{dx}$$

$$3y + 3x\frac{dy}{dx} + 10y\frac{dy}{dx} - 2x = 2\frac{dy}{dx}$$

$$(3x + 10y - 2)\frac{dy}{dx} = 2x - 3y$$

$$\frac{dy}{dx} = \frac{2x - 3y}{3x + 10y - 2}$$

An implicit equation may include a trigonometric function.

Example

Problem:

Find $\frac{dy}{dx}$ for the function of x implicitly defined by $\sin(2x + 3y) = 7x$

Solution:

Differentiating both sides of the equation with respect to x gives:

$\frac{d}{dx}(\sin(2x + 3y)) = \frac{d}{dx}(7x)$

Step 1: Apply the chain rule to $\sin(2x + 3y)$

$\cos(2x + 3y) \times \frac{d}{dx}(2x + 3y) = 7$

Step 2: Apply the chain rule to $3y$ in the last bracket.

$\cos(2x + 3y) \times \left(\frac{d}{dx}(2x) + \frac{d}{dx}(3y)\frac{dy}{dx}\right) = 7$

Step 3: Multiply out the bracket.

$$\cos(2x + 3y) \times \left(2 + 3\frac{dy}{dx}\right) = 7$$

$$2\cos(2x + 3y) + 3\cos(2x + 3y)\frac{dy}{dx} = 7$$

Step 4: Rearrange for $\frac{dy}{dx}$

$$3\cos(2x + 3y)\frac{dy}{dx} = 7 - 2\cos(2x + 3y)$$

$$\frac{dy}{dx} = \frac{7 - 2\cos(2x + 3y)}{3\cos(2x + 3y)}$$

© HERIOT-WATT UNIVERSITY

Here are the steps altogether.

$$\frac{d}{dx}(\sin(2x+3y)) = \frac{d}{dx}(7x)$$
$$\cos(2x+3y) \times \frac{d}{dx}(2x+3y) = 7$$
$$\cos(2x+3y) \times \left(\frac{d}{dx}(2x) + \frac{d}{dx}(3y)\frac{dy}{dx}\right) = 7$$
$$\cos(2x+3y) \times \left(2 + 3\frac{dy}{dx}\right) = 7$$
$$2\cos(2x+3y) + 3\cos(2x+3y)\frac{dy}{dx} = 7$$
$$3\cos(2x+3y)\frac{dy}{dx} = 7 - 2\cos(2x+3y)$$
$$\frac{dy}{dx} = \frac{7 - 2\cos(2x+3y)}{3\cos(2x+3y)}$$

Implicit differentiation exercise   Go online

Using implicit differentiation, find the derivatives of the following implicit functions. Show all your working.

Q83: $y^2 - x^2 = 12$

....

Q84: $x^2 + xy + y^2 = 7$

....

Q85: $3y = 2x^3 + \sin y$

....

Q86: $x \tan y = e^x$

2.8.3 Second derivative

Like regular differentiation, implicit differentiation can be used to find the *second derivative*. The second derivative of y with respect to x is obtained by differentiating twice and is written as: $\frac{d^2y}{dx^2} = \frac{d}{dx}\left(\frac{dy}{dx}\right)$

The second derivative of a function also gives an alternative way of determining the nature of a turning point. This will be covered in more detail in the *Curve sketching* topic.

TOPIC 2. DIFFERENTIATION

Examples

1. Problem:
Find $\frac{d^2y}{dx^2}$ when $3x - 6y + xy = 10$.
Solution:
We have to calculate $\frac{dy}{dx}$ before we can find $\frac{d^2y}{dx^2}$.
Differentiate both sides implicitly: $\frac{d}{dx}(3x) - \frac{d}{dx}(6y) + \frac{d}{dx}(xy) = \frac{d}{dx}(10)$
Apply the product rule to $\frac{d}{dx}(xy)$.

$$\frac{d}{dx}(3x) - \frac{d}{dx}(6y) + \frac{d}{dx}(xy) = \frac{d}{dx}(10)$$

$$3 - 6\frac{dy}{dx} + y + x\frac{dy}{dx} = 0$$

$$(x - 6)\frac{dy}{dx} = -3 - y$$

$$\frac{dy}{dx} = \frac{-3 - y}{x - 6}$$

Now, to find $\frac{d^2y}{dx^2}$, it is easier if we write $\frac{dy}{dx} = \frac{-3-y}{x-6}$ as:
$\frac{dy}{dx} = (-3 - y)(x - 6)^{-1}$
Then we can use the product rule instead of the quotient rule but either could be used.
Apply the product rule to the expression: $(-3 - y)(x - 6)^{-1}$

$$\frac{d}{dx}\left(\frac{dy}{dx}\right) = \frac{dy}{dx}\left((-3 - y)(x - 6)^{-1}\right)$$

$u = (-3 - y) \qquad v = (x - 6)^{-1}$

$\frac{du}{dx} = -\frac{dy}{dx} \qquad \frac{dv}{dx} = -(x - 6)^{-2}$

$$\frac{d^2y}{dx^2} = -(x - 6)^{-1}\frac{dy}{dx} - (-3 - y)(x - 6)^{-2}$$

Substitute $\frac{dy}{dx}$ with $\left(\frac{-3-y}{x-6}\right)$

$$\frac{d^2y}{dx^2} = -(x - 6)^{-1}\left(\frac{-3-y}{x-6}\right) - (-3 - y)(x - 6)^{-2}$$

$$= \frac{3+y}{(x-6)^2} + \frac{3+y}{(x-6)^2}$$

$$= \frac{6+2y}{(x-6)^2}$$

....

2. Problem:
Given $xy - x = 5$ use implicit differentiation to obtain $\frac{dy}{dx}$ in terms of x and y.
Hence obtain $\frac{d^2y}{dx^2}$ in terms of x and y.

116 TOPIC 2. DIFFERENTIATION

Solution:

First derivative:

$\frac{d}{dx}(xy) - \frac{d}{dx}(x) = \frac{d}{dx}(5)$

Remember to apply the product rule to $\frac{d}{dx}(xy)$

$y + x\frac{dy}{dx} - 1 = 0$

$\frac{dy}{dx} = \frac{1+y}{x}$

Second derivative:

$\frac{d}{dx}\left(\frac{dy}{dx}\right) = \frac{d}{dx}\left(\frac{1+y}{x}\right)$

$\frac{d^2y}{dx^2} = \frac{\frac{dy}{dx} \times x - (1+y) \times 1}{x^2}$

Substitute in for $\frac{dy}{dx}$

$\frac{d^2y}{dx^2} = \frac{\left(\frac{1+y}{x}\right) \times x - (1+y)}{x^2}$

Simplify:

$\frac{d^2y}{dx^2} = \frac{1+y-(1+y)}{x^2}$

$\frac{d^2y}{dx^2} = \frac{2y}{x^2}$

....

3. Problem:

Given $4x^2 - 3y^2 = 10$ use implicit differentiation to obtain $\frac{dy}{dx}$ in terms of x and y.

Hence obtain $\frac{d^2y}{dx^2}$ in terms of x and y.

Solution:

First derivative:

$\frac{d}{dx}(4x^2) - \frac{d}{dx}(3y^2) = \frac{d}{dx}(10)$

Remember to apply the chain rule to $3y^2$

$8x - 6y\frac{dy}{dx} = 0$

$\frac{dy}{dx} = \frac{8x}{6y}$

$\frac{dy}{dx} = \frac{4x}{3y}$

Second derivative:

$\frac{d}{dx}\left(\frac{dy}{dx}\right) = \frac{d}{dx}\left(\frac{4x}{3y}\right)$

Apply the quotient rule.

$\frac{d^2y}{dx^2} = \frac{4 \times 3y - 4x \times 3\frac{dy}{dx}}{9y^2}$

Simplify and substitute for $\frac{dy}{dx}$

TOPIC 2. DIFFERENTIATION

$$\frac{d^2y}{dx^2} = \frac{12y - 12x\left(\frac{4x}{3y}\right)}{9y^2}$$

Simplify by multiplying the numerator and the denominator by y:

$$\frac{d^2y}{dx^2} = \frac{12y^2 - 16x^2}{9y^2}$$

....

4. Problem:

Given $3y^2 + xy = 3$ use implicit differentiation to obtain $\frac{dy}{dx}$ in terms of x and y.
Hence obtain $\frac{d^2y}{dx^2}$ in terms of x and y.

Solution:

First derivative:

$$\frac{d}{dx}\left(3y^2\right) + \frac{d}{dx}\left(xy\right) = \frac{d}{dx}\left(3\right)$$

Apply the chain rule to $\frac{d}{dx}\left(3y^2\right)$ and the product rule to $\frac{d}{dx}\left(xy\right)$

$$6y\frac{dy}{dx} + y + x\frac{dy}{dx} = 0$$

Factorise for $\frac{dy}{dx}$

$$\frac{dy}{dx}\left(6y + x\right) + y = 0$$

$$\frac{dy}{dx} = -\frac{y}{6y + x}$$

Second derivative:

$$\frac{d}{dx}\left(\frac{dy}{dx}\right) = \frac{d}{dx}\left(-\frac{y}{6y + x}\right)$$

In this case we will apply the quotient rule:

$$\frac{d^2y}{dx^2} = -\frac{\frac{dy}{dx}(6y+x) - y\left(6\frac{dy}{dx} + 1\right)}{(6y+x)^2}$$

Substitute for $\frac{dy}{dx}$:

$$\frac{d^2y}{dx^2} = -\frac{\left(-\frac{y}{6y+x}\right)(6y+x) - y\left(6\left(-\frac{y}{6y+x}\right) + 1\right)}{(6y+x)^2}$$

Simplify first as much as possible:

$$\frac{d^2y}{dx^2} = -\frac{-y + \frac{6y^2}{6y+x} - 6y}{(6y+x)^2}$$

$$\frac{d^2y}{dx^2} = -\frac{\frac{6y^2}{6y+x} - 7y}{(6y+x)^2}$$

To simplify multiply numerator and denominator by $6y + x$:

$$\frac{d^2y}{dx^2} = -\frac{6y^2 - 7y(6y+x)}{(6y+x)^3}$$

$$\frac{d^2y}{dx^2} = -\frac{-36y^2 - 7xy}{(6y+x)^3}$$

$$\frac{d^2y}{dx^2} = \frac{36y^2 + 7xy}{(6y+x)^3}$$

....

© HERIOT-WATT UNIVERSITY

5. Problem:

Find $\frac{d^2y}{dx^2}$ for the function $3y^2 - xy - x^2 = 2$

Solution:

First derivative:

$$\frac{d}{dx}\left(3y^2\right) - \frac{d}{dx}\left(xy\right) - \frac{d}{dx}\left(x^2\right) = \frac{d}{dx}\left(2\right)$$

Apply the chain rule to $\frac{d}{dx}\left(3y^2\right)$ and the product rule to $\frac{d}{dx}\left(xy\right)$

$$6y\frac{dy}{dx} - \left(y + x\frac{dy}{dx}\right) - 2x = 0$$

Factorise for $\frac{dy}{dx}$:

$$\frac{dy}{dx}\left(6y - x\right) = 2x + y$$

$$\frac{dy}{dx} = \frac{2x + y}{6y - x}$$

Second derivative:

Apply the quotient rule:

$$\frac{d}{dx}\left(\frac{dy}{dx}\right) = \frac{d}{dx}\left(\frac{2x+y}{6y-x}\right)$$

$$\frac{d^2y}{dx^2} = \frac{\left(2 + \frac{dy}{dx}\right)(6y-x) - (2x+y)\left(6\frac{dy}{dx} - 1\right)}{(6y-x)^2}$$

Substitute for $\frac{dy}{dx}$

$$\frac{d^2y}{dx^2} = \frac{\left(2+\left(\frac{2x+y}{6y-x}\right)\right)(6y-x) - (2x+y)\left(6\left(\frac{2x+y}{6y-x}\right) - 1\right)}{(6y-x)^2}$$

Simplify as much as possible:

$$\frac{d^2y}{dx^2} = \frac{(2(6y-x) + 2x+y) - \left(\left(\frac{(2x+y)^2}{6y-x}\right) - (2x+y)\right)}{(6y-x)^2}$$

Multiply the numerator and the denominator by $(6y - x)$:

$$\frac{d^2y}{dx^2} = \frac{2(6y-x)^2 + (2x+y)(6y-x) - \left((2x+y)^2 - (2x+y)(6y-x)\right)}{(6y-x)^3}$$

$$\frac{d^2y}{dx^2} = \frac{2(6y-x)^2 + (2x+y)(6y-x) - (2x+y)^2 + (2x+y)(6y-x)}{(6y-x)^3}$$

Take out a factor of $(6y - x)$:

$$\frac{d^2y}{dx^2} = \frac{(6y-x)\left[12y - 2x + 2x + y + 2x + y\right] - (2x+y)^2}{(6y-x)^3}$$

$$\frac{d^2y}{dx^2} = \frac{(6y-x)\left[14y + 2x\right] - \left(4x^2 + 4xy + y^2\right)}{(6y-x)^3}$$

$$\frac{d^2y}{dx^2} = \frac{84y^2 + 12xy - 14xy - 2x^2 - \left(4x^2 + 4xy + y^2\right)}{(6y-x)^3}$$

$$\frac{d^2y}{dx^2} = \frac{83y^2 - 6xy - 6x^2}{(6y-x)^3}$$

TOPIC 2. DIFFERENTIATION

6. Problem:

Find $\frac{dy}{dx}$ and $\frac{d^2y}{dx^2}$ at the point (-1,-1) when $x^2 + 3xy + y^2 = x + y + 8$

Solution:

We have to calculate $\frac{dy}{dx}$ before we can find $\frac{d^2y}{dx^2}$.

Differentiate both sides implicitly:

$$\frac{d}{dx}\left(x^2 + 3xy + y^2\right) = \frac{d}{dx}(x + y + 8)$$

$$\frac{d}{dx}(x^2) + \frac{d}{dx}(3xy) + \frac{d}{dx}(y^2) = \frac{d}{dx}(x) + \frac{d}{dx}(y) + \frac{d}{dx}(8)$$

Apply the product rule to: $\frac{d}{dx}(3xy)$

$$2x + 3y + 3x\frac{dy}{dx} + 2y\frac{dy}{dx} = 1 + \frac{dy}{dx}$$

$$(3x + 2y - 1)\frac{dy}{dx} = 1 - 2x - 3y$$

$$\frac{dy}{dx} = \frac{1 - 2x - 3y}{3x + 2y - 1}$$

Evaluate $\frac{dy}{dx}$ at (-1,-1).

$$\frac{dy}{dx} = \frac{1 - 2x - 3y}{3x + 2y - 1}$$

$$= \frac{1 - 2(-1) - 3(-1)}{3(-1) + 2(-1) - 1}$$

$$= \frac{6}{-6}$$

$$= -1$$

Now that we know $\frac{dy}{dx}$ we can evaluate $\frac{d^2y}{dx^2}$.

Rearrange and apply the product rule.

$$\frac{dy}{dx} = \frac{1 - 2x - 3y}{3x + 2y - 1}$$

$$\frac{dy}{dx} = (1 - 2x - 3y)(3x + 2y - 1)^{-1}$$

Apply the product rule $y' = u'v + uv'$:

$u = 1 - 2x - 3y$ $\qquad v = (3x + 2y - 1)^{-1}$

$u' = -2 - 3\frac{dy}{dx}$ $\qquad v' = -(3x + 2y - 1)^{-2}\left(3 + 2\frac{dy}{dx}\right)$

$$\frac{d}{dx}\left(\frac{dy}{dx}\right) = \frac{d}{dx}(1 - 2x - 3y)(3x + 2y - 1)^{-1}$$

$$\frac{d^2y}{dx^2} = \left(-2 - 3\frac{dy}{dx}\right)(3x + 2y - 1)^{-1} + (1 - 2x - 3y)\left(-(3x + 2y - 1)^{-2}\left(3 + 2\frac{dy}{dx}\right)\right)$$

Substitute for $x = -1, y = -1, \frac{dy}{dx} = -1$

© HERIOT-WATT UNIVERSITY

120 TOPIC 2. DIFFERENTIATION

$$\begin{aligned}\frac{d^2y}{dx^2} &= \frac{(-2-3(-1))}{(3(-1)+2(-1)-1)} + \frac{(1-2(-1)-3(-1))(-(3+2(-1)))}{(3(-1)+2(-1)-1)^2}\\ &= \frac{1}{-6} - \frac{(6)(-1)}{(-6)^2}\\ &= \frac{1}{-6} - \frac{6}{36}\\ &= -\frac{1}{6} - \frac{1}{6}\\ &= -\frac{1}{3}\end{aligned}$$

Second derivative exercise Go online

Q87: For the function defined implicitly by $\sqrt{y} + x^3 = 1$, find $\frac{dy}{dx}$ and $\frac{d^2y}{dx^2}$ in terms of x and y only.

....

Q88: Find $\frac{d^2y}{dx^2}$ for the function $2y^3 + xy = 3$

....

Q89: Given $5y^2 + x^3 = 4$ use implicit differentiation to obtain $\frac{dy}{dx}$ in terms of x and y. Hence obtain $\frac{d^2y}{dx^2}$ in terms of x and y.

....

Q90: Given $y^2 - x^2y = x$ use implicit differentiation to obtain $\frac{dy}{dx}$ in terms of x and y. Hence obtain $\frac{d^2y}{dx^2}$ in terms of x and y.

....

Q91: Find $\frac{d^2y}{dx^2}$ for the function $2y^2 - xy = 3x^2 + 2$.

....

Q92: For the function defined implicitly by $x^3 - xy + y^2 = 1$, evaluate $\frac{dy}{dx}$ at (1,1) and $\frac{d^2y}{dx^2}$ at (1,0)

2.8.4 Applying implicit differentiation to problems

Implicit differentiation may be needed when finding the tangent to a curve or the nature of a curve.

Examples

1. Problem:

A circle has equation $(x-2)^2 + (y-3)^2 = 25$ which expands to give $x^2 - 4x + y^2 - 6y = 12$.
The point (5,7) lies on the circumference of this circle.

© HERIOT-WATT UNIVERSITY

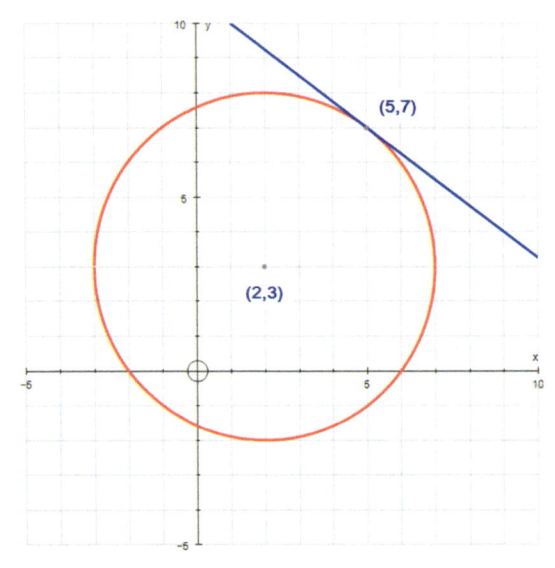

TOPIC 2. DIFFERENTIATION

Find the equation of the tangent to the circle at this point.

Solution:
Remember that the equation of the tangent is given by the straight line equation $y - b = m(x - a)$

Substitute for (5,7): $y - 7 = m(x - 5)$ and the gradient of the tangent, m, is given by $\frac{dy}{dx}$ at (5,7)

Differentiate $x^2 - 4x + y^2 - 6y = 12$:

$$\frac{d}{dx}(x^2) - \frac{d}{dx}(4x) + \frac{d}{dx}(y^2) - \frac{d}{dx}(6y) = \frac{d}{dx}(12)$$

$$2x - 4 + 2y\frac{dy}{dx} - 6\frac{dy}{dx} = 0$$

$$(2y - 6)\frac{dy}{dx} = 4 - 2x$$

$$\frac{dy}{dx} = \frac{4 - 2x}{2y - 6}$$

Now substituting in (5,7) for x and y we get $\frac{dy}{dx}$:

$$\frac{dy}{dx} = \frac{4 - 2(5)}{2(7) - 6}$$

$$= \frac{-6}{8}$$

The gradient of the tangent to the circle at (5,7) is $-\frac{3}{4}$.

© HERIOT-WATT UNIVERSITY

122 TOPIC 2. DIFFERENTIATION

We can now write the equation as:
$$y - 7 = -\frac{3}{4}(x - 5)$$
$$4y - 28 = -3x + 15$$
$$4y + 3x - 43 = 0$$

(Note that there is an alternative method to solve this type of question, that involves using the properties of a circle.)

....

2. Problem:

Given the equation of the curve $3x^2 + 2xy - 5y^2 + 16y = 0$, obtain the x-coordinate of each point at which the curve has a horizontal tangent.

Solution:

Stationary points occur when $\frac{dy}{dx} = 0$, so differentiate and rearrange to get $\frac{dy}{dx}$

$$3x^2 + 2xy - 5y^2 + 16y = 0$$

$$\frac{d}{dx}(3x^2) + \frac{d}{dx}(2xy) - \frac{d}{dx}(5y^2) + \frac{d}{dx}(16y) = \frac{d}{dx}(0)$$

Apply the product rule to $\frac{d}{dx}(2xy)$

$$6x + 2y + 2x\frac{dy}{dx} - 10y\frac{dy}{dx} + 16\frac{dy}{dx} = 0$$

$$(2x - 10y + 16)\frac{dy}{dx} = -6x - 2y$$

$$\frac{dy}{dx} = \frac{-6x - 2y}{2x - 10y + 16}$$

Set $\frac{dy}{dx} = 0$ and solve for x and y

$$\frac{-6x - 2y}{2x - 10y + 16} = 0$$
$$-6x - 2y = 0$$
$$-6x = 2y$$
$$-3x = y$$
$$y = -3x$$

Substitute $y = -3x$ into the original function, collect like terms and solve for x

$$3x^2 + 2x(-3x) - 5(-3x)^2 + 16(-3x) = 0$$
$$3x^2 - 6x^2 - 45x^2 - 48x = 0$$
$$-48x^2 - 48x = 0$$
$$-48x(x + 1) = 0$$
$$-48x = 0 \quad or \quad x + 1 = 0$$
$$x = 0 \qquad\qquad x = -1$$

....

3. Problem:

Calculate the gradient of the curve defined by $\frac{x^3}{y} - 4x = -2 - y$ at the point (2,4).

© HERIOT-WATT UNIVERSITY

Solution:

Differentiate implicitly:

$\frac{d}{dx}\left(x^3 y^{-1}\right) - \frac{d}{dx}\left(4x\right) = \frac{d}{dx}\left(-2\right) - \frac{d}{dx}\left(y\right)$

Apply the product rule to $\frac{d}{dx}\left(x^3 y^{-1}\right)$ and the chain rule to $\frac{d}{dx}\left(y\right)$

$\left(3x^2 y^{-1} - x^3 y^{-2}\frac{dy}{dx}\right) - 4 = -\frac{dy}{dx}$

Rearrange for $\frac{dy}{dx}$:

$\frac{dy}{dx}\left(x^3 y^{-2} - 1\right) = 3x^2 y^{-1} - 4$

Re-write as fractions:

$\frac{dy}{dx}\left(\frac{x^3}{y^2} - 1\right) = \frac{3x^2}{y} - 4$

$\frac{dy}{dx} = \frac{\frac{3x^2}{y} - 4}{\frac{x^3}{y^2} - 1}$

Simplify by multiplying the numerator and denominator by y^2:

$\frac{dy}{dx} = \frac{3x^2 y - 4y^2}{x^3 - y^2}$

Evaluate the gradient at (2,4):

$\frac{dy}{dx} = \frac{3(2)^2 (4) - 4(4)^2}{(2)^3 - (4)^2}$

$\frac{dy}{dx} = \frac{48 - 64}{8 - 16}$

$\frac{dy}{dx} = \frac{-16}{-8}$

$\frac{dy}{dx} = 2$

The gradient at (2,4) is 2.

Applying implicit differentiation to problems exercise   Go online

Q93: A curve has equation $xy - 3y^2 = -1$. Find an equation of the tangent to the curve at the point (2,1).

....

Q94: The equation $y^3 - 2xy = x^2 + 4y$ defines a curve passing through the point B(3,1). Obtain an equation for the tangent to the curve at B.

....

Q95: Find the equation of the tangent at (2,1) for the curve $y(x + y)^2 = 3(x^3 - 5)$

....

Q96: Given the equation $3y^4 - 2x^2 y - y + 5x^2 = 0$ of a curve, obtain the y-coordinate of its stationary points.

....

© HERIOT-WATT UNIVERSITY

124 TOPIC 2. DIFFERENTIATION

Q97: Given the equation $xy^2 + 5x^2y = 10$ for $x > 0$ and $y > 0$ of a curve, obtain the x-coordinate of each point at which the curve has a horizontal tangent.

....

Q98: Find two points on the curve with equation $2x^2 - xy + y^2 = 3$ where the tangent is parallel to the x-axis, and two points where the tangent is parallel to the y-axis.

....

Q99: Find the x-coordinates of the two stationary points for the curve $2x^2 - xy + 3y^2 = 46$.

Astroid   Go online

The following is a curiosity. It is a famous curve first discovered by Roemer (1674). It was found in the search for the best form for gear teeth.

$x^{\frac{2}{3}} + y^{\frac{2}{3}} = 2^{\frac{2}{3}}$ defines a famous curve called the *astroid*.

This curve is drawn in an interesting way. Imagine that a circle of radius $\frac{1}{2}$ is rolling inside another circle of radius 2. The astroid is the curve traced out by a point on the circumference of the smaller circle.

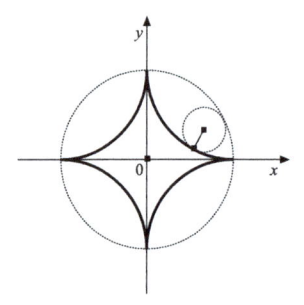

Implicit differentiation can help us to understand the shape of the astroid.

Q100:

a) Calculate $\frac{dy}{dx}$ for $x^{\frac{2}{3}} + y^{\frac{2}{3}} = 2^{\frac{2}{3}}$

b) Determine the gradient of the tangent to the curve at the point $\left(\frac{1}{\sqrt{2}}, \frac{1}{\sqrt{2}}\right)$

....

Q101: Write down the coordinates of the four points on the curve of the astroid where the derivative cannot be defined.

> This can also be checked by looking at the animation. When the blue dot on the small circle comes to one of these points it stops momentarily and then changes direction. There is no tangent at these points.

2.9 Differentiating inverse functions

There are some situations where we do not know what the derivative of a given function is, for example, the natural log, ln x. However, we may know what the derivative of its inverse function is, e.g. $\frac{d}{dx}(e^x) = e^x$. This section will discuss how the inverse function can be used to solve this differentiation problem.

Conditions for inverse functions

> **Key point**
>
> For a function to have an inverse it must be a one-to-one and onto function.
>
> For each $y \in B$ (codomain) there is exactly one element $x \in A$ (domain) such that $f(x) = y$.
>
> The inverse function is defined as $f^{-1}(y) = x$

To make this clearer, the function has to pass some tests:

- every x value must have a y value;
- *vertical ruler test:* if you drag a ruler vertically across the function, one x value must match up with only one y value;
- *horizontal ruler test:* if you drag a ruler horizontally up and down the function, one y value must match up with only one x value.

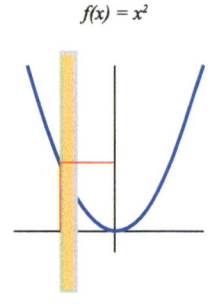

$f(x) = x^2$

Vertical ruler test passed: one y value to one x value.

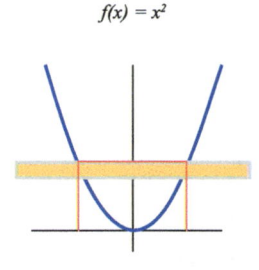

$f(x) = x^2$

Horizontal ruler test failed: more than one x value to one y value so not a one-to-one function.

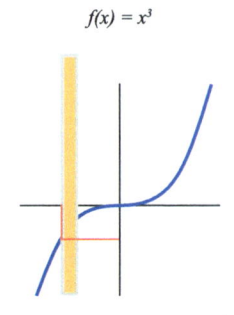

f(x) = x³

f(x) = x³

Vertical ruler test passed: one y value to one x value.

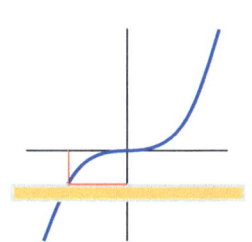

Horizontal ruler test passed: one x value to one y value so is a one-to-one function.

2.9.1 Differentiating inverse functions

Key point

In order to find the derivative of an inverse function, the following result has to be established:
$\frac{dy}{dx} = \frac{1}{\left(\frac{dx}{dy}\right)}$

Establishing the result $\frac{dy}{dx} = \frac{1}{\left(\frac{dx}{dy}\right)}$

Consider $y = f^{-1}(x)$

$f^{-1}(x)$ denotes the inverse function, but $f^{-1}(x) \neq \frac{1}{f(x)}$

so $f(y) = f(f^{-1}(x)) = x$

Now differentiate with respect to x where $f(y)$ is replaced by x:

$\frac{d}{dx}(f(y)) = \frac{d}{dx}(x)$
$\phantom{\frac{d}{dx}(f(y))} = 1$

Now differentiate with respect to x where $f(y)$ is *not* replaced by x:

Using the chain rule:

$\frac{d}{dx}(f(y)) = \frac{d}{dy}(f(y)) \times \frac{dy}{dx}$

So:

$\frac{d}{dx}(f(y)) = \frac{d}{dx}(x)$

$\frac{d}{dy}(f(y)) \times \frac{dy}{dx} = 1$

Since $f(y) = x$ then:

$\frac{d}{dy}(x) \times \frac{dy}{dx} = 1$

$\frac{dx}{dy} \times \frac{dy}{dx} = 1$

Note: Since $f(y) = x$ then x is a function of y. So if we wish to show that we are differentiating x

with respect to y we can write $\frac{d}{dx}(x)$ which can be written as $\frac{dy}{dx}$ in shorthand.
Thus we have the result:

$\frac{dy}{dx} = \frac{1}{\frac{dx}{dy}}$

Rewriting this gives:

$\frac{d}{dx}(y) = \frac{1}{\frac{dx}{dy}}$

Now $y = f^{-1}(x)$, which was stated at the beginning. From this, applying f to both sides we have
$f(y) = f\left(f^{-1}(x)\right) \Rightarrow f(y) = x$
Now $\frac{dx}{dy}$ is equivalent to $\frac{d}{dy}(x)$ where x is a function of y. This means that $\frac{dx}{dy} = f'(y)$
Substituting this into the established result $\frac{d}{dx}(y) = \frac{1}{\frac{dx}{dy}}$ gives: $\frac{d}{dx}f^{-1}(x) = \frac{1}{f'(y)}$

Top tip

A strategy for differentiating inverse functions

The following strategy may help.

1. Let $y = f^{-1}(x)$ and therefore $x = f(y)$
2. Find $f'(y) = \frac{dx}{dy}$ by differentiating with respect to y.
3. To find the derivative of the inverse function use the formula $\frac{dy}{dx} = \frac{1}{\frac{dx}{dy}}$
4. Eliminate y from your answer if required.

Examples

1. Problem:

Consider the function $f(x) = x^3$, it has inverse $f^{-1}(x) = x^{\frac{1}{3}}$
Find the derivative of this inverse function.

Solution:

Let $y = f^{-1}(x)$ so $y = x^{\frac{1}{3}}$ therefore $x = y^3$
Find $f'(y) = 3y^2$
Now $\frac{dy}{dx} = \frac{1}{f'(y)} \Rightarrow \frac{dy}{dx} = \frac{1}{3y^2}$
Lastly substitute y for $x^{\frac{1}{3}}$:
$\frac{dy}{dx} = \frac{1}{3x^{\frac{2}{3}}}$

....

2. Problem:

A function is defined by $f(x) = 3e^{2x} + 4$ Find the derivative of the inverse function.

Solution:

$y = f^{-1}(x)$

Let $y = f^{-1}(x)$ so $x = f(y)$ therefore $x = 3e^{2y} + 4$

Find $f'(y) = 6e^{2y}$

Now $\frac{dy}{dx} = \frac{1}{f'(y)} \Rightarrow \frac{dy}{dx} = \frac{1}{6e^{2y}}$

Lastly substitute to eliminate y. In this case we can substitute for $3e^{2y}$:

$3e^{2y} = x - 4$

$\frac{dy}{dx} = \frac{1}{2 \times 3e^{2y}}$

$\frac{dy}{dx} = \frac{1}{2(x-4)}$

$\frac{dy}{dx} = \frac{1}{2x - 8}$

....

3. Problem:

A function is defined as $f(x) = x^3 + 4x + 2$

a) If $y = f^{-1}(x)$, find the derivative of the inverse, giving your in terms of y.

b) Also find the equation of the tangent to the inverse at the point (7,1).

Solution:

a)

Let $y = f^{-1}(x)$ so $x = f(y)$ therefore $x = y^3 + 4y + 2$

Find $f'(y) = 3y^2 + 4$

Now $\frac{dy}{dx} = \frac{1}{f'(y)} \Rightarrow \frac{dy}{dx} = \frac{1}{3y^2 + 4}$

b)

At $y = 1$:

$\frac{dy}{dx} = \frac{1}{3(1)^2 + 4}$

$\frac{dy}{dx} = \frac{1}{7}$

Substituting $m = \frac{1}{7}$ and (7,1) into the straight line equation:

$y - 1 = \frac{1}{7}(x - 7)$

$7y - 7 = x - 7$

$7y - x = 0$

Differentiating inverse functions exercise

Q102: A function is defined as $f(x) = 3e^{2x} + 4$. Find the derivative of the inverse function.

...

Q103: A function is defined as $f(x) = x^3 + 4x + 2$.

a) If $y = f^{-1}(x)$, find the derivative of the inverse, giving your answer in terms of y.
b) Also find the equation of the tangent to the inverse at the point (7, 1).

...

Q104: A function is defined by $f(x) = x^{\frac{5}{4}}$. Find the derivative of the inverse function in terms of x.

...

Q105: A function is defined by $f(x) = 4x^{-2}$, $x > 0$. Find the derivative of the inverse function in terms of x.

...

Q106: A function is defined by $f(x) = x^2 + 4x - 1$, $x > 0$. Find the derivative of the inverse function in terms of x.

...

Q107: A function is defined by $f(x) = 5e^{3x} - 6$, $x > 0$. Find the derivative of the inverse function in terms of x.

...

Q108: A function is defined by $f(x) = x^3 - 6x + 4$, $x > 2$

a) Find the derivative of the inverse function in terms of y.
b) Find the equation of the tangent to the inverse at the point (2,0). Give your answer in the form $ax + by + c = 0$.

...

Q109: A function is defined by $f(x) = x^3 + 5x^2 - x$.

a) Find the derivative of the inverse function in terms of y.
b) Find the equation of the tangent to the inverse at the point (1,5). Give your answer in the form $ax + by + c = 0$.

...

Q110: A function is defined by $f(x) = 2x^3 - 5x + 11$, $x > 2$

a) Find the derivative of the inverse function in terms of y.
b) Find the equation of the tangent to the inverse at the point (0,11). Give your answer in the form $ax + by + c = 0$.

A strategy for differentiating inverse functions

The following strategy may help.

1. Let $y = f^{-1}(x)$ and therefore $x = f(y)$
2. Find $f'(y) = \frac{dx}{dy}$ by differentiating with respect to y
3. To find the derivative of the inverse function, use the formula $\frac{dy}{dx} = \frac{1}{\left(\frac{dx}{dy}\right)}$.
4. Eliminate y from your answer if required.

2.9.2 Differentiating inverse trigonometric functions

Example : Derivative of $y = \sin^{-1} x$

Problem:

The function $y = \sin^{-1} x$, for $x \in (-1, 1)$ is differentiable.

Note that $\sin^{-1} x$ or arcsin function's domain is restricted so that it is a one-to-one and onto mapping.

Solution:

To differentiate, let $y = \sin^{-1} x$ then $x = \sin y$ then apply the sine function to both sides to give $x = \sin y$.

Using the rule: $\frac{dy}{dx} = \frac{1}{\left(\frac{dx}{dy}\right)}$

$\frac{dx}{dy} = \frac{d}{dy}(\sin y)$ and $\frac{dx}{dy} = \cos y$

We have: $\frac{dy}{dx} = \frac{1}{\cos y}$

We want to write this in terms of x not y:

Now, since $\cos^2 y + \sin^2 y = 1$, then $\cos y = \sqrt{1 - \sin^2 y}$

From the beginning, $x = \sin y$ so $\cos y = \sqrt{1 - x^2}$

Only the positive values of the square root are considered here. This is because of the restricted domain of $\sin^{-1} x$.
The domain of $\sin^{-1} x$ is between -1 and 1.
For this domain the range is $-\frac{\pi}{2} \leqslant y \leqslant \frac{\pi}{2}$.
This range becomes the domain for $\cos y$ and $\cos y$ is positive between $-\frac{\pi}{2} \leqslant y \leqslant \frac{\pi}{2}$.

Therefore: $\frac{dy}{dx}(\sin^{-1} x) = \frac{1}{\sqrt{1 - x^2}}$

The graph for $y = \sin^{-1} x$ is shown here along with the graph of its derivative $\frac{dy}{dx} = \frac{1}{\cos y} = \frac{1}{\sqrt{1 - x^2}}$

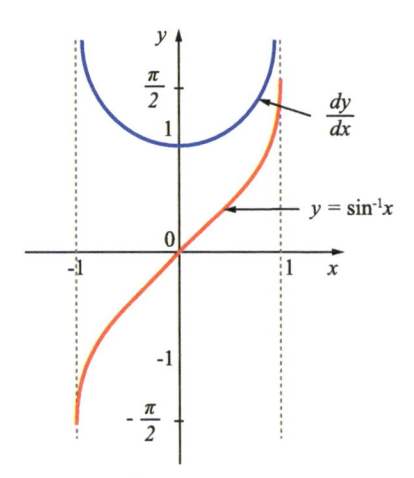

TOPIC 2. DIFFERENTIATION

Notice that $\frac{dy}{dx} = 1$ at $x = 0$ and so the gradient of the curve $y = \sin^{-1} x$ is equal to 1 at $x = 0$
Also note that as $x \to \pm 1$, then $\frac{dy}{dx} \to \infty$

Examples

1. Problem:
The function $y = \cos^{-1} x$ for $x \in (-1, 1)$ is differentiable.
Note that $\cos^{-1} x$ or $arccos$ function's domain is restricted so that it is a one-to-one and onto mapping.

Solution:
To differentiate let $y = \cos^{-1} x$. Now apply the cosine function to both sides to give $x = \cos y$.
Using the rule: $\frac{dy}{dx} = \frac{1}{\frac{dx}{dy}}$
$\frac{dx}{dy} = \frac{d}{dy}(\cos y)$ and $\frac{dx}{dy} = -\sin y$
We have: $\frac{dy}{dx} = -\frac{1}{\sin y}$
We want to write this in terms of x not y.
Now since $\cos^2 y + \sin^2 y = 1$ then $\sin y = \sqrt{1 - \cos^2 y}$
From the beginning $x = \cos y$ so $\sin y = \sqrt{1 - x^2}$
Only the positive values of the square root are considered here. This is because of the restricted domain of $\cos^{-1} x$. The domain of $\cos^{-1} x$ is between -1 and 1. For this domain the range is 0 to π. This range becomes the domain for $\sin y$ and $\sin y$ is positive between 0 to π.
Therefore $\frac{dy}{dx}(\cos^{-1} x) = -\frac{1}{\sqrt{1-x^2}}$
The graph for $y = \cos^{-1} x$ is shown here along with the graph of its derivative $\frac{dy}{dx} = -\frac{1}{\sin y} = -\frac{1}{\sqrt{1-x^2}}$

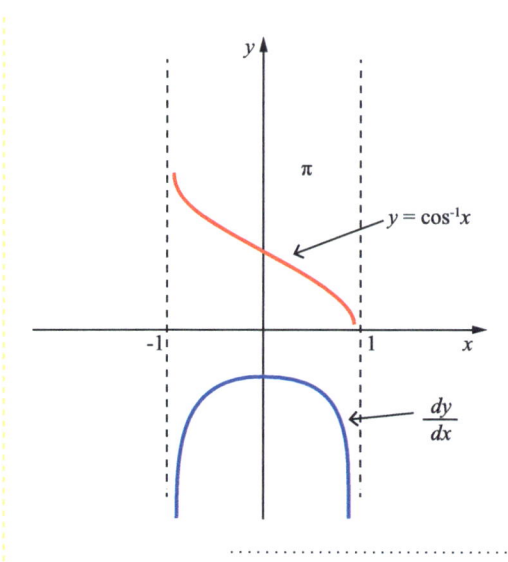

2. Problem:

The function $y = \tan^{-1} x$, for $x \in (-\infty, \infty)$ is differentiable.

Note that $\tan^{-1} x$ or arctan function's domain is restricted so that it is a one-to-one and onto mapping.

Solution:

To differentiate let $y = \tan^{-1} x$. Now apply the tangent function to both sides to give $x = \tan y$.

Using the rule: $\frac{dy}{dx} = \frac{1}{\frac{dx}{dy}}$

$\frac{dx}{dy} = \frac{d}{dy} (\tan y)$ and $\frac{dx}{dy} = \sec^2 y$

We have: $\frac{dy}{dx} = \frac{1}{\sec^2 y}$

We want to write this in terms of x not y.

Now since $\cos^2 y + \sin^2 y = 1$.

Divide through by $\cos^2 y$ gives $1 + \tan^2 y = \sec^2 y$.

From the beginning $x = \tan y$ so $\sec^2 y = 1 + x^2$.

Therefore $\frac{dy}{dx} \left(\tan^{-1} x \right) = \frac{1}{1+x^2}$.

The graph for $y = \tan^{-1} x$ is shown here along with the graph of its derivative $\frac{dy}{dx} = \frac{1}{\sec^2 y} = \frac{1}{1+x^2}$.

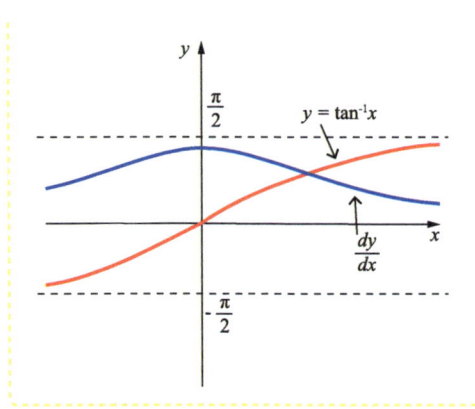

TOPIC 2. DIFFERENTIATION

Key point
This can be summarised as:

$$\frac{d}{dx}\left(\sin^{-1}x\right) = \frac{1}{\sqrt{1-x^2}}$$

$$\frac{d}{dx}\left(\cos^{-1}x\right) = -\frac{1}{\sqrt{1-x^2}}$$

$$\frac{d}{dx}\left(\tan^{-1}x\right) = \frac{1}{1+x^2}$$

Examples

1. Problem:
Differentiate $y = \sin^{-1}4x$ for $\left(-\frac{1}{4}, \frac{1}{4}\right)$

Solution:
Using the standard result of $\frac{d}{dx}\left(\sin^{-1}x\right) = \frac{1}{\sqrt{1-x^2}}$

Now let $u = 4x$ so that $y = \sin^{-1}u$ and use the chain rule $\frac{dy}{dx} = \frac{dy}{du} \times \frac{du}{dx}$

$\frac{dy}{du} = \frac{1}{\sqrt{1-u^2}}$ and $\frac{du}{dx} = 4$

So $\frac{dy}{dx} = \frac{1}{\sqrt{1-u^2}} \times 4$

Substituting $u = 4x$ back in we have:

$$\frac{dy}{dx} = \frac{4}{\sqrt{1-(4x)^2}}$$

$$\frac{dy}{dx} = \frac{4}{\sqrt{1-16x^2}}$$

....

2. Problem:
Differentiate $y = \cos^{-1}\left(\frac{x}{5}\right)$ for (-5,5).

Solution:
Using the standard result of $\frac{d}{dx}\left(\cos^{-1}x\right) = -\frac{1}{\sqrt{1-x^2}}$

Now let $u = \frac{x}{5}$ so that $y = \sin^{-1}u$ and use the chain rule $\frac{dy}{dx} = \frac{dy}{du} \times \frac{du}{dx}$

$\frac{dy}{du} = -\frac{1}{\sqrt{1-u^2}}$ and $\frac{du}{dx} = \frac{1}{5}$

So $\frac{dy}{dx} = -\frac{1}{\sqrt{1-u^2}} \times \frac{1}{5}$

Substituting $u = \frac{x}{5}$ back in we have:

$$\frac{dy}{dx} = -\frac{1}{5\sqrt{1-\left(\frac{x}{5}\right)^2}}$$

$$\frac{dy}{dx} = -\frac{1}{5\sqrt{1-\frac{x^2}{25}}}$$

To simplify get a common denominator for $1 - \frac{x^2}{25}$

This becomes: $\frac{25-x^2}{25}$

We have:

$$\frac{dy}{dx} = -\frac{1}{5\sqrt{\frac{25-x^2}{25}}}$$

Take out the common denominator of 25 outside the square root.

$$\frac{dy}{dx} = -\frac{1}{\frac{5}{5}\sqrt{25-x^2}}$$

$$\frac{dy}{dx} = -\frac{1}{\sqrt{25-x^2}}$$

....

3. Problem:
Differentiate $y = \tan^{-1}\left(\frac{2}{x}\right)$ for $(-\infty, \infty)$

Solution:
Using the standard result of $\frac{d}{dx}\left(\tan^{-1}x\right) = \frac{1}{1+x^2}$

Now let $u = \frac{2}{x}$ so that $y = \sin^{-1}u$ and use the chain rule $\frac{dy}{dx} = \frac{dy}{du} \times \frac{du}{dx}$

$\frac{dy}{du} = \frac{1}{1+u^2}$ and $\frac{du}{dx} = -\frac{2}{x^2}$

So $\frac{dy}{dx} = \frac{1}{1+u^2} \times -\frac{2}{x^2}$

Substituting $u = \frac{2}{x}$ back in we have:

$$\frac{dy}{dx} = -\frac{2}{x^2\left(1+\left(\frac{2}{x}\right)^2\right)}$$

$$\frac{dy}{dx} = -\frac{2}{x^2\left(1+\frac{4}{x^2}\right)}$$

$$\frac{dy}{dx} = -\frac{2}{x^2+4}$$

TOPIC 2. DIFFERENTIATION

Differentiating inverse trigonometric functions exercise   Go online

Q111: Differentiate $y = \sin^{-1}(3x)$ for $\left(-\frac{1}{3}, \frac{1}{3}\right)$

Q112: Find $\frac{dy}{dx}$ when $y = \sin^{-1}\left(\frac{x}{3}\right)$

> **Key point**
>
> The above result can be generalised to give:
>
> $$\frac{d}{dx}\left(\sin^{-1}\left(\frac{x}{a}\right)\right) = \frac{1}{\sqrt{a^2 - x^2}}$$
> $$\frac{d}{dx}\left(\cos^{-1}\left(\frac{x}{a}\right)\right) = -\frac{1}{\sqrt{a^2 - x^2}}$$
> $$\frac{d}{dx}\left(\tan^{-1}\left(\frac{x}{a}\right)\right) = -\frac{a}{a^2 + x^2}$$

Q113: Find $\frac{dy}{dx}$ when $y = \cos^{-1} e^x$

Q114: Differentiate $y = \cos^{-1} 4x$ for $\left(-\frac{1}{4}, \frac{1}{4}\right)$

Q115: Differentiate $y = \tan^{-1} 5x$ for $(-\infty, \infty)$

Q116: Differentiate $y = \sin^{-1}\left(\frac{x}{6}\right)$ for (-6,6).

Q117: Differentiate $y = \cos^{-1}\left(\frac{x}{3}\right)$ for (-3,3).

Q118: Differentiate $y = \tan^{-1}\left(\frac{x}{2}\right)$ for $(-\infty, \infty)$

Q119: Differentiate $y = \sin^{-1}\left(\frac{3}{x}\right)$ for $x < -3$ and $x > 3$

Q120: Differentiate $y = \cos^{-1}\left(\frac{4}{x}\right)$ for $x < -4$ and $x > 4$

Q121: Differentiate $y = \tan^{-1}\left(\frac{5}{x}\right)$ for $(-\infty, \infty)$

Q122: Differentiate $y = \sin^{-1}\left(\frac{3x}{4}\right)$ for $\left(-\frac{4}{3}, \frac{4}{3}\right)$

© HERIOT-WATT UNIVERSITY

Q123: Differentiate $y = \cos^{-1}\left(\frac{2x}{3}\right)$ for $\left(-\frac{3}{2}, \frac{3}{2}\right)$

Q124: Differentiate $y = \tan^{-1}\left(\frac{5x}{2}\right)$ for $(-\infty, \infty)$

Q125: Differentiate $y = \sin^{-1}\left(\frac{5}{2x}\right)$ for $\left(-\frac{5}{2}, \frac{5}{2}\right)$

Q126: Differentiate $y = \cos^{-1}\left(\frac{3}{2x}\right)$ for $\left(-\frac{3}{2}, \frac{3}{2}\right)$

Q127: Differentiate $y = \tan^{-1}\left(\frac{2}{7x}\right)$ for $(-\infty, \infty)$

Q128: Differentiate $y = \sin^{-1}(4x - 1)$ for $\left(0, \frac{1}{2}\right)$

Q129: Differentiate $y = \cos^{-1}(3\sin x)$ for $\left(-\frac{1}{3}, \frac{1}{3}\right)$

Q130: Differentiate $y = \tan^{-1}(\ln|2x|)$ for $(0, \infty)$

2.9.3 Differentiating inverse trigonometric functions implicitly

The results for differentiating $\sin^{-1} x$, $\cos^{-1} x$ and $\tan^{-1} x$ are already known, but it is also possible to differentiate these functions as implicit equations.

Examples

1. Problem:

Find $\frac{dy}{dx}$ when $y = \sin^{-1} x$ for $x \in (-1, 1)$

Solution:

When $y = \sin^{-1} x$ then $\sin y = x$

Now, differentiating this equation implicitly gives:

$$\frac{d}{dy}(\sin y)\frac{dy}{dx} = \frac{d}{dx}(x)$$

$$\cos y \frac{dy}{dx} = 1$$

$$\frac{dy}{dx} = \frac{1}{\cos y}$$

Since $\sin^2 y + \cos^2 y = 1$, then:

$\cos^2 y = 1 - \sin^2 y$

$\cos y = \sqrt{1 - \sin^2 y} = \sqrt{1 - x^2}$ from $\sin y = x$

TOPIC 2. DIFFERENTIATION

(Since $-\frac{\pi}{2} \leqslant y \leqslant \frac{\pi}{2}$ note: only consider the positive root.)
So when $y = \sin^{-1} x$, then:
$\frac{dy}{dx} = \frac{1}{\cos y} = \frac{1}{\sqrt{1 - x^2}}$

....

2. Problem:

Find $\frac{dy}{dx}$ when $y = \tan^{-1}\left(\frac{x}{a}\right)$
The following identity will help: $1 + \tan^2 y = \sec^2 y$

Solution:

When $y = \tan^{-1}\left(\frac{x}{a}\right)$ then $\tan y = \left(\frac{x}{a}\right)$

Reminder: We are assuming that we do not know what the derivative of the inverse is. So instead we rearrange and use implicit differentiation instead.

Differentiating this equation implicitly gives:

$$\frac{d}{dy}(\tan y)\frac{dy}{dx} = \frac{d}{dx}\left(\frac{x}{a}\right)$$

$$\sec^2 y \frac{dy}{dx} = \frac{1}{a}$$

$$\frac{dy}{dx} = \frac{1}{a\sec^2 y}$$

$$= \frac{1}{a(1 + \tan^2 y)}$$

$$= \frac{1}{a\left(1 + \frac{x^2}{a^2}\right)}$$

$$= \frac{1}{a + \frac{x^2}{a}}$$

Multiply the numerator and the denominator by a:

$$= \frac{a}{a^2 + x^2}$$

So when $y = \tan^{-1}\left(\frac{x}{a}\right)$, then:
$\frac{dy}{dx} = \frac{1}{a\sec^2 y} = \frac{a}{a^2 + x^2}$

....

3. Problem:

Find $\frac{dy}{dx}$ when $y = \cos^{-1} x$ for $x \in (-1, 1)$

Solution:

When $y = \cos^{-1} x$ then $\cos y = x$

Now, differentiating this equation implicitly gives,

$$\frac{d}{dy}(\cos y)\frac{dy}{dx} = \frac{d}{dx}(x)$$

$$-\sin y \frac{dy}{dx} = 1$$

$$\frac{dy}{dx} = -\frac{1}{\sin y}$$

© HERIOT-WATT UNIVERSITY

Since $\cos^2 y + \sin^2 y = 1$ then $\sin^2 y = 1 - \cos^2 y$
$\sin y = \sqrt{1 - \cos^2 y} = \sqrt{1 - x^2}$ from $\cos y = x$
(since $0 \leq y \leq \pi$ note: only consider the positive root)
So when $y = \cos^{-1} x$ then
$\frac{dy}{dx} = -\frac{1}{\sin y} = -\frac{1}{\sqrt{1-x^2}}$

Notice that these are the same results as stated in the section on inverse functions.

Differentiating inverse trigonometric functions implicitly exercise   Go online

Using implicit differentiation, find the derivatives of the following inverse trigonometric functions. Show all of your working.

Q131: $y = \cos^{-1}(3x)$

Q132: $y = \sin^{-1}\left(\frac{3x^2}{2}\right)$

Q133: Use implicit differentiation to differentiate $y = \tan^{-1}\left(\frac{x}{6}\right)$

Q134: Use implicit differentiation to differentiate $y = \sin^{-1}\left(\frac{5}{x}\right)$, $-5 < x < 5$

Q135: Use implicit differentiation to differentiate $y = \cos^{-1}\left(\sqrt{2 - 5x}\right)$, $2 \cdot 5 < x < 3$

Q136: Use implicit differentiation to differentiate $y = \tan^{-1}\left(\sqrt{x - 1}\right)$, $x > 0$

2.10 Differentiation using the product, quotient and chain rules

We have looked at the derivatives of a lot of new functions as well as introducing the product and quotient rule. Below is a summary of what we have found:

TOPIC 2. DIFFERENTIATION

> **Key point**
>
> **Product rule:**
>
> If $y = fg$ then $y' = f'g + fg'$
>
> **Quotient rule:**
>
> If $y = \frac{f}{g}$ then $y' = \frac{f'g - fg'}{g^2}$
>
> **Chain rule:**
>
> If $y = f(g)$ then $y' = f'(g) \times g'$

> **Key point**
>
> **Derivatives:**
>
> $\frac{d}{dx}(\sin x) = \cos x$ $\qquad$ $\frac{d}{dx}(\cos x) = -\sin x$ $\qquad$ $\frac{d}{dx}(\tan x) = \sec^2 x$
>
> $\frac{d}{dx}(\csc x) = -\cot x \csc x$ $\qquad$ $\frac{d}{dx}(\sec x) = \sec x \tan x$ $\qquad$ $\frac{d}{dx}(\cot x) = \csc^2 x$
>
> $\frac{d}{dx}\left(\sin^{-1} x\right) = \frac{1}{\sqrt{1-x^2}}$ $\qquad$ $\frac{d}{dx}\left(\cos^{-1} x\right) = -\frac{1}{\sqrt{1-x^2}}$ $\qquad$ $\frac{d}{dx}\left(\tan^{-1} x\right) = \frac{1}{1+x^2}$
>
> $\frac{d}{dx}\left(e^{ax}\right) = ae^{ax}$ $\qquad$ $\frac{d}{dx}(\ln|x|) = \frac{1}{x}$

In this section we will be applying the product and quotient rule to a combination of these functions and more familiar ones as well.

Examples

1. Problem:

Differentiate with respect to x.

$f(x) = (3 - 2x)\sin^{-1}(5x), \quad -\frac{1}{5} < x < \frac{1}{5}$

Solution:

Apply the product rule:

Let $y = fg$ then $f(x) = (3 - 2x)$ and $g(x) = \sin^{-1}(5x)$

So, $f'(x) = -2$ and $g'(x) = \frac{5}{\sqrt{1-25x^2}}$

Then using $y' = f'g + fg'$ gives,

$y' = -2\sin^{-1}(5x) + (3 - 2x)\frac{5}{\sqrt{1-25x^2}}$

So,

$f'(x) = -2\sin^{-1}(5x) + \frac{15-2x}{\sqrt{1-25x^2}}$

....

2. Problem:

Differentiate with respect to x.

$f(x) = e^{\sec 3x}, \quad -\frac{\pi}{6} < x < \frac{\pi}{6}$

© HERIOT-WATT UNIVERSITY

Solution:
To differentiate the exponential, differentiate the power and then multiply by it.
$\frac{d}{dx}(\sec 3x) = 3\tan 3x \sec 3x$
Then,
$f'(x) = 3\tan 3x \sec 3x e^{\sec 3x}$

....

3. Problem:
Differentiate with respect to x.
$y = \frac{2-\ln 4x}{3x}$, $where\ x > 0$
Solution:
Apply the quotient rule.
Let $y = \frac{f}{g}$ then $f(x) = 2 - \ln 4x$ and $g(x) = 3x$
So, $f'(x) = -\frac{1}{x}$ and $g'(x) = 3$
Then,
$y' = \frac{f'g - fg'}{g^2}$
$y' = \frac{-\frac{1}{x} \times 3x - (2 - \ln 4x) \times 3}{(3x)^2}$
$y' = \frac{-x - 6 + 3\ln 4x}{9x^2}$

....

4. Problem:
Differentiate $g(x) = \frac{\tan^{-1} 2x}{1 + 4x^2}$
Solution:
Using the quotient rule: $\frac{d}{dx}\left(\frac{f}{g}\right) = \frac{f'g - fg'}{(g)^2}$
Let $f = \tan^{-1} 2x$ and $g = 1 + 4x^2$
Then $f' = \frac{2}{1+4x^2}$ and $g' = 8x$
Substituting in to the quotient rule:
$\frac{d}{dx}\left(\frac{f}{g}\right) = \frac{\frac{2}{1+4x^2}(1 + 4x^2) - \tan^{-1} 2x\, (8x)}{(1 + 4x^2)^2}$
Simplifying this gives:
$\frac{d}{dx}\left(\frac{f}{g}\right) = \frac{2 - 8x\tan^{-1} 2x}{(1 + 4x^2)^2}$

Differentiation using the product, quotient and chain rules exercise   Go online

Q137: If $f(x) = x^3 \tan 2x$, find $f'(x)$.

....

TOPIC 2. DIFFERENTIATION

Q138: Differentiate $f(x) = \sqrt{3x} \exp(-2x)$, $x \geq 0$

....

Q139: Given $f(x) = \sin^2 x e^{\csc x}$, $0 < x < \pi$, obtain $f'(x)$

....

Q140: For $y = \frac{\cos^{-1}(7x)}{2x-1}$, $-\frac{1}{7} < x < \frac{1}{7}$ find the derivative with respect to x.

....

Q141: Differentiate $\frac{x^5}{1-\tan(2x)}$, $-1 < x < 0$

....

Q142: Differentiate $\frac{6x^3}{\ln|\cos x|}$ with respect to x, for $-\frac{\pi}{4} < x < \frac{\pi}{4}$

....

Q143: Given that $f(x) = \tan^{-1}(\cot 5x)$, $0 < x < \frac{\pi}{6}$, find $f'(x)$

2.11 Parametric differentiation

A situation may be modelled as a function of time. For instance, the position of a child on a merry-go-round. As such, the x and y positions are described as separate functions of time. These are called parametric equations. If we wished to know the rate of change of the child's position for a particular point in time, we need to differentiate. We could find the form of the expression in terms of x and y, but this may be difficult. Given these parametric equations, this section will look at how to find this rate of change without finding the cartesian form. In some cases, it may be easier to differentiate the parametric equations than the cartesian form.

2.11.1 Parametric curves

Parametric curves Go online

When a curve is traced out over time it may sometimes cross itself or double back on itself. In such cases, that curve cannot be described by expressing y directly in terms of x.

The curve is not the graph of a function.

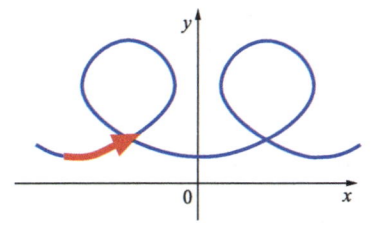

© HERIOT-WATT UNIVERSITY

142 TOPIC 2. DIFFERENTIATION

To cope with this difficulty, we can describe each position along the curve at time t by:
$x = f(t)$
$y = g(t)$

Equations like these are called **parametric equations**.

Parametric equations are equations that are expressed in terms of an independent variable. They are of the form:
$x = f(t)$
$y = g(t)$
where t is the independent variable for x and y and the variable t is called a parameter.

A **parameter** is a variable that is given a series of arbitrary values in order that a relationship between x and y may be established.

The parameter may not always represent time, it might instead denote an angle about the origin or a distance along the curve.

Study the examples in order to understand this more clearly.

Examples

1.

A curve is defined by the following parametric equations:
$x = t^2 - 3t$
$y = 2t$

Notice that the coordinates on the curve can be found by giving the parameter t various values as in the table.

t	0	1	2	3	4
x	0	-2	-2	0	4
y	0	2	4	6	8

This then allows us to plot the points (x, y) and make a sketch of the curve as shown here.

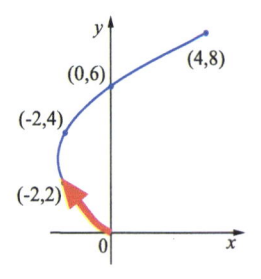

The arrow indicates direction or *orientation* of the curve as t increases.

Note that it is sometimes possible to eliminate t and obtain the cartesian equation of the curve.

In this example, the parametric equations are $x = t^2 - 3t$ and $y = 2t$
From $y = 2t$, then $t = \frac{y}{2}$

Now, substituting this value of t into $x = t^2 - 3t$ gives:

$$x = \left(\frac{y}{2}\right)^2 - 3\left(\frac{y}{2}\right)$$
$$x = \frac{y^2}{4} - \frac{3y}{2}$$
$$4x = y^2 - 6y$$

Therefore, $4x = y^2 - 6y$ is the cartesian equation of the curve and is also called the *constraint equation*.

....

2. Problem:

The parametric equations $x = 4\cos\theta$, $y = 4\sin\theta$, for $0 \leq \theta \leq 2\pi$ describe the position $P(x, y)$ of a particle moving in the plane.
Check that:
$x^2 + y^2 = 16\cos^2\theta + 16\sin^2\theta = 16$

Solution:

The cartesian equation for the curve is $x^2 + y^2 = 16$, which is the equation of a circle centre the origin with radius 4.
The parameter θ is the radian measure of the angle that radius OP makes with positive x-axis.

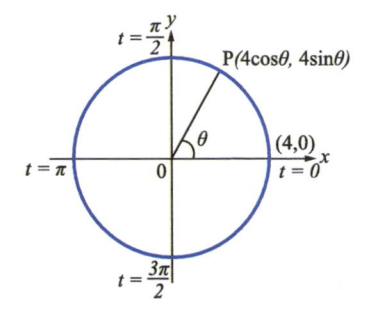

Calculating (x, y) coordinates for some values of θ will gives the direction in which the particle is moving.

θ	0	$\pi/2$	π	$3\pi/2$	2π
x	4	0	-4	0	4
y	0	4	0	-4	0

The table shows that as θ increases the particle moves anticlockwise around the circle starting and ending at (4,0).

The **cartesian equation** of a curve is an equation which links the x-coordinate and the y-coordinate of the general point (x, y) on the curve.

Below are some interesting examples of parametric equations and the curves they generate.

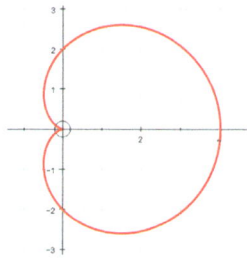

Cardioid
$$x = 2\cos\theta + 2\cos^2\theta$$
$$y = 2\sin\theta + \sin 2\theta$$

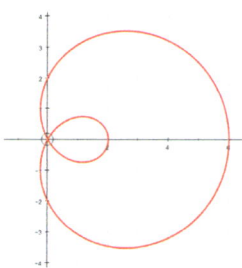

Limaçon
$$x = 2\cos\theta + 4\cos^2\theta$$
$$y = 2\sin\theta + 2\sin 2\theta$$

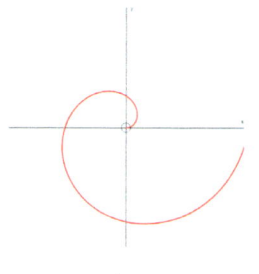

Archimedean spiral
$$x = \theta \cos n\theta$$
$$y = \theta \sin n\theta$$

$$n = 1$$

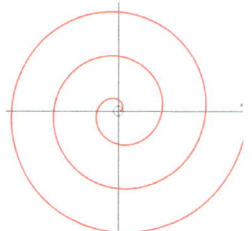

Archimedean spiral
$$x = \theta \cos n\theta$$
$$y = \theta \sin n\theta$$

$$n = 3$$

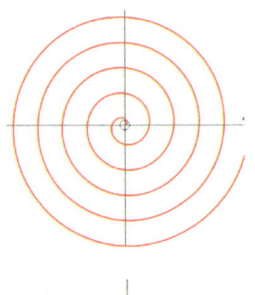

TOPIC 2. DIFFERENTIATION

Archimedean spiral

$x = \theta \cos n\theta$
$y = \theta \sin n\theta$

$n = 5$

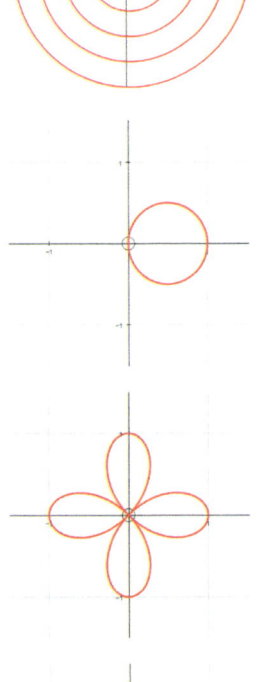

Lobiates

$x = \cos\theta \cos n\theta$
$y = \sin\theta \cos n\theta$

$n = 1$

Lobiates

$x = \cos\theta \cos n\theta$
$y = \sin\theta \cos n\theta$

$n = 2$

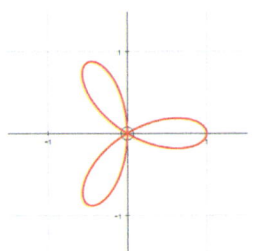

Lobiates

$x = \cos\theta \cos n\theta$
$y = \sin\theta \cos n\theta$

$n = 3$

© HERIOT-WATT UNIVERSITY

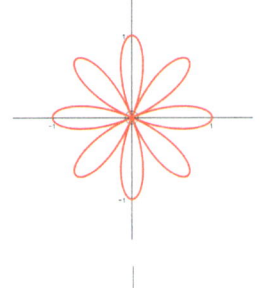

TOPIC 2. DIFFERENTIATION

Lobiates
$x = \cos\theta \cos n\theta$
$y = \sin\theta \cos n\theta$

$n = 4$

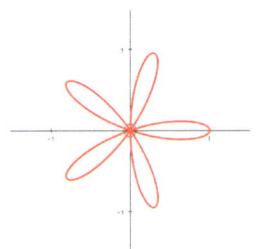

Lobiates
$x = \cos\theta \cos n\theta$
$y = \sin\theta \cos n\theta$

$n = 5$

Parametric curves exercise Go online

The following parametric equations give the position of a general point P(x, y) on a curve.

a) Algebraically find the cartesian form of the curve from the parametric equations.

b) Make a sketch of the curve either from the parametric form or cartesian form. This could be done by plotting points.

Q144: $x = 2\cos t$, $y = 2\sin t$, for $0 \leq t \leq 2\pi$

Q145: $x = 2t$, $y = 4t^2$, for $-\infty < t < \infty$

Q146: $x = 2t - 3$, $y = 4t - 3$, for $-\infty < t < \infty$

Q147: $x = 3\cos t$, $y = -3\sin t$, for $0 \leq t \leq 2\pi$

Q148: $x = \cos(\pi - t)$, $y = \sin(\pi - t)$, for $0 \leq t \leq \pi$

Q149: $x = t$, $y = \sqrt{1 - t^2}$, for $0 \leq t \leq 1$

TOPIC 2. DIFFERENTIATION

Q150: $x = 2t, y = 3 - 3t$, for $0 \leq t \leq 1$

....

Q151: $x = \cos^2 t, y = \sin^2 t$, for $0 \leqslant t \leqslant \frac{\pi}{2}$

2.11.2 First order parametric differentiation

When a curve is determined by parametric equations we may wish to gain more information about the curve by calculating its derivative.

We can do this directly from the parametric equations.

Given that $x = f(t)$ and $y = g(t)$ then, by the chain rule:

$\frac{dy}{dx} = \frac{dy}{dt} \times \frac{dt}{dx}$

Since $\frac{dt}{dx} = \frac{1}{\left(\frac{dx}{dt}\right)}$, then:

$\frac{dy}{dx} = \frac{dy}{dt} \times \frac{1}{\left(\frac{dx}{dt}\right)}$

and the formula for the derivative is:

$\frac{dy}{dx} = \frac{\left(\frac{dy}{dt}\right)}{\left(\frac{dx}{dt}\right)}$

Alternative notation:

$\frac{dy}{dx} = \frac{\dot{y}}{\dot{x}}$

Study the following examples to see how this works.

Examples

1. Problem:

Find $\frac{dy}{dx}$ in terms of the parameter t when $x = t^2 + 6$ and $y = 4t^3$

Solution:

When $x = t^2 + 6$ then $\frac{dx}{dt} = 2t$

When $y = 4t^3$ then $\frac{dy}{dt} = 12t^2$

Now using the formula for parametric differentiation we have:

$\frac{dy}{dx} = \frac{\left(\frac{dy}{dt}\right)}{\left(\frac{dx}{dt}\right)}$

$= \frac{12t^2}{2t}$

$= 6t$

....

2. Problem:

Find $\frac{dy}{dx}$ in terms of the parameter t when $x = \frac{2}{t}$ and $y = \sqrt{t^2 - 3}$

© HERIOT-WATT UNIVERSITY

Solution:

When
$$x = \frac{2}{t} = 2t^{-1}$$
$$\frac{dx}{dt} = -2t^{-2} = -\frac{2}{t^2}$$

When
$$y = \sqrt{t^2 - 3} = (t^2 - 3)^{\frac{1}{2}}$$
$$\frac{dy}{dt} = \frac{1}{2}(t^2 - 3)^{-\frac{1}{2}} \times 2t$$
$$= \frac{t}{(t^2 - 3)^{\frac{1}{2}}}$$

Now, using the formula for parametric differentiation gives:

$$\frac{dy}{dx} = \frac{\left(\frac{dy}{dt}\right)}{\left(\frac{dx}{dt}\right)}$$

$$= \frac{\left(\frac{t}{(t^2 - 3)^{\frac{1}{2}}}\right)}{\left(\frac{-2}{t^2}\right)}$$

$$= \frac{t}{(t^2 - 3)^{\frac{1}{2}}} \times \frac{t^2}{-2}$$

$$= -\frac{t^3}{2\sqrt{t^2 - 3}}$$

First order parametric differentiation exercise   Go online

Q152: Find $\frac{dy}{dx}$ in terms of the parameter t, given $x = \ln|t^2 - 3|$, $y = \ln|t^2|$ use parametric differentiation to find $\frac{dy}{dx}$ in terms of t.

....

Q153: Find $\frac{dy}{dx}$ in terms of the parameter θ, given $x = \csc\theta$, $y = \sec\theta$ use parametric differentiation to find $\frac{dy}{dx}$ in terms of θ.

....

Q154: Find $\frac{dy}{dx}$ in terms of the parameter t, given $x = t^3 + t^2$, $y = t^2 + t$.

....

Q155: Find $\frac{dy}{dx}$ in terms of the parameter θ, given $x = 4\cos^2\theta$, $y = 4\sin^3\theta$.

....

Q156: Find $\frac{dy}{dx}$ in terms of the parameter x, given $x = 4 - 3t$, $y = \frac{3}{t}$.

2.11.3 Motions in a plane

Motions in a plane — Go online

When the position of a particle moving in a plane is given as the parametric equations $x = f(t)$ and $y = g(t)$, then the velocity of the particle can be split into components.

The *horizontal velocity component* is given by $\frac{dx}{dt}$

The *vertical velocity component* is given by $\frac{dy}{dt}$

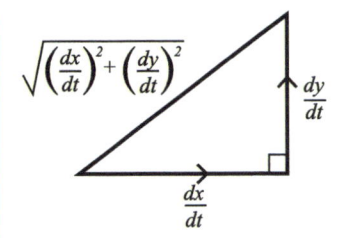

$$\sqrt{\left(\frac{dx}{dt}\right)^2 + \left(\frac{dy}{dt}\right)^2}$$

Therefore, the speed of the particle is:

$$\sqrt{\left(\frac{dx}{dt}\right)^2 + \left(\frac{dy}{dt}\right)^2}$$

The *instantaneous direction of motion* for the particle is given by $\frac{dy}{dx}$

(This concept is similar to the idea of the gradient of a curve at a point.)

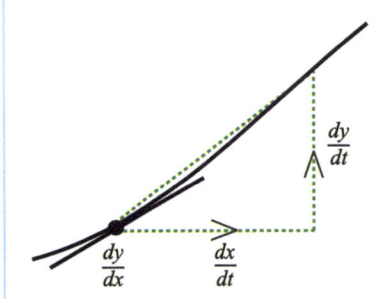

Example

Problem:

A particle is moving along a path determined by the parametric equations $x = \frac{1}{2}t^2 - 3t$ and $y = 4t^3 + t^2$, where t represents time in seconds and distance is measured in metres.

a) How far has the particle travelled from the origin after 2 seconds?

b) What is the speed of the particle at 2 seconds?

c) What is the instantaneous direction of the particle at 2 seconds?

Solution:

a) Horizontal distance travelled when $t = 2$: $x = \frac{1}{2}(2)^2 - 3(2) \Rightarrow x = -4$

Vertical distance travelled when $t = 2$: $y = 4(2)^3 + (2)^2 \Rightarrow y = 40$

Distance travelled from the origin when $t = 2$:

$d = \sqrt{x^2 + y^2}$

$d = \sqrt{(-4)^2 + 40^2}$

$d = 40 \cdot 2$ m

b) Speed is given by $\sqrt{\left(\frac{dx}{dt}\right)^2 + \left(\frac{dy}{dt}\right)^2}$

$\frac{dx}{dt} = t - 3$ and $\frac{dy}{dt} = 12t^2 + 2t$

When $t = 2$ $\frac{dx}{dt} = -1$ and $\frac{dy}{dt} = 52$

Speed $= \sqrt{(-1)^2 + (52)^2}$

$= \sqrt{2705}$

$= 52$ ms^{-1}

c) Instantaneous direction is given by the gradient $\frac{dy}{dx}$

$\frac{dy}{dx} = \frac{\frac{dy}{dt}}{\frac{dx}{dt}}$

$\frac{dy}{dx} = \frac{12t^2 + 2t}{t - 3}$

When $t = 2$

$\frac{dy}{dx} = \frac{52}{-1}$

$\frac{dy}{dx} = -52$

Angle is given by $m = \tan \theta$:

$-52 = \tan \theta$

$\theta = \tan^{-1} 52$

$\theta = 88 \cdot 8°$

Since tan θ is negative the equivalent angle is $180° + 88 \cdot 8° = 268 \cdot 8°$.

The direction of the particle at 2 seconds is $268 \cdot 8°$ anti-clockwise from the x-axis

Motions in a plane exercise

A particle is moving along a path determined by the parametric equations $x = 4t - 1$ and $y = t^2 + 3t$, where t represents the time in seconds and distance is measured in metres.

Q157: How far is the particle from the origin at $t = 0$?

....

Q158: Calculate, in terms of t, the horizontal velocity component for the movement of the particle.

a) $(4t - 1)$ ms^{-1}
b) 4 ms^{-1}
c) $(t^2 + 3t)$ ms^{-1}
d) $(2t + 3)$ ms^{-1}

....

Q159: Calculate, in terms of t, the vertical velocity component for the movement of the particle.

a) $(4t - 1)$ ms^{-1}
b) 4 ms^{-1}
c) $(t^2 + 3t)$ ms^{-1}
d) $(2t + 3)$ ms^{-1}

....

Q160: At $t = 2$, calculate the horizontal velocity.

a) 4 ms^{-1}
b) 6 ms^{-1}
c) 7 ms^{-1}
d) 10 ms^{-1}

....

Q161: At $t = 2$, calculate the vertical velocity.

a) 4 ms^{-1}
b) 6 ms^{-1}
c) 7 ms^{-1}
d) 10 ms^{-1}

....

Q162: At $t = 2$, calculate the speed.

a) $7 \cdot 21$ ms^{-1}
b) $8 \cdot 06$ ms^{-1}
c) $9 \cdot 22$ ms^{-1}
d) $10 \cdot 77$ ms^{-1}

Q163: Derive a formula for the instantaneous direction of motion for the particle.

Q164: In which direction is the particle moving at $t = 2$?

Give your answer in degrees measured relative to the positive direction of the x-axis.

A particle is moving along the path of an ellipse as shown here. The equation of the ellipse is given by the parametric equations:

$x = 5 \cos t, y = 3 \sin t$

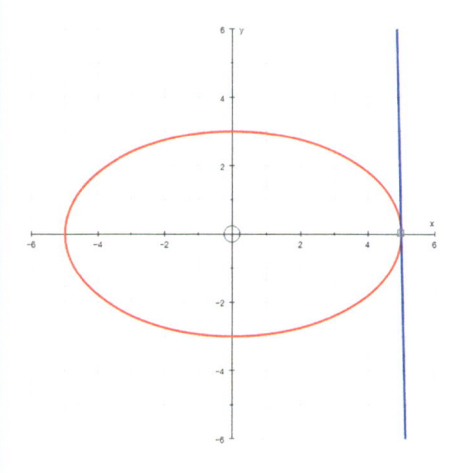

Q165: Calculate the speed of the particle when $t = \frac{\pi}{4}$.

Q166: What is the instantaneous direction of the particle at $t = \frac{\pi}{4}$?

2.11.4 Tangents to parametric curves

Example

Problem:

A curve is defined by the parametric equations: $x = t^2 - 3, y = 2t^3$

Find the equation of the tangent at the point where $t = 1$.

Solution:

When $x = t^2 - 3 \Rightarrow \frac{dx}{dt} = 2t$

TOPIC 2. DIFFERENTIATION

When $y = 2t^3 \Rightarrow \frac{dy}{dt} = 6t^2$

Now, using the formula for differentiating parametric equations, we have:

$\frac{dy}{dx} = \frac{6t^2}{2t} = 3t$

Therefore, at $t = 1$, the gradient of the tangent $\frac{dy}{dx} = 3$

We calculate the corresponding (x, y) at $t = 1$ by substituting into the original parametric equations.

When $t = 1, x = t^2 - 3 \Rightarrow x = -2$

When $t = 1, y = 2t^3 \Rightarrow y = 2$

Therefore, when $t = 1$ we have the coordinate (-2,2).

So now we have $m = 3$ and (-2,2).

We now use the straight line formula $y - b = m(x - a)$ to give the equation of the straight line.

$y - b = m(x - a)$
$y - 2 = 3x + 6$
$y = 3x + 8$

Tangents to parametric curves exercise   Go online

Q167: A curve is defined by the parametric equations $x = \cos 2t$, $y = \sin 2t$, $0 < t < \frac{\pi}{2}$.
Use parametric differentiation to find $\frac{dy}{dx}$.
Hence find the equation of the tangent when $t = \frac{\pi}{3}$.

....

Q168: Given $x = \csc \theta$ and $y = \cos \theta$, use parametric differentiation to find $\frac{dy}{dx}$ in terms of θ.
Hence find the equation of the tangent when $\theta = \frac{\pi}{2}$.

....

Q169: Given $y = t^3 - \frac{3}{2}t^2$ and $x = \sqrt{t}$ for $t > 0$, use parametric differentiation to express $\frac{dy}{dx}$ in terms of t in simplified form.
Hence find the equation of the tangent when $t = 4$.

2.11.5 Second order parametric differentiation

Recall that the second derivative of y with respect to x is obtained by differentiating twice and that $\frac{d^2y}{dx^2} = \frac{d}{dx}\left(\frac{dy}{dx}\right)$

This same method is used when dealing with parametric equations.

Example

Problem:

Find $\frac{dy}{dx}$ and $\frac{d^2y}{dx^2}$ in terms of t given that $x = \frac{2}{t}$ and $y = 2t^3 + 1$.

Solution:

When $x = \frac{2}{t}$ $\Rightarrow$ $\frac{dx}{dt} = -\frac{2}{t^2}$

When $y = 2t^3 + 1$ $\Rightarrow$ $\frac{dy}{dt} = 6t^2$

Now, using the formula for differentiating parametric equations, we have:

$$\begin{aligned}\frac{dy}{dx} &= \frac{6t^2}{\left(-\frac{2}{t^2}\right)} \\ &= 6t^2 \times \left(-\frac{t^2}{2}\right) \\ &= -3t^4\end{aligned}$$

To find the second derivative, we need to differentiate $\frac{dy}{dx}$ with respect to x:

$$\frac{d^2y}{dx^2} = \frac{d}{dx}\left(-3t^4\right)$$

We need to use the chain rule so that we can differentiate $-3t^4$ with respect to x:

$$\begin{aligned}\frac{d^2y}{dx^2} &= \frac{d}{dt}\left(-3t^4\right) \times \frac{dt}{dx} \\ &= \frac{\frac{d}{dt}\left(-3t^4\right)}{\left(\frac{dx}{dt}\right)} \quad \text{Remember } \frac{dt}{dx} = \frac{1}{\left(\frac{dx}{dt}\right)} \\ &= \frac{\frac{d}{dt}\left(-3t^4\right)}{\left(\frac{-2}{t^2}\right)} \\ &= -12t^3 \times \left(\frac{t^2}{-2}\right) \\ &= 6t^5\end{aligned}$$

This gives us $\frac{dy}{dx} = -3t^4$ and $\frac{d^2y}{dx^2} = 6t^5$.

Key point

$$\frac{d^2y}{dx^2} = \frac{\frac{d}{dt}\left(\frac{dy}{dx}\right)}{\frac{dx}{dt}}$$

TOPIC 2. DIFFERENTIATION

Second order parametric differentiation exercise   Go online

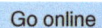

Q170: A curve is defined by the parametric equation $x = \sin t, \ y = \cos t, \ 0 < t < \pi$

Obtain an expression for $\frac{d^2y}{dx^2}$ and hence show that $\cos t \frac{d^2y}{dx^2} + \left(\frac{dy}{dx}\right)^2 = k$ where k is an integer. State the value of k.

....

Q171: Given $y = 5t^2 - \frac{7}{3}t^3$ and $x = 2t^2$, use parametric differentiation to express $\frac{dy}{dx}$ in terms of t in simplified form.

Show that $\frac{d^2y}{dx^2} = a\frac{1}{t}$, determining the values of the constant a.

....

Q172: A curve is defined parametrically, for all t, by the equations $x = 3t + \frac{2}{3}t^3, \ y = \frac{1}{2}t^2 - 4t$

Obtain $\frac{dy}{dx}$ and $\frac{d^2y}{dx^2}$ as functions of t.

2.12 Related rate problems

In this world, nothing is static. Everything is changing all the time. The rate at which a plant grows, the rate at which substances decay, a car slowing down at traffic lights then speeding up on a straight piece of road. All of this can be described through rates of change. These rates are also dependent on the rates of other factors. For instance, a simple example would be to take the area of a rectangle with a constant area. The rate at which the length changes with time will affect the rate at which the breadth changes. The rates are related. This section will look at models in which rates are related and how we can find a particular rate of change.

2.12.1 Explicitly defined rate related problems

How fast does the water level rise when a cylindrical tank is filled at a rate of 2 litres/second?

This question is asking us to calculate the rate at which the *height* of water in the tank is increasing from the rate at which the *volume* of water in the tank is increasing.

This is called a *related rates problem*.

Solutions of related rates problems can often be simplified to an application of the chain rule.

Example

Problem:

A volume of sphere is given by $V = \frac{4}{3}\pi r^3$. The rate of change of the radius of the sphere with respect to time is 2 (i.e. $\frac{dr}{dt} = 2$).

When $r = 3$, find the rate of change of V.

Solution:

We are asked to find the rate of change of V with respect to time, t.

$\frac{dV}{dt} = \frac{dV}{dr} \times \frac{dr}{dt}$ by the chain rule, because V is a function of r and r is a function of time, t.

Now, we substitute for what we know, namely $\frac{dV}{dr}$ and $\frac{dr}{dt}$:

Remembering:

$$V = \frac{4}{3}\pi r^3 \Rightarrow \frac{dV}{dr} = 4\pi r^2$$

$$\frac{dr}{dt} = 2$$

$$\frac{dV}{dt} = 4\pi r^2 \times 2$$

$$= 8\pi r^2$$

So when $\frac{dr}{dt} = 2$ and $r = 3$:

$$\frac{dV}{dt} = 8\pi r^2$$

$$= 8\pi (3)^2$$

$$= 8\pi (9)$$

$$= 72\pi$$

Explicitly defined rate related problems exercise   Go online

Q173: Given that $L = 5h^3 - 2h$ and $h = \frac{3}{2}t^2 - 5t$ find and expression for $\frac{dL}{dt}$

....

Q174: If $K = \sec 2\theta + 3\theta^2$ and $\frac{d\theta}{dt} = t - 4t^3$, find $\frac{dK}{dt}$

....

Q175: If $S = \pi r h$ and we are given $\frac{dh}{dt} = 5$, find $\frac{dS}{dt}$ when $r = 2$

....

Q176: If $B = t^2 + 4\sin 3t$ and $\frac{dt}{dx} = \frac{1}{2}$, find $\frac{dB}{dx}$ when $t = \pi$

2.12.2 Implicitly defined rate related problems

Related rates problems can occur when dealing with implicitly defined functions.

Examples

1.

Problem:

A point moves on the curve $2x^2 - y^2 = 2$ so that the y-coordinate increases at the constant

TOPIC 2. DIFFERENTIATION

rate of 12 ms^{-1}. That is, $\frac{dy}{dt} = 12$

a) At what rate is the x-coordinate changing at the point (3,4)?
b) What is the slope of the curve at the point (3, 4)?

Solution:

a) We are given that $2x^2 - y^2 = 2$ and we are required to find $\frac{dx}{dt}$ at (3,4).
Notice that x is implicitly defined in terms of y.
Differentiate $2x^2 - y^2 = 2$ with respect to t:
$$\frac{d}{dt}(2x^2) - \frac{d}{dt}(y^2) = \frac{d}{dt}(2)$$

Rearrange to express $\frac{dx}{dt}$ in terms of $\frac{dy}{dt}$:
$$4x\frac{dx}{dt} - 2y\frac{dy}{dt} = 0$$
$$4x\frac{dx}{dt} = 2y\frac{dy}{dt}$$
$$\frac{dx}{dt} = \frac{2y}{4x}\frac{dy}{dt}$$

Evaluate $\frac{dx}{dt}$ at (3,4) when $\frac{dy}{dt} = 12$.
$$\frac{dx}{dt} = \frac{2(4)}{4(3)} \times 12$$
$$= \frac{8}{12} \times 12$$
$$= 8$$

b) The slope of the curve is given by $\frac{dy}{dx}$.
At (3,4) $\frac{dx}{dt} = 8$ and $\frac{dy}{dt} = 12$, use the chain rule:
$$\frac{dy}{dx} = \frac{dy}{dt} \times \frac{dt}{dx}$$
$$= \frac{\left(\frac{dy}{dt}\right)}{\left(\frac{dx}{dt}\right)}$$
$$= \frac{12}{8}$$
$$= \frac{3}{2}$$

....

2. Problem:

A particle moves along the curve with equation $5x^2 + xy - y = 6$
If the at the point (2,1) the rate of change of the x coordinate is 1 unit per second, calculate the rate of change of the y coordinate with respect to time.

Solution:

We are given $5x^2 + xy - y = 6$ and we are required to find $\frac{dy}{dt}$ at (2,1).
We can use implicit differentiation to find this.
Differentiate $5x^2 + xy - y = 6$ with respect to t.

$$\frac{d}{dt}\left(5x^2\right) + \frac{d}{dt}\left(xy\right) - \frac{d}{dt}\left(y\right) = \frac{d}{dt}\left(6\right)$$

Apply the chain rule to $\frac{d}{dt}\left(5x^2\right)$ and the product rule to $\frac{d}{dt}\left(xy\right)$

$$10x\frac{dx}{dt} + \left(\frac{dx}{dt}y + x\frac{dy}{dt}\right) - \frac{dy}{dt} = 0$$

Rearrange to express $\frac{dy}{dt}$ in terms of $\frac{dx}{dt}$:

$$(x-1)\frac{dy}{dt} = (-10x - y)\frac{dx}{dt}$$
$$\frac{dy}{dt} = \frac{(-10x - y)}{(x-1)}\frac{dx}{dt}$$

Evaluate $\frac{dy}{dt}$ at (2,1) when $\frac{dx}{dt} = 1$

$$\frac{dy}{dt} = \frac{(-10(2) - (1))}{((2) - 1)} \times 1$$
$$\frac{dy}{dt} = -21$$

The y coordinate changes at a rate of -21 units per second.

Implicitly defined rate related problems exercise   Go online

Q177:

A particle moves along the curve with equation $x^2 - 3xy + 2y^2 = 4$

If, at the point (2,3), the rate of change of the y-coordinate is 3 units per second, calculate the rate of change of the x-coordinate with respect to time.

....

Q178: Given that $x^2 - 5xy - 3y^2 = -7$ and $\frac{dy}{dt} = 2$. Find $\frac{dx}{dt}$ at the point (1,1).

....

Q179:

What is the instantaneous direction of motion of the particle in the previous question at (1,1)?

Give your answer in degrees, to one decimal place, relative to the positive direction of the x-axis.

2.12.3 Related rate problems in practice

> **Key point**
>
> **Strategy for solving related rates problems in practice**
>
> 1. Draw a diagram and label the variables and constants. Use t for time.
> 2. Write down any additional numerical information.
> 3. Write down what you are asked to find. Usually this is a rate and you should write this as a derivative.
> 4. Write down an equation that describes the relationship between the variables.
> 5. Differentiate (usually with respect to time).

Examples

1. Problem:

When a stone is dropped into a still pond, ripples move out from the point where it hits in the form of concentric circles.

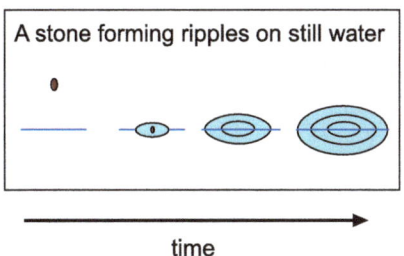

A stone forming ripples on still water

time

Find the rate at which the *area* of the disturbed water is increasing when the *radius* reaches 10 metres.

At this stage, the radius is increasing at a rate of 2 ms^{-1}, i.e. $\frac{dr}{dt} = 2$ when $r = 10$

Solution:

Try to follow the strategy as detailed previously.

1. Make a diagram if it helps.

   Area of disturbed water
   $A = \pi r^2$

   $r = 10$ m

160 TOPIC 2. DIFFERENTIATION

Note that both the area, A, and the radius, r, change with time. Think of A and r as differentiable functions of time and use t to represent time. The derivatives $\frac{dA}{dt}$ and $\frac{dr}{dt}$ give the rates at which A and r change.

2. Note that the area of the circle is $A = \pi r^2$ and that when $r = 10$, then $\frac{dr}{dt} = 2$
3. The requirement is to find $\frac{dA}{dt}$
4. The equation to connect the variables is $\frac{dA}{dt} = \frac{dA}{dr} \times \frac{dr}{dt}$
5. Note that when $A = \pi r^2$, then $\frac{dA}{dr} = 2\pi r$
Therefore $\frac{dA}{dt} = \frac{dA}{dr} \times \frac{dr}{dt} = 2\pi r \times \frac{dr}{dt}$
When $r = 10$, $\frac{dA}{dt} = 2\pi \times 10 \times 2 = 40\pi$
Therefore, the area of disturbed water is increasing at a rate of 40π ms^{-2} when the radius of the outermost ripple reaches 10 metres.

....

2. Problem:

Sand runs into a conical tank at a rate of 5π litres/s. The tank stands vertex down and has a height of 4 m and a base radius of 2 m.

Find the rate at which the depth of the sand is increasing when the sand is 1 metre deep.

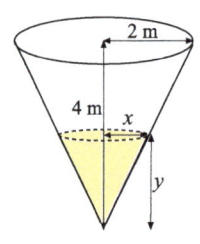

Solution:

The variables in the problem are:

V = the volume of sand in the tank at time t
x = the radius of the surface of the sand at time t
y = the depth of the sand in the tank at time t

The constants are the dimensions of the tank and the rate at which the tank fills with sand = $\frac{dV}{dt}$

To find $\frac{dy}{dt}$ when $y = 1$ metre (= 100 cm), calculate this from $\frac{dV}{dt} = \frac{dV}{dy} \times \frac{dy}{dt}$

We change the units from metres to centimetres here because the volume will be converted from litres to cubic centimetres.

The volume of sand in the tank is given by the equation $V = \frac{1}{3}\pi x^2 y$

However, this last equation involves both the variables x and y, and before we can find $\frac{dV}{dy}$ we should eliminate x.

TOPIC 2. DIFFERENTIATION

We can do this by observing similar triangles:

$$\frac{x}{200} = \frac{y}{400}$$
$$x = \frac{1}{2}y$$

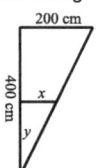

Rewrite V in terms of y alone:

$$V = \frac{1}{3}\pi x^2 y$$
$$= \frac{1}{3}\pi \left(\frac{1}{2}y\right)^2 y$$
$$= \frac{1}{12}\pi y^3$$

Then calculate $\frac{dV}{dy} = \frac{1}{4}\pi y^2$

The sand is filling the cone at a rate of $\frac{dV}{dt} = 5\pi l s^{-1} = 5000\pi cm^{-3} s^{-1}$

Therefore $\frac{dV}{dt} = \frac{dV}{dy} \times \frac{dy}{dt}$ becomes:

$$5000\pi = \frac{1}{4}\pi y^2 \frac{dy}{dt}$$
$$\frac{dy}{dt} = \frac{20000}{y^2}$$

and when $y = 1$ metre (= 100 cm) then $\frac{dy}{dt} = 2 \, cms^{-1}$

When the sand is 1 metre deep in the tank, the level is rising at a rate of 2 cms⁻¹.

....

3. Problem:

A right angled triangle has a fixed base of 20 feet and an unknown altitude. The length of the base increases at a rate of 2·5 feet per second when travelling towards this altitude. At one point the height of the triangle is 6 feet tall.

How fast is the height of the triangle changing when the distance from the altitude is 8 feet?

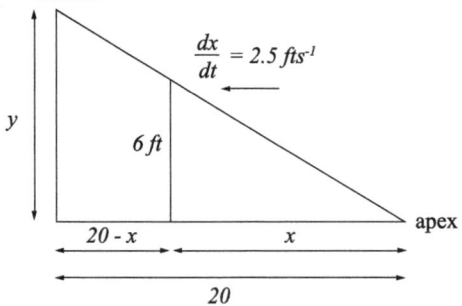

Solution:

The variables in the problem are:

- x = distance from the apex
- y = altitude of triangle

We need to calculate $\frac{dy}{dt}$. This means we need to find an expression for y in terms of x and we know that $\frac{dx}{dt} = 2 \cdot 5$ when the base is 12 feet.

Using similar triangles we have:

$$\frac{y}{6} = \frac{20}{x}$$

$$y = \frac{120}{x}$$

Now differentiate and evaluate when $\frac{dx}{dt} = 2 \cdot 5$ and $20 - x = 12$

$$\frac{dy}{dt} = \frac{dy}{dx} \times \frac{dx}{dt}$$

$$\frac{dy}{dt} = -\frac{120}{(12)^2} \times \frac{5}{2}$$

$$= -\frac{600}{288}$$

$$= -21 \, fts^{-1}$$

Related rate problems exercise   Go online

Q180: A snowball, in the shape of a sphere, is melting at a uniform rate of 0·5 cms⁻¹. How fast if the volume decreasing when the radius is 4 cm?

....

Q181: Let S be the surface area of a cube whose edges have length x at time t. Write an equation that relates $\frac{dS}{dt}$ to $\frac{dx}{dt}$

....

Q182: If the velocity of a particle is given by $v = 2s^3 + 5s$, where s is the displacement of the particle from the origin at time t, find an expression for the acceleration of the particle in terms of s.

....

Q183: A huge block of ice is in the shape of a cube.

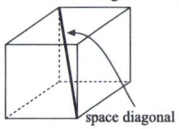

At what rate is the space diagonal of the cube decreasing if the edges of the cube are melting at a rate of 2 cm s⁻¹?

2.13 Displacement, velocity and acceleration

Before looking at the applications of **displacement**, **velocity** and **acceleration**, it is important to understand the difference between these and **distance** and **speed**.

Distance and speed are **scalar** quantities. This means that they are described by a magnitude only. For instance, how far someone has travelled or how fast someone is travelling.

Displacement, velocity and acceleration are **vector** quantities. This means that they have a magnitude and a direction. Since direction is now important, we need to tell the difference between movement to the right from that to the left, for instance. In order to do this, we state that movement in one direction is positive and movement in the opposite direction is negative. As long as this reference point is consistent within the worked solution it does not matter which is labelled which. By convention, movement to the right is positive and to the left is negative. It is good practice to state this before a solution; however, you can assume that this is the case in this section unless otherwise stated.

Distance and displacement

An athlete is training for the 100 metres. He runs 100 metres in a straight line from the starting line and then turns around and jogs back 10 metres towards the starting line. What is the distance travelled and what is the displacement?

Distance

Distance is only concerned about the total distance that the athlete has covered.

He travelled 100 metres in the first instance, turned around and then travelled another 10 metres.

A distance/time graph can be drawn to visually represent the athlete's progress, assuming that he travels at a constant speed.

Distance/time graph

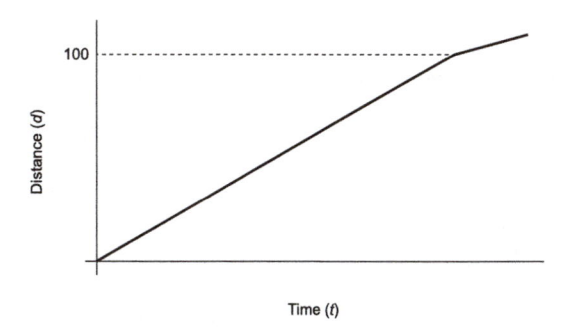

Distance is often denoted by the letter d and is usually measured in metres.

In total, he has travelled:

$d = 100 + 10$
$d = 110$ m

Displacement

Displacement is concerned about how far the athlete has travelled relative to his starting point. Generally, it is the case that:

- movement to the right is positive;
- movement to the left is negative.

He travelled 100 metres in the first instance to the right (+100 m), turned around and then travelled another 10 metres to the left (-10 m).

A displacement/time graph can be drawn to visually represent the athlete's progress, assuming that he travels at a constant speed.

Displacement/time graph

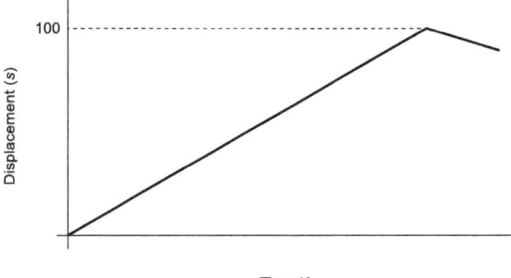

Displacement is often denoted by the letter s or x and is usually measured in metres.

In total, he has travelled:

$s = 100 - 10$
$s = 90$ m

Since 90 m is a positive number, the athlete has travelled 90 m to the right of his starting point.

TOPIC 2. DIFFERENTIATION

Velocity and speed

The same athlete runs the 100 metres in 10 seconds.

A displacement/time graph can be drawn to visually represent the athlete's progress, assuming that he travels at a constant speed.

Displacement/time graph

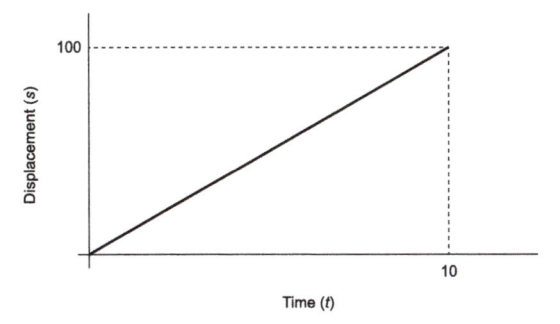

The displacement can be represented by the function: $s(t) = 10t$, which can be derived from the graph.

Velocity

Velocity, by definition, is the rate of change of displacement with respect to time. It is usually denoted by the letter v or $\dot{x}$ and is measured in m/s.

We therefore have: $v = \frac{ds}{dt}$ or $\dot{x} = \frac{dx}{dt}$ which is the gradient of the line $s(t)$.

In the above example, the velocity is: v = 10 m/s to the right.

The velocity is positive because the displacement was in the positive direction.

Speed

Speed is a scalar quantity and is the magnitude of the velocity $|v|$. It is often denoted by the letter s, but should not be confused with displacement.

Note: $|v|$ means that the answer to $|v|$ is v when $v \geq 0$ and $-v$ when $v < 0$. In other words, by taking the modulus, v becomes positive.

In this case, the speed is $|v|$ = 10 m/s

Note that a direction is not given in the answer.

© HERIOT-WATT UNIVERSITY

Importantly, notice that in all cases when finding speed and velocity, it is the rate of change of displacement, *not distance*, that is used.

Acceleration

The same athlete decides to try running a longer distance. Whilst he is running, he speeds up and slows down as shown in the velocity/time graph.

Velocity/time graph

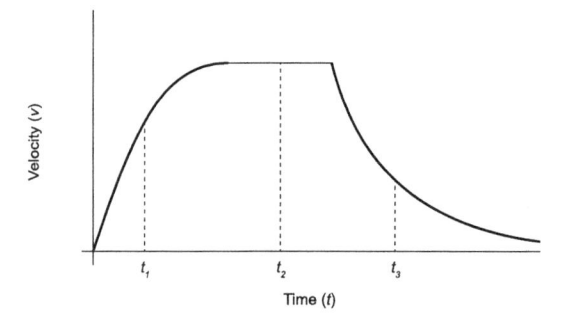

Velocity is given by: $v(t)$

Acceleration is given by: $\frac{dv}{dt}$ which is the gradient of $v(t)$.

Acceleration is a vector quantity and is the rate of change of velocity with respect to time. It is usually denoted by the letter a or $\ddot{x}$ and measured in m/s^2 where $a = \frac{dv}{dt}$ or $\ddot{x} = \frac{d^2x}{dt^2}$

Since acceleration is a vector quantity and direction is important, it takes some thought to interpret what a positive and negative answer means. There are two cases to consider.

1. When the velocity is *positive* (i.e. the object travelling to the right):

   - if the acceleration is *positive* in this direction, it is *accelerating*. To be clearer, the acceleration is acting in the same direction as the velocity so the object is speeding up.
   - if the acceleration is *negative* in this direction, it is *decelerating*. The acceleration this time is acting in the opposite direction of the velocity. It is acting against velocity and is therefore slowing the object down.

2. When the velocity is *negative* (i.e. travelling to the left):

   - if the acceleration is *positive* in this direction, it is *decelerating*. That is to say, the acceleration is acting in the opposite direction to the velocity. It is acting against velocity and is therefore slowing the object down.
   - if the acceleration is *negative* it is *accelerating*. To be clearer, the acceleration is acting in the same direction as the velocity so the object is speeding up.

TOPIC 2. DIFFERENTIATION

Examples

1. Velocity is positive

Imagine that you are an observer watching a car. From the starting point it travels to the right (velocity is positive). The car then increases its speed in this direction so the velocity is increasing and therefore the car is accelerating. Since it is travelling to the right and accelerating to the right as well, the acceleration is positive. If the driver of the car suddenly brakes so that the speed of the car is decreasing, but it is still travelling to the right, the velocity is decreasing and therefore the car is decelerating, i.e. its acceleration is negative.

....

2. Velocity is negative

Again, imagine that you are observing a car, but this time it is travelling to the left from its starting point. Since it is travelling to the left, the velocity is negative. The car then increases its speed in this direction so the velocity is increasing and therefore the car is accelerating. Since it is travelling to the left and accelerating to the left as well, the acceleration is negative. If the driver of the car suddenly brakes so that the speed of the car is decreasing, but it is still travelling to the left, the velocity is decreasing and therefore the car is decelerating, i.e. its acceleration is positive. Since a negative acceleration shows an increase in speed to the left, deceleration must be positive to show a reduction in speed to the left.

In summary, the sign of the velocity determines the direction of travel and the sign of the acceleration determines whether the object is increasing or decreasing in speed in that direction.

Key point

If we have been given the acceleration we can integrate to find the velocity: $v = \int a \, dt$

If we have been given the velocity we can integrate to find the displacement: $s = \int v \, dt$

Displacement, velocity and acceleration exercise   Go online

Q184: The distance (in metres) that a rocket has travelled at time t (seconds) in the initial stages of lift-off is calculated using the formula $s(t) = 2t^3$

a) Find expressions for the speed and acceleration of the rocket.
b) Calculate the velocity and acceleration of the rocket after 10 seconds.

....

Q185: A stone is thrown upwards. Its height above the ground, s metres, after t seconds is given by $s = 2 + 9t - 5t^2$

a) Find the speed of the stone after 1·5 seconds.
b) Find the maximum height of the stone and the time at which this occurs.
c) At what speed is the stone travelling when it hits the ground?

© HERIOT-WATT UNIVERSITY

2.14 Curve sketching in closed intervals

A **closed interval** is denoted by square brackets, e.g. [3,6]. This means that the numbers 3 and 6, as well as the numbers in between, are included in the interval. The numbers 3 and 6 at the end of the interval are often called the end points.

An open interval is denoted by round brackets, e.g. (3,6). This means that the numbers 3 and 6 are not included in the interval, but all numbers in between are included.

If we are to identify maximum and minimum values of a graph, then we must identify if the end points are included in the interval or not.

Study the following graphs and notice that the maximum and minimum values can occur at stationary points or end points.

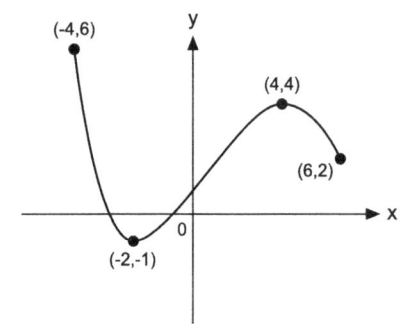

For the interval [-4, 6], the maximum value is 6 at (-4,6) which occurs at an end point of the interval, and the minimum value is -1 at (-2,-1) which occurs at a stationary point.

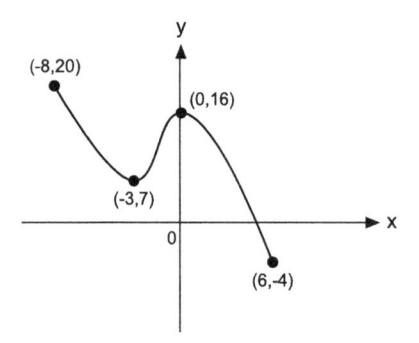

For the interval [-8, 6], the maximum value is 20 at (-8,20) which occurs at an end point of the interval, and the minimum value is -4 at (6,-4) which occurs at the other end point of the interval.

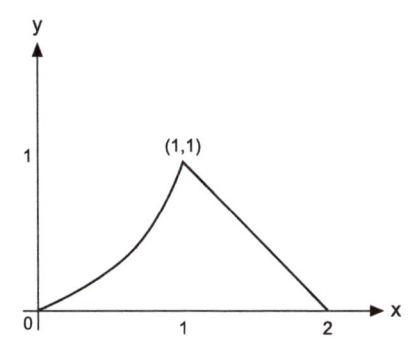

The previous graph is defined as $f(x) = \begin{cases} x^2 \text{ for } 0 \leqslant x < 1 \\ 2 - x \text{ for } 1 \leqslant x \leqslant 2 \end{cases}$ which has maximum value of 1.

However, at $x = 1$ the graph has a corner (as opposed to being smooth) which means that $f'(x)$ does not exist.

Therefore (1,1) is not a stationary point.

Key point

In a closed interval, the maximum and minimum values of a function can occur at a stationary point, an end point of the interval or where $f'(x)$ does not exist.

2.15 Optimisation and other applications

If a situation can be represented by a mathematical formula, then it is possible to find the maximum or minimum values of this model, if they exist. For instance, when manufacturing a cuboid-shaped packaging, a company may want to minimise the quantity of material used but maximise the volume, the aim being to pay the minimum cost for material for their product.

Some or all of these steps may be necessary:

1. label any unknown quantities;
2. identify the quantity that is to be maximised or minimised, in terms of the variables;
3. use the information in the question to establish equations connecting the variables;
4. express the unknown quantity in terms of only one variable;
5. calculate the extrema of the function, i.e. calculate the turning points and their nature, and, if appropriate, consider any end points.

Examples

1. Problem

Two workers A and B share an 8 hour shift. A works for x hours and B works for y hours.

Their work output varies with time (in hours) from the start of the shift.
A has work output equal to $x - \frac{x^2}{16}$ units per hour and B has work output equal to $\frac{1}{2}\ln(2y)$ units per hour.

Calculate the hours each should be employed for maximum output.

Solution

Since they share the eight hour shift, $x + y = 8$ and so $y = 8 - x$
The total output of the two workers is:

$$T = x - \frac{x^2}{16} + \frac{1}{2}\ln(2y)$$
$$= x - \frac{x^2}{16} + \frac{1}{2}\ln(16 - 2x) \quad \text{(replacing } 2y \text{ with } 16 - 2x\text{)}$$

with T written in terms of one variable, any stationary points can now be determined.

$$\frac{dT}{dx} = 1 - \frac{2x}{16} - \frac{1}{16 - 2x} = 0 \text{ (at a stationary point)}$$
$$16(16 - 2x) - (16 - 2x)2x - 16 = 0 \text{ (multiplied by } 16(16 - 2x)\text{)}$$
$$(16 - 2x)(16 - 2x) = 16$$
$$(16 - 2x)^2 = 16$$
$$16 - 2x = \pm 4$$
$$2x = 12 \text{ or } 20$$
$$x = 6 \text{ or } 10$$

$x = 10$ is impossible since the shift has a maximum of eight hours, so the stationary point occurs at $x = 6$
From the table shown here we can see that a maximum occurs at $x = 6$

x	6^-	6	6^+
$\frac{dT}{dx}$	+	0	-
Slope	↗	→	↘

The total output when $x = 6$ is:

$$T(6) = 6 - \frac{6^2}{16} + \frac{1}{2}\ln(16 - 2(6))$$
$$T(6) = 4 \cdot 44 \text{ units per hour}$$

TOPIC 2. DIFFERENTIATION

We need to check if there are any points where the derivative does not exist. Looking at the following graph, there are none, so we do not need to consider this.

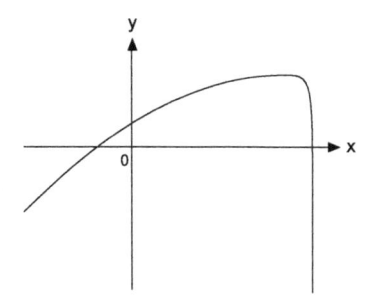

For the context of this question, we should consider the end points: when $x = 0$ and $x = 8$

The total output when $x = 0$ is:

$$T(0) = 0 - \frac{0^2}{16} + \frac{1}{2}\ln(16 - 2(0))$$

$T(0) = 1.39$ output of work

$T(0) < T(6)$ so it is not a maximum.

The total output when $x = 8$ is:

$$T(8) = 8 - \frac{8^2}{16} + \frac{1}{2}\ln(16 - 2(8))$$

$T(8) =$ undefined output of work

$T(8)$ is undefined so it is not a maximum.

So for maximum total output, the shift should be split so that A works 6 hours and B works 2 hours.

....

2. Problem

A sheet of metal 3 m long and 2 m wide is to be made into a rectangular box by cutting a square of side x metres from each corner and turning up the edges. If V represents the volume of the box, find V in terms of x and determine the depth of the box so that it may have a maximum volume.

Solution

We start with the rectangular sheet, which then has the corners cut out.

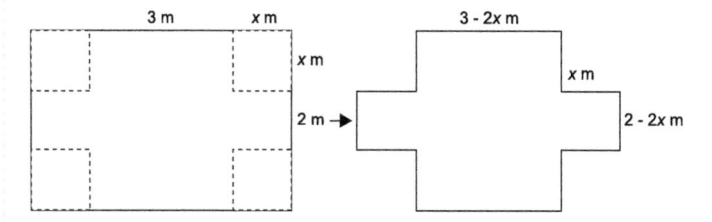

© HERIOT-WATT UNIVERSITY

172 TOPIC 2. DIFFERENTIATION

Note that there are limitations on what x can be.

Using common sense, $x > 0$ since we cannot cut out negative lengths and $x < 1$ otherwise the measure $2 - 2x$ would not exist. Therefore $x \in \mathbb{R} : 0 < x < 1$

When this is folded up it will create a cuboid with lengths $3 - 2x$, $2 - 2x$ and x m

So $V = x(3 - 2x)(2 - 2x)$

To maximise the volume we find the stationary points. We can either multiply out the brackets and differentiate or apply the product rule.

In this case, multiplying out and differentiating may be faster.

$V(x) = 6x - 10x^2 + 4x^3$

So $V'(x) = 6 - 20x + 12x^2$

For stationary points, $V'(x) = 0$:

$6 - 20x + 12x^2 = 0$

$3 - 10x + 6x^2 = 0$

Using the quadratic formula: $x = \frac{-b \pm \sqrt{b^2 - 4ac}}{2a}$ for the equation $ax^2 + bx + c = 0$:

$x = \frac{10 \pm \sqrt{100 - 4(6)(3)}}{2(6)}$

$x = \frac{10 \pm \sqrt{4 \times 7}}{2(6)}$

$x = \frac{5 \pm \sqrt{7}}{6}$

$x = 0 \cdot 392$ or $x = 1 \cdot 274$

Since we have already determined that $0 < x < 1$, the only possible solution is $x = 0 \cdot 392$

The nature needs to be determined as to whether these are maximum turning points or not.

x	$\rightarrow$	0·392	$\rightarrow$	1·274	$\rightarrow$
$V'(x)$	+	0	-	0	+
Slope	↗	$\rightarrow$	↘	$\rightarrow$	↗

There is a maximum turning point at 0·392 m.

The corresponding volume is:

$V(0 \cdot 392) = (0 \cdot 392)(3 - 2(0 \cdot 392))(2 - 2(0 \cdot 392))$

$V(0 \cdot 392) = 1 \cdot 056$ m^3

So the volume is maximised when the depth of the box is 0·392 m.

....

TOPIC 2. DIFFERENTIATION

3. Problem:

A water container is made into the following shape:

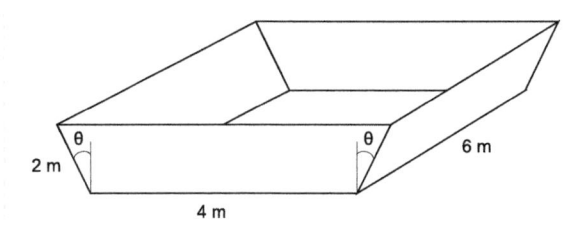

a) Express the volume V (in cubic metres) of the container in terms of the angle θ where $0 \leqslant \theta \leqslant \frac{\pi}{2}$
b) Find the value of θ which gives the maximum volume for this container and justify your answer.
c) Write down how many litres of water the container will hold.

Solution:

a) Consider a cross section of the container as follows:

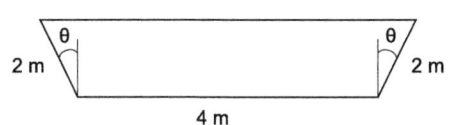

The area of the rectangular part of the cross section is $4 \times 2\cos\theta = 8\cos\theta$ m².
The area of the triangular parts of the cross section is $2\sin\theta \times 2\cos\theta = 2 \times 2\sin\theta\cos\theta = 2\sin(2\theta)$ m²
We do not need to multiply by $\frac{1}{2}$ since there are two triangles with the same area, i.e. $2 \times \frac{1}{2} = 1$.
$V = 6 \times (8\cos\theta + 2\sin(2\theta))$
$V = 48\cos\theta + 12\sin(2\theta)$

b) Maximum volume occurs when $\frac{dV}{d\theta} = 0$ where $0 \leqslant \theta \leqslant \frac{\pi}{2}$.
$\frac{d}{d\theta}(48\cos\theta + 12\sin(2\theta)) = 0$
$-48\sin\theta + 24\cos(2\theta) = 0$
To solve, we need to replace $\cos(2\theta)$ with $1 - 2\sin^2\theta$:
$-48\sin\theta + 24 - 48\sin^2\theta = 0$
$-48\sin^2\theta - 48\sin\theta + 24 = 0$

© HERIOT-WATT UNIVERSITY

Solving using the quadratic formula:

$$\theta = \frac{48 \pm \sqrt{(-48)^2 - 4(-48)(24)}}{2(-48)}$$

$$\theta = -\frac{48 \pm 48\sqrt{3}}{96}$$

$$\theta = -\frac{1 \pm \sqrt{3}}{2}$$

$\theta = 0 \cdot 366$ or $\theta = -1 \cdot 366$

We will discard the solution $\theta = -1 \cdot 366$ since this value is not within $0 \leqslant \theta \leqslant \frac{\pi}{2}$

Now, to determine the nature of the stationary point when $\theta = 0 \cdot 366$, we will use the second derivative:

$V''\theta = -48cos\theta - 48sin(2\theta)$
$V''(0 \cdot 366) = -76 \cdot 9 < 0$ therefore, a maximum

Evaluating:
$V(0 \cdot 366) = 24cos(0 \cdot 366) \times (2 + \sin(0 \cdot 366)) = 52 \cdot 8$ m^3

c) $V(0 \cdot 366) = 52 \cdot 8$ m^3
The container will hold 52,800 litres

2.16 Learning points

Differentiation
Conditions for differentiability

- A function is:
  - continuous if $\lim_{h \to 0} f(x) = f(a)$, otherwise it is discontinuous, in other words if you can draw a curve without lifting the pen from the paper it is continuous.;
  - differentiable at a point if it has a tangent at that point.
- $f(x)$ is differentiable over the interval $[a, b]$ if $f'(x)$ exists for each $x \in [a, b]$
- Higher derivatives are denoted using either $f^{(n)}(x)$ or $\frac{d^n y}{dx^n}$ where n is the nth derivative.

The product rule

- When $k(x) = f(x)\,g(x)$, then $k'(x) = f'(x)\,g(x) + f(x)\,g'(x)$

The quotient rule

- When $k(x) = \frac{f(x)}{g(x)}$, then $k'(x) = \frac{f'(x)g(x) - f(x)g'(x)}{[g(x)]^2}$

Differentiating cot(x), sec(x), cosec(x) and tan(x)

Secant Function	Cosecant Function	Cotangent Function
$\sec x = \frac{1}{\cos x}$	$\operatorname{cosec} x = \frac{1}{\sin x}$ or $\csc x = \frac{1}{\sin x}$	$\cot x = \frac{1}{\tan x}$

- $\frac{d}{dx}(\sec x) = \sec x \tan x$
- $\frac{d}{dx}(\csc x) = -\csc x \cot x$
- $\frac{d}{dx}(\cot x) = -\csc^2 x$
- $\frac{d}{dx}(\tan x) = \sec^2 x$

Differentiate exp(x)

- $\frac{d}{dx} e^{ax} = ae^{ax}$

Implicit differentiation

- For two variables x and y, y is an explicit function of x if y can be written as an expression in which the only variable is x and we obtain only one value for y. Otherwise it is implicit.
- When differentiating an implicit function with variables x and y, differentiate both sides with respect to x, then rearrange to make $\frac{dy}{dx}$ the subject.
- When differentiating implicit functions of y with respect to x, apply the chain rule, e.g.
$$\frac{d}{dx}(y^2) = \frac{d}{dy}(y^2)\frac{dy}{dx}$$
$$\frac{d}{dx}(y^2) = 2y\frac{dy}{dx}$$

Logarithmic differentiation

- The natural logarithm is the logarithm with base e.
- The notation is $\log_e |x| = \ln |x|$
- $\frac{dy}{dx} = \frac{1}{\frac{dx}{dy}}$
- $\frac{d}{dx} \ln(x) = \frac{1}{x}$
- $\frac{d}{dx} \ln |f(x)| = f'(x) \times \frac{1}{f(x)}$
- When differentiating a function that is either a product or quotient of two or more functions or has a variable in the power:
   - apply the natural logarithm to both the LHS and the RHS;
   - logarithmic rules may need to be applied to simplify before differentiation at this stage;
   - differentiate the LHS and RHS implicitly;
   - re-arrange to give $\frac{dy}{dx} =$.
   - replace y with the original function to give the final solution in terms of only x.

Differentiating inverse functions

- For a function to have an inverse, it must be a one-to-one and onto function.
- Given a function, we can use the standard result $\frac{d}{dx}\left(f^{-1}(x)\right) = \frac{1}{f'(y)}$ to find the derivative of the inverse.
- $\frac{dy}{dx}\left(\sin^{-1} x\right) = \frac{1}{\sqrt{1-x^2}}$ and $\frac{dy}{dx}\left(\sin^{-1}\frac{x}{a}\right) = \frac{1}{\sqrt{a^2-x^2}}$
- $\frac{dy}{dx}\left(\cos^{-1} x\right) = -\frac{1}{\sqrt{1-x^2}}$ and $\frac{dy}{dx}\left(\cos^{-1}\frac{x}{a}\right) = -\frac{1}{\sqrt{a^2-x^2}}$
- $\frac{dy}{dx}\left(\tan^{-1} x\right) = \frac{1}{1+x^2}$ and $\frac{dy}{dx}\left(\tan^{-1}\frac{x}{a}\right) = \frac{a}{\sqrt{a^2+x^2}}$

TOPIC 2. DIFFERENTIATION

Parametric differentiation

- Parametric equations are equations that are expressed in terms of an independent variable. They are of the form $x = f(t)$ and $y = g(t)$ where t is the independent variable for x and y, and the variable t is called a parameter.
- To sketch a parametric equation we can either:
  - eliminate the independent variable t, resulting in an equation in terms of x and y which can be drawn.
  - or set up a table of values for t, x and y and then plot the values x and y.
- Given $x = f(t)$ and $y = g(t)$, then $\frac{dy}{dx} = \frac{\left(\frac{dy}{dt}\right)}{\left(\frac{dx}{dt}\right)}$
- When the position of a particle moving in a plane is given by the parametric equations $x = f(t)$ and $y = g(t)$, then the velocity of the particle can be split into:
  - a horizontal velocity component $\frac{dx}{dt}$
  - a vertical velocity component $\frac{dy}{dt}$
- The speed of the particle is given by $\sqrt{\left(\frac{dx}{dt}\right)^2 + \left(\frac{dy}{dt}\right)^2}$
- The instantaneous direction of motion is given by $\frac{dy}{dx}$
- When a curve is defined parametrically, to find the equation of a tangent:
  - find an expression for the gradient using $\frac{dy}{dx} = \frac{\left(\frac{dy}{dt}\right)}{\left(\frac{dx}{dt}\right)}$
  - substitute the value of t in $\frac{dy}{dx}$ to evaluate the gradient.
  - evaluate $x = f(t)$ and $y = g(t)$ for the value of t to find a Cartesian coordinate.
  - substitute the coordinate and gradient into the equation of a straight line $y - b = m(x - a)$
- Given $x = f(t)$ and $y = g(t)$, then $\frac{d^2y}{dx^2} = \frac{\frac{d}{dt}\left(\frac{dy}{dx}\right)}{\frac{dx}{dt}}$

Related rate problems

- Strategy for solving related rate problems:
  - draw a diagram and label the variables and constants. Use t for time;
  - write down:
    - any additional numerical information;
    - the rate of change you are asked to find as a derivative;
    - an equation that describes the relationship between the variables;
  - differentiate (usually with respect to time).

Displacement, velocity and acceleration

- Distance is a scalar quantity. It is the total distance travelled. It is usually denoted by the letter d.

- Displacement is a vector quantity. It is the shortest distance between the starting point and the end point. It is usually denoted by the letter s or x.

- Velocity is a vector quantity. It is the rate of change of displacement with respect to time. It is usually denoted by $v = \frac{ds}{dt}$ or $\dot{x} = \frac{dx}{dt}$.

- Speed is a scalar quantity. It is the magnitude of the velocity and usually denoted by s. Not to be confused with displacement.

- Acceleration is a vector quantity. It is the rate of change of velocity with respect to time. It is usually denoted by $a = \frac{dv}{dt}$ or $\ddot{x} = \frac{d^2x}{dt^2}$.

Curve Sketching in closed intervals

- A closed interval $[a, b]$ is the set of numbers $x \in \mathbb{R}$ where $a \leq x \leq b$, i.e. the end points are included.

- Within a closed interval, the maximum and minimum may occur at stationary points and end points.

Optimisation and other applications

- To solve an optimisation problem:

  1. find the stationary points by differentiating and equating the derivative to zero;
  2. determine their nature using a nature table or by using the second derivative;
  3. evaluate the maximum/minimum depending on which is required;
  4. if the domain is a closed interval, evaluate the end points;
  5. from the stationary points and end points, choose the one that maximises or minimises the problem.

TOPIC 2. DIFFERENTIATION

2.17 Proofs

Proof 1: Justification of connection between differentiability and continuity

Prove that $\lim_{h \to 0} g(x + h) - \lim_{h \to 0} g(x) = 0 \Rightarrow \lim_{h \to 0} g(x + h) = g(x)$

$$\lim_{h \to 0} [g(x + h) - g(x)] = \lim_{h \to 0} \left[\frac{g(x + h) - g(x)}{h} \times h \right]$$

$$= \lim_{h \to 0} \frac{g(x + h) - g(x)}{h} \times \lim_{h \to 0} h$$

$$= g'(x) \times 0$$

$$= 0$$

Thus, $\lim_{h \to 0} [g(x + h) - g(x)] = 0$

Hence, $\lim_{h \to 0} g(x + h) - \lim_{h \to 0} g(x) = 0 \Rightarrow \lim_{h \to 0} g(x + h) = g(x)$

Proof 2: The product rule

Let $k(x) = f(x) \, g(x)$

then $k'(x) = \lim_{h \to 0} \dfrac{k(x+h) - k(x)}{h}$

$= \lim_{h \to 0} \dfrac{f(x+h) \, g(x+h) - f(x) \, g(x)}{h}$

$= \lim_{h \to 0} \dfrac{f(x+h) \, g(x+h) - f(x) \, g(x+h) + f(x) \, g(x+h) - f(x) \, g(x)}{h}$

$= \lim_{h \to 0} \dfrac{[f(x+h) - f(x)] \, g(x+h) + f(x) \, [g(x+h) - g(x)]}{h}$

$= \lim_{h \to 0} \left\{ \dfrac{f(x+h) - f(x)}{h} \times g(x+h) \right\} + \lim_{h \to 0} \left\{ \dfrac{g(x+h) - g(x)}{h} \times f(x) \right\}$

$= \lim_{h \to 0} \dfrac{f(x+h) - f(x)}{h} \times \lim_{h \to 0} g(x+h) + \lim_{h \to 0} \dfrac{g(x+h) - g(x)}{h} \times \lim_{h \to 0} f(x)$

$= f'(x) \, g(x) + g'(x) \, f(x)$

$= f'(x) \, g(x) + f(x) \, g'(x)$

Proof 3: The quotient rule

Let $k(x) = \dfrac{f(x)}{g(x)}$ then $k(x) = f(x) \, [g(x)]^{-1}$

$k'(x) = f'(x) \, [g(x)]^{-1} + f(x)(-1)[g(x)]^{-2} \, g'(x)$

$= [g(x)]^{-2} \, [f'(x) \, g(x) - f(x) \, g'(x)]$

$= \dfrac{f'(x) g(x) - f(x) g'(x)}{[g(x)]^2}$

© HERIOT-WATT UNIVERSITY

Proof 4: Derivative of exp(x)

When $f(x) = e^x$ then,

$$\begin{aligned} f'(x) &= \lim_{h \to 0} \frac{e^{x+h} - e^x}{h} \\ &= \lim_{h \to 0} \frac{e^x(e^h - 1)}{h} \\ &= \lim_{h \to 0} e^x \lim_{h \to 0} \frac{e^h - 1}{h} \\ &= e^x \times 1 \\ &= e^x \end{aligned}$$

Proof 5: Rule connecting $\frac{dy}{dx}$ and $\frac{dx}{dy}$

Prove that $\frac{dy}{dx} = \frac{1}{\left(\frac{dx}{dy}\right)}$

By definition:
$$\begin{aligned} \frac{dy}{dx} &= \lim_{\Delta x \to 0} \left(\frac{\Delta y}{\Delta x}\right) \\ &= \lim_{\Delta x \to 0} \left(\frac{1}{\frac{\Delta x}{\Delta y}}\right) \end{aligned}$$

But, as $\delta x \Rightarrow 0$, then so also $\delta y \Rightarrow 0$ so:
$$\begin{aligned} \frac{dy}{dx} &= \lim_{\delta x \to 0} \left(\frac{1}{\frac{\Delta x}{\Delta y}}\right) \\ &= \frac{1}{\left(\frac{dx}{dy}\right)} \end{aligned}$$

2.18 Extended information

The following links should serve as an insight into the wealth of information on the internet and encourage readers to explore the subject further.

The authors do not maintain these web links and no guarantee can be given as to their effectiveness at a particular date.

http://www.univie.ac.at/future.media/moe/galerie.html

This site consists of interactive learning areas. Of specific interest to this topic are the sites under the headings:

- Differentiation 1
- Applications of differential calculus
- Differentiation 2

http://www.quickmath.com/

Look under for the Differentiate link under the heading Calculus. This provides a tool to check your answers.

http://www-history.mcs.st-andrews.ac.uk/history/index.html

The MacTutor History of Mathematics archive. This site provides a very comprehensive directory of biographies for hundreds of important mathematicians. This was the source for the following information on Newton and Leibniz.

http://xahlee.org/SpecialPlaneCurves_dir/specialPlaneCurves.html

This site gives some historical information and methods of construction for many curves along with many colourful diagrams.

Of particular interest might be the special plane curves: astroid, cardoid, cycloid, epicycloid and hypocycloid.

http://web.mit.edu/wwmath/calculus/index.html

Here you will find further information on implicit differentiation and differentiation of inverse functions.

The following song might amuse you! Hopefully you will not feel this way after working your way through this section.

A Calculus Carol:

Written by Denis Gannon (1940-1991)

Sung to the tune of 'Oh, Christmas Tree' also known as 'The Red Flag'

Oh, Calculus; Oh, Calculus, How tough are both your branches. Oh, Calculus; Oh, Calculus, To pass what are my chances? Derivatives I cannot take, At integrals my fingers shake. Oh, Calculus; Oh, Calculus, How tough are both your branches.

Oh, Calculus; Oh, Calculus, Your theorems I can't master. Oh, Calculus; Oh, Calculus, My Proofs are a disaster. You pull a trick out of the air, Or find a reason, God knows where. Oh, Calculus; Oh, Calculus, Your theorems I can't master.

Oh, Calculus; Oh, Calculus, Your problems do distress me. Oh, Calculus; Oh, Calculus, Related rates depress me. I walk toward lampposts in my sleep, And running water makes me weep. Oh, Calculus; Oh, Calculus, Your problems do distress me.

Oh, Calculus; Oh,Calculus, My limit I am reaching. Oh, Calculus; Oh, Calculus, For mercy I'm beseeching. My grades do not approach a B, They're just an epsilon from D. Oh, Calculus; Oh,Calculus, My limit I am reaching.

Sir Isaac Newton

- Born on 4th January 1643 in Woolsthorpe, England.
- Died on 31st March 1727 in London.

"If I have been able to see further, it was because I stood on the shoulders of giants."

Newton came from a family of farmers but never knew his father who died three months before he was born. Although a wealthy man, Newton's father was uneducated and could not sign his own name. His mother, Hannah Ayscough, remarried when Newton was two years old. Newton was then left in the care of his grandmother and he had a rather unhappy childhood.

In 1653, he attended the Free Grammar School in Grantham. However, his school reports described him as idle and inattentive and he was taken away from school to manage his mothers estate. He showed little interest for this and, due to the influence of an uncle, he was allowed to return to the Free Grammar School in 1660. This time he was able to demonstrate his academic promise and passion for learning and on 5th June 1661 he entered Trinity College, Cambridge.

His ambition at Cambridge was to obtain a law degree, but he also studied philosophy, mechanics and optics. His interest in mathematics began in 1663 when he bought an astrology book at a fair and found that he could not understand the mathematics in it. This spurred him on to read several mathematical texts and to make further deep mathematical studies.

Newton was elected a scholar at Cambridge on 28th April 1664 and received his bachelors degree in April 1665. In the summer of 1665, the University was closed due to the plague and Newton had to return to Lincolnshire. There, while still less than 25 years old, he made revolutionary advances in mathematics, physics, astronomy and optics. While at home, Newton established the foundations for differential and integral calculus, several years before the independent discovery by Leibniz. The method of fluxions as he named it was based on his crucial insight that integration is merely the inverse procedure to differentiating a function.

In 1672, he was elected a fellow of the Royal Society after donating a reflecting telescope. In that year, he also published his first scientific paper on light and colour. However, he came in for some criticism from other academics who objected with some of his methods of proof and from then on Newton was torn between wanting fame and recognition, and the fear of criticism. He found the easiest way to avoid this was to publish nothing.

Newton's greatest achievement was his work in physics and celestial mechanics that lead to his theory of universal gravitation. He was persuaded to write a full account of his new physics and its application to astronomy. In 1687, he published the *Philosophiae naturalis principia mathematica* or *Principia* as it is known. This is recognised as the greatest scientific book ever written. It made him an international leader in scientific research.

On 15th January, Newton was elected by the University of Cambridge as one of their two members to the Convention Parliament in London. This may have led him to see that there was a life in London which might appeal more to him than that of the academic world in Cambridge.

After suffering a nervous breakdown in 1693, Newton retired from research and decided to leave Cambridge to take up a government position in London as Warden, and then later as Master, of the Royal Mint. He made an effective contribution to the work of the Mint particularly on measures to prevent counterfeiting of the coinage.

In 1703, he was elected as president of the Royal Society, a position he retained until his death. He was knighted by Queen Anne in 1705, the first scientist to be honoured in this way for his work.

However, his last years were not easy, dominated in many ways over the controversy with Leibniz as to who had first invented calculus.

Gottfried Leibniz

- Born on 1st July 1646 in Leipzig, Germany.
- Died on 14th November 1716 in Hannover.

"The soul is the mirror of an indestructible universe."

His father Friedrich was a professor of moral philosophy and his mother Catharina Schmuck was Friedrich's third wife. Friedrich died when Leibniz was only six, so he was brought up by his mother and it was her influence that played an important role in his life and philosophy.

In 1661, Leibniz entered the University of Leipzig. He was only fourteen, which nowadays would be considered highly unusual; however, at that time, there would be others of a similar age. He studied philosophy and mathematics, and graduated with a bachelors degree in 1663. Further studies took him on to a Masters Degree in philosophy, and a bachelors degree and doctorate in Law.

By November 1667, Leibniz was living in Frankfurt where he investigated various different projects: scientific, literary and political. He also continued his law career.

In 1672, Leibniz went to Paris with the aim of contacting the French government and dissuading them from attacking German land. While there, he made contact with mathematicians and philosophers and began construction of a calculating machine. On the January of the following year, he went to England to try the same peace mission, the French one having failed, and he visited The Royal Society of London and presented his incomplete calculating machine while there. The Royal Society elected him as a fellow on 19th April 1673, but by 1674 he had not kept his promise to finish his mechanical calculating machine and so he fell out of favour.

It was during his time in Paris that Leibniz developed his version of calculus. However, the English mathematician Sir Isaac Newton had already laid the foundations for differential and integral calculus several years before Leibniz. This lead to much controversy over who had invented calculus and caused Newton to fly into an irrational temper directed against Leibniz. Neither Leibniz nor Newton thought in terms of functions, both always worked in terms of graphs. Leibniz concentrated on finding a good notation for calculus and spent a lot of time thinking about it, whereas Newton wrote more for himself and tended to use whatever notation he thought of on the day.

Amongst Leibniz's other achievements in mathematics were his development of the binary system of arithmetic and his work on determinants, which arose from his developing methods to solve systems of linear equations. He also produced an important piece of work on dynamics.

Leibniz is described as "a man of medium height with a stoop, broad shouldered but bandy-legged, as capable of thinking for several days sitting in the same chair as of travelling the roads of Europe summer and winter. He was an indefatigable worker, a universal letter writer (he had more than 600 correspondents), a patriot and cosmopolitan, a great scientist, and one of the most powerful spirits of Western civilisation."

It was also said about him that "it is rare to find learned men who are clean, do not stink and have a sense of humour."

2.19 End of topic test

End of topic 2 test — Go online

Implicit differentiation

Q186: Use implicit differentiation to find an expression for $\frac{dy}{dx}$ when $3x^2 - y^2 = 9$.

....

Q187: If $xy^2 + x^3y = 7$, use implicit differentiation to find $\frac{dy}{dx}$.

....

Q188: The equation $y^3 + 3xy = 3x^2 - 5$ defines a curve passing through the point $A(2,1)$. Obtain an equation for the tangent to the curve at A.
Give your answer in standard form i.e. $Ax + By + C = 0$

....

Q189: Find the two points where the curve $x^2 + 3xy + y^2 = 9$ crosses the x-axis, and show that the tangents to the curve at these points are parallel. What is the common slope of these tangents?

....

Q190: Given $2xy - x = 6$, use implicit differentiation to obtain $\frac{dy}{dx}$ in terms of x and y. Hence obtain $\frac{d^2y}{dx^2}$ in terms of x and y.

Logarithmic differentiation

Q191: Given $y = 4^x$, use logarithmic differentiation to obtain $\frac{dy}{dx}$ in terms of x.

....

Q192: Given $y = (\cos x)^{5x}$, use logarithmic differentiation to obtain $\frac{dy}{dx}$ in terms of x.

....

Q193: Using logarithmic differentiation find the values of x for which the graph of $y = \dfrac{x^{\frac{1}{2}}(3-x)^{\frac{1}{6}}}{(2x+1)^{\frac{2}{3}}}$ is stationary.

Differentiating inverse functions

Q194:

Starting with $\cos^2 \theta + \sin^2 \theta = 1$, what do you divide by to obtain $1 + \cot^2 \theta = \csc^2 \theta$ for $0 < \theta < \frac{\pi}{2}$?
By expressing $y = \cot^{-1} x$ as $x = \cot y$, obtain $\frac{dy}{dx}$ in terms of x.

....

TOPIC 2. DIFFERENTIATION

Q195: By expressing $y = \sin^{-1} x$ as $x = \sin y$ and using $\cos^2 x + \sin^2 x = 1$, obtain $\frac{dy}{dx}$ in terms of x.

Q196: Let $f(x) = x^2 - 4x - 3$.
Find the value of $\frac{d}{dx}(f^{-1}(x))$ when $y = 5$.

Differentiation using the product, quotient and chain rules

Find the derivatives of:

Q197: $3x^4 \ln x$

Q198: $(x^3 - 2x + 1) \exp(x)$.

Q199: $(x^3 - 2x) \sin x$

Q200: $e^x \sin x$

Q201: $\sin x \ln x$

Q202: $\frac{3x+2}{4x+1}$

Q203: $\frac{1 + \ln x}{4x}$

Q204: $\frac{\cos x}{1 + 4x^2}$

Q205: $\exp(\cos x)$

Q206: $\cos^{-1}(x^3)$, $-1 \leq x \leq 1$

Q207: $\exp(\sin^{-1} x)$

Q208: $\exp(x^3 - 2x)$

© HERIOT-WATT UNIVERSITY

Q209: ln($\sin x$)

Q210: ln($x^3 - 2x$)

Q211: $x^2 e^{-x^3}$

Q212: $\frac{\sin x}{2 + \cos x}$, $-\pi \leq x \leq \pi$

Q213: $\cos(x^2) \sin 3x$

Q214: $\frac{\ln(x + 4)}{x + 4}$, $x > -4$

Q215: exp ($\tan 2x$)

Q216: $x^3 \ln |\cos x|$

Q217: $2\sin^{-1}\sqrt{1 - x}$

Q218: $(3x^2 + 5)(2x - 1)^3$

Q219: A function f is defined by

$f(x) = \frac{3x}{(x - 2)}$, $x \neq 2$

Find $f'(x)$ and deduce that the derivative is always negative.

Q220: Use calculus to find all the values of x for which the function

$f(x) = (1 - x)^2 e^x$, $x \in \mathbb{R}$ is decreasing.

Q221: Given that $y = e^x \sin x$, find the value of x in the interval $0 < x < \pi$ such that $\frac{dy}{dx} = 0$.

Q222: Show that the function $y = \frac{\cos kx}{x}$, where $x \neq 0$ and k is a non-zero constant, satisfies the differential equation $\frac{d^2y}{dx^2} + \frac{2}{x}\frac{dy}{dx} + k^2 y = 0$

TOPIC 2. DIFFERENTIATION

Parametric differentiation

Q223:
A curve is given by the parametric equations $x = 4t^2 + 3$, $y = 2t^3$.
Find $\frac{dy}{dx}$ in terms of t.

....

Q224: The parametric equations $x = 3\cos t$ and $y = 4\sin t$ describe a curve in the plane. Find $\frac{dy}{dx}$.

....

Q225: A curve is defined by the parametric equations $x = 3t^2 - t$, $y = \sqrt{t}$ for $t \geq 0$. Work out the equation of the tangent when $t = 4$.

....

Q226: Find $\frac{dy}{dx}$ and $\frac{d^2y}{dx^2}$ when $x = \cos 4t$ and $y = \sin^2 4t$.

....

Q227: A particle moves on the curve $2x^2 - y^2 = 2$ with a vertical velocity of 9 m/s.

a) What is the horizontal velocity when the particle is at the point (3, 4)?

b) What is the slope of the curve at this point?

Related rate problems

Q228: A spherical football has a puncture. Its volume is decreasing at a rate of $0 \cdot 02$. m^3 s^{-1} Find the rate of change of the radius of the football when the radius is $0 \cdot 1$ m.
(Recall that the volume of a sphere is given by: $V = \frac{4}{3}\pi r^3$)

....

Q229: Two parallel sides of a rectangle are being lengthened at a rate of 2 cm s^{-1}, while the other two sides are shortened in such a way that the figure remains a rectangle with constant area of 50 cm^2.
What is the rate of change of the perimeter when the length of an increasing side is 5 cm?
Is the perimeter increasing or decreasing?

Displacement, velocity and acceleration

Q230: A train travels along straight tracks from rest. Its velocity, $v(t)$ metres per second, is given by $v(t) = \frac{150t}{12+25t}$

a) Write down the unsimplified numerator of the expression for the acceleration.

b) Find the acceleration of the train at 5 seconds. Give your answer to two decimal places.

....

© HERIOT-WATT UNIVERSITY

Q231: A body moves in a straight line with velocity $v = 3t^2 - 4t + 5$ at time t.

a) Obtain the value of its acceleration when $t = 0$
b) At time $t = 0$ the body is at the origin. Write down an expression for the displacement.

....

Q232: A body moves in a straight line with acceleration, $a = 3t + 1$ at time t.

a) At time $t = 0$ the velocity is zero. Obtain the value of its velocity when $t = 2$
b) At time $t = 0$ the displacement is zero. Obtain the value of its displacement when $t = 4$

....

Q233: A body moves in a straight line with displacement $s = 4t^3 - t^2 + 5t - 2$ at time t.

a) Obtain the value of its velocity when $t = 1$
b) Obtain the value of its acceleration when $t = 3$

....

Q234: A ball is thrown vertically upwards. It's height, h metres, after t seconds is given by $h = 2 - 4t + 6t^2$.

a) Find its velocity after 1 second.
b) When does it reach its greatest height? What is this greatest height?
c) What is the acceleration of the ball?

....

Q235: The displacement of a weight on a spring after t seconds is given by $s(t) = 2\sin(3t)$ where $t \in [0, 2\pi]$.

a) Find its velocity after 1 second. Write the answer to one decimal place.
b) After what time is the velocity first zero?
c) At what time, t, belonging to the interval $[0, 2\pi]$ is the acceleration first at a maximum?

Optimisation and other applications

Q236: Two numbers vary but their sum is always 18. Find the maximum value of the sum of their product and difference.

a) Write down an expression in its simplest form for the sum of the product and difference.
b) What is the value of x that maximises the sum?
c) What is the value of y that maximises the sum?

....

TOPIC 2. DIFFERENTIATION

Q237: Find the point (x, y) on the graph of $y = \sqrt{x}$ nearest the point (4,0).

a) Write down the co-ordinates of the point to be found in terms of x alone.
b) Write down an expression for the square of the distance from (4,0) to the curve $y = \sqrt{x}$ in terms of x. Do not simplify.
c) What point of (x, y) is closest to (4, 0)? Give your answer to one decimal place.

....

Q238: Of all lines tangent to the graph $y = \frac{6}{x^2+3}$ find the tangent lines of minimum slope.

a) Write down an expression for the gradient of the tangent line.
b) Write down, if it exists, the point of contact where the gradient of the tangent is a minimum.

....

Q239: A movie screen on a wall is 20 metres high and 10 metres above the floor.

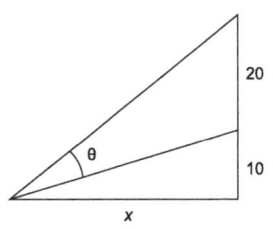

The viewing angle θ is given by: $\theta = \tan^{-1}\left(\frac{30}{x}\right) - \tan^{-1}\left(\frac{10}{x}\right)$

At what distance x from the wall should you position yourself so that the viewing angle of the movie screen is as big as possible?
What is this angle?

a) Write down the derivative of θ with respect to x. Simplify the fraction in each term in your answer.
b) What is the value of x^2 that maximises the viewing angle?
c) What is the maximum viewing angle in radians? Write the answer as a multiple of π.

Topic 3

Integration

Contents

- 3.1 Looking back ... 195
  - 3.1.1 Completing the square ... 199
  - 3.1.2 Integrating expressions ... 201
  - 3.1.3 Integration of sin and cos ... 205
  - 3.1.4 Integrating composite functions ... 208
  - 3.1.5 Finding the area under a curve ... 210
  - 3.1.6 Finding the area between two curves ... 212
  - 3.1.7 Terminology ... 215
- 3.2 Integrate using standard results ... 216
- 3.3 Integration by substitution ... 221
  - 3.3.1 Integration by substitution (substitution given) ... 221
  - 3.3.2 Definite integrals with substitution ... 225
  - 3.3.3 Integration by substitution (substitution not given) ... 230
- 3.4 Integration involving inverse trigonometric functions and completing the square ... 235
  - 3.4.1 Integration involving inverse trigonometric functions ... 235
  - 3.4.2 Integration involving inverse tangent and logarithms ... 238
  - 3.4.3 Integration involving inverse tangent and completing the square ... 240
- 3.5 Partial fractions and integration ... 243
  - 3.5.1 Distinct linear factors ... 243
  - 3.5.2 Repeated linear factors ... 244
  - 3.5.3 Irreducible quadratic factors ... 248
  - 3.5.4 Improper rational functions ... 250
- 3.6 Integration by parts ... 252
  - 3.6.1 One application ... 252
  - 3.6.2 Repeated application ... 255
  - 3.6.3 Cyclical (Getting back to the original function) ... 258
  - 3.6.4 Introducing 1 as a 'dummy variable' ... 259
- 3.7 Area between a curve and an axis ... 260
  - 3.7.1 Area between a curve and the x-axis ... 261
  - 3.7.2 Area between a curve and the y-axis ... 267
- 3.8 Volume of revolution ... 269
- 3.9 Learning points ... 276
- 3.10 Extended information ... 279

3.11 End of topic test 281

TOPIC 3. INTEGRATION

> **Learning objective**
>
> By the end of this topic, you should be able to:
>
> - use and understand the terminology:
>   - anti-derivative;
>   - integration;
>   - constant of integration;
>   - general indefinite integral of f(x);
>   - integrand;
>   - particular integral;
>   - definite integral;
> - integrate the functions:
>   - x^n and $(ax + b)^n$;
>   - $\cos(x)$ and $\cos(ax + b)$;
>   - $\sin(x)$ and $\sin(ax + b)$;
>   - $\sec^2(x)$ and $\sec^2(ax + b)$;
>   - e^x and e^{ax+b};
>   - $\frac{1}{x}$ and $\frac{1}{ax+b}$;
> - use the substitution method of integration to integrate:
>   - when the substitution is given;
>   - definite integrals;
>   - when the substitution is not given;
> - recognise and use the standard integrals that give inverse sine and tangent functions;
> - use a combinations of techniques to solve integrals that may include an inverse trigonometric function in the answer;
> - integrate a rational function where the denominator:
>   - can be factorised into distinct linear factors;
>   - can be factorised into repeated linear factors;
>   - has an irreducible quadratic factor;
> - integrate improper rational functions;
> - use integration by parts:
>   - to find the integral of the product of two functions;
>   - more than once, to find the integral of the product of two functions;
>   - repeatedly, to get back to the original function;

Learning objective continued

- use a 'dummy variable' of 1 to integrate specific functions using integration by parts e.g. $\ln |x|$, $\tan^{-1}(x)$;
- to gain an elementary understanding of the integral as the area under a curve;
- understand how to calculate the area between a curve and the y-axis;
- to calculate the volume of a solid of revolution about the x and y-axes.

3.1 Looking back

Pre-requisites from Advanced Higher

You should have covered the following in the *Partial fractions* topic. If you need to reinforce your learning go back and study this topic.

Algebraic long division

- The method of partial fractions is applied to proper rational functions only. If the function is improper, algebraic long division must be carried out first.

$$
\begin{array}{r}
x - 2 \\
x^2 + 2x - 3 \overline{\smash{\big)} x^3 + 0x^2 - 2x + 5} \\
\underline{x^3 + 2x^2 - 3x} \\
-2x^2 + x + 5 \\
\underline{-2x^2 - 4x + 6} \\
5x - 1
\end{array}
$$

where $x - 2$ is the Quotient, $x^3 + 0x^2 - 2x + 5$ is the Dividend, and $5x - 1$ is the Remainder.

- When setting up algebraic long division each power of the variable must be accounted for in the dividend. See the example above.

Partial fractions: Linear factors

- Any rational functions that have distinct linear factors on the denominator take the following form:

$$\frac{\cdots}{(x+a)(x+b)(x+c)} \equiv \frac{A}{(x+a)} + \frac{B}{(x+b)} + \frac{C}{(x+c)}$$

- Note that this rule can be extended to any number of distinct linear factors on the denominator.

Partial fractions: Repeated linear factors

- Any rational functions that have repeated linear factors on the denominator take the following form:

$$\frac{\cdots}{(x+a)^3} \equiv \frac{A}{(x+a)} + \frac{B}{(x+a)^2} + \frac{C}{(x+a)^3}$$

- Note that this rule can be extended to any number of distinct linear factors on the denominator.

Partial fractions: Irreducible quadratic factors

- An *Irreducible Quadratic* is a quadratic polynomial that cannot be factorised to give real roots.

- Any rational functions that have irreducible quadratic factors on the denominator take the following form:

$$\frac{\cdots}{x^2 + bx + c} \equiv \frac{Ax + B}{x^2 + bx + c}$$

- Note that this rule can be extended to any irreducible polynomial factor on the denominator.

Summary of prior knowledge
Completing the square: National 5
After completing the square, a quadratic expression has the form $a(x+k)^2 + m$

Integration: Higher

- $\int a\, dx = ax + C$, where a is a constant and C is the constant of integration.
- $\int x^n\, dx = \frac{x^{n+1}}{n+1} + C$, where $n \neq -1$.
- $\int ax^n\, dx = a \int x^n\, dx = \frac{ax^{n+1}}{n+1} + C$, where $n \neq -1$ and a is a constant.
- To integrate expressions with multiple terms, each term must first be expressed in the form ax^n.
- Products and quotients must first be expressed as the sum of terms given in the form ax^n.
- $\int \sin x\, dx = -\cos x + C$
- $\int \cos x\, dx = \sin x + C$
- $\int \sin(ax+b)\, dx = -\frac{1}{a}\cos(ax+b) + C$
- $\int \cos(ax+b)\, dx = \frac{1}{a}\sin(ax+b) + C$
- $\int (ax+b)^n\, dx = \frac{(ax+b)^{n+1}}{a(n+1)} + C$, where $n \neq -1$.

Definite integrals: Higher

- If $F(x)$ is an anti-derivative or integral of $f(x)$ then $\int_a^b f(x)\, dx = F(b) - F(a)$, where $a \leq x \leq b$.

The area under a curve

- The area between the graph $y = f(x)$ and the x-axis from $x = a$ to $x = b$ is given by the definite integral, $\int_a^b f(x)\, dx$

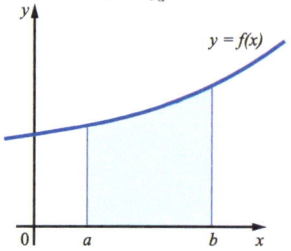

TOPIC 3. INTEGRATION

- An area above the x-axis will give a positive value for the integral.

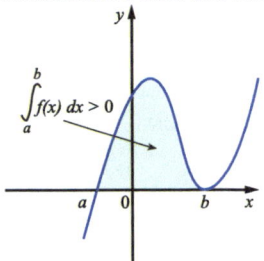

$$\int_a^b f(x)\, dx > 0$$

- An area below the x-axis will give a negative value for the integral.

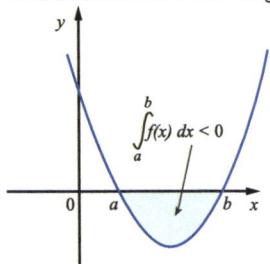

$$\int_a^b f(x)\, dx < 0$$

Area above and below the x-axis: Higher

- To calculate an area between a curve and the x-axis:
    - calculate the areas above and below the axis separately;
    - ignore the negative sign for areas below the axis;
    - add the areas together.

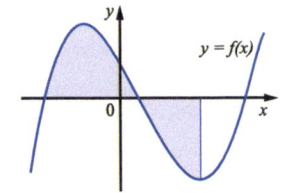

$y = f(x)$

© HERIOT-WATT UNIVERSITY

The area between two curves: Higher

- To calculate an area between two graphs:
  - calculate points of intersection;
  - note which graph is above the other between the points of intersection.

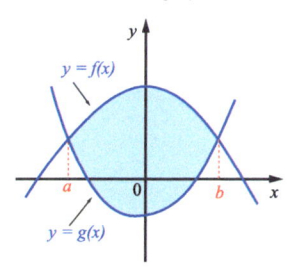

- For the diagram shown here $g(x) \leq f(x)$ for $a \leq x \leq b$ and thus the area enclosed by the curves is given by $\int_a^b (f(x) - g(x))\, dx$
- When calculating definite integrals of trigonometric functions, the limits must be in radians.

TOPIC 3. INTEGRATION

3.1.1 Completing the square

These examples demonstrate the method for completing the square and reminds us how to find the turning point of a parabola by using this method.

Examples

1. Problem:

The diagram below shows part of the graph of $y = x^2 - 4x + 5$.

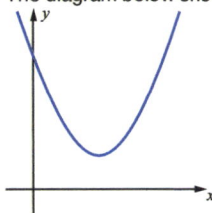

What are the coordinates of the turning point?

Solution:

Since the parabola has no roots (i.e. does not cross the x-axis) we must use the method of completing the square.

$$y = x^2 - 4x + 5$$
$$= (x^2 - 4x) + 5$$
$$= (x - 2)^2 - 4 + 5$$
$$y = (x - 2)^2 + 1$$

- use the first two terms to find a perfect square (half -4)
- $(x - 2)^2 = x^2 - 4x + 4$ (we don't want the +4 so subtract it)
- collect like terms

To find the x coordinate make the bracket equal to zero, $x - 2 = 0$ so $x = 2$.
When the bracket equals zero $y = 0^2 + 1$ so $y = 1$.
The coordinates of the minimum turning point are (2,1).

....

2. Problem:

Express the $2x^2 + 8x - 7$ in the form $a(x + p)^2 + q$.

Solution:

$$y = 2x^2 + 8x - 7$$
$$= 2\left[x^2 + 4x\right] - 7$$
$$= 2\left[(x + 2)^2 - 4\right] - 7$$
$$= 2(x + 2)^2 - 8 - 7$$
$$y = 2(x + 2)^2 - 15$$

- use the first two terms to find a perfect square (half 4)
- $(x + 2)^2 = x^2 + 4x + 4$ (we don't want the +4 so subtract it)
- expand the square brackets [ ]
- collect like terms

....

© HERIOT-WATT UNIVERSITY

3. Problem:

Find the coordinates of the turning point of $y = -3x^2 - 24x - 55$.

Solution:

$$y = -3x^2 - 24x - 55$$
$$= -3\left[x^2 + 8x\right] - 55$$
$$= -3\left[(x+4)^2 - 16\right] - 55$$
$$= -3(x+4)^2 + 48 - 55$$
$$y = -3(x+4)^2 - 7$$

- factorise the first 2 terms
- use the first two terms to find a perfect square (half 8)
- $(x+4)^2 = x^2 + 8x + 16$ (we don't want the +16 so subtract it)
- expand the square brackets [ ] (be careful: $-3 \times -16 = 48$)
- collect like terms

The graph has a sad face shape because x^2 is negative so the turning point is a maximum.
To find the x coordinate make the bracket equal to zero. $x + 4 = 0$ so $x = -4$.
When the bracket equals zero $y = -3 \times 0^2 - 7$ so $y = -7$.
The coordinates of the maximum turning point are (-4,-7).

> **Key point**
>
> Make sure that the equation of the quadratic is in the form $y = ax^2 + bx + c$ before completing the square.

Example

Problem:

Express the $1 - 12x - x^2$ in the form $q - (x+p)^2$.

Solution:

$$y = 1 - 12x - x^2$$
$$= -x^2 - 12x + 1$$
$$= -\left[x^2 + 12x\right] + 1$$
$$= -\left[(x+6)^2 - 36\right] + 1$$
$$= -(x+6)^2 + 36 + 1$$
$$= -(x+6)^2 + 37$$
$$y = 37 - (x+6)^2$$

- swap the order of the first and last terms
- factorise the first 2 terms
- use the first two terms to find a perfect square (half 12)
- $(x+6)^2 = x^2 + 12x + 36$ (we don't want the +4 so subtract it)
- expand the square brackets [ ]
- collect like terms
- swap the order of the first and last terms

TOPIC 3. INTEGRATION

> **Completing the square exercise**   Go online
>
> **Q1:** If $x^2 - 8x + 7$ is written in the form $(x + p)^2 + q$.
> What is the value of q?
>
> ....
>
> **Q2:** If $2x^2 - 12x + 8$ is written in the form $a(x + p)^2 + q$.
> What is the value of q?
>
> ....
>
> **Q3:** If $-5x^2 + 20x + 3$ is written in the form $a(x + p)^2 + q$.
> What is the value of q?
>
> ....
>
> **Q4:** Give $x^2 + 5x - 1$ in the form $(x + p)^2 + q$.

3.1.2 Integrating expressions

> **Key point**
>
> The notation for finding an integral is $\int x^n dx = \frac{x^{n+1}}{n+1} + C$ where $n \neq -1$ and C is the constant of integration.
>
> The German mathematician Gottfried Leibniz devised the notation used for integration.
>
> $\int f(x)\, dx = F(x) + C$
>
> - $f(x)$ is the **integrand**;
> - $F(x)$ is the **anti-derivative** of $f(x)$ and is called the **integral**;
> - C is the **constant of integration**.

> **Key point**
>
> We can take a common factor of a constant to the front of the integral sign.
>
> $$\int ax^n \, dx = a \int x^n$$
> $$= \frac{ax^{n+1}}{n+1} + C$$
>
> where $n \neq -1$, a is a constant and C is the constant of integration.

© HERIOT-WATT UNIVERSITY

Examples

1. Problem:

Find $\int 4\, dx$

Solution:

$$\int 4\, dx = \int 4x^0\, dx \quad \text{Remember } x^0 = 1$$
$$= 4\int x^0\, dx$$
$$= 4\frac{x^1}{1} + C$$
$$= 4x + C$$

....

2. Problem:

Find $\int x^{-3}\, dx$

Solution:

$$\int x^{-3}\, dx = \frac{x^{-3+1}}{-3+1} + C$$
$$= -\frac{x^{-2}}{2} + C$$
$$= -\frac{1}{2x^2} + C$$

....

3. Problem:

Find $\int 3x^2 + 2x - 1\, dx$

Solution:

Integrate each term in turn.

$$\int 3x^2\, dx + \int 2x\, dx - \int 1\, dx = \frac{3x^3}{3} + \frac{2x^2}{2} - 1x + C$$
$$= x^3 + x^2 - x + C$$

Key point

To integrate more complex functions, the integrand must be expressed as the sum of individual terms each given in the form ax^n.

Examples

1. Problem:

Find $\int (2p - 3)^2\, dp$

TOPIC 3. INTEGRATION

Solution:
Step 1: We need to simplify the expression before we can integrate.
$$\int (2p - 3)^2 \, dp = \int (2p - 3)(2p - 3) \, dp$$
$$= \int 4p^2 - 12p + 9 \, dp$$
Step 2: Integrate each term in turn.
$$\int (2p - 3)^2 \, dp = \int 4p^2 - 12p + 9 \, dp$$
$$= \frac{4p^3}{3} - 6p^2 + 9p + C$$

2. Problem:
Find $\int \frac{u^2 + 2}{\sqrt{u}} \, du$

Solution:
Step 1: Before integrating, separate the terms and express each one in the form ax^n.
$$\int \frac{u^2 + 2}{\sqrt{u}} \, du = \int \frac{u^2}{\sqrt{u}} + \frac{2}{\sqrt{u}} \, du$$
$$= \int \frac{u^2}{u^{\frac{1}{2}}} + \frac{2}{u^{\frac{1}{2}}} \, du$$
$$= \int u^{\frac{3}{2}} + 2u^{-\frac{1}{2}} \, du$$

Step 2: Integrate each term in turn.
$$\int \frac{u^2 + 2}{\sqrt{u}} \, du = \int u^{\frac{3}{2}} + 2u^{-\frac{1}{2}} \, du$$
$$= \frac{u^{\frac{3}{2} + 1}}{\frac{3}{2} + 1} + \frac{2u^{-\frac{1}{2} + 1}}{-\frac{1}{2} + 1} + C$$
$$= \frac{2u^{\frac{5}{2}}}{5} + 4u^{\frac{1}{2}} + C$$

Step 3: Simplify the fractional indices.
$$\int \frac{u^2 + 2}{\sqrt{u}} \, du = \frac{2u^{\frac{5}{2}}}{5} + 4u^{\frac{1}{2}} + C$$
$$= \frac{2\sqrt{u^5}}{5} + 4\sqrt{u} + C$$

3. Problem:
Find $\int \frac{1}{x^2} + \sqrt{x} \, dx, \; x \neq 0$

Solution:
Step 1: Before integrating, express each term in the form ax^n.
$$\int \frac{1}{x^2} + \sqrt{x} \, dx = \int x^{-2} + x^{\frac{1}{2}} \, dx$$

Step 2: Integrate each term in turn.

$$\int \frac{1}{x^2} + \sqrt{x}\,dx = \int x^{-2} + x^{\frac{1}{2}}\,dx$$

$$= \frac{x^{-2+1}}{-2+1} + \frac{x^{\frac{1}{2}+1}}{\frac{1}{2}+1} + C$$

$$= \frac{x^{-1}}{-1} + \frac{x^{\frac{3}{2}}}{\frac{3}{2}} + C$$

$$= -\frac{1}{x} + \frac{2}{3}x^{\frac{3}{2}} + C$$

Step 3: Simplify the fractional indices.

$$\int \frac{1}{x^2} + \sqrt{x}\,dx = -\frac{1}{x} + \frac{2}{3}x^{\frac{3}{2}} + C$$

$$= -\frac{1}{x} + \frac{2\sqrt{x^3}}{3} + C$$

....

4. Problem:

Find $\int \frac{9}{\sqrt[8]{t^5}} - \frac{1}{2}\,dt$, $t \neq 0$

Solution:

Step 1: Before integrating, express each term in the form ax^n.

$\int \frac{9}{\sqrt[8]{t^5}} - \frac{1}{2}dt = \int 9t^{-\frac{5}{8}} - \frac{1}{2}dt$

Step 2: Integrate each term in turn.

$$\int \frac{9}{\sqrt[8]{t^5}} - \frac{1}{2}dt = \int 9t^{-\frac{5}{8}} - \frac{1}{2}t\,dt$$

$$= \frac{9t^{-\frac{5}{8}+1}}{-\frac{5}{8}+1} - \frac{1}{2}t + C$$

$$= \frac{9t^{\frac{3}{8}}}{\frac{3}{8}} - \frac{1}{2}t + C$$

$$= \frac{8}{3} \times 9t^{\frac{3}{8}} - \frac{1}{2}t + C$$

$$= 24t^{\frac{3}{8}} - \frac{1}{2}t + C$$

Step 3: Simplify the fractional indices.

$$\int \frac{9}{\sqrt[8]{t^5}} - \frac{1}{2}dt = 24t^{\frac{3}{8}} - \frac{1}{2}t + C$$

$$= 24\sqrt[8]{t^3} - \frac{t}{2} + C$$

TOPIC 3. INTEGRATION

Integrating expressions exercise   Go online

Q5: Find $\int 3x^2 - 6x - 5\, dx$.

....

Q6: Find $\int 3x^{\frac{1}{2}} + x^{-4} dx$, $x \neq 0$

....

Q7: Find $\int \frac{1}{\sqrt[3]{r}} - \frac{1}{2r^3}\, dr$, $r \neq 0$

....

Q8: Find $\int \frac{p^5 + 1}{\sqrt{p^3}}\, dp$, $p \neq 0$

3.1.3 Integration of sin and cos

Remember just as with differentiation x must be measured in radians.

Key point

$$\int \sin x\, dx = -\cos x + C$$

$$\int \cos x\, dx = \sin x + C$$

Examples

1. Problem:

Find $\int 8 + 3\sin x\, dx$

Solution:

$\int 8 + 3\sin x\, dx = 8x - 3\cos x + C$

....

2. Problem:

Find $\int 4\sin\theta + 5\cos\theta\, d\theta$

Solution:

$$\begin{aligned}
\int 4\sin\theta + 5\cos\theta\, d\theta &= 4\int \sin\theta + 5\int \cos\theta\, d\theta \\
&= 4 \times (-\cos\theta) + 5 \times \sin\theta + C \\
&= -4\cos\theta + 5\sin\theta + C
\end{aligned}$$

© HERIOT-WATT UNIVERSITY

Key point

$$\int \sin(ax+b)\,dx = -\frac{1}{a}\cos(ax+b)$$
$$\int \cos(ax+b)\,dx = \frac{1}{a}\sin(ax+b)$$

Examples

1. Problem:
Find $\int \cos(3x+5)\,dx$
Solution:
$\int \cos(3x+5)\,dx = \frac{1}{3}\sin(3x+5) + C$

....

2. Problem:
Find $\int 2\sin(1-6\theta)\,d\theta$
Solution:
$$\begin{aligned}\int 2\sin(1-6\theta)\,d\theta &= 2\int \sin(1-6\theta)\,d\theta \\ &= 2 \times -\frac{1}{-6}\cos(1-6\theta) + C \\ &= \frac{2}{6}\cos(1-6\theta) + C \\ &= \frac{1}{3}\cos(1-6\theta) + C\end{aligned}$$

Key point

$$\sin^2 x = \frac{1}{2}(1-\cos(2x))$$
$$\cos^2 x = \frac{1}{2}(1+\cos(2x))$$

TOPIC 3. INTEGRATION

Example

Problem:

Knowing that $\cos^2 x = \frac{1}{2}(1 + \cos 2x)$, find $\int \cos^2 x \, dx$

Solution:

$$\begin{aligned}\int \cos^2 x \, dx &= \int \frac{1}{2}(1 + \cos 2x) \, dx \\ &= \frac{1}{2}\int (1 + \cos 2x) \, dx \\ &= \frac{1}{2}\left(x + \frac{1}{2}\sin 2x\right) + C \\ &= \frac{1}{2}x + \frac{1}{4}\sin 2x + C\end{aligned}$$

Integration of sin and cos exercise Go online

Q9: What is $\int \frac{6}{\sqrt{u}} - \sin u \, du$?

....

Q10: What is $\int \frac{\cos x}{4} + \sqrt{7}\sin x - \pi \, dx$?

....

Q11: What is $\int \sin 2x \, dx$?

....

Q12: What is $\int \cos(7x + 6) \, dx$?

....

Q13: What is $\int \sin(4x + 5) \, dx$?

....

Q14: What is $\int 6\cos\left(4x + \frac{\pi}{4}\right) dx$?

....

Q15: What is $\int 5\sin\left(\frac{x}{5} - \frac{\pi}{3}\right) dx$?

....

Q16: Use the double angle formula $\cos 2x = 1 - 2\sin^2 x$ to express $\sin 2x$ in terms of $\cos 2x$.

a) What is $\sin^2 x$ in terms of $\cos 2x$?
b) Hence what is $\int \sin^2 x \, dx$?

3.1.4 Integrating composite functions

Top tip

It is easier to remember how to find $\int (ax+b)^n$ by following a few simple steps:

1. write down the bracket $(ax+b)$
2. increase the power by 1 $(ax+b)^{n+1}$
3. divide by the value of a multiplied by the new power $\frac{(ax+b)^{n+1}}{a(n+1)}$
4. remember to add C $\frac{(ax+b)^{n+1}}{a(n+1)} + C$

Examples

1. Problem:

Find $\int (4x+3)^4 \, dx$

Solution:

$$\int (4x+3)^4 \, dx = \frac{(4x+3)^5}{4 \times 5} + C$$
$$= \frac{(4x+3)^5}{20} + C$$

2. Problem:

Find $\int (2x+5)^{\frac{3}{4}} \, dx$

Solution:

$$\int (2x+5)^{\frac{3}{4}} \, dx = \frac{(2x+5)^{\frac{7}{4}}}{2 \times \frac{7}{4}} + C$$
$$= \frac{(2x+5)^{\frac{7}{4}}}{\frac{14}{4}} + C$$
$$= \frac{4}{14}(2x+5)^{\frac{7}{4}} + C$$
$$= \frac{2}{7}(2x+5)^{\frac{7}{4}} + C$$

Key point

Remember your expression must be in the form $(ax+b)^n$ before you can use this method of integration:

$$\int (ax+b)^n \, dx = \frac{(ax+b)^{n+1}}{a(n+1)} + C$$

TOPIC 3. INTEGRATION

Example

Problem:

Find $\int \frac{1}{\sqrt{8t-3}}\, dt$

Solution:

$$\int \frac{1}{\sqrt{8t-3}}\, dt = \int (8t-3)^{-\frac{1}{2}}\, dt$$

$$= \frac{(8t-3)^{\frac{1}{2}}}{8 \times \frac{1}{2}} + C$$

$$= \frac{\sqrt{8t-3}}{4} + C$$

Integrating composite functions exercise

Go online

Q17: What is $\int (8x+7)^{\frac{1}{4}}\, dx$?

Q18: What is $\int (8-3x)^{-\frac{2}{3}}\, dx$?

Q19: What is $\int \sqrt{2x+7}\, dx$?

Q20: What is $\int \frac{1}{(4x+1)^2}\, dx$?

Q21: What is $\int \frac{1}{\left(\sqrt{4x+5}\right)^3}\, dx$?

Q22: When $\frac{ds}{dt} = (1-8t)^{-7}$, what is s?

3.1.5 Finding the area under a curve

The area between a curve and the x-axis   Go online

Example

Problem:

Calculate the total shaded area in the graph shown here.

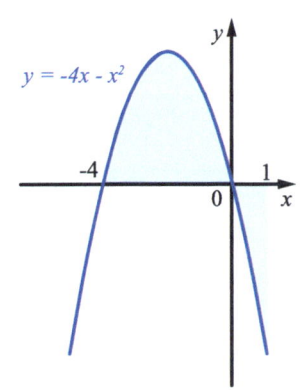

$y = -4x - x^2$

Solution:

Note from the graph that the area is in two sections, one part above the x-axis and the other part below the x-axis. These areas should be calculated separately as follows.

The area above the x-axis is given by the integral,

$$\int_{-4}^{0} \left(-4x - x^2\right) dx = \left[-2x^2 - \frac{x^3}{3}\right]_{-4}^{0}$$

$$= [0] - \left[-32 + \frac{64}{3}\right]$$

$$= 32 - 21\frac{1}{3}$$

$$= 10\frac{2}{3}$$

Thus the area above the x-axis is $10\frac{2}{3}\ units^2$.

TOPIC 3. INTEGRATION

The area below the x-axis is given by the integral

$$\int_0^1 \left(-4x - x^2\right) dx = \left[-2x^2 - \frac{x^3}{3}\right]_0^1$$
$$= \left[-2 - \frac{1}{3}\right] - [0]$$
$$= -2\frac{1}{3}$$

The integral gives a negative answer since the area is below the x-axis thus the area is $2\frac{1}{3}\ units^2$.

Hence,

Total shaded area $= 10\frac{2}{3} + 2\frac{1}{3} = 13$ units^2

Finding the area under a curve exercise   Go online

The graph for $y = 3x^2 - 12x + 9$ is shown here.

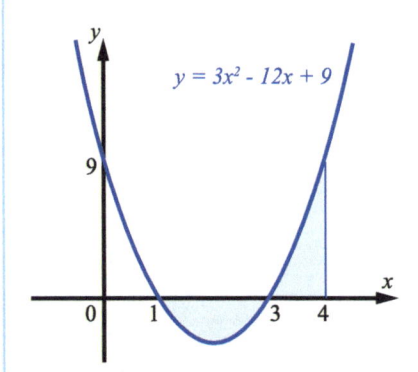

Evaluate the following definite integrals for this graph.

Q23: $\int_1^3 \left(3x^2 - 12x + 9\right) dx$
....

Q24: $\int_3^4 \left(3x^2 - 12x + 9\right) dx$
....

Q25: What is the total shaded area between the curve and the x-axis for the function above?

3.1.6 Finding the area between two curves

To calculate an area between two graphs:

- calculate the points of intersection where $f(x) = g(x)$;
- note which graph is above the other between the points of intersection;
- evaluate $\int_a^b f(x) - g(x)\, dx$ for $f(x)$ above $g(x)$ and $a \leq x \leq b$.

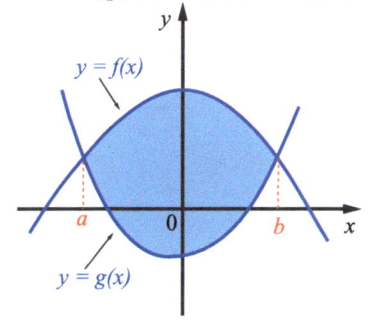

Example

Problem:

The parabolas shown in the diagram have equations $y = 22 - x^2$ and $y = x^2 + 4$. Calculate the shaded areas between the two curves.

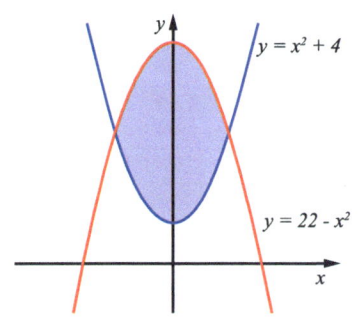

Solution:

Step 1: Find the points of intersection of the parabolas.

$$x^2 + 4 = 22 - x^2$$
$$2x^2 - 18 = 0$$
$$2(x^2 - 9) = 0$$
$$2(x - 3)(x + 3) = 0$$
$$x = 3 \text{ or } x = -3$$

TOPIC 3. INTEGRATION

Step 2: Determine which curve is above the other.

$y = 22 - x^2$ is above $y = x^2 + 4$ when $-3 \leq x \leq 3$

Step 3: Calculate the area.

$$\int_{-3}^{3} (22 - x^2) - (x^2 + 4) \, dx = \int_{-3}^{3} 18 - 2x^2 \, dx$$

$$= \left[18x - \frac{2x^3}{3} \right]_{-3}^{3}$$

$$= \left(18 \times 3 - \frac{2 \times 3^3}{3} \right) - \left(18 \times (-3) - \frac{2 \times (-3)^3}{3} \right)$$

$$= 72 \text{ units}^2$$

Finding the area between two curves exercise — Go online

Q26: Calculate the shaded area in the diagram shown here.

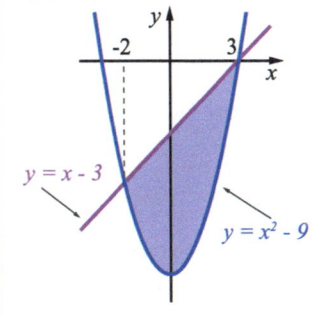

$y = x - 3$

$y = x^2 - 9$

....

Q27: Find the points of intersection between the two graphs shown here and hence calculate the shaded area.

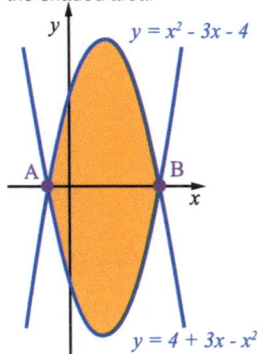

$y = x^2 - 3x - 4$

$y = 4 + 3x - x^2$

Q28: Calculate the area enclosed by the graphs $y = x - 4$ and $y = x^2 - 5x + 4$.

Q29: Calculate the area enclosed by the graphs $y = x^2 - 20x + 17$ and $y = -1 - x^2$

Q30: The graphs $y = x^3 - 3x - 2$ and $y = x - 2$ intersect in three places as shown in the diagram.
Find the x-coordinates of the three points of intersection and hence calculate the total area enclosed between these two graphs.

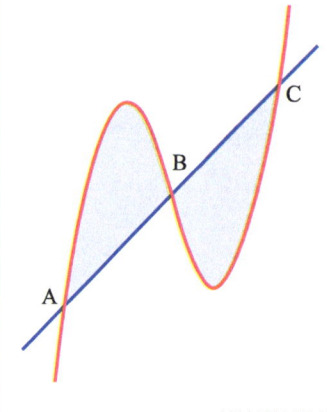

Q31: The diagram shows the curve with equation $y = 2x^2 - 7x + 7$ and the line with equation $y = x + 7$.
Find the integral which represents the shaded area.

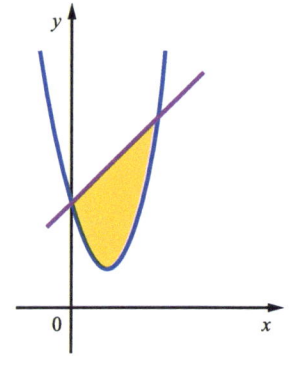

TOPIC 3. INTEGRATION

The concrete on the 20 feet by 28 feet rectangular facing of the entrance to an underground cavern is to be repainted.
Coordinate axes are chosen as shown in the diagram with a scale of 1 unit equal to 1 foot.
The roof is in the form of a parabola with equation $y = 16 - \frac{x^2}{4}$.

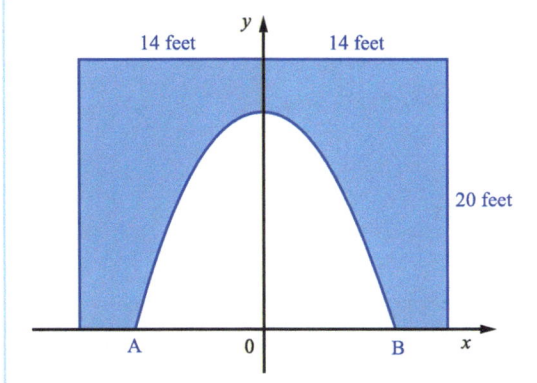

Q32: What are the coordinates of points A and B?

....

Q33: Calculate the total cost of repainting the facing at £5 per square foot.

3.1.7 Terminology

Recall the following definitions.

> **Anti-differentiation** is the reverse process of differentiation.
>
> **Integration** is the method used to find anti-derivatives.

> **Key point**
>
> $\int f(x)\, dx = F(x) + C$
>
> - $f(x)$ is the **integrand**;
> - $F(x)$ is the **anti-derivative** of $f(x)$ and is called the **integral**;
> - C is the **constant of integration**;
> - $F(x) + C$ is the **general indefinite integral of f(x)** with respect to x;
> - When the value of C is evaluated, $F(x) + C$ is called the **particular integral**.

216 TOPIC 3. INTEGRATION

> **Key point**
>
> When the limits of integration are known, we have a definite integral and
> $\int_a^b f(x)dx = [F(x)]_a^b = F(b) - F(a)$ for $a \leq x \leq b$.
>
> This is known as the **Fundamental Theorem of Calculus**.

So, for example, $F(x) = x^3 + C$ is the indefinite integral of $f(x) = 3x^2$ whereas $F(x) = x^3 + 6$ is a particular integral of $f(x) = 3x^2$

3.2 Integrate using standard results

We have already seen the standard derivatives.

> $$\frac{d}{dx}(\tan(x)) = \sec^2(x)$$
> $$\frac{d}{dx}(e^x) = e^x$$
> $$\frac{d}{dx}(\ln(x)) = \frac{1}{x}$$

The standard integrals then follow.

> **Key point**
>
> $$\int \sec^2(x)dx = \tan(x) + C$$
> $$\int e^x dx = e^x + C$$
> $$\int \frac{1}{x}dx = \ln|x| + C$$

Notice that, in this last integral, we take the modulus of x as the ln function does not exist for negative values of x.

It may be useful to have a summary of the results:

> **Key point**
>
> $$\int x^n dx = \frac{x^{n+1}}{n+1} + C \qquad \int (ax+b)^n dx = \frac{(ax+b)^{n+1}}{a(n+1)} + C$$

> **Key point**
>
> $$\int \cos x\, dx = \sin x + C \qquad \int \cos(ax+b)dx = \frac{1}{a}\sin(ax+b) + C$$

© HERIOT-WATT UNIVERSITY

TOPIC 3. INTEGRATION

> **Key point**
>
> $$\int \sin x \, dx = -\cos x + C \qquad \int \sin(ax+b) \, dx = -\frac{1}{a}\cos(ax+b) + C$$

> **Key point**
>
> $$\int \sec^2 x \, dx = \tan x + C \qquad \int \sec^2(ax+b) \, dx = \frac{1}{a}\tan(ax+b) + C$$

> **Key point**
>
> $$\int e^x \, dx = e^x + C \qquad \int e^{(ax+b)} \, dx = \frac{1}{a}e^{(ax+b)} + C$$

> **Key point**
>
> $$\int \frac{1}{x} \, dx = \ln|x| + C \qquad \int \frac{dx}{ax+b} = \frac{1}{a}\ln|ax+b| + C$$

Examples

1.

$$\int \frac{2}{(3x+7)^3} dx = 2 \int (3x+7)^{-3} dx$$
$$= 2\frac{(3x+7)^{-2}}{3 \times (-2)} + C$$
$$= -\frac{1}{3(3x+7)^2} + C$$

....

2.

$\int 3\cos(2x-1) \, dx = \frac{3}{2}\sin(2x-1) + C$

....

3.

$$\int 6e^{3x+2} dx = \frac{6}{3}e^{3x+2} + C$$
$$= 2e^{3x+2} + C$$

....

© HERIOT-WATT UNIVERSITY

4.

$$\int_0^{\frac{\pi}{4}} \sec^2(2x)dx = \left[\frac{1}{2}\tan(2x)\right]_0^{\frac{\pi}{4}}$$
$$= \frac{1}{2}\left(\tan(\frac{\pi}{4}) - \tan 0\right)$$
$$= \frac{1}{2}(1 - 0)$$
$$= \frac{1}{2}$$

5.

$$\int_0^1 \frac{dx}{3x + 5} = \left[\frac{1}{3}\ln|3x + 5|\right]_0^1$$
$$= \frac{1}{3}(\ln 8 - \ln 5)$$
$$= \frac{1}{3}\ln\left(\frac{8}{5}\right)$$
$$= 0 \cdot 157 \text{ to 3 d.p.}$$

6.

Find $\int \frac{1}{3x} dx$

This question can be done in two ways:

Method 1: Take the constant to the front of the integral and then integrate to get the natural logarithm.

$$\int \frac{1}{3x} dx = \frac{1}{3} \int \frac{1}{x} dx$$
$$= \frac{1}{3}\ln|x| + C_1$$

Method 2: Integrate to get the natural logarithm and divide by the derivative of $3x$.

$\int \frac{1}{3x} dx = \frac{1}{3}\ln|3x| + C_2$

Both solutions are correct even though at first sight they look different, and in fact:

$$\frac{1}{3}\ln|3x| + C_2 = \frac{1}{3}\ln|x| + \frac{1}{3}\ln 3 + C_2 \quad \text{(since } \ln|ab| = \ln|a| + \ln|b|\text{)}$$
$$= \frac{1}{3}\ln|x| + C_1 \qquad \text{(where } C_1 = \frac{1}{3}\ln 3 + C_2\text{)}$$

A tricky integration

Now we consider a tricky integration. It is more difficult because the integrand is an improper rational function, and it does not fit into our standard results above. We met improper rational functions in the topic of partial fractions. In the cases below, the numerators and denominators both have a degree of 1.

TOPIC 3. INTEGRATION

Examples

1. Problem:

Integrate $\int \frac{x}{x+2} \, dx$

Solution:

Before integrating this function we note first that it is an improper rational function since the degree of the numerator is the same as the degree of the denominator. We must therefore re-write it as a proper rational function.

To do this we note that the denominator takes the form $x + 2$. We need part of the numerator to look like this. The question we ask is: "What must I do to the denominator to make it the same as the numerator?". In this case we subtract 2 i.e. $(x + 2) - 2$. This is now the new numerator.

$$\int \frac{x}{x+2} \, dx = \int \left(\frac{x+2-2}{x+2} \right) dx \quad \text{(Rearranging to obtain proper fractions)}$$

$$= \int \left(\frac{x+2}{x+2} - \frac{2}{x+2} \right) dx$$

$$= \int \left(1 - \frac{2}{x+2} \right) dx$$

$$= x - 2\ln|x+2| + C$$

2. Problem:

Integrate $\int \frac{2x}{6-8x} \, dx$

Solution:

Take the denominator $6 - 8x$, the coefficient of x on the numerator is positive $2x$. We therefore need to multiply $(6 - 8x)$ by $-1/4$ i.e. $-\frac{1}{4}(6 - 8x)$, but when multiplied out this give a constant of $-\frac{6}{4}$ we don't want, so we must add $\frac{6}{4}$ back on i.e. $-\frac{1}{4}(6-8x) + \frac{6}{4}$. This now becomes the numerator of the rational function.

$$\int \frac{2x}{6-8x} dx = \int \left(\frac{-\frac{1}{4}(6-8x) + \frac{6}{4}}{6-8x} \right) dx \quad \text{(Write the numerator in terms of 6 - 8x)}$$

$$= \int \left(\frac{-\frac{1}{4}(6-8x)}{6-8x} + \frac{\frac{6}{4}}{6-8x} \right) dx \quad \text{(Separate the two terms)}$$

$$= -\frac{1}{4} \int \frac{6-8x}{6-8x} dx + \frac{3}{2} \int \frac{1}{6-8x} dx$$

$$= -\frac{1}{4} \int 1 \, dx + \frac{3}{2} \int \frac{1}{6-8x} dx \quad \text{(Simplify each term before integrating)}$$

$$= -\frac{1}{4}x + \frac{3}{2}(-\frac{1}{8}) \ln|6-8x| + C$$

$$= -\frac{1}{4}x - \frac{3}{16} \ln|6-8x| + C$$

The method used above to transform an improper rational function into a proper rational functions before integration is not the only method. In the topic of partial fractions you saw the method of algebraic long division.

© HERIOT-WATT UNIVERSITY

The next example will use the method of algebraic long division.

Example

Problem:

Integrate $\int \frac{2x-3}{x+4} dx$ by first using algebraic long division to write the integrand as a proper rational function.

Solution:

Transform the improper rational function $\frac{2x-3}{x+4}$ into a proper rational function using algebraic long division.

$$x+4 \overline{\smash{\big)}\begin{array}{c} 2 \\ 2x-3 \\ \underline{2x+8} \\ -11 \end{array}}$$

The improper rational function now becomes $\frac{2x-3}{x+4} = 2 - \frac{11}{x+4}$

This is the function that we now integrate.

$$\int \frac{2x-3}{x+4} dx = \int 2 - \frac{11}{x+4} dx$$
$$= \int 2 dx - \int \frac{11}{x+4} dx$$
$$= 2x - 11 ln|x+4| + C$$

Integrate using standard results exercise   Go online

Q34: Integrate $\int 3(2x+5)^4 \, dx$

....

Q35: Integrate $\int 2\cos(3x-1) \, dx$

....

Q36: Integrate $\int 5\sin(2x+3) \, dx$

....

Q37: Integrate $\int 2\sec^2(3x-7) \, dx$

....

Q38: Integrate $\int -7e^{(2x-4)} \, dx$

....

Q39: Integrate $\int \frac{-5}{3x+2} \, dx$

....

Q40: Integrate $\int \frac{x}{x-6} \, dx$

....

Q41: Integrate $\int \frac{r+3}{r+7} dr$

Q42: Integrate $\int \frac{2x}{6x+3} dx$

3.3 Integration by substitution

In the previous section we used standard results to integrate functions. However, it is not always the case that standard results can be used in the first instance. This section demonstrates the method of integration by substitution, which transforms the integrand into a form where standard results can be applied.

3.3.1 Integration by substitution (substitution given)

Consider how to integrate the function $x^3(2x^4 - 3)^4$. It would be possible to expand the bracket, multiply by x^3 and then integrate each term separately. Not an enviable task!

This integration is a lot easier if the variable is changed from x to some other suitably chosen variable.

Examples

1. Problem:

Find $\int x^3 (2x^4 - 3)^4 dx$

Solution:

Note that this integrand is a product of two functions: x^3 and $(2x^4 - 3)^4$.

First we select the function that is to be change to another variable, u. This function is chosen so that when it is differentiated it resembles the other function in the integrand.

Let $u = 2x^4 - 3$ then $\frac{du}{dx} = 8x^3$ and $\frac{dx}{du} = \frac{1}{8x^3}$

$$\int x^3(2x^4 - 3)^4 dx = \int x^3(2x^4 - 3)^4 \frac{dx}{du} du$$

$$= \int x^3 u^4 \frac{1}{8x^3} du \quad \text{Substitute } u = (2x^4 - 3)^4 \text{ and } \frac{dx}{du} = \frac{1}{8x^3}$$

$$= \int \frac{x^3}{8x^3} u^4 du$$

$$= \int \frac{1}{8} u^4 du \quad \text{Now integrate with respect to the variable u.}$$

$$= \frac{1}{40} u^5 + C$$

$$= \frac{1}{40}(2x^4 - 3)^5 + C$$

2. Problem:

Find $\int \sin^2(x) \cos(x) \, dx$

Solution:

Let $u = \sin(x)$ then $\frac{du}{dx} = \cos(x)$ and $\frac{dx}{du} = \frac{1}{\cos(x)}$

$$\begin{aligned}
\int \sin^2(x) \cos(x) \, dx &= \int \sin^2(x) \cos(x) \frac{dx}{du} \, du \\
&= \int u^2 \cos(x) \frac{1}{\cos(x)} \, du \\
&= \int u^2 \, du \\
&= \frac{1}{3} u^3 + C \\
&= \frac{1}{3} \sin^3(x) + C
\end{aligned}$$

Integration by substitution (substitution given) exercise 1   Go online

Q43: Find $\int 5x\sqrt{2 + 3x^2} \, dx$ using the substitution $u = 2 + 3x^2$

....

Q44: Find $\int \sin^3(x) \cos(x) \, dx$ using the substitution $u = \sin(x)$

....

Q45: Find $\int \frac{5x}{3x^2 + 7} \, dx$ using the substitution $u = 3x^2 + 7$

The previous examples have been straight forward in the sense that the substitution was given and it was clear how to simplify the integrand before integration.

This is not always the case and an extra step is needed before integrating. Study the next examples carefully and try to follow the steps of working.

Examples

1. Problem:

Find $\int 3x(x + 2)^5 \, dx$

Solution:

Let $u = x + 2$ then $\frac{du}{dx} = 1$ and $\frac{dx}{du} = 1$
Also since $u = x + 2$ then $x = u - 2$
This substitution makes it easier to multiply out the brackets.

TOPIC 3. INTEGRATION

Thus,
$$\int 3x(x+2)^5 dx = \int 3x(x+2)^5 \frac{dx}{du} du$$
$$= \int 3xu^5 du$$
$$= \int 3(u-2)u^5 du \quad \text{(u - 2 has been substituted for x)}$$
$$= \int 3u^6 - 6u^5 du$$
$$= \frac{3}{7}u^7 - u^6 + C$$
$$= \frac{1}{7}u^6(3u-7) + C$$
$$= \frac{1}{7}(x+2)^6(3(x+2) - 7) + C$$
$$= \frac{1}{7}(x+2)^6(3x-1) + C$$

2. Problem:

Find $\int 7x(3x-4)^{-2} dx$

Solution:

Let $u = 3x - 4$ then $\frac{du}{dx} = 3$ and $\frac{dx}{du} = \frac{1}{3}$
Also since $u = 3x - 4$ then $x = \frac{1}{3}(u + 4)$

Thus,
$$\int 7x(3x-4)^{-2} dx = \int 7x(3x-4)^{-2} \frac{dx}{du} du$$
$$= \int 7 \times \frac{1}{3}(u+4) \times u^{-2} \times \frac{1}{3} du$$
$$= \frac{7}{9} \int (u+4) u^{-2} du$$
$$= \frac{7}{9} \int u^{-1} + 4u^{-2} du$$
$$= \frac{7}{9} \int \frac{1}{u} + 4u^{-2} du$$
$$= \frac{7}{9}(\ln|u| + \frac{4u^{-1}}{-1}) + C$$
$$= \frac{7}{9}(\ln|u| - \frac{4}{u}) + C$$
$$= \frac{7}{9}(\ln|3x-4| - \frac{4}{3x-4}) + C$$

3. Problem:

Find $\int 4x(3x+2)^{-3} dx$ using the substitution $u = 3x - 2$.

Solution:

Let $u = 3x + 2$ then $\frac{du}{dx} = 3$ and $\frac{dx}{du} = \frac{1}{3}$

© HERIOT-WATT UNIVERSITY

Since $u = 3x + 2$, then $x = \frac{1}{3}(u-2)$

Thus,

$$\int 4x(3x+2)^{-3}dx = \int 4x(3x+2)^{-3}\frac{dx}{du}du$$

$$= \int 4 \times \frac{1}{3}(u-2) \times u^{-3} \times \frac{1}{3}du$$

$$= \frac{4}{9}\int (u-2)u^{-3}du$$

$$= \frac{4}{9}\int u^{-2} - 2u^{-3}du$$

$$= \frac{4}{9}\left(\frac{u^{-1}}{-1} - \frac{2u^{-2}}{-2}\right) + C$$

$$= \frac{4}{9}\left(-\frac{1}{u} + \frac{2}{u^2}\right) + C$$

$$= \frac{4}{9}\left(\frac{2}{u^2} - \frac{1}{u}\right) + C$$

$$= \frac{4}{9}\left(\frac{2}{u^2} - \frac{u}{u^2}\right) + C$$

$$= \frac{4}{9}\left(\frac{2-u}{u^2}\right) + C$$

Now substitute $u = 3x + 2$ to obtain the solution in terms of x.

$$\int 4x(3x+2)^{-3}dx = \frac{4}{9}\left(\frac{2-u}{u^2}\right) + C$$

$$= \frac{4}{9}\left(\frac{2-(3x+2)}{(3x+2)^2}\right) + C$$

$$= \frac{-4x}{3(3x+2)^2} + C$$

To integrate by the method of substitution we take the following steps:

1. Select the expression that is to be substituted for and write it equal to u.
2. Differentiate u with respect to x to get $\frac{du}{dx}$.
3. Invert $\frac{du}{dx}$ to get $\frac{dx}{du}$.
4. Substitute u and $\frac{dx}{du}$ into the integrand.
5. If possible simplify the integrand to eliminate all variables other than u.
6. If variables other than u appear in the integrand after simplifying, rearrange u for that variable and substitute back into the integrand so that only u appears.
7. Integrate

TOPIC 3. INTEGRATION

> **Integration by substitution (substitution given) exercise 2**   Go online
>
> Now try the following questions using the method of substitution. Remember to rearrange the substituted expression to allow simplification of the integrand.
>
> **Q46:** Find $\int 4x(x-5)^3\,dx$ using the substitution $u = x - 5$
>
> ....
>
> **Q47:** Find $\int 5x(2x+1)^{-4}\,dx$ using the substitution $u = 2x + 1$
>
> ....
>
> **Q48:** Find $\int x(x+6)^{-2}\,dx$ using the substitution $u = x + 6$

3.3.2 Definite integrals with substitution

Substitution can also be used for definite integrals. The limits of these integrals, however, are in terms of x not of the new variable. We therefore must change the limits and write them in terms of the new variable.

> **Key point**
>
> You must be careful to pay attention to the limits of the integral as these will change with a substitution.

Work through the following example, paying close attention to how the new limits are written in terms of the new variable u.

Example

Problem:

Find $\int_0^1 15x^2(x^3+1)^4\,dx$

Solution:

First select the expression that is to be replaced by u and find an expression for $\frac{dx}{du}$.

Let $u = x^3 + 1$ then $\frac{du}{dx} = 3x^2$ and $\frac{dx}{du} = \frac{1}{3x^2}$.

$$\int_0^1 15x^2(x^3+1)^4\,dx = \int 15x^2(x^3+1)^4 \frac{dx}{du}\,du$$
$$= \int 15x^2(u)^4 \frac{1}{3x^2}\,du$$
$$= \int 5u^4\,du$$

The limits of the original function are from 0 to 1.

When $x = 0$ then $u = 0^3 + 1 = 1$
When $x = 1$ then $u = 1^3 + 1 = 2$

So the limits after substitution are from $u = 1$ to $u = 2$.

Now integrate,

$$\int_0^1 15x^2(x^3+1)^4 dx = \int_1^2 5u^4 du$$
$$= \left[\frac{5u^5}{5}\right]_1^2$$
$$= \left[u^5\right]_1^2$$
$$= (2)^5 - (1)^5$$
$$= 32 - 1$$
$$= 31$$

> **Key point**
>
> Note that when evaluating a definite integral we do not replace u with the original expression in x.

Example

Problem:

Find $\int_0^{\frac{\pi}{4}} \tan(4x) dx$, using the substitution $u = \cos(4x)$

Solution:

To solve $\int_0^{\frac{\pi}{4}} \tan(4x) dx$ we need to recall the trigonometric identity $\tan(x) = \frac{\sin(x)}{\cos(x)}$

Before starting we will re-write the integrand using the trigonometric identity for $\tan(x)$.

$\int_0^{\frac{\pi}{4}} \frac{\sin(4x)}{\cos(4x)} dx$

Now select the expression that is to be replaced by u, then find an expression for $\frac{dx}{du}$.

Let $u = \cos(4x)$, then $\frac{du}{dx} = -4\sin(4x)$ and $\frac{dx}{du} = -\frac{1}{4\sin(4x)}$

Next we must work out the new limits for the new integrand.

The limits for the original function are from 0 to $\frac{\pi}{4}$

When $x = 0$ then $u = \cos(x) = 1$
When $x = \frac{\pi}{4}$ then $u = \cos\left(\frac{\pi}{4}\right) = \frac{1}{\sqrt{2}}$

Note here that the lower limit value is greater than the upper limit value i.e. $1 > \frac{1}{\sqrt{2}}$. This is substituted into the integral and evaluated as you have done before. So the limits after substitution are from $u = 1$ to $u = \frac{1}{\sqrt{2}}$.

TOPIC 3. INTEGRATION

$$\int_0^{\frac{\pi}{4}} \frac{\sin(4x)}{\cos(4x)} dx = \int_0^{\frac{1}{\sqrt{2}}} \frac{\sin(4x)}{u} \times \left(-\frac{1}{4\sin(4x)}\right) dx$$

$$= -\frac{1}{4} \int_0^{\frac{1}{\sqrt{2}}} \frac{1}{u} dx$$

$$= -\frac{1}{4} \left[\ln|u|\right]_1^{\frac{1}{\sqrt{2}}}$$

$$= -\frac{1}{4} \left(\ln\left|\frac{1}{\sqrt{2}}\right| - \ln|1|\right) \text{ Recall that } \ln|1| = 0$$

$$= -\frac{1}{4} \left(\ln\left|\frac{1}{\sqrt{2}}\right|\right) \text{ Recall the logarithmic rule } \ln\left|\frac{x}{y}\right| = \ln|x| - \ln|y|$$

$$= -\frac{1}{4} \left(\ln|1| - \ln\left|\sqrt{2}\right|\right)$$

$$= \frac{1}{4} \ln\left|\sqrt{2}\right| \text{ Recall the logarithmic rule } \ln|x|^n = n\ln|x|$$

$$= \frac{1}{4} \ln|2|^{\frac{1}{2}}$$

$$= \frac{1}{8} \ln|2|$$

Definite integrals with substitution exercise   Go online

Q49: Find $\int_0^{\frac{1}{4}} \frac{(1+2\sqrt{x})^3}{\sqrt{x}} dx$, using the substitution $u = 1 + 2\sqrt{x}$

....

Q50: Find $\int_0^2 (4x-3)e(2x^2 - 3x)dx$, using the substitution $u = 2x^2 - 3x$

....

Q51: Find $\int_0^{\frac{\pi}{4}} \sec^2(x)\tan^5(x)dx$, using the substitution $u = \tan(x)$
Remember: $\frac{d}{dx}\tan(ax+b) = \frac{1}{a}\sec^2(ax+b)$

....

Q52: Find $\int_0^{\frac{\pi}{3}} \tan(x)dx$, using the substitution $u = \cos(x)$

Definite integrals with substitution and trigonometric identity

Often a trigonometric substitution will make an integral much easier, especially when used along with a trigonometric identity. The following examples illustrate this.

Examples

1. Problem:

Find $\int_0^{\frac{3}{4}} \frac{x}{\sqrt{9-4x^2}} dx$ using $x = \frac{3}{2}\sin(u)$ for $0 \leqslant u \leqslant \frac{\pi}{2}$.

© HERIOT-WATT UNIVERSITY

Solution:

Notice that we have a different type of substitution this time.

Let $x = \frac{3}{2}\sin(u)$ then $\frac{dx}{du} = \frac{3}{2}\cos(u)$ and $x^2 = \frac{9}{4}\sin^2(u)$

The limits of the integral will also change,

When $x = 0$ then $\frac{3}{2}\sin(u) = 0$ and so $u = 0$
When $x = 3/4$ then $\frac{3}{2}\sin(u) = \frac{3}{4}$ and so $u = \frac{\pi}{6}$

This is found by,
$$\frac{3}{2}\sin(u) = \frac{3}{4}$$
$$\sin(u) = \frac{1}{2}$$
$$u = \sin^{-1}\left(\frac{1}{2}\right)$$
$$u = \frac{\pi}{6}$$

$$\int_0^{\frac{3}{4}} \frac{x}{\sqrt{9-4x^2}} dx = \int_0^{\frac{\pi}{6}} \frac{\frac{3}{2}\sin(u)}{\sqrt{9-4\left(\frac{3}{2}\sin(u)\right)^2}} \frac{3}{2}\sin(u) du$$

$$= \int_0^{\frac{\pi}{6}} \frac{\frac{3}{2}\sin(u)}{\sqrt{9-4\left(\frac{9}{4}\sin^2(u)\right)}} \frac{3}{2}\sin(u) du$$

$$= \int_0^{\frac{\pi}{6}} \frac{\frac{3}{2}\sin(u)}{\sqrt{9-9\sin^2(u)}} \frac{3}{2}\cos(u) du$$

$$= \int_0^{\frac{\pi}{6}} \frac{\frac{3}{2}\sin(u)}{\sqrt{9}\sqrt{1-\sin^2(u)}} \frac{3}{2}\cos(u) du$$

Remember $1 - \sin^2(u) = \cos^2(u)$

$$\int_0^{\frac{3}{4}} \frac{x}{\sqrt{9-4x^2}} dx = \int_0^{\frac{\pi}{6}} \frac{\frac{3}{2}\sin(u)}{\sqrt{3\cos^2(u)}} \frac{3}{2}\cos(u) du$$

$$= \int_0^{\frac{\pi}{6}} \frac{\frac{3}{2}\sin(u)}{3\cos(u)} \frac{3}{2}\cos(u) du$$

$$= \frac{9}{4}\int_0^{\frac{\pi}{6}} \frac{\sin(u)}{3\cos(u)} \cos(u) du$$

$$= \frac{3}{4}\int_0^{\frac{\pi}{6}} \frac{\sin(u)}{\cos(u)} \cos(u) du$$

$$= \frac{3}{4}\int_0^{\frac{\pi}{6}} \sin(u) du$$

TOPIC 3. INTEGRATION

Now solve the integral,

$$\int_0^{\frac{3}{4}} \frac{x}{\sqrt{9-4x^2}} dx = \frac{3}{4} \int_0^{\frac{\pi}{6}} \sin(u) du$$
$$= \frac{3}{4} [-\cos(u)]_0^{\frac{\pi}{6}}$$
$$= \frac{3}{4} \left(-\frac{\sqrt{3}}{2} + 1\right)$$
$$= \frac{3}{4} \left(1 - \frac{\sqrt{3}}{2}\right)$$

....

2. Problem:

Find $\int_0^{\frac{4}{\sqrt{2}}} \sqrt{16 - x^2} dx$ using $x = 4\sin(u)$ for $0 \leqslant u \leqslant \frac{\pi}{2}$

Solution:

Notice that we have a different type of substitution this time.

Let $x = 4\sin(u)$ then $\frac{dx}{du} = 4\cos(u)$ and $x^2 = 16\sin^2(u)$

The limits of the integral will also change,

When $x = 0$ then $4\sin(u) = 0$ and so $u = 0$
When $x = \frac{4}{\sqrt{2}}$ then $4\sin(u) = \frac{4}{\sqrt{2}}$ and so $u = \frac{\pi}{4}$

This is found by,

$$4\sin(u) = \frac{4}{\sqrt{2}}$$
$$\sin(u) = \frac{1}{\sqrt{2}}$$
$$u = \sin^{-1}\left(\frac{1}{\sqrt{2}}\right)$$
$$u = \frac{\pi}{4}$$

$$\int_0^{\frac{4}{\sqrt{2}}} \sqrt{16-x^2} \int_0 dx = \int_0^{\frac{\pi}{4}} \sqrt{16 - 16\sin^2(u)} \times 4\cos(u) du$$
$$= \int_0^{\frac{\pi}{4}} \sqrt{16}\sqrt{1 - \sin^2(u)} \times 4\cos(u) du$$
$$= 16 \int_0^{\frac{\pi}{4}} \sqrt{1 - \sin^2(u)} \times \cos(u) du$$

Remember $1 - \sin^2(u) = \cos^2(u)$

$$\int_0^{\frac{4}{\sqrt{2}}} \sqrt{16 - x^2} dx = 16 \int_0^{\frac{\pi}{4}} \sqrt{\cos^2(u)} \times \cos(u) du$$
$$= 16 \int_0^{\frac{\pi}{4}} \cos^2(u) du$$

© HERIOT-WATT UNIVERSITY

To be able to integrate $\cos^2 u$ we use the following double angle formula.
$\cos 2u = 2\cos^2 u - 1$ so $\cos^2 u = \frac{1}{2}(\cos 2u + 1)$

$\int_0^{\frac{4}{\sqrt{2}}} \sqrt{16 - x^2}\,dx = 16 \int_0^{\frac{\pi}{4}} \frac{1}{2}(\cos(2u) + 1)\,du$

Now solve the integral,

$$\int_0^{\frac{4}{\sqrt{2}}} \sqrt{16 - x^2}\,dx = 8 \int_0^{\frac{\pi}{4}} (\cos(2u) + 1)\,du$$

$$= 8 \left[\frac{1}{2}\sin(2u) + u\right]_0^{\frac{\pi}{4}}$$

$$= 8 \left\{\left(\frac{1}{2}\sin\left(\frac{\pi}{2}\right) + \frac{\pi}{4}\right) - \left(\frac{1}{2}\sin(0) + 0\right)\right\}$$

$$= 8\left(\frac{1}{2}(1) + \frac{\pi}{4}\right)$$

$$= 4 + 2\pi$$

Below is a reminder of the list of trigonometric identities that may be useful with the type of integration covered in this section. At National 5 and Higher you would have met the first trigonometric identity, but not the other two.

Trigonometric identities	Double angle formulae
$\cos^2(x) + \sin^2(x) = 1$ $1 + \tan^2(x) = \sec^2(x)$ $\cot^2(x) + 1 = \text{cosec}^2(x)$	$\cos(2x) = \cos^2(x) - \sin^2(x)$ $\quad\quad\quad = 1 - \sin^2(x)$ $\quad\quad\quad = 2\cos^2(x) - 1$ $\sin(2x) = 2\sin(x)\cos(x)$

Definite integrals with substitution and trigonometric identity exercise   Go online

Q53: Find $\int_0^{\frac{6}{\sqrt{5}}} \frac{1}{\sqrt{36-5x^2}}\,dx$, using the substitution $x = \frac{6}{\sqrt{5}}\sin(u)$ for $0 \leq u \leq \frac{\pi}{2}$.

....

Q54: Find $\int_0^{\frac{3}{2}} \frac{1}{9+4x^2}\,dx$, using the substitution $x = \frac{3}{2}\tan(u)$.

....

Q55: Find $\int_0^{\frac{1}{4}} \sqrt{1 - 4x^2}\,dx$, using the substitution $x = \frac{1}{2}\sin(t)$.

3.3.3 Integration by substitution (substitution not given)

In the previous sections, the substitution for integration was given, however, with practice it becomes easier to recognise which substitution to use. Becoming familiar with the different types of integration will make it easier.

TOPIC 3. INTEGRATION

The following examples may help:

Example: When $\int f(ax + b) \, dx$ **substitute** $u = ax + b$

Problem:

Calculate $\int \sin(3x + 2) \, dx$

Solution:

Let $u = 3x + 2$ then $\frac{du}{dx} = 3$ and $\frac{dx}{du} = \frac{1}{3}$

Hence

$$\int \sin(3x + 2) \, dx = \int \sin u \frac{dx}{du} \, du$$
$$= \int \frac{\sin u}{3} \, du$$
$$= -\frac{1}{3} \cos u + C$$
$$= -\frac{1}{3} \cos(3x + 2) + C$$

Note that the example above could have been integrated using the 'reverse chain rule' method as you would have seen at Higher. We did not have to use substitution to work out the answer. The same is true for: $\int (5x - 2)^4 \, dx$

Examples

1. When $\int f'(x) \, [f(x)]^n \, dx$ **substitute** $u = f(x)$, where the derivative of u can be a multiple of $f(x)$

In words this rule says that when the function being raised to a power, $f(x)$, is multiplied by its derivative, $f'(x)$, or a multiple of its derivative, then substitute for $f(x)$ i.e. let $u = f(x)$.

Problem:

Calculate $\int x^4 (x^5 - 6)^3 \, dx$

Solution:

Let $u = x^5 - 6$ then $\frac{du}{dx} = 5x^4$ and $\frac{dx}{du} = \frac{1}{5x^4}$

Hence

$$\int x^4 (x^5 - 6)^3 \, dx = \int x^4 u^3 \frac{dx}{du} \, du$$
$$= \int x^4 u^3 \frac{1}{5x^4} \, du$$
$$= \int \frac{1}{5} u^3 \, du$$
$$= \frac{u^4}{20} + C$$
$$= \frac{1}{20}(x^5 - 6)^4 + C$$

© HERIOT-WATT UNIVERSITY

2. When $\int (ax + b)(cx + d)^n \, dx$ substitute $u = cx + d$

Problem:

Calculate $\int x(4x - 3)^5 \, dx$

Solution:

In this example the substitution makes it easier to multiply out the brackets.

Let $u = 4x - 3$ then $x = \frac{1}{4}(u + 3)$

Also, $\frac{du}{dx} = 4$ and $\frac{dx}{du} = \frac{1}{4}$

Hence,

$$\begin{aligned}
\int x(4x - 3)^5 \, dx &= \int \frac{1}{4}(u+3) u^5 \frac{dx}{du} du \\
&= \int \frac{1}{4}(u+3) u^5 \frac{1}{4} du \\
&= \frac{1}{16} \int u^6 + 3u^5 \, du \\
&= \frac{1}{16} \left(\frac{u^7}{7} + \frac{3u^6}{6} \right) + C \\
&= \frac{1}{16} \left(\frac{u^7}{7} + \frac{u^6}{2} \right) + C \\
&= \frac{1}{16} \left(\frac{2u^7}{14} + \frac{7u^6}{14} \right) + C \\
&= \frac{1}{224} u^6 (2u + 7) + C \\
&= \frac{1}{224} (4x - 3)^6 (2(4x - 3) + 7) + C \\
&= \frac{1}{224} (4x - 3)^6 (8x + 1) + C
\end{aligned}$$

....

3. When $\int f'(x) \, e^{f(x)} \, dx$ substitute $u = f(x)$

In words this rule says that if the exponential is multiplied by the derivative of $f(x)$, or a multiple of the derivative of $f(x)$, then substitute for $f(x)$ i.e. let $u = f(x)$.

Problem:

Calculate $\int x e^{(x^2 - 5)} \, dx$

Solution:

Let $u = x^2 - 5$ then $\frac{du}{dx} = 2x$ and $\frac{dx}{du} = \frac{1}{2x}$

Hence

$$\begin{aligned}
\int x e^{(x^2 - 5)} \, dx &= \int x e^u \frac{dx}{du} du \\
&= \int x e^u \frac{1}{2x} du \\
&= \int \frac{1}{2} e^u \, du \\
&= \frac{1}{2} e^u + C \\
&= \frac{1}{2} e^{(x^2 - 5)} + C
\end{aligned}$$

TOPIC 3. INTEGRATION

4. When $\int {f'(x)}/{f(x)} \, dx$ **substitute** $u = f(x)$

In words this rule says that if the numerator can be written as the derivative of the denominator, or a multiple of the derivative then substitute for the denominator, $f(x)$ i.e. let $u = f(x)$.

Problem:

Calculate $\int \frac{6x}{x^2 + 3} \, dx$

Solution:

Let $u = x^2 + 3$, then $\frac{du}{dx} = 2x$ and $\frac{dx}{du} = \frac{1}{2x}$

Hence

$$\int \frac{6x}{x^2+3} \, dx = \int \frac{6x}{u} \frac{dx}{du} \, du$$
$$= \int \frac{6x}{u} \frac{1}{2x} \, du$$
$$= \int \frac{3}{u} \, du$$
$$= 3\ln|u| + C$$
$$= 3\ln\left|x^2 + 3\right| + C$$

Key point

Integration by substitution:

- When $\int f(ax + b) \, dx$ substitute using $u = ax + b$
- When $\int f'(x) [f(x)]^n dx$ substitute using $u = f(x)$
- When $\int (ax + b)(cx + d)^n dx$ substitute using $u = cx + d$
- When $\int f'(x) e^{f(x)} dx$ substitute using $u = f(x)$
- When $\int \frac{f'(x)}{f(x)} dx$ substitute using $u = f(x)$

Integration by substitution (substitution not given) exercise   Go online

For the integrals given identify the correct type of integrand.

Q56: $\int x^2 (4x^3 + 3)^2 \, dx$

a) $\int f(ax+b) \, dx$
b) $\int f'(x)[f(x)]^n \, dx$
c) $\int (ax+b)(cx+d)^n \, dx$
d) $\int f'(x) e^{f(x)} \, dx$
e) $\int \frac{f'(x)}{f(x)} \, dx$

© HERIOT-WATT UNIVERSITY

Q57: $\int \frac{2x+3}{2x^2+6x-1} \, dx$

a) $\int f(ax+b) \, dx$
b) $\int f'(x)[f(x)]^n \, dx$
c) $\int (ax+b)(cx+d)^n \, dx$
d) $\int f'(x)e^{f(x)} \, dx$
e) $\int \frac{f'(x)}{f(x)} \, dx$

....

Q58: $\int 5\cos(4x+1) \, dx$

a) $\int f(ax+b) \, dx$
b) $\int f'(x)[f(x)]^n \, dx$
c) $\int (ax+b)(cx+d)^n \, dx$
d) $\int f'(x)e^{f(x)} \, dx$
e) $\int \frac{f'(x)}{f(x)} \, dx$

....

Q59: $\int 3xe^{(x^2+4)} \, dx$

a) $\int f(ax+b) \, dx$
b) $\int f'(x)[f(x)]^n \, dx$
c) $\int (ax+b)(cx+d)^n \, dx$
d) $\int f'(x)e^{f(x)} \, dx$
e) $\int \frac{f'(x)}{f(x)} \, dx$

....

Q60: $\int (3x+1)(2x-7)^3 \, dx$

a) $\int f(ax+b) \, dx$
b) $\int f'(x)[f(x)]^n \, dx$
c) $\int (ax+b)(cx+d)^n \, dx$
d) $\int f'(x)e^{f(x)} \, dx$
e) $\int \frac{f'(x)}{f(x)} \, dx$

Perform the following integrations, using a suitable substitution.

Q61: $\int 2\sec^2(4x-5) \, dx$

....

Q62: $\int (2x^2+x)(4x^3+3x^2+5)^5 \, dx$

....

Q63: $\int (x+4)(2x-3)^4 \, dx$

....

Q64: $\int 4xe^{(x^2-3)} \, dx$

....

Q65: $\int \frac{2x-4}{2x^2-8x+3} \, dx$

3.4 Integration involving inverse trigonometric functions and completing the square

In the topic of differentiation we saw the derivatives of the inverse trigonometric functions. This section will look at the anti-derivatives and the process of integrating to get back to the inverse trigonometric functions. In some cases, completing the square is necessary before integrating. Each case will be explored.

3.4.1 Integration involving inverse trigonometric functions

From the topic of differentiation we learned the standard derivative results for the inverse trigonometric functions $\sin^{-1}(x)$ and $\tan^{-1}(x)$.

These are given on the formula sheet.

$\frac{d}{dx}(\sin^{-1} x) = \frac{1}{\sqrt{1-x^2}}$

$\frac{d}{dx}(\tan^{-1} x) = \frac{1}{1+x^2}$

$\frac{d}{dx}\left(\sin^{-1}\left(\frac{x}{a}\right)\right) = \frac{1}{\sqrt{a^2-x^2}} + C$

$\frac{d}{dx}\left(\tan^{-1}\left(\frac{x}{a}\right)\right) = \frac{a}{a^2+x^2} + C$

It therefore follows that:

$\int \frac{1}{\sqrt{1-x^2}}\, dx = \sin^{-1} x + C$

$\int \frac{1}{1+x^2}\, dx = \tan^{-1} x + C$

$\int \frac{1}{\sqrt{a^2-x^2}}\, dx = \sin^{-1}\left(\frac{x}{a}\right) + C$

$\int \frac{1}{a^2+x^2}\, dx = \frac{1}{a}\tan^{-1}\left(\frac{x}{a}\right) + C$

These examples demonstrate the use of the standard results. Note that often a little algebra may be necessary before applying the above results.

Read through the examples carefully, paying close attention to how the denominator is factorised by taking a common factor.

Examples

1. Problem:

Find $\int \frac{dx}{4x^2+1}$

Solution:

$$\int \frac{dx}{4x^2+1} = \int \frac{dx}{4(x^2+\frac{1}{4})}$$

Take a common factor so that the constant in front of x^2, on the denominator, is 1

$$= \frac{1}{4} \int \frac{dx}{x^2+(\frac{1}{2})^2}$$

Now apply the standard result and simplify

$$= \frac{1}{4} \frac{1}{\frac{1}{2}} \tan^{-1}\left(\frac{x}{\frac{1}{2}}\right) + C$$

$$= \frac{1}{2} \tan^{-1}(2x) + C$$

2. Problem:

Find $\int \frac{dx}{\sqrt{9-25x^2}}$

Solution:

$$\int \frac{dx}{\sqrt{9-25x^2}} = \int \frac{dx}{\sqrt{25\left(\frac{9}{25}-x^2\right)}}$$

Take a common factor so that the constant in front of x^2, on the denominator, is 1

$$= \frac{1}{5} \int \frac{dx}{\sqrt{\left(\frac{3}{5}\right)^2 - x^2}}$$

...

$$\int \frac{dx}{\sqrt{9-25x^2}} = \cdots$$

Now apply the standard result and simplify

$$= \frac{1}{5} \sin^{-1}\left(\frac{x}{\frac{3}{5}}\right) + C$$

$$= \frac{1}{5} \sin^{-1}\left(\frac{5x}{3}\right) + C$$

3.
Problem:

Find $\int_0^4 \frac{9}{3x^2+16} dx$

Just as in previous integration questions we can evaluate definite integrals as shown in this example.

Solution:

$$\int_0^4 \frac{9}{3x^2+16}dx = \int_0^4 \frac{9}{3\left(x^2+\frac{16}{3}\right)}dx$$

Take a common factor so that the constant in front of x^2, on the denominator, is 1

$$=3\int_0^4 \frac{1}{x^2+\left(\frac{4}{\sqrt{3}}\right)^2}dx$$

Now apply the standard result and simplify

$$=\left[3\times\frac{1}{\frac{4}{\sqrt{3}}}tan^{-1}\left(\frac{x}{\frac{4}{\sqrt{3}}}\right)\right]_0^4$$

$$=\left[3\frac{\sqrt{3}}{4}\tan^{-1}\left(\frac{x}{\frac{4}{\sqrt{3}}}\right)\right]_0^4$$

$$=\left[\frac{3\sqrt{3}}{4}\tan^{-1}\left(\frac{\sqrt{3}x}{4}\right)\right]_0^4$$

$$=\left(\frac{3\sqrt{3}}{4}\tan^{-1}\left(\frac{\sqrt{3}(4)}{4}\right)\right)-\left(\frac{3\sqrt{3}}{4}\tan^{-1}\left(\frac{\sqrt{3}(0)}{4}\right)\right)$$

$$=\left(\frac{3\sqrt{3}}{4}\tan^{-1}\left(\sqrt{3}\right)\right)-\left(\frac{3\sqrt{3}}{4}\tan^{-1}(0)\right)$$

$$=\frac{3\sqrt{3}}{4}\frac{\pi}{3}-0$$

$$=\frac{\sqrt{3}}{4}\pi$$

Key point

$\int \frac{1}{\sqrt{1-x^2}}\,dx = \sin^{-1}x + C$ and $\int \frac{1}{\sqrt{a^2-x^2}}\,dx = \sin^{-1}\left(\frac{x}{a}\right) + C$

$\int \frac{1}{1+x^2}\,dx = \tan^{-1}x + C$ and $\int \frac{1}{a^2+x^2}\,dx = \frac{1}{a}\tan^{-1}\left(\frac{x}{a}\right) + C$

Integration involving inverse trigonometric functions exercise   Go online

Q66: Find $\int \frac{4}{\sqrt{64-x^2}}\,dx$

....

Q67: Find $\int \frac{6}{36+4x^2}\,dx$

....

Q68: Find $\int_0^{\frac{3}{4}} \frac{dx}{\sqrt{9-4x^2}}$

....

238 TOPIC 3. INTEGRATION

Q69: For the following question recall that to rationalise a surd we multiply by the surd from the denominator e.g. $\frac{2}{3\sqrt{5}} \times \frac{\sqrt{5}}{\sqrt{5}} = \frac{2\sqrt{5}}{15}$

Find $\int_0^3 \frac{16}{3x^2+9} dx$

3.4.2 Integration involving inverse tangent and logarithms

We will now consider further examples which often look to have inverse tangent solutions but instead have logarithmic solutions or a combination of both.

First we examine how to find any integral of the form: $\int \frac{bx+c}{x^2+a^2} dx$

> **Top tip**
>
> Note that the numerator is a polynomial with a degree of one less than the denominator.

To integrate we make use of the formula: $\int \frac{f'(x)}{f(x)} dx = \ln|f(x)| + C$

In particular, for example: $\int \frac{2x}{x^2+4} dx = \ln|x^2+4| + C$

We can now find any integral of the form $\int \frac{bx+c}{x^2+a^2} dx$ by writing the integrand as a multiple of $\frac{2x}{x^2+a^2}$ + a multiple of $\frac{1}{x^2+a^2}$.

Examples

1. Problem:

Find $\int \frac{x-3}{x^2+4} dx$

Solution:

To make this integration possible, split the original expression into two separate fractions.

$$\int \frac{x-3}{x^2+4} dx = \int \frac{x}{x^2+4} dx - \int \frac{3}{x^2+4} dx$$
$$= \frac{1}{2} \int \frac{2x}{x^2+4} dx - 3 \int \frac{1}{x^2+2^2} dx$$
$$= \frac{1}{2} \ln|x^2+4| - \frac{3}{2} \tan^{-1}\left(\frac{x}{2}\right) + C$$

2. Problem:

Find $\int \frac{x+5}{x^2+9} dx$

Solution:

$$\int \frac{x+5}{x^2+9} dx = \int \frac{x}{x^2+9} dx + \int \frac{5}{x^2+9} dx$$
$$= \frac{1}{2} \int \frac{2x}{x^2+9} dx + 5 \int \frac{1}{x^2+3^2} dx$$
$$= \frac{1}{2} \ln|x^2+9| + \frac{5}{3} \tan^{-1}\left(\frac{x}{3}\right) + C$$

© HERIOT-WATT UNIVERSITY

TOPIC 3. INTEGRATION

As was observed in the previous two examples, when the integrand was split up the first term integrated to a logarithm and the second integrated to an inverse tangent function.

> **Key point**
>
> For the term that integrates to a logarithm it is necessary for the numerator to become the derivative of the denominator.

The following examples demonstrate how this is done.

Examples

1. Problem:

Find $\int \frac{2x+3}{9x^2+4}\,dx$

Solution:

Once split into two separate fractions, the first term will integrate to a logarithm so the numerator must become the derivative of the denominator.

Take the denominator of the first integrand: $9x^2 + 4$ and differentiate to get $18x$.

We must transform the numerator of this integrand, $2x$, into $18x$.

We do this by multiplying the integrand by $\frac{18}{18}$ and taking a factor of $\frac{2}{18}$ to the front of the integral sign.

$$\int \frac{2x+3}{9x^2+4}\,dx = \int \frac{2x}{9x^2+4}\,dx + \int \frac{3}{9x^2+4}\,dx$$

$$= \frac{2}{18}\int \frac{18x}{9x^2+4}\,dx + 3\int \frac{1}{9x^2+2^2}\,dx$$

$$= \frac{1}{9}\int \frac{18x}{9x^2+4}\,dx + \frac{3}{9}\int \frac{1}{x^2+\left(\frac{2}{3}\right)^2}\,dx$$

$$= \frac{1}{9}\ln\left|9x^2+4\right| + \frac{1}{3}\times\frac{3}{2}\tan^{-1}\left(\frac{3x}{2}\right) + C$$

$$= \frac{1}{9}\ln\left|9x^2+4\right| + \frac{1}{2}\tan^{-1}\left(\frac{3x}{2}\right) + C$$

....

2. Problem:

Find $\int \frac{5x-2}{3x^2+25}\,dx$

Solution:

Once split into two separate fractions, the first term will integrate to a logarithm so the numerator must become the derivative of the denominator.

Take the denominator of the first integrand: $3x^2 + 25$ and differentiate to get $6x$.

We must transform the numerator of this integrand, $5x$, into $6x$.

We do this by multiplying the integrand by $\frac{6}{6}$ and taking a factor of $\frac{5}{6}$ to the front of the integral sign.

© HERIOT-WATT UNIVERSITY

$$\int \frac{5x-2}{3x^2+25}dx = \int \frac{5x}{3x^2+25}dx - \int \frac{2}{3x^2+25}dx$$
$$= \frac{5}{6}\int \frac{6x}{3x^2+25}dx - \frac{2}{3}\int \frac{1}{x^2+\frac{25}{3}}dx$$
$$= \frac{5}{6}\int \frac{6x}{3x^2+25}dx - \frac{2}{3}\int \frac{1}{x^2+\left(\frac{5}{\sqrt{3}}\right)^2}dx$$
$$= \frac{5}{6}\ln|3x^2+25| - \frac{2}{3} \times \frac{\sqrt{3}}{5}\tan^{-1}\left(\frac{\sqrt{3}x}{5}\right) + C$$
$$= \frac{5}{6}\ln|3x^2+25| - \frac{2\sqrt{3}}{15}\tan^{-1}\left(\frac{\sqrt{3}x}{5}\right) + C$$

Integration involving inverse tangent and logarithms exercise Go online

Q70: Find $\int \frac{x-1}{x^2+4} dx$

Q71: Find $\int \frac{x+3}{x^2+16} dx$

Q72: Find $\int \frac{x-7}{x^2+1} dx$

Q73: Find $\int \frac{x-1}{4x^2+36} dx$

Q74: Find $\int \frac{3x+2}{5x^2+4} dx$

3.4.3 Integration involving inverse tangent and completing the square

Lastly, in this section we see how to find any integral of the form $\int \frac{dx}{ax^2+bx+c}$ where the quadratic in the denominator does not factorise (i.e. $b^2 - 4ac < 0$).

We do this by 'completing the square' in the denominator and then making an easy use of substitution to obtain an integral which we can identify as an inverse trigonometric function.

Examples

1. Problem:

Evaluate $\int_2^7 \frac{dx}{x^2-4x+29}$

TOPIC 3. INTEGRATION

Solution:
Step 1: First we complete the square for the denominator.
$$x^2 - 4x + 29 = (x^2 - 4x) + 29$$
$$= (x - 2)^2 - (-2)^2 + 29$$
$$= (x - 2)^2 - 4 + 29$$
$$= (x - 2)^2 + 25$$

Therefore the integral becomes: $\int \frac{dx}{x^2 - 4x + 29} = \int \frac{dx}{(x-2)^2 + 25}$

Step 2: Apply the method of substitution, remembering to change the limits to the new variable.

Let $u = x - 2$ then $\frac{du}{dx} = 1$ and $\frac{dx}{du} = 1$

When $x = 2$ then $u = 0$.
When $x = 7$ then $u = 5$.

Therefore the integral becomes:

$$\int_2^7 \frac{dx}{x^2 - 4x + 29} = \int_2^7 \frac{dx}{(x-2)^2 + 25}$$
$$= \int_0^5 \frac{1}{(x-2)^2 + 25} \frac{dx}{du} du$$
$$= \int_0^5 \frac{1}{u^2 + 25} \times 1 \, du$$
$$= \int_0^5 \frac{du}{u^2 + 25}$$

Step 3: Integrate with respect to u.

$$\int_2^7 \frac{dx}{x^2 - 4x + 29} = \int_0^5 \frac{du}{u^2 + 25}$$
$$= \left[\frac{1}{5} \tan^{-1} \left(\frac{u}{5} \right) \right]_0^5$$
$$= \left(\frac{1}{5} \tan^{-1} \left(\frac{5}{5} \right) \right) - \left(\frac{1}{5} \tan^{-1} \left(\frac{0}{5} \right) \right)$$
$$= \frac{1}{5} \times \frac{\pi}{4}$$
$$= \frac{\pi}{20}$$

....

2. Problem:
Find $\int \frac{dx}{x^2 + 2x + 4}$

Solution:
Step 1: First we complete the square for the denominator.
$$x^2 + 2x + 4 = (x^2 + 2x) + 4$$
$$= (x + 1)^2 - 1^2 + 4$$
$$= (x + 1)^2 - 1 + 4$$
$$= (x + 1)^2 + 3$$

Therefore the integral becomes: $\int \frac{dx}{x^2 + 2x + 4} = \int \frac{dx}{(x+1)^2 + 3}$

Step 2: Now apply the method of substitution.
Let $u = x + 1$ then $\frac{du}{dx} = 1$ and $\frac{dx}{du} = 1$
Therefore the integral becomes:

$$\int \frac{dx}{x^2 + 2x + 4} = \int \frac{dx}{(x+1)^2 + 3}$$
$$= \int \frac{1}{(x+1)^2 + 3} \frac{dx}{du} du$$
$$= \int \frac{1}{u^2 + 3} \times 1 \, du$$
$$= \int \frac{du}{u^2 + 3}$$

Step 3: Now integrate with respect to u using the standard results.

$$\int \frac{dx}{x^2 + 2x + 4} = \int \frac{du}{u^2 + 3}$$
$$= \frac{1}{\sqrt{3}} \tan^{-1}\left(\frac{u}{\sqrt{3}}\right)$$

Step 4: Change back to the original variable x.

$$\int \frac{dx}{x^2 + 2x + 4} = \frac{1}{\sqrt{3}} \tan^{-1}\left(\frac{u}{\sqrt{3}}\right)$$
$$= \frac{1}{\sqrt{3}} \tan^{-1}\left(\frac{x+1}{\sqrt{3}}\right)$$

Integration involving inverse tangent and completing the square Go online
exercise

Q75: Find $\int_3^4 \frac{1}{x^2 - 6x + 12} dx$

....

Q76: Find $\int_{-4}^{-3} \frac{7}{x^2 + 8x + 19} dx$

Challenge question 1

This is a challenge activity. It involves a combination of the techniques given in this section.
Find $\int \frac{x-2}{x^2 + 2x + 5} \, dx$

TOPIC 3. INTEGRATION

3.5 Partial fractions and integration

This section examines how certain integrals can be found by using the sum of partial fractions.

In general partial fractions are useful for evaluating integrals of the form $\int \frac{f(x)}{g(x)} dx$.

Where $f(x)$ and $g(x)$ are polynomials and $g(x)$ can be factorised into the product of linear and quadratic factors.

In this case expressing $\frac{f(x)}{g(x)}$ in partial fractions will enable us to carry out the integration. When $f(x)$ and $g(x)$ are polynomials then $\frac{f(x)}{g(x)}$ is called a rational function.

To be able to proceed with this section a good working knowledge of Partial Fractions is needed. To revisit these techniques, go to the Partial Fractions topic before continuing.

3.5.1 Distinct linear factors

In this section techniques to integrate functions such as $\int \frac{x+7}{(x-1)(x+3)} dx$ are examined.

The integrand has a denominator which is the product of linear factors.

For $\int f(x) dx = F(x) + C$, $f(x)$ is the integrand.

> **Key point**
>
> To integrate a rational function with linear factors on the denominator and a constant on the numerator we use the following standard result after writing the function as the sum of partial fractions:
>
> $$\int \frac{dx}{ax+b} = \frac{1}{a} \ln|ax+b| + C$$

The example below demonstrates the use of partial fractions and the standard result above to integrate a rational function with linear factors on the denominator.

Example

Problem:

Find the indefinite integral $\int \frac{x+7}{(x-1)(x+3)} dx$

Solution:

Step 1: Rewrite $\frac{x+7}{(x-1)(x+3)}$ as partial fractions.

$\frac{x+7}{(x-1)(x+3)} = \frac{A}{(x-1)} + \frac{B}{(x+3)}$

Step 2: Obtain a common denominator and equate the numerators.

$x + 7 = A(x+3) + B(x-1)$

Step 3: Find the values of A and B.

Let $x = -3$ then $4 = 0 \times A + (-4) \times B$
$B = -1$

© HERIOT-WATT UNIVERSITY

Let $x = 1$ then $8 = 4 \times A + 0 \times B$
$A = 2$

Step 4: Substitute the values of A and B into the equation.
$$\frac{x+7}{(x-1)(x+3)} = \frac{2}{(x-1)} - \frac{1}{(x+3)}$$

Step 5: Now find the integrand.
$$\int \frac{x+7}{(x-1)(x+3)} dx = \int \frac{2}{(x-1)} dx - \int \frac{1}{(x+3)} dx$$
$$= 2\ln|x-1| - \ln|x+3| + C$$

Key point

Note that:

i. This method will work for any integral of the form $\int \frac{cx+d}{ax^2+bx+c} dx$, where $ax^2 + bx + c$ can be factorised.

ii. The method can be extended to integrals such as $\int \frac{9}{(x-1)(x+2)(2x+1)} dx$, where the denominator is a cubic function written as three separate linear factors.

Distinct linear factors exercise   Go online

Q77: Find $\int \frac{9}{2x^3 + 3x^2 - 3x - 2} dx$

....

Q78: Find $\int \frac{-x^2+x+22}{x^3+3x^2-4x-12} dx$

3.5.2 Repeated linear factors

This section examines how to integrate rational functions which have repeated linear factors in the denominator, such as $\int \frac{x+5}{(2x+1)(x-1)^2} dx$

Key point

When a rational function is split into the sum of partial fractions with repeated linear factors on the denominator the following result is used:

$$\int \frac{1}{(ax+b)^n} dx = \frac{1}{a(1-n)} \times \frac{1}{(ax+b)^{n-1}} + C$$

The next example should make this clear.

TOPIC 3. INTEGRATION

Examples

1. Problem:
Find the indefinite integral $\int \frac{x+5}{(2x+1)(x-1)^2} dx$

Solution:
Step 1: Rewrite the indefinite integral as partial fractions.
$$\frac{x+5}{(2x+1)(x-1)^2} = \frac{A}{(2x+1)} + \frac{B}{(x-1)} + \frac{C}{(x-1)^2}$$

Step 2: Obtain a common denominator and equate numerators.
$$\frac{x+5}{(2x+1)(x-1)^2} = \frac{A(x-1)^2}{(2x+1)(x-1)^2} + \frac{B(2x+1)(x-1)}{(2x+1)(x-1)^2} + \frac{C(2x+1)}{(2x+1)(x-1)^2}$$
$$x+5 = A(x-1)^2 + B(2x+1)(x-1) + C(2x+1)$$

Step 3: Find the values of A, B and C by selecting values for x.

Let $x = 1$ then,
$$1 + 5 = A(1-1)^2 + B(2(1)+1)(1-1) + C(2(1)+1)$$
$$6 = C(3)$$
$$C = 2$$

Given $C = 2$, let $x = -\frac{1}{2}$ then,
$$-\frac{1}{2} + 5 = A\left(-\frac{1}{2}-1\right)^2 + B\left(2\left(-\frac{1}{2}\right)+1\right)\left(-\frac{1}{2}-1\right) + 2\left(2\left(-\frac{1}{2}\right)+1\right)$$
$$\frac{9}{2} = A\left(-\frac{3}{2}\right)^2$$
$$\frac{9}{2} = \frac{9}{4}A$$
$$A = 2$$

Given $A = 2$ and $C = 2$, let $x = 0$ then,
$$0 + 5 = 2(0-1)^2 + B(2(0)+1)(0-1) + 2(2(0)+1)$$
$$5 = 2 + B(1)(-1) + 2(1)$$
$$5 = 2 - B + 2$$
$$1 = -B$$
$$B = -1$$

Step 4: Substitute for $A = 2$, $B = -1$ and $C = 2$ back into the sum of partial fractions.
$$\frac{x+5}{(2x+1)(x-1)^2} = \frac{2}{2x+1} - \frac{1}{x-1} + \frac{2}{(x-1)^2}$$

Step 5: Integrate, begin by rewriting into a form that can be integrated.
$$\int \frac{x+5}{(2x+1)(x-1)^2} dx = \int \frac{2}{(2x+1)} dx - \int \frac{1}{(x-1)} dx + \int \frac{2}{(x-1)^2} dx$$
$$= 2\int \frac{1}{(2x+1)} dx - \int \frac{1}{(x-1)} dx + 2\int (x-1)^{-2} dx$$
$$= \frac{2}{2} \ln|2x+1| - \ln|x-1| + \frac{2}{(-1)}(x-1)^{-1} + C$$
$$= \ln|2x+1| - \ln|x-1| - \frac{2}{(x-1)} + C$$

© HERIOT-WATT UNIVERSITY

2. Problem:

Evaluate $\int_0^1 \frac{2x^2+7x+9}{x^3+6x^2+12x+8}dx$

Solution:

Step 1: Factorise the denominator using synthetic division.

Try $x = -2$

```
-2 | 1   6   12   8
   | 0  -2   -8  -8
   ----------------
     1   4    4   0
```

Since the remainder is zero, this means that $x = -2$ is a root of the cubic and so $(x+2)$ is a factor.

We therefore have,

$$x^3 + 6x^2 + 12x + 8 = (x+2)\left(x^2 + 4x + 4\right)$$
$$= (x+2)(x+2)^2$$
$$= (x+2)^3$$

Step 2: Rewrite the integrand as the sum of partial fractions.

$\frac{2x^2+7x+9}{(x+2)^3} = \frac{A}{x+2} + \frac{B}{(x+2)^2} + \frac{C}{(x+2)^3}$

Step 3: Obtain a common denominator and equate numerators.

$$\frac{2x^2+7x+9}{(x+2)^3} = \frac{A(x+2)^2}{(x+2)^3} + \frac{B(x+2)}{(x+2)^3} + \frac{C}{(x+2)^3}$$

$$2x^2 + 7x + 9 = A(x+2)^2 + B(x+2) + C$$

Step 4: Find the values of A, B and C by selecting values for x.

Let $x = -2$ then,

$$2(-2)^2 + 7(-2) + 9 = A(-2+2)^2 + B(-2+2) + C$$
$$8 - 14 + 9 = C$$
$$3 = C$$
$$C = 3$$

Given $C = 3$, let $x = 0$

$$2(0)^2 + 7(0) + 9 = A(0+2)^2 + B(0+2) + 3$$
$$9 = 4A + 2B + 3$$
$$6 = 4A + 2B$$
$$3 = 2A + B$$

We could leave this equation as it is and use simultaneous equations after we generate the second equation, or we could rearrange and substitute. For this example we will rearrange and substitute.

TOPIC 3. INTEGRATION

$B = 3 - 2A$
Given $B = 3 - 2A, C = 3$, let $x = 1$
$2(1)^2 + 7(1) + 9 = A(1+2)^2 + (3-2A)(1+2) + 3$
$\qquad 2 + 7 + 9 = 9A + (3-2A)(3) + 3$
$\qquad\qquad 18 = 9A + 9 - 6A + 3$
$\qquad\qquad 18 = 3A + 12$
$\qquad\qquad 6 = 3A$
$\qquad\qquad A = 2$
Substitute $A = 2$ back into $B = 3 - 2A$:
$B = 3 - 2(2)$
$B = -1$

Step 4: Substitute $A = 2, B = -1$ and $C = 3$ back into the sum of partial fractions
$\frac{2x^2+7x+9}{(x+2)^3} = \frac{2}{x+2} - \frac{1}{(x+2)^2} + \frac{3}{(x+2)^3}$

Step 5: Integrate, begin by rewriting in a form that can be integrated.

$\int_0^1 \frac{2x^2+7x+9}{x^3+6x^2+12x+8} dx = \int_0^1 \frac{2}{x+2} dx - \int_0^1 \frac{1}{(x+2)^2} dx + \int_0^1 \frac{3}{(x+2)^3} dx$

$\qquad = 2\int_0^1 \frac{1}{x+2} dx - \int_0^1 (x+2)^{-2} dx + 3\int_0^1 (x+2)^{-3} dx$

$\qquad = 2\left[\ln|x+2|\right]_0^1 + \left[(x+2)^{-1}\right]_0^1 + 3\left[-\frac{1}{2}(x+2)^{-2}\right]_0^1$

$\qquad = 2\left[\ln|x+2|\right]_0^1 + \left[(x+2)^{-1}\right]_0^1 - \frac{3}{2}\left[(x+2)^{-2}\right]_0^1$

$\qquad = 2\left[\ln|x+2|\right]_0^1 + \left[\frac{1}{x+2}\right]_0^1 - \frac{3}{2}\left[\frac{1}{(x+2)^2}\right]_0^1$

$\qquad = 2(\ln|3| - \ln|2|) + \left(\frac{1}{3} - \frac{1}{2}\right) - \frac{3}{2}\left(\frac{1}{9} - \frac{1}{4}\right)$

$\qquad = 2\ln\left|\frac{3}{2}\right| - \frac{1}{6} + \frac{5}{24}$

$\qquad = 2\ln\left|\frac{3}{2}\right| + \frac{1}{24}$

Repeated linear factors exercise   Go online

Q79: Find $\int \frac{x^2-4}{(x+1)^3} dx$

....

Q80: Find $\int_3^4 \frac{2x^2-7x+5}{x^3-6x^2+12x-8} dx$

3.5.3 Irreducible quadratic factors

We will now see how to deal with the case where the denominator includes an irreducible quadratic factor.

A quadratic is irreducible when it has no real roots i.e. $b^2 - 4ac < 0$.

The integral $\int \frac{3x - 1}{(x - 2)(x^2 + 1)} \, dx$ is an example of this type with $x^2 + 1$ the irreducible quadratic factor in the denominator.

> **Key point**
>
> When a rational function is split into the sum of partial fractions and has an irreducible quadratic factor of the form $x^2 + a^2$ on the denominator the following result is used:
>
> $$\int \frac{dx}{x^2 + a^2} = \frac{1}{a}\tan^{-1}\left(\frac{x}{a}\right) + C$$

Examples

1. Problem:
Find the indefinite integral $\int \frac{3x - 1}{(x - 2)(x^2 + 1)} \, dx$

Solution:
Step 1: Rewrite the integral as the sum of partial fractions.
$\frac{3x - 1}{(x - 2)(x^2 + 1)} = \frac{A}{x - 2} + \frac{Bx + C}{x^2 + 1}$

Step 2: Obtain a common denominator and equate numerators.
$3x - 1 = A(x^2 + 1) + (Bx + C)(x - 2)$

Step 3: Find the values of A, B and C.
Let $x = 2$ then $5 = A(5) + (2B + C)(0)$
so $A = 1$
Let $x = 0$ then $-1 = A(1) + (Bx0 + C)(-2)$ and $A = 1$
so $C = 1$
Let $x = 1$ then $2 = A(2) + (B + C)(-1)$, and $A = 1$ and $C = 1$
so $B = -1$

Step 4: Substitute the values, $A = 1$, $B = -1$ and $C = 1$ into the original equation.
$\frac{3x - 1}{(x - 2)(x^2 + 1)} = \frac{1}{x - 2} + \frac{-x + 1}{x^2 + 1}$

TOPIC 3. INTEGRATION

Step 5: Integrate, begin by rewriting into a form that can be integrated.

$$\int \frac{3x-1}{(x-2)(x^2+1)} dx = \int \frac{1}{x-2} dx + \int \frac{-x+1}{x^2+1} dx$$

$$= \int \frac{1}{x-2} dx - \int \frac{x}{x^2+1} dx + \int \frac{1}{x^2+1} dx$$

$$= \int \frac{1}{x-2} dx - \frac{1}{2} \int \frac{2x}{x^2+1} dx + \int \frac{1}{x^2+1} dx$$

$$= \ln|x-2| - \frac{1}{2} \ln|x^2+1| + \tan^{-1}(x) + C$$

Remember: $\int \frac{f'(x)}{f(x)} dx = \ln|f(x)| + C$ and $\int \frac{1}{a^2+x^2} dx = \frac{1}{a} \tan^{-1}\left(\frac{x}{a}\right)$

....

2. Problem:

Calculate $\int_0^2 \frac{5x^2+10x+11}{(x+3)(x^2+4)} dx$

Solution:

Step 1: Rewrite the integral as the sum of partial fractions.

$\frac{5x^2+10x+11}{(x+3)(x^2+4)} = \frac{A}{x+3} + \frac{Bx+C}{x^2+4}$

Step 2: Obtain a common denominator and equate numerators.

$5x^2 + 10x + 11 = A\left(x^2+4\right) + (Bx+C)(x+3)$

Step 3: Find the values of A, B and C.

Let $x = -3$ then

$5(-3)^2 + 10(-3) + 11 = A\left((-3)^2+4\right) + (B(-3)+C)(-3+3)$

$$26 = 13A$$

$$A = 2$$

Given $A = 2$ let $x = 0$ then

$5(0)^2 + 10(0) + 11 = 2\left(0^2+4\right) + (B(0)+C)(0+3)$

$$11 = 8 + 3C$$

$$3 = 3C$$

$$C = 1$$

Given $C = 1$ let $x = 1$ then

$5(1)^2 + 10(1) + 11 = 2\left(1^2+4\right) + (B(1)+1)(1+3)$

$$26 = 10 + 4B + 4$$

$$12 = 4B$$

$$B = 3$$

Step 4: substitute the values of $A = 2$, $B = 3$ and $C = 1$ into the original equation.

$\frac{5x^2+10x+11}{(x+3)(x^2+4)} = \frac{2}{x+3} + \frac{3x+1}{x^2+4}$

© HERIOT-WATT UNIVERSITY

Step 5: Integrate, begin by writing into a form that can be integrated.

$$\int_0^2 \frac{5x^2 + 10x + 11}{(x+3)(x^2+4)}dx = \int_0^2 \frac{2}{x+3}dx + \int_0^2 \frac{3x+1}{x^2+4}dx$$

$$= 2\int_0^2 \frac{1}{x+3}dx + 3\int_0^2 \frac{x}{x^2+4}dx + \int_0^2 \frac{1}{x^2+4}dx$$

$$= 2\int_0^2 \frac{1}{x+3}dx + \frac{3}{2}\int_0^2 \frac{2x}{x^2+4}dx + \int_0^2 \frac{1}{x^2+2^2}dx$$

$$= 2\left[ln|x+3|\right]_0^2 + \frac{3}{2}\left[ln|x^2+4|\right]_0^2 + \left[\frac{1}{2}tan^{-1}\left(\frac{x}{2}\right)\right]_0^2$$

$$= 2\left[ln|x+3|\right]_0^2 + \frac{3}{2}\left[ln|x^2+4|\right]_0^2 + \frac{1}{2}\left[tan^{-1}\left(\frac{x}{2}\right)\right]_0^2$$

$$= 2\left(ln|5| - ln|3|\right) + \frac{3}{2}\left(ln|8| - ln|4|\right) + \frac{1}{2}\left(tan^{-1}(1) - tan^{-1}(0)\right)$$

$$= 2\left(ln|5| - ln|3|\right) + \frac{3}{2}\left(ln|8| - ln|4|\right) + \frac{1}{2}\left(tan^{-1}(1) - tan^{-1}(0)\right)$$

$$= ln\left|\frac{5}{3}\right|^2 + ln|2|^{\frac{3}{2}} + \frac{\pi}{8}$$

$$= ln\left|\frac{25}{9}\right| + ln\left|\sqrt{8}\right| + \frac{\pi}{8}$$

$$= ln\left|\frac{25\sqrt{8}}{9}\right| + \frac{\pi}{8}$$

Irreducible quadratic factors exercise Go online

Q81: Find $\int \frac{4x^2 - 9x + 13}{x^3 - 3x^2 + 2x - 6} dx$

....

Q82: Find $\int_0^1 \frac{6x^2+x+11}{(x+1)(x^2+3)} dx$

Challenge question 2

Try to find the following indefinite integral $\int \frac{dx}{x(x^2 + x + 1)}$

Be warned, this is very nasty.
(You will definitely *not* be asked to do anything as nasty as this in an exam.)

3.5.4 Improper rational functions

Let $f(x)$ be a polynomial of degree n and $g(x)$ be a polynomial of degree m; when $n \geq m$ then $\frac{f(x)}{g(x)}$ is an improper rational function.

For a polynomial, the degree is the value of the highest power.

To integrate an **improper rational function** we take the following steps:

1. Use algebraic long division to obtain a polynomial plus a proper rational function.
2. Apply the method of partial fractions to the proper rational function.
3. Integrate using the methods described in previous sections.

Example

Problem:

Find the indefinite integral $\int \frac{x^3 + 3x^2 + 4}{x^2 + 2x - 3} \, dx$

Solution:

Since the degree of the numerator is greater than the degree of the denominator then the integrand is an improper rational function.

Step 1: Divide by the denominator in the following way.

Note that the term $0x$ is used to position the x terms.

```
                  x    +1
x² +2x -3 | x³  +3x²  +0x   +4
            x³  +2x²  -3x
            ─────────────
                 x²   +3x   +4
                 x²   +2x   -3
                 ─────────────
                       x    +7
```

So, $\frac{x^3 + 3x + 4}{x^2 + 2x - 3} = x + 1 + \frac{x+7}{x^2 + 2x - 3}$

$x + 1$ is a polynomial and $\frac{x+7}{x^2 + 2x - 3}$ is a proper rational function: both expressions can be integrated.

Step 2: Before integrating we must apply the method of partial fractions to the proper rational function.

Factorise the denominator: $x^2 + 2x - 3 = (x - 1)(x + 3)$

Substitute back into the proper rational function: $\frac{x+7}{x^2+2x-3} = \frac{x+7}{(x-1)(x+3)}$

Select the correct form of the sum of partial fractions: $\frac{x+7}{(x-1)(x+3)} = \frac{A}{x-1} + \frac{B}{x+3}$

Obtain a common denominator and equate the numerators.

$$\frac{x+7}{(x-1)(x+3)} = \frac{A(x+3)}{(x-1)(x+3)} + \frac{B(x-1)}{(x-1)(x+3)}$$

$$x + 7 = A(x+3) + B(x-1)$$

Work out the values of A and B by selecting values of x.

Let $x = 1$ then,

$1 + 7 = A(1+3) + B(1-1)$

$8 = 4A$

$A = 2$

Given $A = 2$, let $x = -3$ then,

$-3 + 7 = 2(-3 + 3) + B(-3 - 1)$
$\quad 4 = -4B$
$\quad B = -1$

Substitute $A = 2$ and $B = -1$ back into the sum of partial fractions.

$\frac{x+7}{x^2+2x-3} = \frac{2}{x-1} - \frac{1}{x+3}$

Step 3: Integrate

$$\int \frac{x^3 + 3x + 4}{x^2 + 2x - 3} = \int (x + 1)\, dx + \int \frac{2}{x-1}\, dx - \int \frac{1}{x+3}\, dx$$
$$= \frac{1}{2}x^2 + x + 2\ln|x-1| - \ln|x+3| + C$$

Improper rational functions exercise   Go online

Q83: Find $\int \frac{2x^3 + 5x^2 - 9x - 18}{x^2 + x - 6}\, dx$

....

Q84: Find $\int \frac{4x^3 + 6x^2 - 17x - 36}{x^3 + x^2 - 8x - 12}\, dx$

....

Q85: Find $\int \frac{x^3 + 4x^2 + 3x + 2}{x^3 - x^2 + 4x - 4}\, dx$

3.6 Integration by parts

Integration by parts lets us integrate a product of two functions. This section will explore four cases: single application, repeated application, cyclical application and the use of a dummy variable.

3.6.1 One application

The product rule for differentiation gives rise to a very useful rule for integration known as integration by parts. Remember that the product rule can be stated as:

$\frac{d}{dx}(f(x)g(x)) = f'(x)g(x) + f(x)g'(x)$

Integrating both sides of the above equation with respect to x gives the following:

$\int \frac{d}{dx}(f(x)g(x)) = \int f'(x)g(x) + f(x)g'(x))dx$

$f(x)g(x) = \int f'(x)g(x)\, dx + \int f(x)g'(x)\, dx$

TOPIC 3. INTEGRATION

> **Key point**
>
> Rearranging this last equation gives the integration by parts formula:
>
> $\int f(x) g'(x)\, dx = f(x) g(x) - \int f'(x) g(x)\, dx$

The shortened form may be easier to remember: $\int f g' = fg - \int f' g$

This formula provides a means of finding $\int f(x) g'(x)\, dx$ as long as it is possible to find $\int f'(x) g(x)\, dx$ and so is useful if $\int f'(x) g(x)\, dx$ is simpler than $\int f(x) g'(x)\, dx$

Alternative notation

Let $u = f(x)$ and $v = g(x)$
then $\frac{du}{dx} = f'(x)$ and $\frac{dv}{dx} = g'(x)$

Therefore in terms of u and v the formula for integration by parts is:

$\int u \frac{dv}{dx}\, dx = uv - \int v \frac{du}{dx}\, dx$

Example

Problem:

Integrate $\int x \sin 3x\, dx$

Solution:

This integral is the product of two functions x and $\sin 3x$.

If $f(x) = x$ and $g'(x) = \sin 3x$
Then $f'(x) = 1$ and $g(x) = \int g'(x)\, dx = -\frac{1}{3} \cos 3x$
Note that $f'(x)$ is simpler than $f(x)$.

The formula for integration by parts $\int f g'\, dx = fg - \int f' g\, dx$ then gives:

$$\int x \sin 3x\, dx = x \times \left(-\frac{1}{3} \cos 3x\right) - \int 1 \times \left(-\frac{1}{3} \cos 3x\right) dx$$

$$= -\frac{1}{3} x \cos 3x + \frac{1}{3} \int \cos 3x\, dx$$

$$= -\frac{1}{3} x \cos 3x + \frac{1}{9} \sin 3x + C$$

Note that by choosing $f(x) = x$ and $g'(x) = \sin 3x$ then the integral $\int f'(x) g(x)\, dx$ reduces to a simple integral.

This is the aim when integrating by parts.

The following strategy may help with the choices for $f(x)$ and $g'(x)$.

© HERIOT-WATT UNIVERSITY

Key point

Strategy for choosing $f(x)$ and $g'(x)$:

- For $f(x)$ choose the function that becomes simpler when differentiated. Often this is a polynomial (but not always).

- For $g'(x)$ choose the function so that $g(x)$ can be easily determined by integrating. Often this is either a trigonometric or an exponential function.

Example

Problem:

Find $\int (x+1) e^{-x}\, dx$

Solution:

Choose $f(x) = x + 1$ then $f'(x) = 1$ and $g'(x) = e^{-x}$ then $g(x) = -e^{-x}$

Now, using the formula for integration by parts $\int f g'\, dx = fg - \int f'g\, dx$ gives

$$\begin{aligned}
\int (x+1)e^{-x}\, dx &= -(x+1)e^{-x} - \int 1 \times (-e^{-x})\, dx \\
&= -(x+1)e^{-x} + \int e^{-x}\, dx \\
&= -(x+1)e^{-x} - e^{-x} + C \\
&= -e^{-x}(x+2) + C
\end{aligned}$$

Definite integrals using integration by parts

It is also possible to evaluate definite integrals using integration by parts.

Key point

The integration by parts formula for definite integrals is:
$\int_a^b f(x)g'(x)\,dx = [f(x)g(x)]_a^b - \int_a^b f'(x)g(x)\,dx$

This formula is used in the following example.

Example

Problem:

Evaluate $\int_0^{\frac{\pi}{2}} x \cos x\, dx$

Solution:

Choose $f(x) = x$ then $f'(x) = 1$
and $g'(x) = \cos x$ then $g(x) = \sin x$

TOPIC 3. INTEGRATION

Using the formula for integration by parts... $\int_a^b fg'dx = [fg]_a^b - \int_a^b f'g\,dx$ gives,

$$\int_0^{\frac{\pi}{2}} x \cos x\, dx = [x \sin x]_0^{\frac{\pi}{2}} - \int_0^{\frac{\pi}{2}} \sin x\, dx$$
$$= [x \sin x]_0^{\frac{\pi}{2}} - [\cos x]_0^{\frac{\pi}{2}}$$
$$= (\frac{\pi}{2} - 0) - (0 - 1)$$
$$= \frac{\pi}{2} + 1$$

One application exercise   Go online

Q86: Find $\int (3x + 1)e^{2x}\, dx$

....

Q87: Find $\int (2x - 3)\sin(2x)\, dx$

....

Q88: Find $\int_1^2 \frac{2\ln|x|}{x^4}dx$

3.6.2 Repeated application

Sometimes we may have to use integration by parts more that once to obtain an answer.

Examples

1. Problem:

Find the indefinite integral $x^2 \cos x\, dx$

Solution:

Step 1: The integrand is the product of two functions so we must use integration by parts.

Let $f(x) = x^2$ then $f'(x) = 2x$
$g'(x) = \cos x$ then $g(x) = \sin x$

The formula for integration by parts $\int fg'dx = fg - \int f'g\,dx$ gives,

$$\int x^2 \cos x\, dx = x^2 \sin x - \int 2x \sin x\, dx$$
$$= x^2 \sin x - 2\int x \sin x\, dx \quad (1)$$

The integrand on the RHS is still the product of two functions, $x \sin x$. Use integration by parts a second time for $\int x \sin x\, dx$

Step 2: Use integration by parts a second time to integrate $\int x \sin x \, dx$

Let $f(x) = x$ then $f'(x) = 1$
and $g'(x) = \sin x$ then $g(x) = -\cos x$
The formula for integration by parts $\int fg' dx = fg - \int f'g \, dx$ gives,

$$\int x \sin x \, dx = x(-\cos x) - \int 1 \times (-\cos x) dx$$
$$= -x \cos x + \int \cos x \, dx + C$$
$$= -x \cos x + \sin x + C$$

Step 3: Substitute the solution for $\int x \sin x \, dx$ back into (1) the solution from Step 1.

$$\int x^2 \cos x \, dx = x^2 \sin x - 2 \int x \sin x \, dx$$
$$= x^2 \sin x - 2(-x \cos x + \sin x) + C$$
$$= x^2 \sin x + 2x \cos x - 2 \sin x + C$$
$$= (x^2 - 2) \sin x + 2x \cos x + C$$

....

2. Problem:

Evaluate $\int_0^3 x^2 e^{2x} dx$

Solution:

Step 1: The integrand is the product of two functions so we must use integration by parts.
Let $f(x) = x^2$ then $f'(x) = 2x$
and $g'(x) = e^{2x}$ then $g(x) = \frac{1}{2} e^{2x}$
The formula for integration by parts $\int_a^b fg' dx = [fg]_a^b - \int_a^b f'g \, dx$ gives,

$$\int_0^3 x^2 e^{2x} dx = \left[x^2 \times \frac{1}{2} e^{2x} \right]_0^3 - \int_0^3 2x \times \frac{1}{2} e^{2x} dx$$
$$= \frac{1}{2} \left[x^2 e^{2x} \right]_0^3 - \int_0^3 x e^{2x} dx$$

At this stage we can evaluate the first term of the solution.

$$\int_0^3 x^2 e^{2x} dx = \frac{1}{2} \left[x^2 e^{2x} \right]_0^3 - \int_0^3 x e^{2x} dx$$
$$= \frac{1}{2} \left(3^2 e^{2(3)} - 0^2 e^{2(0)} \right) - \int_0^3 x e^{2x} dx$$
$$= \frac{1}{2} \left(9 e^6 \right) - \int_0^3 x e^{2x} dx$$
$$= \frac{9}{2} e^6 - \int_0^3 x e^{2x} dx \quad (1)$$

The integrand on the RHS is still the product of two functions, $x e^{2x}$. Use integration by parts a second time for $\int_0^3 x e^{2x} dx$

TOPIC 3. INTEGRATION

Step 2: Use integration by parts a second time to integrate $\int_0^3 xe^{2x}dx$

Let $f(x) = x$ then $f'(x) = 1$
and $g'(x) = e^{2x}$ then $g(x) = \frac{1}{2}e^{2x}$

The formula for integration by parts is, $\int_a^b fg'dx = [fg]_a^b - \int_a^b f'gdx$ gives,

$$\int_0^3 xe^{2x}dx = \left[x \times \frac{1}{2}e^{2x}\right]_0^3 - \int_0^3 1 \times \frac{1}{2}e^{2x}dx$$

$$= \frac{1}{2}\left[xe^{2x}\right]_0^3 - \frac{1}{2}\int_0^3 e^{2x}dx$$

$$= \frac{1}{2}\left[xe^{2x}\right]_0^3 - \frac{1}{2}\left[\frac{1}{2}e^{2x}\right]_0^3$$

$$= \frac{1}{2}\left[xe^{2x}\right]_0^3 - \frac{1}{4}\left[e^{2x}\right]_0^3$$

$$= \frac{1}{2}\left(3e^{2(3)} - 0e^0\right) - \frac{1}{4}\left(e^{2(3)} - e^0\right)$$

$$= \frac{1}{2}(3e^6) - \frac{1}{4}(e^6 - 1)$$

$$= \frac{3}{2}e^6 - \frac{1}{4}e^6 + \frac{1}{4}$$

$$= \frac{5}{4}e^6 + \frac{1}{4}$$

$$= \frac{1}{4}(5e^6 + 1)$$

Step 3: Substitute the solution for $\int_0^3 xe^{2x}dx$ back into (1), the solution from Step 1.

$$\int_0^3 x^2e^{2x}dx = \frac{9}{2}e^6 - \int_0^3 xe^{2x}dx$$

$$= \frac{9}{2}e^6 - \frac{1}{4}(5e^6 + 1)$$

$$= \frac{9}{2}e^6 - \frac{5}{4}e^6 - \frac{1}{4}$$

$$= \frac{13}{4}e^6 - \frac{1}{4}$$

$$= \frac{1}{4}(13e^6 - 1)$$

Repeated application exercise   Go online

Q89: Find $\int 3x^2 e^{-x}\, dx$

....

Q90: Find $\int_0^{\frac{\pi}{6}} x^2 \sin(3x)dx$

3.6.3 Cyclical (Getting back to the original function)

This section looks at finding an integral of the form $I = \int e^{ax} \sin(bx)\, dx$ or similar.

Sometimes an integration by parts can lead round in a circle. The solution is to integrate by parts twice to obtain a simple equation for I which is then easy to solve.

Example

Problem:

Find the indefinite integral $\int e^x \sin(x)\, dx$

Solution:

Step 1:

Let $f(x) = e^x$ then $f'(x) = e^x$
and $g'(x) = \sin(x)$ then $g(x) = -\cos(x)$

The formula for integration by parts $\int f g' = fg - \int f'g$ gives,

$\int e^x \sin(x)\, dx = -e^x \cos(x) + \int e^x \cos(x)\, dx$

Letting $I = \int e^x \sin(x)\, dx$ we have, $I = -e^x \cos(x) + \int e^x \cos(x)\, dx$

Step 2: Apply integration by parts to $\int e^x \cos(x)\, dx$

Let $f(x) = e^x$ then $f'(x) = e^x$
and $g'(x) = \cos(x)$ then $g(x) = \sin(x)$

Step 3: Apply integration by parts a second time.

$\int f g' = fg - \int f'g$

$\int e^x \cos(x)\, dx = e^x \sin(x) - \int e^x \sin(x)\, dx$

Step 4: Substitute the solution of $\int e^x \cos(x)\, dx$ back into the first application of integration by parts in step 1.

$\int e^x \sin(x)\, dx = -e^x \cos(x) + \int e^x \cos(x)\, dx$

$\int e^x \sin(x)\, dx = -e^x \cos(x) + e^x \sin(x) - \int e^x \sin(x)\, dx$

Notice that $\int e^x \sin(x)\, dx$ appears on the LHS and the RHS.

Since $I = \int e^x \sin(x)\, dx$ we can now write: $I = -e^x \cos(x) + e^x \sin(x) - I$

Step 5: Rearrange the equation to give,

$2I = e^x \cos(x) - e^x \sin(x) + C$

$I = \frac{1}{2}(e^x \cos(x) - e^x \sin(x)) + C$

Therefore, $\int e^x \sin(x)\, dx = \frac{1}{2}(e^x \cos(x) - e^x \sin(x)) + C$

Note that since C is an arbitrary constant we leave $1/2\,C$ as C in the working.

TOPIC 3. INTEGRATION

Getting back to the original function exercise

Q91: Find $\int e^{3x} \cos(2x)\, dx$

Q92: Find $\int_1^3 \frac{\ln|x|}{x}\, dx$

3.6.4 Introducing 1 as a 'dummy variable'

At first sight some integrands do not appear to be suitable to integrate by parts, as there is only one function e.g. $\ln|x|$ and $\tan^{-1}(x)$.

To integrate such functions we introduce the 'dummy variable' 1 and write as a product. The example below will illustrate this.

Integral ln x

Examples

1. Problem:
Integrate $\int \ln|x|\, dx$
Solution:

Rewrite the integrand as $\ln|x| \times 1$ then find the integral by using integration by parts as follows.

Let $f(x) = \ln|x|$ then $f'(x) = \frac{1}{x}$
and $g'(x) = 1$ then $g(x) = x$

The formula for integration by parts $\int fg'\, dx = fg - \int f'g\, dx$ then gives us,

$$\int (\ln|x| \times 1)\, dx = x\ln|x| - \int \frac{1}{x} x\, dx$$
$$= x\ln|x| - \int 1\, dx$$
$$= x\ln|x| - x + C$$

2. Problem:
Integrate $\int \tan^{-1}(3x)\, dx$
Solution:

Rewrite the integrand as $\tan^{-1}(3x) \times 1$ then find the integral by using integration by parts as follows.

Let $f(x) = \tan^{-1}(3x)$ then $f'(x) = \frac{3}{1+9x^2}$
and $g'(x) = 1$ then $g(x) = x$

© HERIOT-WATT UNIVERSITY

The formula for integration by parts $\int fg'\,dx = fg - \int f'g\,dx$ then gives us,

$$\int \tan^{-1}(3x) \times 1\,dx = x\tan^{-1}(3x) - \int \frac{3x}{1+9x^2}\,dx$$
$$= x\tan^{-1}(3x) - 3\int \frac{x}{1+9x^2}\,dx$$

To integrate the integrand on the RHS recall the standard result $\int \frac{f'(x)}{f(x)}\,dx = \ln|f(x)| + C$.

We need to make the numerator look like the derivative of the denominator
i.e. $\frac{d}{dx}\{1+9x^2\} = 18x$.

We therefore multiply the integrand by $\frac{18}{18}$.

$$\int \tan^{-1}(3x) \times 1\,dx = x\tan^{-1}(3x) - \int \frac{3x}{1+9x^2}\,dx$$
$$= x\tan^{-1}(3x) - 3\int \frac{x}{1+9x^2}\,dx$$
$$= x\tan^{-1}(3x) - \frac{3}{18}\int \frac{18x}{1+9x^2}\,dx$$
$$= x\tan^{-1}(3x) - \frac{3}{18}\ln|1+9x^2| + C$$

Introducing 1 as a 'dummy variable' exercise   Go online

Q93: Find $\int \ln|3x|\,dx$

Q94: Find $\int \tan^{-1}(2x)\,dx$

Q95: Find $\int [\ln|x|]^2\,dx$

Challenge question 3   Go online

Use integration by parts to find the indefinite integral $\int \sin^{-1}x\,dx$

For this challenge question you will need to recall the standard result $\frac{d}{dx}\{\sin^{-1}x\} = \frac{1}{\sqrt{1-x^2}}$

3.7 Area between a curve and an axis

Working out the area under a curve was covered at Higher. These next subsections describe, in more detail, the theory behind the derivation of the integral to find the area under the curve.

We will look at how to find the area between a curve and the x-axis and the area between a curve and the y-axis.

3.7.1 Area between a curve and the x-axis

In the diagram A is the area under the curve $y = f(x)$ from $x = a$ to $x = b$

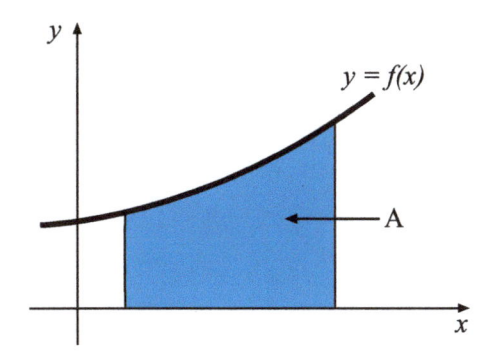

This area can be estimated by dividing it into strips which approximate to rectangles and then totalling the areas of these rectangles.
The total area of the rectangles gives an approximate value for A.
Generally the more rectangles used, the closer the result will be to the actual value for A.

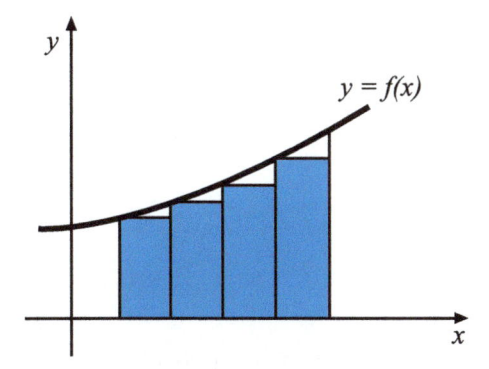

Consider a rectangle, width δx and let δs be the shaded area under the curve $y = f(x)$.

Where δ is the Greek letter delta. It represents an infinitesimal change. In this case δx will represent the infinitesimal change in the width of each rectangle underneath the curve. The widths, δx, will become smaller and smaller so that more rectangles are drawn underneath the curve. δs represents the infinitesimal change in the area of the function, $f(x)$, at a given point on the curve.

Take one shaded section underneath the curve as illustrated below in blue. The area of two rectangles with width δx can be calculated. One rectangle, with height y and area $y\delta x$, underestimates the area because the height of the rectangle is less than the height of the curve to the right of it. The second rectangle, with height $y + \delta y$ and area $(y + \delta y)\delta x$, over estimates the area because the height of the rectangle is greater than the height of the curve to the left of it.

By considering the area of the two rectangles in the diagram, we can see that the area underneath

© HERIOT-WATT UNIVERSITY

the curve, δs, at a given point on the curve, is greater than $y\delta x$, the rectangle that underestimates the area and less than $(y + \delta y)\delta x$, the area of the rectangle that overestimates that area.

This can be written mathematically as: $y\delta x \leqslant \delta s \leqslant (y + \delta y)\delta x$

and then, $y \leqslant \frac{\delta s}{\delta x} \leqslant y + \delta y$

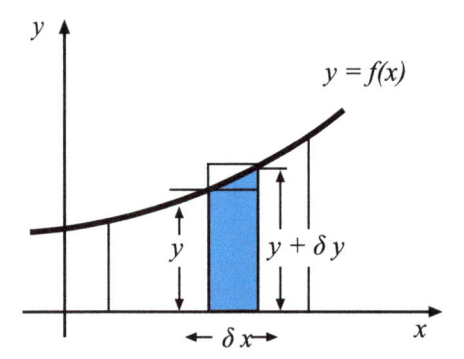

Ideally, we want the difference between the underestimated area and the overestimated area to be zero. That way both would give the same area, and hence, the exact area underneath the curve. To make this happen we decrease the width of each rectangle so that it tends to zero i.e. let $\delta x \to 0$. The number of rectangles underneath the curve would increase and $\delta x \to 0$ i.e. the height of the underestimated rectangle is the same height as the overestimated rectangle. This in turn allows us to write $\frac{\delta s}{\delta x} \to \frac{dS}{dx}$ i.e. the infinitesimal change in the height of the curve over the infinitesimal change in the width of the rectangle tends to the derivative $\frac{dS}{dx}$.

As $\delta x \to 0$ we can write:

$$y \leqslant \frac{\delta s}{\delta x} \leqslant y$$
$$\Rightarrow \frac{dS}{dx} = y$$

Integrating both sides with respect to x gives: $S = \int y dx$

This integration will give an area function $S(x)$ and will involve a constant of integration C.

Key point

If the right- and left-hand boundaries are defined for x it is possible to obtain the area function as a definite integral, and in fact the original area under the curve is,

$$A = \lim_{\delta x \to 0} \sum_{x=a}^{x=b} y\delta x = \int_a^b y dx = S(b) - S(a)$$

This result has been shown to be true for a **monotonically** increasing function; however, note that this same result is also true for more complicated functions.

TOPIC 3. INTEGRATION

Area under a curve   Go online

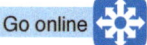

We can approximate the area under a curve using rectangles. The greater the number of rectangles under the curve, the smaller the margin of error and the area of the rectangles is closer to the accurately calculated area.

Remember that when finding the area under a curve:

- find the roots to identify where the graph cuts the x-axis;
- make a sketch of the curve;
- areas above the x-axis give a positive value for the definite integral;
- areas below the x-axis give a negative value for the definite integral.

Examples

1. Problem:

Find the area between the curve $y = 4x - 2x^2$ and the x-axis from $x = 0$ to $x = 3$.

Solution:

A sketch of the curve shows that the area is in two parts.
One part is above the x-axis and has a positive value for the definite integral and the other part is below the x-axis and has a negative value for the definite integral.
So the two areas need to be calculated separately.

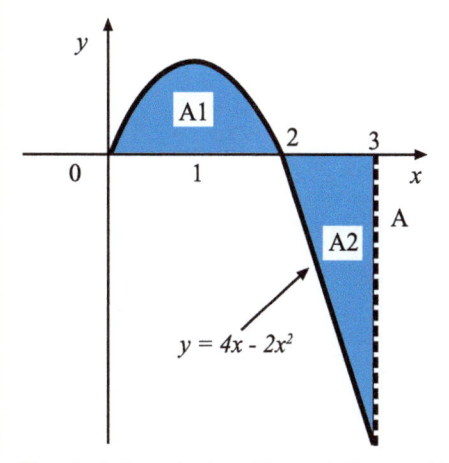

The shaded area is above the x-axis for $x = 0$ to $x = 2$.

$$\int_0^2 (4x - 2x^2)dx = \left[2x^2 - \frac{2}{3}x^3\right]_0^2$$
$$= 8 - \frac{16}{3}$$
$$= 2\frac{2}{3}$$

The shaded area is below the x-axis for $x = 2$ to $x = 3$.

$$\int_2^3 (4x - 2x^2)dx = \left[2x^2 - \frac{2}{3}x^3\right]_2^3$$
$$= (18 - 18) - \left(8 - \frac{16}{3}\right)$$
$$= -2\frac{2}{3}$$

The answer is negative as the area is below the x-axis. So the actual area is $2^2/_3$

The total shaded area is therefore $2^2/_3 + 2^2/_3 = 5^1/_3 \; units^2$

....

2. Problem:

Find the area between the curve $y = \tan^{-1} x$ and the x-axis from $x = -\frac{\pi}{2}$ to $x = \frac{\pi}{2}$.

Solution:

A sketch of the curve shows that the area is in two parts.
One part is above the x-axis and has a positive value for the definite integral and the others part is below the x-axis and has a negative value for the definite integral.
So the two areas need to be calculated separately.

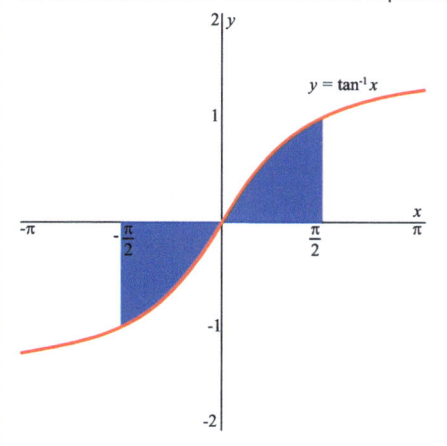

The shaded area is above the x-axis for $x = 0$ to $x = \frac{\pi}{2}$. So to calculate this area we work out: $\int_0^{\frac{\pi}{2}} \tan^{-1} x \; dx$

TOPIC 3. INTEGRATION

Recall from the section on integration by parts that there is not a standard result that allows us to integrate $tan^{-1}x$, but there is a standard result to differentiate it: $\frac{d}{dx}tan^{-1}x = \frac{1}{1+x^2}$
We must therefore use one of the previously learned strategies, specifically integration by parts, using the dummy variable '1' and the integration by parts formula
$\int_a^b fg' \, dx = [fg]_a^b - \int_a^b f'g \, dx$

Step 1:

Given $\int_0^{\frac{\pi}{2}} tan^{-1}x \times 1 dx$

Let $f(x) = \tan^{-1}x$ then $f'(x) = \frac{1}{1+x^2}$
and $g'(x) = 1$ then $g(x) = x$

Step 2: Substitute into the integration by parts formula $\int_a^b fg' \, dx = [fg]_a^b - \int_a^b f'g \, dx$

$$\int_0^{\frac{\pi}{2}} tan^{-1}x \times 1 dx = \left[(\tan^{-1}x)(x)\right]_0^{\frac{\pi}{2}} - \int_0^{\frac{\pi}{2}} \frac{1}{2(1+x^2)} \times x dx$$

$$= \left[x\tan^{-1}x\right]_0^{\frac{\pi}{2}} - \frac{1}{2}\int_0^{\frac{\pi}{2}} \frac{x}{1+x^2} dx$$

$$= \left(\frac{\pi}{2}\tan^{-1}\frac{\pi}{2} - 0\tan^{-1}0\right) - \frac{1}{2}\int_0^{\frac{\pi}{2}} \frac{x}{1+x^2} dx$$

$$= \frac{\pi}{2}\tan^{-1}\frac{\pi}{2} - \frac{1}{2}\int_0^{\frac{\pi}{2}} \frac{x}{1+x^2} dx$$

To integrate the integrand on the RHS we use the standard result $\int \frac{f'(x)}{f(x)} dx = \ln|f(x)| + C$. To use this we need to make the numerator look like the derivative of the denominator i.e. $\frac{d}{dx}\{1+x^2\} = 2x$. We therefore multiply the integrand by $\frac{2}{2}$.

Step 3: Integrate the integrand on the RHS.

$$\int_0^{\frac{\pi}{2}} \frac{x}{1+x^2} dx = \frac{1}{2}\int_0^{\frac{\pi}{2}} \frac{2x}{1+x^2} dx$$

$$= \frac{1}{2}\left[\ln|1+x^2|\right]_0^{\frac{\pi}{2}}$$

$$= \frac{1}{2}\left(\ln\left|1+\frac{\pi}{2}^2\right| - \ln|1+0^2|\right)$$

$$= \frac{1}{2}\left(\ln\left|1+\frac{\pi}{2}^2\right| - \ln|1|\right)$$

Recall that $\ln|1| = 0$

So, $\int_0^{\frac{\pi}{2}} \frac{x}{1+x^2} dx = \frac{1}{2}\ln\left|1+\frac{\pi}{2}^2\right|$

Step 4: Substitute this result back into the application of integration by parts in step 2.

$$\int_0^{\frac{\pi}{2}} \tan^{-1}x \times 1 dx = \frac{\pi}{2}\tan^{-1}\frac{\pi}{2} - \frac{1}{2}\int_0^{\frac{\pi}{2}} \frac{x}{1+x^2} dx$$

$$= \frac{\pi}{2}\tan^{-1}\frac{\pi}{2} - \frac{1}{2} \times \frac{1}{2}\ln\left|1+\frac{\pi}{2}^2\right|$$

$$= \frac{\pi}{2}\tan^{-1}\frac{\pi}{2} - \frac{1}{4}\ln\left|1+\frac{\pi}{2}^2\right|$$

© HERIOT-WATT UNIVERSITY

Step 5: Work out the area below the x-axis for $x = -\frac{\pi}{2}$ to $x = 0$.

To calculate the area below the x-axis evaluate the integral $\int_{-\frac{\pi}{2}}^{0} \tan^{-1} x \, dx$. We have integrated for the integrand $\tan^{-1} x$ in steps 1-4 above. We will do this again but for the limits from $x = -\frac{\pi}{2}$ to $x = 0$.

Using integration by parts as done in step 2 we have:

$\int_{-\frac{\pi}{2}}^{0} \tan^{-1} x \times 1 \, dx = \left[x \tan^{-1} x \right]_{-\frac{\pi}{2}}^{0} - \frac{1}{2} \int_{-\frac{\pi}{2}}^{0} \frac{x}{1+x^2} dx$

We can also use the process in step 3 to integrate $\int_{-\frac{\pi}{2}}^{0} \frac{x}{1+x^2} dx$

$\int_{-\frac{\pi}{2}}^{0} \tan^{-1} x \times 1 \, dx = \left[x \tan^{-1} x \right]_{-\frac{\pi}{2}}^{0} - \frac{1}{2} \times \frac{1}{2} \left[\ln \left| 1 + x^2 \right| \right]_{-\frac{\pi}{2}}^{0}$

$= \left[x \tan^{-1} x \right]_{-\frac{\pi}{2}}^{0} - \frac{1}{4} \left[\ln \left| 1 + x^2 \right| \right]_{-\frac{\pi}{2}}^{0}$

$= \left(0 \tan^{-1} 0 - \left(-\frac{\pi}{2} \tan^{-1} \left(-\frac{\pi}{2} \right) \right) \right) - \frac{1}{4} \left(\ln \left| 1 + 0^2 \right| - \ln \left| 1 + \left(-\frac{\pi}{2} \right)^2 \right| \right)$

$= \frac{\pi}{2} \tan^{-1} \left(-\frac{\pi}{2} \right) - \frac{1}{4} \left(\ln |1| - \ln \left| 1 + \frac{\pi}{4} \right| \right)$

Recall that $ln\,|1| = 0$

$\int_{-\frac{\pi}{2}}^{0} \tan^{-1} x \times 1 \, dx = \frac{\pi}{2} \tan^{-1} \left(-\frac{\pi}{2} \right) + \frac{1}{4} \ln \left| 1 + \frac{\pi}{4} \right|$

$= -\frac{\pi}{2} \tan^{-1} \frac{\pi}{2} + \frac{1}{4} \ln \left| 1 + \frac{\pi}{4} \right|$

Ignoring the coefficient $\frac{\pi}{2}$, we have said that $\tan^{-1} \left(-\frac{\pi}{2} \right)$ is equal to $-\tan^{-1} \left(\frac{\pi}{2} \right)$. This is because $\tan^{-1} x$ is an odd function. Odd functions have the property $f(-x) = -f(x)$.

Also recall that this answer is the area below the x-axis and so is negative. The actual area is therefore $\frac{\pi}{2} \tan^{-1} \left(\frac{\pi}{2} \right) - \frac{1}{4} ln \left| 1 + \frac{\pi}{4} \right|$

Step 6: Work out the total area by adding the two areas together.

Total area $= \frac{\pi}{2} \tan^{-1} \frac{\pi}{2} - \frac{1}{4} \ln \left| 1 + \frac{\pi^2}{2} \right| + \frac{\pi}{2} \tan^{-1} \frac{\pi}{2} - \frac{1}{4} \ln \left| 1 + \frac{\pi}{4} \right|$

$= 2 \times \frac{\pi}{2} \tan^{-1} \frac{\pi}{2} - 2 \times \frac{1}{4} \ln \left| 1 + \frac{\pi^2}{2} \right|$

$= \pi \tan^{-1} \frac{\pi}{2} - \frac{1}{2} \ln \left| 1 + \frac{\pi^2}{2} \right|$

Area between a curve and the x-axis exercise   Go online

Q96: Find the area between the curve $f(x) = x^2 - 3x + 2$ and the x-axis, from $x = 0$ and $x = 3$. Remember to sketch the curve.

....

Q97: Find the area between the curve $f(x) = 2 \sin x + 1$ and the x-axis for $0° \leqslant x \leqslant 360°$.

3.7.2 Area between a curve and the y-axis

In the previous section we integrated to work out the area between the curve and the x-axis. We did this by drawing rectangles underneath the curve down to the x-axis. By decreasing the width of these rectangles and summing their areas together we got a better and better approximation to the exact area. When the width of the rectangles tended to zero the sum of the areas tended to the exact area underneath the curve.

We can follow the same process to work out the area between the curve and the y-axis. The rectangles are drawn from the curve to the y-axis and their areas are summed together as the width of each one tends to zero, giving the exact area between the curve and the y-axis.

> **Key point**
>
> To work out the area between the curve and the y-axis integrate the function with respect to y between the limits $y = a$ and $y = b$.
>
> $$A = \int_a^b x\,dy, \quad a \leqslant y \leqslant b$$

This is illustrated in the diagram below.

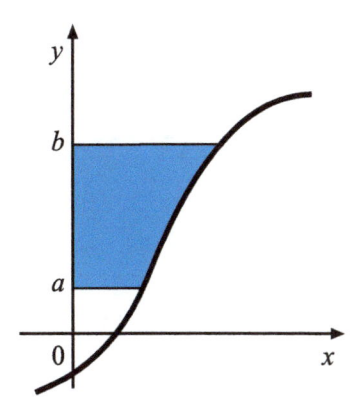

> To integrate a function between the curve and y-axis:
>
> - sketch the graph of the curve.
> - rearrange the function and write in terms of y.
> - integrate the rearranged equation and evaluate for the limits.

The following example should make this process clear.

Example

Problem:

Find the shaded area enclosed by the y-axis and the curve $y = (x - 4)^3$ between $y = 1$ and $y = 8$.

Solution:

Step 1: Sketch the graph of the curve.

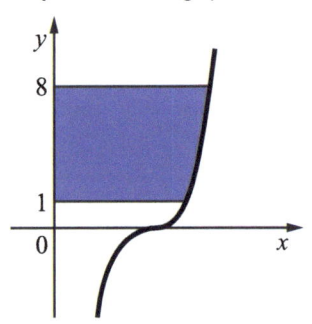

Step 2: Find the area.

Area = $\int_1^8 x \, dy$

We cannot integrate x with respect to y, so we must rearrange the equation of the curve in terms of y.

$$y = (x - 4)^3$$
$$y^{\frac{1}{3}} = x - 4$$
$$x = y^{\frac{1}{3}} + 4$$

Step 3: Integrate the rearranged equation.

$$\text{Area} = \int_1^8 y^{\frac{1}{3}} + 4 \, dy$$
$$= \left[\frac{3}{4} y^{\frac{4}{3}} + 4y \right]_1^8$$
$$= \left(\frac{3}{4}(8)^{\frac{4}{3}} + 4(8) \right) - \left(\frac{3}{4}(1)^{\frac{4}{3}} + 4(1) \right)$$
$$= 44 - 4\frac{3}{4}$$
$$= 39\frac{1}{4} \text{ square units}$$

TOPIC 3. INTEGRATION

Area between a curve and the y-axis exercise　　Go online

Q98: Find the area enclosed by the curve $y = x^2 + 1$ and the y-axis between $y = 1$ and $y = 4$.

....

Q99: What is the area enclosed by the curve $y = \ln |x| + 5$ and the y-axis between $y = 0$ and $y = 3$? Write your answer in terms of e with negative powers.

....

Q100: Find the area enclosed by the curve $xy - y = 2$ and the y-axis between $y = 1$ and $y = 3$.

3.8 Volume of revolution

Volume of revolution about the x-axis

The shaded area enclosed by the curve $y = f(x)$ and the x-axis between $x = a$ and $x = b$ is shown in the diagram.

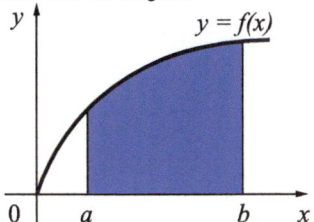

Suppose this area is rotated 360° about the x-axis, then the solid that is formed is known as a solid of revolution.

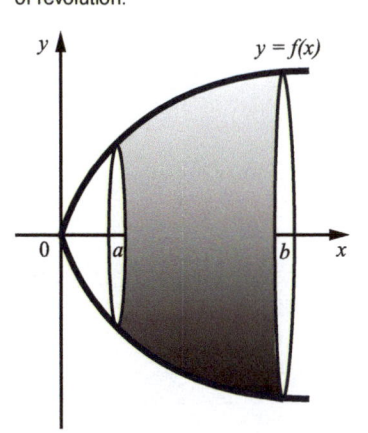

Calculate its volume, V.

Note that if we look at the volume of the solid from the end i.e. looking from b towards the origin, we would see a circle. No matter which cross-section we take the shape we see will always be a circle. The radius of this circle is the height of the function from the x-axis i.e. $y = f(x)$. The radius of each circle changes as the value of x changes. We can work out the area of each circle using $A = \pi r^2$, where the radius is given by $f(x)$. By adding the area of each circle between the limits $x = a$ and $x = b$ we obtain the volume of the solid of revolution.

The description given below is a more mathematical explanation of this process.

Now, consider the strip with thickness δx and height y.
When this is rotated about the x-axis it produces a disc of radius y and thickness δx. Let δV be the volume of this disc, then $\delta V \approx \pi y^2 \delta x$

(Compare with the volume of a cylinder = $\pi r^2 h$)

Now as $\delta x \rightarrow 0$ then,

$$V = \lim_{\delta x \to 0} \sum_{x=a}^{x=b} \pi y^2 dx$$

$$= \int_a^b \pi y^2 dx$$

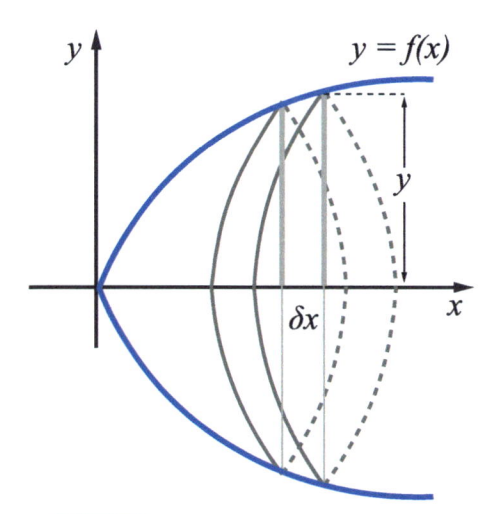

Key point

The volume of a solid of revolution about the x-axis, for $y = f(x)$, is given by:

$$V = \int_a^b \pi y^2 dx \quad a \leqslant y \leqslant b$$

© HERIOT-WATT UNIVERSITY

Examples

1. Problem:

The shaded region shown in the diagram is the area between the curve $y = 2x + x^2$ and the x-axis from $x = 2$ to $x = 3$.

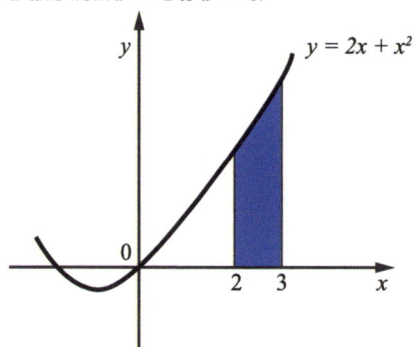

Calculate the volume of the solid of revolution formed when this area is given a full turn about the x-axis.

Solution:

The volume is given by: $V = \int_a^b \pi y^2 dx$ where $y = 2x + x^2$, when $a = 2$ and $b = 3$.

$$V = \pi \int_2^3 (2x + x^2)^2 dx$$

$$= \pi \int_2^3 (4x^2 + 4x^3 + x^4) dx$$

$$= \pi \left[\frac{4}{3}x^3 + \frac{4}{4}x^4 + \frac{1}{5}x^5 \right]_2^3$$

$$= \pi \left[\left(\frac{4}{3}(3)^3 + (3)^4 + \frac{1}{5}(3)^5 \right) - \left(\frac{4}{3}(2)^3 + (2)^4 + \frac{1}{5}(2)^5 \right) \right]$$

$$= \pi \left[\left(36 + 81 + \frac{243}{5} \right) - \left(\frac{32}{3} + 16 + \frac{32}{5} \right) \right]$$

$$= 132\frac{8}{15} \pi \text{ units}^3$$

....

2. Problem:

Calculate the volume of the solid formed when the semi-circle with equation $x^2 + y^2 = r^2$, $y \geq 0$ is given a full turn about the x-axis.

Solution:

Note that $x^2 + y^2 = r^2$ is the equation of a circle centred at (0,0) with radius r. We have restricted the functions such that $y \geq 0$.

This produces a semi-circle, centred at the origin with radius r. This means that we are integrating the volume of revolution between $x = -r$ and $x = r$.

Before integrating we must rearrange the equation to $y^2 = r^2 - x^2$
Using the formula,

$$V = \int_{-r}^{r} \pi y^2 dx$$
$$= \int_{-r}^{r} \pi \left(r^2 - x^2\right) dx$$
$$= \int_{-r}^{r} \left(\pi r^2 - \pi x^2\right) dx$$
$$= \left[\pi r^2 x - \frac{1}{3}\pi x^3\right]_{-r}^{r}$$
$$= \pi r^3 - \frac{1}{3}\pi r^3 + \pi r^3 - \frac{1}{3}\pi r^3$$
$$= \frac{4}{3}\pi r^3$$

Which gives the formula for the volume of a sphere.

The shaded area shows the area enclosed by the curve $x^2 + y^2 = r^2$ and the x-axis between $-r$ and r.

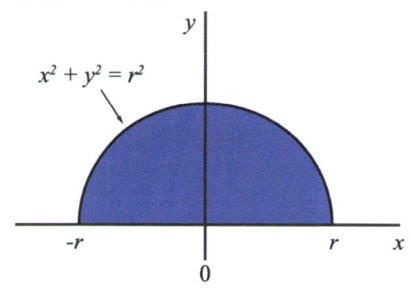

Rotating this area 360° about the x-axis forms a solid of revolution.

One of the first mathematicians to perfect a method for computing the volume of a sphere was Archimedes of Syracuse. There is more about this mathematician in the Extended information section of this topic.

Volume of revolution about the x-axis exercise   Go online

Q101: Calculate the volume of revolution formed when the area between the curve $y^2 = 4x$ and the x-axis from $x = 0$ to $x = 3$ is rotated about the x-axis.

....

Q102: Calculate the volume of revolution formed when the area between the curve $y = \sqrt{4 - x^2}$ and the x-axis from $x = -2$ to $x = 2$ is rotated about the x-axis.

....

Q103: By revolving the curve $y = \frac{r}{h}x$ for $0 \leqslant x \leqslant h$ about the x-axis show that we can generate a cone of base r radius and height h.

TOPIC 3. INTEGRATION

Volume of revolution about the y-axis

Above we have considered the volume of revolution about the x-axis. We can now apply a similar process for working out the volume of revolution about the y-axis.

> **Key point**
>
> The volume of a solid of revolution about the y-axis, for $x = f(y)$, is given by:
>
> $$V = \int_a^b \pi x^2 dy \quad a \leqslant x \leqslant b$$

Examples

1. Problem:

Calculate the volume of the solid formed when the equation $y = \frac{x}{4}$, bounded by $y = 0$ and $y = 2$, is rotated about the y-axis.

Solution:

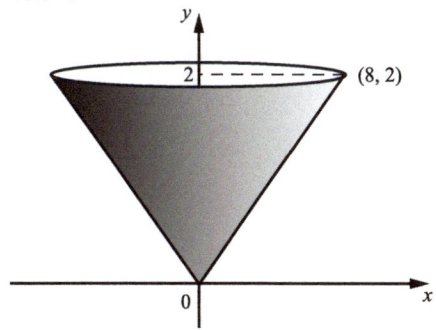

Before integrating we must rearrange the equation:

$y = \dfrac{x}{4}$

$x = 4y$

$x^2 = (4y)^2$

$x^2 = 16y^2$

Using the formula:

$$\begin{aligned}
V &= \int_0^2 \pi x^2 dy \\
&= \pi \int_0^2 16y^2 dy \\
&= \pi \left[\frac{16}{3} y^3 \right]_0^2 \\
&= \pi \left[\frac{16}{3}(2)^3 - \frac{16}{3}(0)^3 \right] \\
&= \frac{128\pi}{3} \text{ units}^3
\end{aligned}$$

© HERIOT-WATT UNIVERSITY

2. Problem:

Find the volume of the solid of revolution formed when the area enclosed by the y-axis and $y = \cos^{-1}(x)$ from $y = 0$ to $y = \frac{\pi}{2}$

Solution:

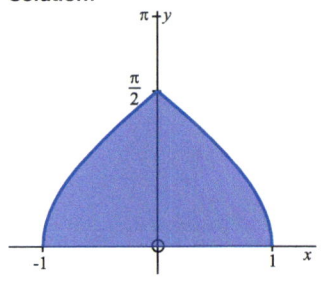

Before integrating we must rearrange the equation:
$$y = \cos^{-1}(x)$$
$$\cos(y) = x$$
$$x = \cos(y)$$

Using the formula:
$$V = \int_0^{\frac{\pi}{2}} \pi \cos^2 y \, dy$$
$$= \pi \int_0^{\frac{\pi}{2}} \cos^2 y \, dy$$

To integrate $\cos^2 y$ we use $\cos^2 y = \frac{1}{2}(\cos(2y) + 1)$

$$V = \pi \int_0^{\frac{\pi}{2}} \frac{1}{2}(\cos(2y) + 1) dy$$
$$= \frac{\pi}{2} \int_0^{\frac{\pi}{2}} \cos(2y) + 1 \, dy$$
$$= \frac{\pi}{2} \left[\frac{1}{2}\sin(2y) + y \right]_0^{\frac{\pi}{2}}$$
$$= \frac{\pi}{2} \left\{ \left(\frac{1}{2}\sin(\pi) + \frac{\pi}{2}\right) - \left(\frac{1}{2}\sin(0) + 0\right) \right\}$$
$$= \frac{\pi}{2} \left(\frac{\pi}{2}\right)$$
$$= \frac{\pi^2}{2} \text{ units}^2$$

TOPIC 3. INTEGRATION

Volume of revolution about the y-axis exercise   Go online

Q104: Calculate the volume of revolution formed when the area between the curve $y = 4x^2 - 1$ and the y-axis from $y = -1$ to $y = 4$ is rotated about the y-axis.

....

Q105: Calculate the volume of revolution formed when the area between the curve $xy^2 = 2$ and the y-axis from $y = 1$ to $y = 3$ is rotated about the y-axis.

....

Q106: Calculate the volume of revolution formed when the area between the curve $y = \ln |x|$ and the y-axis from $y = 1$ to $y = 4$ is rotated about the y-axis.
By taking out a common factor of $\frac{1}{2}\pi^2$ write your answer in factorised form.

3.9 Learning points

Integration
Terminology

- *Anti-differentiation* is the reverse process of differentiation.

- *Integration* is the method used to find the anti-derivatives.

- In general if $\frac{d}{dx}(F(x)) = f(x)$ then $F(x)$ is called an anti-derivative, or integral, of $f(x)$, written as $\int f(x)\,dx = F(x) + C$ and C is the *constant of integration*.

- For $\int f(x)\,dx = F(x) + C$, $F(x) + C$ is the general indefinite integral of $f(x)$ with respect to x of the integral $\int f(x)\,dx$

- For $\int f(x)\,dx = F(x) + C$, $f(x)$ is the *integrand*.

- For $\int f(x)\,dx = F(x) + C$ an anti-derivative given by a particular value of C is a *particular integral*.

- $\int_a^b f(x)\,dx$ is a definite integral because the limits of integration are known and $\int_a^b f(x)\,dx = [F(x)]_a^b = F(b) - F(a)$, $a \leqslant x \leqslant b$

Integration using standard results

- Learn and use the following standard integrals:
    - $\int x^n\,dx = \frac{1}{n+1}x^{n+1} + C$ and $\int (ax+b)^n\,dx = \frac{(ax+b)^{n+1}}{a(n+1)} + C$
    - $\int \cos(x)\,dx = \sin(x) + C$ and $\int \cos(ax+b)\,dx = \frac{1}{a}\sin(ax+b) + C$
    - $\int \sin(x)\,dx = -\cos(x) + C$ and $\int \sin(ax+b)\,dx = -\frac{1}{a}(ax+b) + C$
    - $\int \sec^2(x)\,dx = \tan(x) + C$ and $\int \sec^2(ax+b)\,dx = \frac{1}{a}\tan(ax+b) + C$
    - $\int \exp(x)\,dx = \exp(x) + C$ and $\int \exp(ax+b)\,dx = \frac{1}{a}\exp(ax+b) + C$
    - $\int \frac{1}{x}\,dx = \ln|x| + C$ and $\int \frac{dx}{ax+b} = \frac{1}{a}\ln|ax+b| + C$

Integration by substitution

- When $\int f(ax+b)\,dx$ substitute $u = ax+b$.
- When $\int f'(x)[f(x)]^n\,dx$ substitute $u = f(x)$.
- When $\int (ax+b)(cx+d)^n\,dx$ substitute $u = (cx+d)$.
- When $\int f'(x)e^{f(x)}\,dx$ substitute $u = f(x)$.
- When $\int \frac{f'(x)}{f(x)}\,dx$ substitute $u = f(x)$.

Integrating inverse trigonometric functions

- $\int \frac{1}{\sqrt{1-x^2}}\,dx = \sin^{-1}(x) + C$ and $\int \frac{1}{\sqrt{a^2-x^2}}\,dx = \sin^{-1}\left(\frac{x}{a}\right) + C$
- $\int \frac{1}{1+x^2}\,dx = \tan^{-1}(x) + C$ and $\int \frac{1}{a^2+x^2}\,dx = \frac{1}{a}\tan^{-1}\left(\frac{x}{a}\right) + C$

© HERIOT-WATT UNIVERSITY

TOPIC 3. INTEGRATION

- To integrate $\int \frac{bx + c}{x^2 + a^2} \, dx$ we separate the fraction to give two integrals, then use $\int \frac{f'(x)}{f(x)} \, dx = \ln|f(x)| + C$ and standard results to integrate. The solution will contain an inverse tangent.

- To integrate $\int \frac{dx}{ax^2 + bx + c}$, where $b^2 - 4ac < 0$, we complete the square on the denominator, then after using substitution apply $\int \frac{f'(x)}{f(x)} \, dx = \ln|f(x)| + C$ and standard results to integrate.

Partial fractions and integration

- To integrate proper rational functions, we must first, split them into partial fractions with either linear, repeated linear, irreducible quadratic factors or a combination of these in the denominator. We then integrate the partial fractions.

Integration by parts

- Integration by parts is used to integrate the product of two functions and is given by:
$\int fg' = fg - \int f'g$

- For definite integrals, integration by parts is given by: $\int_a^b fg' = [fg]_a^b - \int_a^b f'g$

- We select f and g' so that the integral on the RHS is simpler than the one we started with on the LHS.

- Sometimes we must apply integration by parts repeatedly to get to a solution.

- When integrating cyclical functions we may use integration by parts several times in order to obtain the original integrand on the RHS. From here we use algebra to rearrange the equation and solve the integral.

- When integrating $\ln|x|$, $\tan^{-1}(x)$, $\sin^{-1}(x)$ we must integrate each as a product with 1 e.g. $\int 1 \times \ln|x| \, dx$, $\int 1 \times \tan^{-1}(x) \, dx$, $\int 1 \times \sin^{-1}(x) \, dx$. In all cases $g' = 1$.

Area between a curve and the x-axis

- To find the area between a curve and the x-axis we must:
  - make a sketch of the curve to determine where the area is above and below the x-axis.
  - calculate each area separately and take the modulus of the negative areas.
  - add all the areas together to get the total area between the curve and the x-axis.

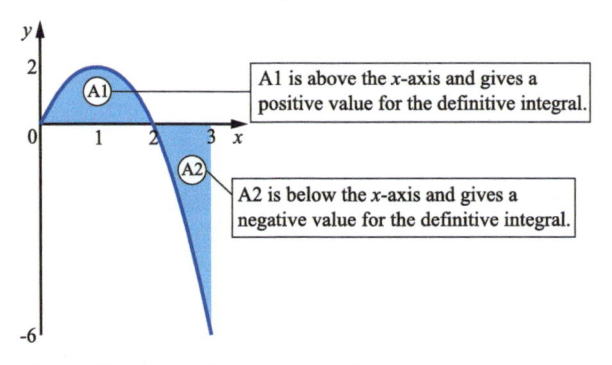

A1 is above the x-axis and gives a positive value for the definitive integral.

A2 is below the x-axis and gives a negative value for the definitive integral.

© HERIOT-WATT UNIVERSITY

Area between a curve and the y-axis

- To find the area between a curve and the y-axis we must integrate with respect to y between the limits $y = a$ and $y = b$. The integral will take the form $A = \int_a^b x\,dy$. We must rearrange the equation of the curve in terms of y.

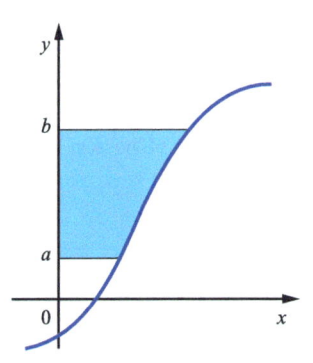

Volume of revolution

- The volume of revolution around the x-axis is given by $V = \int_a^b \pi y^2 dx$ where $y = f(x)$ and $a \leqslant x \leqslant b$.

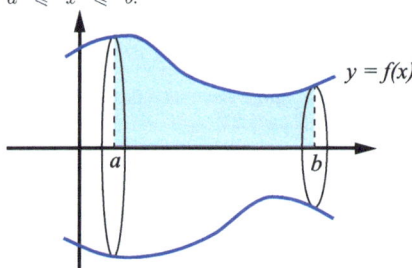

- The volume of revolution around the y-axis is given by $V = \int_a^b \pi x^2 dy$ where $x = f(y)$ and $a \leqslant y \leqslant b$.

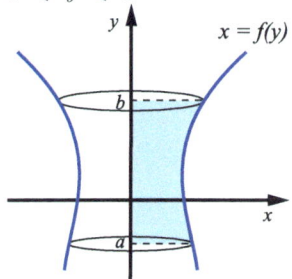

3.10 Extended information

The authors do not maintain these web links and no guarantee can be given as to their effectiveness at a particular date.
They should serve as an insight to the wealth of information from the World Wide Web and encourage readers to explore the subject further.

http://www.univie.ac.at/future.media/moe/galerie.html
This site consists of interactive learning units. Of specific interest to this topic is the site under the heading: Integration, then Intuitively understanding the integral.

http://math.furman.edu/~dcs/java/change.html
This site shows an animation of integration by change of variable for a particular function.

http://xanadu.math.utah.edu/java/ApproxArea.html
You are able to see an animation for approximating the area under various curves at this site.

http://www.quickmath.com/
Look under Calculus, then Integrate. This provides a tool to calculate your answers.

http://www-history.mcs.st-andrews.ac.uk/history/index.html
The MacTutor History of Mathematics archive. This site provides a very comprehensive directory of biographies for hundreds of important mathematicians. This was the source for the following information on Archimedes.

Archimedes of Syracuse

Archimedes was born in 287 BC and died in 212 BC in Syracuse, Scilly.

Eureka!

He achieved fame in his own time through his mechanical inventions. Many of these were war machines, which proved to be particularly effective in the defence of Syracuse when attacked by the Roman general Marcellus.

On a visit to Egypt he also invented a type of pump known as Archimedes' screw, which is still used today in many parts of the world.

Another of his inventions, which brought him great fame, was the compound pulley. It was written that using this small device he was able to move the great weight of 'a ship of burden out of the king's arsenal, which could not be drawn out of the dock without great labour and many men' and 'holding the head of the pulley in his hand and drawing the cords by degrees, he drew the ship in a straight line, as smoothly and evenly as if she had been in the sea.'

He is noted as having stated: 'give me a place to stand and I will move the earth.' This was in the context of his work on levers.

give me a place to stand and I will move the earth...

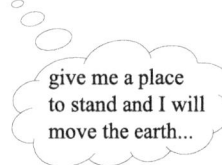

Although he achieved fame through his mechanical inventions, Archimedes believed that pure mathematics was the only worthy pursuit. He is considered by most historians to be one of the greatest mathematicians of all time. He demonstrated that he could approximate square roots accurately. He invented a system of expressing large numbers, and also perfected an integration method that allowed him to calculate the areas, volumes and surface areas of many solids. He found the volume and surface area of a sphere. Archimedes considered his most significant accomplishments were those concerning a cylinder circumscribing a sphere, and he gave instructions that his tombstone should have displayed on it a diagram consisting of a sphere with a circumscribing cylinder.

It is also claimed that his work on integration 'gave birth to the calculus of the infinite' conceived and brought to perfection by Kepler, Cavalieri, Fermat, Leibniz and Newton.

In mechanics, Archimedes discovered theorems for the centre of gravity of plane figures and solids. His most famous theorem gives the weight of a solid immersed in a liquid, called Archimedes' principle. This discovery is said to have inspired his famous shout 'Eureka!' ('I have found it!').

Archimedes was killed in 212 BC by a Roman soldier during the capture of Syracuse by the Romans in the Second Punic War. Cicero, a Roman statesman, while serving in Sicily, had Archimedes' tombstone restored. It has been claimed that 'the Romans had so little interest in pure mathematics that this action by Cicero was probably the greatest single contribution of any Roman to the history of mathematics'.

TOPIC 3. INTEGRATION

3.11 End of topic test

End of topic 3 test — Go online

Integration using standard results

Q107: Find $\int \frac{5}{\sqrt{1-(2x)^2}}\,dx$.

Q108: Find $\int 3\sec^2(5x)\,dx$.

Q109: Find $\int 3e^{4x-1}\,dx$.

Q110: Evaluate $\int_3^6 \frac{3}{2x-5}\,dx$.

Q111: Evaluate $\int_0^{\frac{\pi}{3}} \frac{5}{1+(3x)^2}\,dx$.

Integration by substitution (given)

Q112: Find $\int 3x^2(2x^3+5)^3\,dx$ using the substitution $u = 2x^3 + 5$.

Q113: Find $\int \cos^4 x \sin x\,dx$ using the substitution $u = \cos(x)$.
Note that $\cos^4(x)$ means $\cos(x)$ to the power of 4 **not** $\cos(x^4)$.

Q114: Find $\int \frac{5x}{2x^2+1}\,dx$ using the substitution $u = 2x^2 + 1$.

Q115: Find $\int 2x^3 e^{4x^4+3}\,dx$ using the substitution $u = 4x^4 + 3$.

Q116: Find $\int \tan x\,dx$ using the substitution $u = \cos x$.

Q117: Evaluate $\int_0^{\sqrt{2}} 4x\sqrt{1+3x^2}\,dx$ using the substitution $u = 1 + 3x^2$.

Q118: Evaluate $\int_0^{\frac{\pi}{2}} 2x^2 \sin(x^3+1)\,dx$ using the substitution $u = x^3 + 1$.

Q119: Evaluate $\int_0^{\frac{\pi}{4}} \sec^2 x \tan^3 x\,dx$ using the substitution $u = \tan x$.

282 TOPIC 3. INTEGRATION

Q120: A/B level

Evaluate $\int_{\frac{1}{5}}^{\frac{2}{5}} 3x(5x+2)^3 dx$ using the substitution $u = 5x + 2$.

Q121: A/B level

Evaluate $\int_{2}^{15} \frac{3x}{\sqrt{4x+1}} dx$ using the substitution $u = 4x + 1$.

Integration by substitution (not given)

Q122: A/B level

Using a suitable substitution, find $\int x^3 (3x^4 - 1)^2 dx$.

Q123: A/B level

Using a suitable substitution, find $\int (2x + 1)(3x + 4)^2 dx$.

Q124: A/B level

Using a suitable substitution, find $\int \frac{7x^2}{3x^3+1} dx$.
Write your answer as a multiple of the natural logarithm.

Q125: A/B level

Evaluate $\int_{0}^{\frac{\pi}{2}} \sin\left(4x - \frac{\pi}{2}\right) dx$ using the substitution $u = 4x - \frac{\pi}{2}$.

Q126: A/B level

Using a suitable substitution, evaluate $\int_{-1}^{1} 5x^2 e^{(4x^3 + 1)} dx$.

Integration involving inverse trigonometric functions

Q127: Find $\int \frac{2}{\sqrt{9-16x^2}} dx$.

Q128: Find $\int \frac{2}{9+16x^2} dx$.

Q129: A/B level

Find $\int \frac{2}{x^2-6x+25} dx$.

TOPIC 3. INTEGRATION

Q130: Evaluate $\int_0^2 \frac{2}{3x^2+4}dx$. Give your answer in terms of π.

....

Q131: Evaluate $\int_0^{\frac{2}{\sqrt{3}}} \frac{1}{\sqrt{16-9x^2}}dx$. Give your answer in terms of π.

....

Q132: A/B level

Evaluate $\int_{-2}^{1} \frac{2}{x^2+4x+13}\,dx$. Give your answer in terms of π.

Integration involving partial fractions

Q133: Find the sum of partial fractions for the expression $f(x) = \frac{-9}{(x+1)(x-2)}$. Hence determine $\int f(x)\,dx$.

....

Q134: Find the sum of partial fractions for the expression $f(x) = \frac{8x-8}{4x^2+4x-3}$. Hence determine $\int f(x)\,dx$.

....

Q135: A/B level

Find the sum of partial fractions for the expression $f(x) = \frac{5x+7}{(x-1)(x+2)(x+3)}$. Hence determine $\int f(x)\,dx$.

....

Q136: Find the sum of partial fractions for the expression $f(x) = \frac{-6}{x^3-2x^2-x+2}$. Hence determine $\int f(x)\,dx$.

....

Q137: Find the sum of partial fractions for the expression $f(x) = \frac{-3x-5}{(x+2)^2}$. Write down the anti-derivative.

....

Q138: A/B level

Find the sum of partial fractions for the expression $f(x) = \frac{x^2-4x+5}{(x-1)^3}$. Write down the anti-derivative.

....

Q139: A/B level

Find the sum of partial fractions for the expression $f(x) = \frac{9x-9}{(2x+1)(x-2)^2}$. Write down the anti-derivative.

....

Q140: A/B level

Find the sum of partial fractions for the expression $f(x) = \frac{1}{(x-1)(x^2+4)}$. Write down the anti-derivative.

....

© HERIOT-WATT UNIVERSITY

284 TOPIC 3. INTEGRATION

Q141: A/B level

Integrate the improper rational function $f(x) = \frac{x^3 + 2x^2 + 3x + 1}{x^2 + 3x + 2}$.
Hence determine $\int f(x)\, dx$.

....

Q142: Find the sum of partial fractions for the expression $f(x) = \frac{5x - 2}{(x - 1)(x + 2)}$.
Hence determine $\int_2^3 f(x)\, dx$.

....

Q143: Find the sum of partial fractions for the expression $f(x) = \frac{5}{2x^2 + 7x + 3}$.
Hence determine $\int_0^3 f(x)\, dx$.

....

Q144: Find the sum of partial fractions for the expression $f(x) = \frac{2}{(x - 1)(x - 2)(x - 3)}$.
Hence determine $\int_4^5 f(x)\, dx$.

....

Q145: A/B level

Find the sum of partial fractions for the expression $f(x) = \frac{11x^2 - 46x + 47}{2x^3 - 12x^2 + 22x - 12}$.
Hence determine $\int_4^5 f(x)\, dx$.

....

Q146: Find the sum of partial fractions for the expression $f(x) = \frac{3x - 2}{x^2 - 2x + 1}$.
Hence determine $\int_3^5 f(x)\, dx$.

....

Q147: A/B level

Find the sum of partial fractions for the expression $f(x) = \frac{3x^2 - 13x + 18}{x^3 - 6x^2 + 12x - 8}$.
Hence determine $\int_4^5 f(x)\, dx$.

....

Q148: A/B level

Find the sum of partial fractions for the expression $f(x) = \frac{4x^2 - 22x - 8}{(2x + 3)(2x - 1)^2}$.
Hence determine $\int_1^2 f(x)\, dx$.

....

Q149: Find the sum of partial fractions for the expression $f(x) = \frac{25}{x^3 - x^2 - 8x - 12}$.
Hence determine $\int_4^5 f(x)\, dx$.

....

Q150: A/B level

Find the sum of partial fractions for the expression $f(x) = \frac{2x^2 + 2x + 7}{2x^3 - 7x^2 + 8x - 3}$.
Hence determine $\int_2^4 f(x)\, dx$.

....

© HERIOT-WATT UNIVERSITY

TOPIC 3. INTEGRATION

Q151: A/B level
Find the sum of partial fractions for the expression $f(x) = \frac{2x^2 - x + 5}{(x + 1)(x^2 + 3)}$.
Hence determine $\int_1^{\sqrt{3}} f(x)dx$.

Integration by parts

Q152: Use integration by parts to find $\int (3x + 4) e^x dx$.

Q153: A/B level
Using integration by parts find $\int (4x^2 + 1) \cos 3x dx$.

Q154: A/B level
Using integration by parts integrate the function $\int 2 \sin 3x \cos 2x dx$.

Q155: A/B level
Using integration by parts integrate the function $\int e^x \cos x dx$.

Q156: A/B level
Using integration by parts find $\int [\ln |x|]^2 dx$.

Q157: A/B level
Using integration by parts find $\int \cos^{-1} 2x dx$.

Q158: Using integration by parts evaluate $\int_{\pi/6}^{\pi/2} 5x \cos 2x \, dx$.

Q159: A/B level
Using integration by parts evaluate $\int_0^2 x^2 e^{3x} dx$.

Q160: A/B level
Using integration by parts evaluate $\int_0^{\pi} 5x^2 \sin 2x dx$.

Q161: A/B level
Using integration by parts evaluate $\int_1^3 \ln |4x| dx$.

© HERIOT-WATT UNIVERSITY

Area between a curve and the y-axis

Q162: *A/B level*
Find the area enclosed between the y-axis and the curve $y = (x - 2)^3$ between $y = 2$ and $y = 8$.

....

Q163: *A/B level*
Find the area enclosed between the y-axis and the curve $y = \frac{1}{x^4}$ between $y = \frac{1}{16}$ and $y = 1$.

....

Q164: *A/B level*
Find the area enclosed between the y-axis and the curve $y^2 = 16 - x$ between $y = 4$ and $y = -4$.

Volume of revolution about the x-axis

Q165: The area under the curve $y = 2x - \frac{3}{x}$ between $x = 1$ and $x = 2$ is rotated through 2π radians about the x-axis.
Determine the exact value of the volume of the solid formed.

....

Q166: The area under the curve $y^2 = 6 - x^2$ in the first quadrant is rotated through 2π radians about the x-axis.
Determine the exact value of the volume of the solid formed.

....

Q167: The area under the curve $y = \cos \theta$ between $x = 0$ and $x = \frac{\pi}{4}$ is rotated through 2π radians about the x-axis.
Determine the exact value of the volume of the solid formed.

....

Q168: *A/B level*
A semi-circle with centre (3,0) and radius 2, lies on the x-axis as shown.

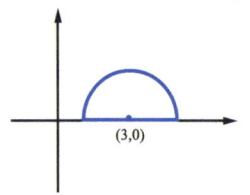

Find the volume of the solid of revolution formed when it is rotated completely about the x-axis.
Determine the exact value of the volume of the solid formed.

TOPIC 3. INTEGRATION

Volume of revolution about the y-axis

Q169: A/B level

The area under the curve $3y = x^2$ between $y = -1$ and $y = 2$ is rotated through 2π radians about the y-axis.

Determine the exact value of the volume of the solid formed.

....

Q170: A/B level

The area under the curve $y = 5 - x^2$ between $y = -5$ and $y = 5$ is rotated through 2π radians about the y-axis.

Determine the exact value of the volume of the solid formed.

....

Q171: A/B level

The area under the curve $3y = \ln|x|$ between $y = 2$ and $y = 4$ is rotated through 2π radians about the y-axis.

Determine the exact value of the volume of the solid formed.

....

Q172: A/B level

The area under the curve $xy^2 = 3$ between $y = 1$ and $y = 3$ is rotated through 2π radians about the y-axis.

Determine the exact value of the volume of the solid formed.

....

Q173: A/B level

The area under the curve $y = \sin \theta^2$ between $y = 0$ and $y = \frac{1}{2}$ is rotated through 2π radians about the y-axis.

Determine the exact value of the volume of the solid formed.

Topic 4

Differential equations

Contents

- 4.1 Looking back 291
  - 4.1.1 Laws of exponentials 292
  - 4.1.2 Laws of logarithms 295
  - 4.1.3 The chain rule 300
- 4.2 First order linear differential equations introduction 301
- 4.3 First order linear differential equations with separable variables 308
  - 4.3.1 Separable variables 308
  - 4.3.2 Growth and decay 319
  - 4.3.3 Further applications of differential equations 326
- 4.4 Integrating factor 330
  - 4.4.1 Method of integrating factor 331
  - 4.4.2 Integrating factor applications 340
- 4.5 Second order linear differential equations with constant coefficients 344
  - 4.5.1 Homogeneous second order linear differential equations 345
  - 4.5.2 Initial value problems 356
  - 4.5.3 Non-homogeneous second order linear differential equations 360
- 4.6 Learning points 374
- 4.7 Proofs 377
- 4.8 Extended information 379
- 4.9 End of topic test 385

TOPIC 4. DIFFERENTIAL EQUATIONS

Learning objective

By the end of this topic, you should be able to:

- use the method of separate variables to solve first order linear differential equations;
- formulate differential equations for growth and decay and solve them by separating variables;
- formulate differential equations according to Newton's Law and solve them by separating variables;
- use the integrating factor to solve first order linear differential equations;
- solve second order linear differential equations.

4.1 Looking back

Pre-requisites from Advanced Higher

You should have covered the following in the *Partial fractions* topic. If you need to reinforce your learning go back and study this topic.

Algebraic long division

- The method of partial fractions is applied to proper rational functions only. If the function is improper, algebraic long division must be carried out first.

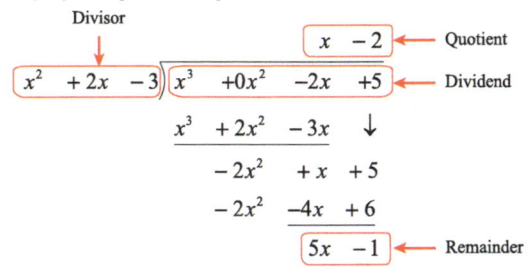

$$
\begin{array}{r}
x - 2 \\
x^2 + 2x - 3 \overline{\smash{\big)}\, x^3 + 0x^2 - 2x + 5} \\
\underline{x^3 + 2x^2 - 3x} \\
-2x^2 + x + 5 \\
\underline{-2x^2 - 4x + 6} \\
5x - 1
\end{array}
$$

where $x - 2$ is the Quotient, $x^2 + 2x - 3$ is the Divisor, $x^3 + 0x^2 - 2x + 5$ is the Dividend, and $5x - 1$ is the Remainder.

- When setting up algebraic long division each power of the variable must be accounted for in the dividend. See the example above.

The quotient rule

The quotient rule gives us a method that allows us to differentiate algebraic fractions.

It states that, in function notation:

when $k(x) = \frac{f(x)}{g(x)}$, then $k'(x) = \frac{f'(x)g(x) - f(x)g'(x)}{[g(x)]^2}$

In Leibnitz notation:

when $y = \frac{u}{v}$, where u and v are functions of x, then $\frac{dy}{dx} = \frac{\frac{du}{dx}v - u\frac{dv}{dx}}{v^2}$

Summary of prior knowledge
Laws of exponentials: National 5

1. $a^m \times a^n = a^{m+n}$
2. $\frac{a^m}{a^n} = a^{m-n}$
3. $(a^m)^n = a^{m \times n}$
4. $a^0 = 1$
5. $a^{-m} = \frac{1}{a^m}$
6. $a^{\frac{m}{n}} = \sqrt[n]{a^m}$

Laws of logarithms: Higher

1. $\log_a m + \log_a n = \log_a(m \times n)$
2. $\log_a m - \log_a n = \log_a \frac{m}{n}$
3. $\log_a x^n = n \times \log_a x$
4. $\log_a 1 = 0$
5. $\log_a a = 1$

The chain rule: Higher

Let $y(x)$ be the composite function $g(f(x))$, then the chain rule is:

- Function notation: $y'(x) = g'(f(x)) \times f'(x)$
- Leibniz notation: $\frac{dy}{dx} = \frac{dg}{df} \times \frac{df}{dx}$

4.1.1 Laws of exponentials

Using the laws of indices   Go online

Multiply the expression below by a x a

$a^1 \times a^9 = $ a x a x a x a x a x a x a x a x a
$= a^{10}$
$= a^{1+9}$

General Rule

$a^n \times a^m = a^{n+m}$

$a^6 \times a^4 = $ a x a x a x a x a x a x a x a x a x a
$= a^{10}$
$= a^{6+4}$

Multiply the expression below by $\frac{a}{a}$

$\dfrac{a^4}{a^1} = \dfrac{a \times a \times a \times a}{a}$

$= \dfrac{a \times a \times a}{1}$

$= a^3$
$= a^{4-1}$

General Rule

$\dfrac{a^n}{a^m} = a^{n-m}$

$\dfrac{a^7}{a^4} = \dfrac{a \times a \times a \times a \times a \times a \times a}{a \times a \times a \times a}$

$= a^3$
$= a^{7-4}$

Examples

1. Problem:

Simplify $\frac{x^5 \times x^4}{x^2}$, $x \neq 0$

Solution:

$\frac{x^5 \times x^4}{x^2} = x^{5+4-2} = x^7$

....

TOPIC 4. DIFFERENTIAL EQUATIONS

2. Problem:

Simplify $3x^{\frac{1}{3}} \times 4x^{\frac{2}{3}}$

Solution:

$$\begin{aligned}
3x^{\frac{1}{3}} \times 4x^{\frac{2}{3}} &= 3 \times 4 \times x^{\frac{1}{3}} \times x^{\frac{2}{3}} \\
&= 3 \times 4 \times x^{\frac{1}{3}+\frac{2}{3}} \\
&= 12 \times x^{\frac{3}{3}} \\
&= 12x^1 \\
&= 12x
\end{aligned}$$

How to raise a power to a power   Go online

$(a^x)^y \quad if \quad x=1 \quad and \quad y=1$
$= (a^1)^1$
$= (a)^1$
$= a$

$(a^x)^y \quad if \quad x=4 \quad and \quad y=2$
$= (a^4)^2$
$= (a \times a \times a \times a)^2$
$= (a \times a \times a \times a)(a \times a \times a \times a)$
$= a^8$
$= a^{4 \times 2}$

Examples

1. Problem:

Simplify $(2y^4)^2$.

Solution:

$(2y^4)^2 = 2^2 \times y^{4 \times 2} = 4y^8$

....

2. Problem:

Simplify $(2g^{\frac{1}{2}})^6$.

Solution:

$(2g^{\frac{1}{2}})^6 = 2^6 \times g^{\frac{1}{2} \times 6} = 64g^{\frac{6}{2}} = 64g^3$

Negative and zero indices   Go online

$If \; x = 0$
$a^0 = 1$

$If \quad x = -2$
$a^{-2} = \dfrac{1}{a \times a}$

$If \quad x = -5$
$a^{-5} = \dfrac{1}{a \times a \times a \times a \times a}$

© HERIOT-WATT UNIVERSITY

Examples

1. Problem:

Simplify, giving your answer with a positive index $\frac{y^7}{y^{10}}$

Solution:

$\frac{y^7}{y^{10}} = y^{7-10} = y^{-3} = \frac{1}{y^3}$

....

Problem:

Simplify $(2g^{-2})^3$

Solution:

$(2g^{-2})^3 = 2^3 \times g^{-2 \times 3} = 8g^{-6} = \frac{8}{g^6}$

The purpose of a fractional index is to define a surd.

$a^{\frac{1}{2}} = \sqrt{a}$ $a^{\frac{1}{3}} = \sqrt[3]{a}$ $a^{\frac{1}{4}} = \sqrt[4]{a}$ $a^{\frac{3}{2}} = \sqrt{a^3}$ $a^{\frac{2}{3}} = \sqrt[3]{a^2}$

In essence the numerator of the index is the power and the denominator is the root.

$a^{\frac{x}{y}} = \sqrt[y]{a^x}$

If $x = 1$ and $y = 2$ then $a^{\frac{1}{2}} = \sqrt{a}$

If $x = 3$ and $y = 2$ then $a^{\frac{3}{2}} = \sqrt{a^3}$

If $x = 4$ and $y = 3$ then $a^{\frac{4}{3}} = \sqrt[3]{a^4}$

If $x = 4$ and $y = 4$ then $a^{\frac{4}{4}} = \sqrt[4]{a^4} = a$

Examples

1. Problem:

Evaluate $25^{\frac{3}{2}}$

Solution:

$25^{\frac{3}{2}} = \sqrt[2]{25^3} = 5^3 = 125$

It does not matter whether you square root or cube the term first and since you probably don't know 25^3 it is easier to find $\sqrt{25}$ then cube the answer.

....

2. Problem:

Evaluate $8^{-\frac{1}{3}}$

Solution:

$8^{-\frac{1}{3}} = \frac{1}{8^{\frac{1}{3}}} = \frac{1}{\sqrt[3]{8}} = \frac{1}{2}$

- Remember a negative index moves the term onto the denominator.
- The power a third means the cube root.
- The cube root of 8 is 2 because $2^3 = 8$

TOPIC 4. DIFFERENTIAL EQUATIONS

Laws of exponentials exercise Go online

Q1: Simplify $\frac{x^6 \times x^2}{x^3}$, $x \neq 0$

Q2: Simplify $2x^{\frac{1}{4}} \times 5x^{\frac{1}{4}}$

Q3: Simplify $(3y^4)^3$

Q4: Simplify $(5m^{\frac{3}{2}})^2$

Q5: Simplify, giving your answer with a positive index $(3m^{-4})^2$

Q6: Simplify, giving your answer with a positive index $\frac{y^{-10} \times y^{\frac{3}{2}}}{y^{\frac{1}{2}}}$

Q7: Evaluate $16^{\frac{1}{4}}$

Q8: Evaluate $27^{\frac{2}{3}}$

Q9: Evaluate $4^{\frac{3}{2}}$

4.1.2 Laws of logarithms

Key point

The first law of indices states that $a^m \times a^n = a^{m+n}$ and leads us to the first law of logarithms.

$$\log_a m + \log_a n = \log_a(m \times n)$$

Examples

1. Problem:

Evaluate $\log_6 4 + \log_6 9$

Solution:

$$\log_6 4 + \log_6 9 = \log_6 (4 \times 9)$$
$$= \log_6 36$$

© HERIOT-WATT UNIVERSITY

2. Problem:

What is $\log_2 32 + \log_2 64$?

Solution:

$$\log_2 32 + \log_2 64 = \log_2 (32 \times 64)$$
$$= \log_2 2048$$

We can simplify this further with a calculator but we must also be able to evaluate it without a calculator..

Let $y = \log_2 2048$ turn this into an exponential

$2^y = 2048$

2 to the power is easy to find by repeated doubling (use your fingers to keep track of the power).

2	4	8	16	32	64	128	256	512	1024	2048
2^1	2^2	2^3	2^4	2^5	2^6	2^7	2^8	2^9	2^{10}	2^{11}

and $\log_2 2048 = 11$

Hence $\log_2 32 + \log_2 64 = 11$

Key point

The second law of indices states that $\frac{a^m}{a^n} = a^{m-n}$ and leads us to the second law of logarithms.

$$\log_a m - \log_a n = \log_a \frac{m}{n}$$

Example

Problem:

What is $\log_3 81 - \log_3 9$?

Solution:

$$\log_3 81 - \log_3 9 = \log_3 \frac{81}{9}$$
$$= \log_3 9$$

We can evaluate this with a calculator but again we must be able to work out the answer without a calculator.

Let $y = \log_3 9$
$3^y = 9$

we know that $y = 2$ because $3^2 = 9$

so $\log_3 9 = 2$

Hence $\log_3 81 - \log_3 9 = 2$.

TOPIC 4. DIFFERENTIAL EQUATIONS

> **Key point**
>
> The third law of logarithms is:
>
> $$\log_a x^n = n\log_a x$$
>
> Notice that the power becomes a scalar.

Example

Problem:

What is $\log_5 125^4$?

Solution:

$\log_5 125^4 = 4 \log_5 125$

We can simplify this further with a calculator but we must also be able to evaluate it without a calculator.

Let $y = \log_5 125$
$5^y = 125$

we know that $y = 3$ because $5^3 = 125$

so $4\log_5 125 = 4 \times 3 = 12$

Hence $\log_5 125^4 = 12$.

> **Key point**
>
> The fourth law of indices states that $a^0 = 1$ and leads us to the fourth law of logarithms.
>
> $$\log_a 1 = 0$$

> **Key point**
>
> We also know that $a^1 = a$ and that leads us to the fifth law of logarithms.
>
> $$\log_a a = 1$$

Example

Problem:

What is $\log_8 4 + \log_8 6 - \log_8 3$?

© HERIOT-WATT UNIVERSITY

Solution:

$$\log_8 4 + \log_8 6 - \log_8 3 = \log_8 \left(\frac{4 \times 6}{3}\right)$$
$$= \log_8 8$$
$$= 1$$

A word of warning $\log_a m - \log_a n = \log_a \frac{m}{n}$ and should not be confused with $\frac{\log_a m}{\log_a n}$ which can only be evaluated as it stands.

> **Key point**
>
> The laws of logarithms only apply when the bases of the logs are equal.

These laws are required for manipulation of expressions involving logs and exponentials. Although it is important to learn these laws, familiarity with them will increase with practice.

> **Key point**
>
> The laws of logarithms are:
>
> 1. $\log_a m + \log_a n = \log_a (m \times n)$
> 2. $\log_a m - \log_a n = \log_a \frac{m}{n}$
> 3. $\log_a x^n = n \log_a x$
> 4. $\log_a 1 = 0$
> 5. $\log_a a = 1$

Examples

1. Problem:

Simplify $\log_3 (p + 2) - 4 \log_3 2$.

Solution:

$$\log_3 (p + 2) - 4 \log_3 2 = \log_3 (p + 2) - \log_3 2^4$$
$$= \log_3 (p + 2) - \log_3 16$$
$$= \log_3 \frac{p + 2}{16}$$

....

2. Problem:

What is the value of $\frac{\log_2 16}{\log_2 2}$?

TOPIC 4. DIFFERENTIAL EQUATIONS

Solution:

$$\frac{\log_2 16}{\log_2 2} = \frac{\log_2 2^4}{\log_2 2}$$
$$= \frac{4\log_2 2}{\log_2 2}$$
$$= 4$$

because $\frac{\log_2 2}{\log_2 2} = 1$ and $4 \times 1 = 4$

3. Problem:

If $\log_8 r = \frac{1}{3}$, what is the value of r?

Solution:

$\log_8 r = \frac{1}{3}$ can be simplified to,

$r = 8^{\frac{1}{3}}$
$r = \sqrt[3]{8}$
$r = 2$

4. Problem:

What is the exact value of $\log_4 64$?

Solution:

$\log_4 64 = \log_4 4^3$
$= 3\log_4 4$
$= 3 \times 1$
$= 3$

Laws of logarithms exercise   Go online

Without using a calculator answer these questions.

Q10: What is $\log_6 18 + \log_6 2$?

Q11: What is $5\log_5 25$?

Q12: What is $\frac{1}{4}\log_2 16$?

Without using a calculator, simplify these expressions.

Q13: $3\log_{10} 4 - \log_{10} 8$ giving your answer as a log.

© HERIOT-WATT UNIVERSITY

Q14: $\log_{10}\left(5 \times 10^{1000}\right) - \log_{10} 5$

Find the value of x in the following by using the log laws.

Q15: $\ln x^2 - \ln 36 = \ln e$

....

Q16: $3\log_5 x - \log_5 6^3 = 0$

4.1.3 The chain rule

It will be useful to remember both forms of the chain rule. (However, you are not required to be able to prove either of them.)

Key point

Function notation:
$h\,'(x) = g\,'(f(x)) \times f\,'(x)$

Leibniz notation:
$dy/dx = dy/du \times du/dx$

Many types of composite functions can be differentiated using the chain rule. Note that either function or Leibniz notation can be used.

Examples

1. Problem:

Find $\frac{dy}{dx}$ when $y = \sqrt{x^3 - 4x^2 - 5}$.

Solution:

It is often easier to follow these steps.

Step 1:	Turn the square root into a power.	$y = \left(x^3 - 4x^2 - 5\right)^{\frac{1}{2}}$
Step 2:	Bring down the power.	$\frac{1}{2}$
Step 3:	Write down the bracket.	$\frac{1}{2}\left(x^3 - 4x^2 - 5\right)$
Step 4:	Reduce the power by 1.	$\frac{1}{2}\left(x^3 - 4x^2 - 5\right)^{-\frac{1}{2}}$
Step 5:	Differentiate the bracket.	$\frac{1}{2}\left(x^3 - 4x^2 - 5\right)^{-\frac{1}{2}} \times \left(3x^2 - 8x\right)$
Step 6:	Simplify the answer.	$\frac{dy}{dx} = \frac{3x^2 - 8x}{2\sqrt{x^3 - 4x^2 - 5}}$

....

2. Problem:

Find $f\,'(x)$ when $f(x) = \cos(2x + 1)$.

TOPIC 4. DIFFERENTIAL EQUATIONS

Solution:

Step 1:	Differentiate cos.	$-sin$
Step 2:	Write down the bracket.	$-sin(2x+1)$
Step 3:	Differentiate the bracket.	$-sin(2x+1) \times 2$
Step 4:	Simplify the answer.	$-2sin(2x+1) \times 2$

3. Problem:

Find the derivative of $\sin^2(3-2x)$.

Solution:

Step 1:	Remember:	$\sin^2(3-2x) = (sin(3-2x))^2$
Step 2:	Bring down the power.	2
Step 3:	Write down the bracket.	$2sin(3-2x)$
Step 4:	Reduce the power by 1.	$2sin(3-2x)^1$
Step 5:	Differentiate the bracket.	
	Now the bracket is:	$sin(3-2x)$
	We have to use the chain rule here too.	$cos(3-2x) \times -2 = -2cos(3-2x)$
	Giving:	$2sin(3-2x) \times -2cos(3-2x)$
Step 6:	Simplify the answer.	$\frac{d}{dx} = -4\sin(3-2x)\cos(3-2x)$

Differentiating composite functions exercise Go online

Try these questions on paper and then check your answers.

Q17: Find $h'(x)$ when $h(x) = \sqrt{x^2 + 6x}$

Q18: If $f(x) = \cos^3 x$, find $f'(x)$.

4.2 First order linear differential equations introduction

A differential equation is an equation for a function that relates the values of the function to the values of its derivatives.

They are used in a wide variety of disciplines and can be used to model various situations such as exponential growth and decay, chemical reactions, optimum investment strategies, the oscillation of a spring and many more. Any situation that involved a rate of change can be described by a

© HERIOT-WATT UNIVERSITY

differential equation.

In the first section we will consider first order linear differential equations. This will be followed by second order linear differential equations.

> **Key point**
>
> **Differential equation**
>
> A differential equation is an equation involving an unknown function and its derivatives.

> **Key point**
>
> **Order**
>
> The order of a differential equation is that of the highest-order derivative appearing in the equation.

Examples

1. First order differential equation

$\frac{dy}{dx} = x^2 y$

This is a first order differential equation because $\frac{dy}{dx}$ is the first derivative. It has order 1.

....

2. Second order differential equation

$\frac{d^2y}{dx^2} + \frac{dy}{dx} + y^3 = \sin x$

This is a second order differential equation because $\frac{d^2y}{dx^2}$ (the highest-order derivative in the equation) is the second derivative. It has order 2.

> **Key point**
>
> **First order linear differential equation**
>
> A first order linear differential equation is an equation that can be expressed in the standard linear form:
>
> $$\frac{dy}{dx} + P(x)y = f(x)$$

The equation is termed linear because it involves only first order terms in $\frac{dy}{dx}$ and y, and no terms such as $y^2, y\frac{dy}{dx}, y^3, \cos(y)$ etc.

Examples

1. $(4 - x^2)\frac{dy}{dx} + 3y = (4 - x^2)^2$ is a first order linear differential equation.

....

2. $x\frac{dy}{dx} + y^2 = 0$ is a first order differential equation which is not linear because it contains a y term of degree 2.

3. $\frac{dy}{dx} + 3xy = x$ where $P(x) = 3x$, $f(x) = x$

4. $\frac{dy}{dx} - y = e^x$ where $P(x) = -1$, $f(x) = e^x$

5. $\frac{dy}{dt} + y\sin t = t^2$ where $P(t) = \sin t$, $f(t) = t^2$

Some algebraic manipulation may be necessary in order to express an equation in standard linear form. This is explored in these examples.

Examples

1. Problem:

Write $x\frac{dy}{dx} - 3x^2 = y$ in standard linear form.

Solution:

To rearrange this first order linear differential equation into standard linear form use the normal algebraic manipulation rules you would use to rearrange any other algebraic equation. In this case, $\frac{dy}{dx}$ must have a coefficient of one, and $\frac{dy}{dx}$ and y must both be on the LHS.

Given $x\frac{dy}{dx} - 3x^2 = y$

Add $3x^2$ to both sides and subtract y from both sides:

$x\frac{dy}{dx} - y = 3x^2$

Divide both sides by x:

$\frac{dy}{dx} - \frac{y}{x} = \frac{3x^2}{x}$

Simplify:

$\frac{dy}{dx} - \frac{y}{x} = 3x$

2. Problem:

Write $\frac{dy}{dx} = x^2 + xy - 3$ in standard linear form.

Solution:

To rearrange this first order linear differential equation into standard linear form use the normal algebraic manipulation rules you would use to rearrange any other algebraic equation. In this case, $\frac{dy}{dx}$ and the y term must be on the LHS. Everything else must be on the RHS.

Given: $\frac{dy}{dx} = x^2 + xy - 3$

Subtract xy from both sides:

$\frac{dy}{dx} - xy = x^2 - 3$

304 TOPIC 4. DIFFERENTIAL EQUATIONS

Differential equations arise in many situations and they are usually solved by integration. A first order differential equation is usually solved by a single integration and the solution obtained contains an arbitrary constant of integration.

> **Key point**
>
> An **arbitrary constant** is a constant that occurs in the general solution of a differential equation.

> **Key point**
>
> The solution of a differential equation containing a constant of integration is known as a **general solution** of the equation.

Since the constant can assume any value, the differential equation actually has infinitely many solutions.

In applications of differential equations, the solution often has to satisfy an additional condition - usually called the **initial condition**. The initial condition can be used to find the value of the constant in the general solution.

> **Key point**
>
> When the value of the constant of integration is known in the general solution, then the resulting solution is called the **particular solution**.

For a differential equation an initial condition is additional information required to determine a particular solution. This could be a coordinate on a curve, a velocity at $t = 0$, the amount of money in a bank account on 1st January 2000, etc.

The particular solution of a differential equation is a solution which is often obtained from the general solution when an initial condition is known.

This is illustrated in the following examples.

Equation of a curve

Example

Problem:

The gradient of the tangent to a curve is given by the differential equation $\frac{dy}{dx} = 2x - 4$

Find the equation of the curve.

Solution:

We solve this problem by integrating:

$$\int \frac{dy}{dx}\, dx = \int 2x - 4\, dx$$
$$y = x^2 - 4x + C$$

The above is a general solution which gives a family of curves.

TOPIC 4. DIFFERENTIAL EQUATIONS

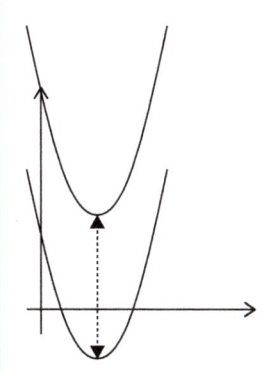

The curves all have the same gradient given by $\frac{dy}{dx} = 2x - 4$

If we are given more information, then we can identify one curve in particular.

For example, if we know that the point $(1, 0)$ lies in the curve then we can calculate C.

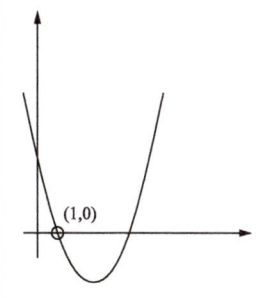

$y = x^2 - 4x + C$
$0 = 1 - 4 + C$
$C = 3$

In this case, $y(1) = 0$ is the initial condition for the differential equation.

Using this initial condition, we obtain the particular solution $y = x^2 - 4x + 3$

Example

Problem:

A stone is dropped into a still pond. Ripples move out from the point where the stone hits in the form of concentric circles.

The area, A m², of disturbed water increases at a rate of $10t$ m/s, where t is the time in seconds.

Calculate the area of disturbed water after 2 seconds.

Solution:

The rate of change of area is given by the differential equation $\frac{dA}{dt} = 10t$

Integrating this with respect to t gives:

$$\int \frac{dA}{dt} \, dt = \int 10t \, dt$$
$$A = 5t^2 + C$$

This is the general solution of the equation.

To find the appropriate particular solution, use a suitable initial condition. In this case, use the fact that $A = 0$ when $t = 0$ (this corresponds to the fact that the area equals zero at the instant that the stone is dropped).

Substituting these values into:

$A = 5t^2 + C$
$0 = 5(0)^2 + C$
$C = 0$

Therefore, the particular solution to the differential equation is $A = 5t^2$

Use this particular solution to calculate the area of disturbed water after 2 seconds.

When $t = 2$, then:

$A = 5t^2$
$A = 5(2)^2$
$A = 20$ m^2

First order linear differential equations introduction exercise   Go online

Match the following differential equations to their correct titles.

Q19: $\frac{d^2y}{dx^2} - 4\frac{dy}{dx} + 3 = 0$
a) First order linear
b) Second order linear
c) Neither

....

TOPIC 4. DIFFERENTIAL EQUATIONS

Q20: $\frac{d^2y}{dx^2} - xy^2 = 0$
a) First order linear
b) Second order linear
c) Neither

....

Q21: $3\frac{dy}{dx} - 4y = 10$
a) First order linear
b) Second order linear
c) Neither

....

Q22: $3 - x^4y = \frac{dy}{dx}$
a) First order linear
b) Second order linear
c) Neither

....

Q23: $\frac{dy}{dx} - 4x^2 = \cos(y)$
a) First order linear
b) Second order linear
c) Neither

....

Q24: $\frac{d^2y}{dx^2} - 6x^2y = 12$
a) First order linear
b) Second order linear
c) Neither

....

Q25: A curve $f(x)$ has a tangent given by $f'(x) = 3x^2 + 2$
The curve passes through the point $(2, 11)$.

  a) Integrate $f'(x) = 3x^2 + 2$ to find the general solution $f(x)$.

  b) Evaluate the constant of integration and write down the equation of the curve $f(x)$.

....

Q26: A ball is thrown vertically upwards from a point 6m above the ground. The velocity of the ball is given by the differential equation $V = \frac{dh}{dt} = 13 - 10t$

  a) Write down the function, $h(t)$, which describes the height of the ball above the ground at time t.

  b) Calculate the height of the ball above the ground after 1·5 seconds.

© HERIOT-WATT UNIVERSITY

Q27: Bacteria are cultured in a lab. The number of bacteria, $B(t)$, increases at a rate of $12e^{0.6t}$ bacteria per hour, where t is the time in hours. There are 20 bacteria to start with.

a) Write down the function $B(t)$, which describes the number of bacteria at time t.
b) How many bacteria are present after 10 hours? (Give your answer to the nearest unit.)

4.3 First order linear differential equations with separable variables

The previous section explored the format of first order linear differential equations and how they could be written in standard linear form. It also looked at differential equations that we could apply straightforward integration to in order to solve them. This is not always the case and other methods of solving first order linear differential equations must be used. Examples illustrating this are the equations for bacterial growth and decay and radioactive decay. Both of which will be explored in the following sections.

4.3.1 Separable variables

First order linear variable separable equations are made up of two different variables, usually x and y, which can be separated through algebraic manipulation. The method is to separate like variables, say the x variables, onto the RHS and the y variable onto the LHS.

The general form of the first order separable variable equation is given in the key point below.

Key point

First order separable variable equations can be written in the form:

$$\frac{dy}{dx} = f(x)g(y)$$

We can rewrite $\frac{dy}{dx} = f(x)g(y)$ as $\frac{1}{g(y)} dy = f(x) dx$

Integrating both sides with respect to x gives:

$$\int \frac{1}{g(y)} dy\, dx = \int f(x) dx\, dx$$

$$\int \frac{1}{g(y)} dy \frac{dx}{dx} = \int f(x) dx$$

$$\int \frac{1}{g(y)} dy = \int f(x) dx$$

If both integrations can be carried out in this equation, then it is possible to find a general solution.

TOPIC 4. DIFFERENTIAL EQUATIONS

Examples

1. Problem:
Find the general solution for the differential equation $\frac{dy}{dx} = \frac{3x^2}{y}$

Solution:
Separating variables gives: $y\, dy = 3x^2\, dx$
Integrating both sides gives:
$$\int y\, dy = \int 3x^2\, dx$$
$$\frac{y^2}{2} = \frac{3}{3}x^3 + C$$
$$y^2 = 2x^3 + C$$
$$y = \pm\sqrt{2x^3 + C}$$

(Note that, in the third line of the working above, $2C$ was rewritten as C. This is because C is just an arbitrary constant.)

....

2. Problem:
Find the general solution for the differential equation $\frac{dy}{dx} = x^3 y$

Solution:
Separating the variables gives: $\frac{1}{y} dy = x^3\, dx$
Integrating both sides gives:
$$\int \frac{1}{y} dy = \int x^3\, dx$$
$$\ln|y| = \frac{1}{4}x^4 + C$$
$$y = e^{\left(\frac{1}{4}x^4 + C\right)} \text{ use the exponential rule } e^{(x+y)} = e^x e^y$$
$$= e^{\left(\frac{1}{4}x^4\right)} e^C$$
$$= A e^{\left(\frac{1}{4}x^4\right)} \text{ where } A = e^C \text{ because } e^C \text{ is a constant}$$

....

3. Problem:
Given that $y = -1$ when $x = 0$, find the particular solution for the differential equation $x \frac{dy}{dx} = y^2 - \frac{dy}{dx}$

Solution:
First of all, rearrange the equation so that the variables can be separated:
$$x\frac{dy}{dx} = y^2 - \frac{dy}{dx}$$
$$x\frac{dy}{dx} + \frac{dy}{dx} = y^2$$
$$(x+1)\frac{dy}{dx} = y^2$$

Then, separating variables gives: $\frac{1}{y^2} dy = \frac{1}{x+1} dx$

© HERIOT-WATT UNIVERSITY

Integrating both sides gives:

$$\int \frac{1}{y^2}\, dy = \int \frac{1}{x+1}\, dx$$

$$-\frac{1}{y} = \ln|x+1| + C$$

Substitute in $x = 0$ and $y = -1$:

$1 = \ln 1 + C$
$C = 1$

Therefore, the particular solution is:

$$-\frac{1}{y} = \ln(x+1) + 1$$

$$y = -\frac{1}{\ln(x+1)+1}$$

When writing your solution to this type of question always rearrange to $y = \ldots$

....

4. Problem:

Integrate to obtain the general solution for the differential equations $\frac{dy}{dx} = \frac{xy}{1+x^2}$

Solution:

First separate the variables by dividing both sides by y and multiplying by dx: $\frac{dy}{y} = \frac{x}{1+x^2}\, dx$

Integrate both sides: $\int \frac{dy}{y} = \int \frac{x}{1+x^2}\, dx$

We can use standard results to integrate both the LHS and RHS. Both will integrate to logarithms. Recall the standard result:

$\int \frac{f'(x)}{f(x)}\, dx = \ln|f(x)| + C$

Note that the derivative of the denominator, $1 + x^2$, is $2x$. To make it obvious that the numerator is the derivative of the denominator we multiply it by 2, then compensate by multiplying the integrand by $\frac{1}{2}$: $\ln|y| = \frac{1}{2}\int \frac{2x}{1+x^2}\, dx$

Now integrate: $\ln|y| = \frac{1}{2}\ln\left|1+x^2\right| + C$

Now rearrange for y. Start by using the logarithmic rule $a\ \ln|x| = \ln|x|^a$: $\ln|y| = \ln\left|1+x^2\right|^{\frac{1}{2}} + C$

Take the exponential of both sides: $e^{\ln|y|} = e^{\ln\left|1+x^2\right|^{\frac{1}{2}} + C}$

Using the exponential rule $x^{(a+b)} = x^a x^b$: $y = e^{\ln\left|1+x^2\right|^{\frac{1}{2}}} e^C$

Recall that e^C is a constant and can be replaced by A in the general solution:

$y = A e^{\ln\left|1+x^2\right|^{\frac{1}{2}}}$

$y = A\left|1+x^2\right|^{\frac{1}{2}}$

$y = A\sqrt{1+x^2}$

Note that we do not need the modulus signs for the expression $1 + x^2$ as it will always be positive and so we do not need to use them in the final line of the general solution.

....

TOPIC 4. DIFFERENTIAL EQUATIONS

5. Problem:

Find the particular solution for the differential equation $x\frac{dy}{dx} - 2 = 2y - 3\frac{dy}{dx}$, when $x = 1$ and $y = 3$.

Solution:

First we must rearrange the differential equation to collect terms involving $\frac{dy}{dx}$ to the LHS. To do this add $3\frac{dy}{dx}$ and 2 to both sides:

$x\frac{dy}{dx} + 3\frac{dy}{dx} = 2y + 2$

Take a common factor of $\frac{dy}{dx}$ on the LHS: $(x+3)\frac{dy}{dx} = 2y + 2$

Separate variables, x terms on the RHS and y terms on the LHS: $\frac{dy}{2y+2} = \frac{dx}{x+3}$

Now integrate both sides: $\int \frac{dy}{2y+2} = \int \frac{dx}{x+3}$

We can use standard results to integrate both the LHS and RHS. Both will integrate to logarithms. Recall the standard result:

$\int \frac{f'(x)}{f(x)} dx = \ln|f(x)| + C$

Note that the derivative of the denominator on the LHS, $2y + 2$, is 2. To make it obvious that the numerator is the derivative of the denominator we multiply it by 2, then compensate by multiplying the integrand by $\frac{1}{2}$:

$\frac{1}{2} \int \frac{2dy}{2y+2} = \int \frac{dx}{x+3}$

Now integrate: $\frac{1}{2} \ln|2y+2| = \ln|x+3| + C$

Now rearrange for y. Start by using the logarithmic rule $a \ln|x| = \ln|x|^a$.

$\ln|2y+2|^{\frac{1}{2}} = \ln|x+3| + C$

Take the exponential of both sides and apply the exponential rule $x^{(a+b)} = x^a x^b$:

$e^{\ln|2y+2|^{\frac{1}{2}}} = e^{\ln|x+3|+C}$

$e^{\ln|2y+2|^{\frac{1}{2}}} = e^{\ln|x+3|} e^C$

Recall that e^C is a constant and so we can replace e^C by A:

$|2y+2|^{\frac{1}{2}} = A|x+3|$

Square both sides: $2y + 2 = [A(x+3)]^2$ (*)

Before rearranging for y and obtaining a general solution, we first want to find the value of the constant of integration A. We do this by substituting for the initial conditions $x = 1$ and $y = 3$:

$2(3) + 2 = [A(1+3)]^2$

$8 = 16A^2$

$A^2 = \frac{1}{2}$

$A = \frac{1}{\sqrt{2}}$

Substituting $A = \frac{1}{\sqrt{2}}$ back into (*) we therefore have: $2y + 2 = \left[\frac{1}{\sqrt{2}}(x+3)\right]^2$

© HERIOT-WATT UNIVERSITY

Now expand the brackets and rearrange for y:

$$2y + 2 = \frac{1}{2}(x+3)^2$$

$$2y = \frac{1}{2}(x+3)^2 - 2$$

$$y = \frac{1}{4}(x+3)^2 - 1$$

This is the particular solution.

These examples have demonstrated the solving of first order linear differential equations by using separation of variables. In each case, standard results have been used to integrate each function. It may be the case, however, that previously learned methods of integration must be used to complete the integration, such as writing the function as a sum of partial fractions or using integration by parts. The following examples will look at these two cases.

Examples

1. Problem:

Find the general solution to the differential equation $\frac{dy}{dx} = \frac{x^2(y^2-4)}{2y}$

Solution:

Separate the variables by arranging the y terms on the LHS and the x terms on the RHS:

$\frac{2y}{y^2-4} dy = x^2 dx$

Now integrate both sides: $\int \frac{2y}{y^2-4} dy = \int x^2 dx$

Note that the denominator of the LHS is $y^2 - 4$. This can be factorised using the difference of two squares:

$\int \frac{2y}{(y-2)(y+2)} dy = \int x^2 dx$

The LHS is still not in integrable form. We need to re-write it as the sum of partial fractions.

Set up the form of the partial fraction for the integrand on the LHS: $\frac{2y}{(y-2)(y+2)} \equiv \frac{A}{y-2} + \frac{B}{y+2}$

Obtain a common denominator: $\frac{2y}{(y-2)(y+2)} \equiv \frac{A(y+2)}{(y-2)(y+2)} + \frac{B(y-2)}{(y-2)(y+2)}$

Equate the numerators: $2y = A(y+2) + B(y-2)$

Select values of y to evaluate for A and B:

Let $y = 2$: $2(2) = A(4) + B(0) \Rightarrow A = 1$

Let $y = -2$: $2(-2) = A(0) + B(-4) \Rightarrow B = 1$

Substitute for $A = 1$ and $B = 1$ into the sum of partial fractions: $\frac{2y}{(y-2)(y+2)} \equiv \frac{1}{y-2} + \frac{1}{y+2}$

Now substitute for the sum of partial fractions for the integrand on the LHS and integrate:

$$\int \frac{1}{y-2} + \frac{1}{y+2} dy = \int x^2 dx$$

$$\ln|y-2| + \ln|y+2| = \frac{x^3}{3} + C$$

TOPIC 4. DIFFERENTIAL EQUATIONS

Using the logarithmic rule $\ln|x|+\ln|y| = \ln|xy|$, we can simplify the LHS: $\ln|(y-2)(y+2)| = \frac{x^3}{3} + C$

Taking the exponential of both sides we get:

$$e^{\ln|(y-2)(y+2)|} = e^{\frac{x^3}{3}+C}$$

$$(y-2)(y+2) = e^{\frac{x^3}{3}+C}$$

To continue to simplify we must recall the exponential rule $e^{(x+y)} = e^x e^y$ and the fact that e^C is a constant and can be replaced by the arbitrary value A:

$$y^2 - 4 = e^{\frac{x^3}{3}} e^C$$

$$y^2 - 4 = A e^{\frac{x^3}{3}}$$

Now continue to simply by rearranging the equation for y, recalling that when square rooting there are always two solutions: one positive and one negative:

$$y^2 = A e^{\frac{x^3}{3}} + 4$$

$$y = \pm \sqrt{A e^{\frac{x^3}{3}} + 4}$$

....

2. Problem:

Find the general solution to the first order linear differential equation.

$$-3\frac{dy}{dx} = (y+2)(y-1)$$

Solution:

The first step is to separate the variables by arranging the y terms on the LHS and the x terms on the RHS:

$$\frac{-3}{(y+2)(y-1)} dy = dx$$

Now integrate both sides: $\int \frac{-3}{(y+2)(y-1)} dy = \int 1 \, dx$

It is clear that the RHS can be integrated using standard results, but the LHS is not in integrable form. It must be rewritten as the sum of partial fractions.

Taking the integrand of the LHS set up the form of the sum of partial fractions: $\frac{-3}{(y+2)(y-1)} \equiv \frac{A}{y+2} + \frac{B}{y-1}$ (*)

Obtain a common denominator and equate the numerators:

$$\frac{-3}{(y+2)(y-1)} \equiv \frac{A(y-1)}{(y+2)(y-1)} + \frac{B(y+2)}{(y+2)(y-1)}$$

$$-3 = A(y-1) + B(y+2)$$

We can now evaluate the constants A and B by selecting values for y:

Let $y = -2$: $-3 = A(-3) + B(0) \Rightarrow A = 1$

Let $y = 1$: $-3 = A(0) + B(3) \Rightarrow B = -1$

© HERIOT-WATT UNIVERSITY

Now substitute $A = 1$ and $B = -1$ back into (*) to obtain the integrand for the LHS. The differential equation that we will solve now becomes:

$$\int \frac{1}{y+2} - \frac{1}{y-1} \, dy = \int 1 \, dx$$

$\ln|y+2| - \ln|y-1| = x + C$

From here we want to rewrite the general solution in terms of y. To start this process use the logarithmic rule $\ln|a| - \ln|b| = \ln\left|\frac{a}{b}\right|$ to rewrite the LHS:

$\ln\left|\frac{y+2}{y-1}\right| = x + C$

Now take the exponential of both sides: $e^{\ln\left|\frac{y+2}{y-1}\right|} = e^{x+C}$

To continue to simplify the RHS we must recall the exponential rule $e^{(x+y)} = e^x e^y$ and the fact that e^C is a constant and can be replaced by the arbitrary value A:

$$\frac{y+2}{y-1} = e^x e^C$$

$$\frac{y+2}{y-1} = Ae^x$$

To simplify the LHS, we can rewrite the numerator in the form $y + 2 = (y - 1) + 3$, then simplify, as shown below:

$$\frac{(y-1)+3}{y-1} = Ae^x$$

Now separate the LHS into the sum of two fractions and simplify:

$$\frac{y-1}{y-1} + \frac{3}{y-1} = Ae^x$$

$$1 + \frac{3}{y-1} = Ae^x$$

Continue to simplify by arranging the y terms on the LHS and everything else on the RHS:

$\frac{3}{y-1} = Ae^x - 1$

Invert both sides and continue to rearrange:

$$\frac{y-1}{3} = \frac{1}{Ae^x - 1}$$

$$y - 1 = \frac{3}{Ae^x - 1}$$

$$y = \frac{3}{Ae^x - 1} + 1$$

Now write the RHS as one fraction by getting a common denominator:

$$y = \frac{3}{Ae^x - 1} + \frac{Ae^x - 1}{Ae^x - 1}$$

$$y = \frac{3 + Ae^x - 1}{Ae^x - 1}$$

$$y = \frac{Ae^x + 2}{Ae^x - 1}$$

3. Problem:

Solve the first order linear differential equation $\frac{dy}{dx} = \frac{\ln|x|}{y^2}$

Solution:

First rearrange the differential equation by collecting all y terms on the LHS and x terms on the RHS:

$y^2 dy = \ln|x|\, dx$

Integrate both sides: $\int y^2\, dy = \int \ln|x|\, dx$

Now that the variables have been separated it is clear that we can used standard results to integrate the LHS, but not the RHS. To integrate the RHS, we need to use integration by parts, a technique that we learned about in the previous topic. Integration by parts allows us to integrate the product of two functions. To make the RHS integrand a product of two functions we multiply it by 1:

$\int y^2\, dy = \int 1 \times \ln|x|\, dx$

Integrating the LHS gives: $\frac{y^3}{3} = \int 1 \times \ln|x|\, dx$ (*)

To integrate the RHS, recall the integration by parts rule: $\int fg' = fg - \int f'g$

Since we can differentiate ln|x| and integrate 1, we can set up the integration as follows:

Let $f = \ln|x|$, then $f' = \frac{1}{x}$

Let $g' = 1$, then $g = x$

The RHS integrand then becomes:

$$\int 1 \times \ln|x|\, dx = \ln|x| \times x - \int \frac{1}{x} \times x\, dx$$

$$= x\ln|x| - \int 1\, dx$$

$$= x\ln|x| - x + C$$

Now substituting back into (*) we get:

$\frac{y^3}{3} = x\ln|x| - x + C$

The last step is to rearrange for y:

$y^3 = 3(x\ln|x| - x + C)$

$y^3 = 3x\ln|x| - 3x + 3C$

$y^3 = 3x(\ln|x| - 1) + C$

Note that after expanding the bracket, the constant of integration becomes $3C$. Since this is an arbitrary constant, it can replace it by C.

The last step is to take the cube root of the RHS in order to get the general solution in the form $y = \ldots$

$y = \sqrt[3]{3x(\ln|x| - 1) + C}$

Key point

To solve a first order separable variable differential equation of the form $\frac{dy}{dx} = f(x)g(y)$:

1. Separate the variables: $\frac{1}{g(y)} dy = f(x) \, dx$
2. Integrate both sides: $\int \frac{1}{g(y)} dy = \int f(x) \, dx$

First order linear differential equations with separable variables extra help

First order linear differential equations with separable variables can be rearranged into the form: $g(y) \, dy = f(x) \, dx$

Once you have arranged it in this form you may need to use any known process to integrate i.e. algebraic manipulation to produce a standard integral, partial fractions, substitution or integration by parts.

The general solution will contain an arbitrary constant. If you are given an initial condition in order to determine the constant, then you will obtain a particular solution.

Find the general solution of $x \frac{dy}{dx} = y$ by answering the following set of multiple choice questions.

Q28:

Multiplying both sides of $x \frac{dy}{dx} = y$ by the differential dx gives:

a) $x \, dy = y + dx$
b) $x \, dy = y \, dx$
c) $x \, dx = y \, dy$

....

Q29: Separate x and dx to one side and y and dy to the other side.

In this case, we need to divide both sides by x and also divide both sides by y, i.e. divide both sides by xy which gives:

a) $\frac{dy}{y} = \frac{dx}{x}$
b) $\frac{dy}{x} = \frac{dx}{y}$
c) $y \, dy = x \, dx$

....

Q30: Integrating the separated expression gives:

a) $y^{-2} = x^{-2} + C$
b) $y = x + C$
c) $\ln y = \ln x + C$

TOPIC 4. DIFFERENTIAL EQUATIONS

Q31: Express the answer as "$y =$"
As a start, we will need to replace C with $\ln(A)$
$\ln(y) = \ln(x) + \ln(A)$
Continuing to simplify gives:

a) $y = x + A$
b) $\ln y = \ln(Ax) \Rightarrow y = e^{Ax}$
c) $\ln y = \ln(Ax) \Rightarrow y = Ax$

Find the general solution of $x \frac{dy}{dx} = 1$ by answering the following set of multiple choice questions.

Q32: Multiplying both sides of $x \frac{dy}{dx} = 1$ by the differential dx gives:

a) $dy = x\, dx$
b) $x\, dy = 1 + dx$
c) $x\, dy = dx$

Q33: Separate x and dx to one side and y and dy to the other side.
In this case, we need to divide both sides by x, which gives.

a) $dy = \frac{dx}{x}$
b) $dy = x\, dx$
c) $dx = x\, dy$

Q34: Integrating the separated expression gives:

a) $y = \ln(x) + C$
b) $\ln(y) = \ln(x) + C$
c) $y^2 = \frac{1}{x} + C$

Q35: Express the answer as "$y =$"
As a start, we will need to replace C with $\ln(A)$
$y = \ln(x) + \ln(A)$
Continuing to simplify gives:

a) $y = x + A$
b) $y = \ln(Ax)$
c) $y = e^{Ax}$

First order linear differential equations with separable variables Go online
exercise

Not all linear differential equations can be written in the form: $\frac{dy}{dx} = f(x)g(y)$

Decide which of the following six differential equations can be solved by the separable variable method. Recall that you may have to rearrange the differential equation using standard balancing rules including: adding, subtracting, multiplying, dividing and factorising.

Q36: $\frac{dy}{dx} = \sqrt{xy}$

....

Q37: $x\frac{dy}{dx} - \frac{y}{x} = 3x$

....

Q38: $\frac{dy}{dx} = 1 - x + y - xy$

....

Q39: $\frac{dy}{dx} = x^2 + xy - 3$

....

Q40: $x\frac{dy}{dx} = \frac{dy}{dx} + y^2 + xy$

....

Q41: $x\frac{dy}{dx} = \frac{dy}{dx} + y^2$

....

Q42:

A curve passes through the point $(3, 2)$ and its gradient function is $\frac{y+2}{x-2}$

a) Keeping your answer in logarithmic form, write down the general solution.
b) Using the coordinate $(3, 2)$, work out the value of the constant of integration, C.
c) Write down the equation of the curve y.

....

Q43:

Water leaks from a tank so that the rate of change of the depth of the water h metres is modelled by $\frac{dh}{dt} = -\frac{3h^2}{4}$ where t is the time in minutes. The initial depth of water in the tank is 1 metre.

a) Find the particular solution of the differential equation in terms of h.
b) How long does it take the depth of the water in the tank to decrease to 10 cm?

....

TOPIC 4. DIFFERENTIAL EQUATIONS 319

Q44:

Find the general solution to the differential equation $\frac{dy}{dx} = 1 - x + y - xy$

a) Factorise $1 - x + y - xy$
b) Keeping your answer in logarithmic form, write down the general solution.
c) By taking the exponential of both sides and using the exponential rule $e^{x+y} = e^x e^y$, write down the general solution in terms of y. Use A for the constant of integration.

....

Q45:

Find the particular solution for the differential equation $(1 + x)\frac{dy}{dx} - y(y + 1) = 0$ for $y = 1$ when $x = 1$.

a) Rearrange the differential equation into integrable form. Leave all expressions factorised.
b) Before integrating, the LHS must be split into its sum of partial fractions of the form $\frac{A}{y} + \frac{B}{y+1}$. State the value of A and B
c) Integrate and work out the constant of integration, C, for the initial conditions $y = 1$ when $x = 1$. Use the logarithmic rule $a \times \ln x = \ln(x^a)$ to write down your simplified answer.
d) Write down the function y.

4.3.2 Growth and decay

Many real-life situations can be modelled by differential equations. This section covers the forming and solving of these equations.

Consider the following situations.

Bacterial growth

In an experiment, the number of bacteria $B(t)$ present in a culture after t days is found to be increasing at a rate proportional to the number of bacteria present.

From this statement, it is possible to write down a differential equation for $B(t)$.

The rate of change of B is $\frac{dB}{dt}$, which is proportional to B.

Hence, $\frac{dB}{dt} = kB$ (i.e. the rate of change of number of bacteria is equal to the constant of proportionality multiplied by the number of bacteria at time t).

We can solve this to find B in terms of t. This is done in the following way.

Separating the variables gives $\frac{dB}{B} = k\,dt$

© HERIOT-WATT UNIVERSITY

Hence,

$$\int \frac{1}{B} dB = \int k \, dt$$
$$\ln B = kt + C$$
$$B = e^{kt+C}$$
$$B = Ae^{kt} \text{ where } A = e^C$$

$B = Ae^{kt}$ is the general solution of the equation.

If it is known that there are B_0 bacteria present at the start of the experiment, then $B = B_0$ when $t = 0$

Hence $B_0 = Ae^{0k} \Rightarrow A = B_0$ and we have the particular solution $B = B_0 e^{kt}$

Radioactive decay

When a radioactive atom emits some of its mass as radiation, the remainder of the atom reforms to make an atom of some new substance. This process is called radioactive decay. For example, radium decays into lead and carbon-14 decays into nitrogen-14.

For a radioactive substance, the mass m (grams) at time t (years) decreases at a rate proportional to the mass at that time.

In a similar way to the above example for growth, we can obtain an equation for decay. This time the constant of proportionality is negative because the mass is decreasing so $\frac{dm}{dt} = -km$ (i.e. the rate of change of mass is equal to the negative of the constant of proportionality multiplied by the mass at time t).

We can solve this differential equation to get the particular solution for radioactive decay. This is done as follows:

$\frac{dm}{dt} = -km$

Separate the variables, m terms on the LHS and t terms on the RHS, then integrate:

$$\int \frac{dm}{m} = \int -k \, dt$$
$$\ln |m| = -kt + C$$
$$m = e^{-kt+C}$$
$$m = e^{-kt} e^C$$
$$m = Ae^{-kt} \quad \text{where } A = e^C$$

$m = Ae^{-kt}$ is the general solution of the equation.

If it is known that there is m_0 mass to begin with, then $m = m_0$ when $t = 0$.

Hence, $m_0 = Ae^{-k(0)} \Rightarrow A = m_0$.

Hence we have the particular solution $m = m_0 e^{-kt}$

The examples of bacterial growth and radioactive decay above illustrate that **exponential growth** and **exponential decay** are of fundamental importance.

TOPIC 4. DIFFERENTIAL EQUATIONS

> **Key point**
>
> **Exponential growth**
>
> Exponential growth occurs when the population increases at a rate proportional to the size of the population and is written as:
>
> $$\frac{dP}{dt} = kP$$
>
> The general solution is: $P(t) = Ae^{kt}$

> **Key point**
>
> **Exponential decay**
>
> Exponential decay occurs when the population decreases at a rate proportional to the size of the population and is written as:
>
> $$\frac{dP}{dt} = -kP$$
>
> The general solution is: $P(t) = Ae^{-kt}$

Examples

1. Problem:

In an experiment, the number of bacteria $B(t)$ present in a culture after t days is increasing at a rate proportional to the number of bacteria present.

Suppose that there are B_0 bacteria present at the start of the experiment, i.e. $B(0) = B_0$.

If it is found that the bacteria population doubles in 10 hours, i.e. $B = 2B_0$ when $t = 10$, then calculate k, the constant of proportionality, and hence find a formula for $B(t)$ in terms of t.

Solution:

The population growth is described by the differential equation $\frac{dB}{dt} = kB$

Hence,

$$\int \frac{dB}{B} = \int k \, dt$$

$$\ln B = kt + C$$

$$B = e^{kt+C}$$

Since $B(0) = B_0$, then $B = B_0 e^{kt}$

Also, since $B(10) = 2B_0$, then substituting 10 for t we have:
$$2B_0 = B_0 e^{10k}$$
$$e^{10k} = 2$$
$$10k = \ln 2$$
$$k = \frac{\ln 2}{10} = 0 \cdot 069 \text{ (3 d.p.)}$$
Hence, the number of bacteria present in the culture at time t is given by: $B(t) = B_0 e^{0 \cdot 069t}$

....

2. Problem:

For a radioactive substance, the mass m (grams) at time t (years) decreases at a rate proportional to the mass at that time.

The half-life of any radioactive material is the time taken for half of the mass to decay. Given that the original mass of a particular radioactive substance is 800 grams and after 10 years it has decayed to 600 grams, find the half-life for this substance.

Solution:

The radioactive decay is described by the differential equation $\frac{dm}{dt} = -km$, hence:
$$\int \frac{dm}{m} = \int -k \, dt$$
$$\ln m = -kt + C$$
$$m = e^{-kt+C}$$
Since $m(0) = 800$, then:
$$800 = e^{-k(0) + C}$$
$$800 = e^C$$
Therefore: $m = 800e^{-kt}$

Also $m = 600$ when $t = 10$, hence: $600 = 800e^{-10k}$

Thus:
$$e^{-10k} = 0 \cdot 75$$
$$-10k = \ln(0 \cdot 75)$$
$$k = \frac{\ln(0 \cdot 75)}{-10} = 0 \cdot 0288 \text{ (3 s.f.)}$$
Therefore, the mass of the radioactive substance at time t is given by: $m(t) = 800e^{-0 \cdot 0288t}$

To find the half life for this substance, we let $m(t) = \frac{1}{2}800 = 400$ and solve the above equation for t.
$$400 = 800e^{-0 \cdot 0288t}$$
$$e^{-0 \cdot 0288t} = 0 \cdot 5$$
$$-0 \cdot 0288t = \ln 0 \cdot 5$$
$$t = 24 \cdot 1 \text{ years (approximately)}$$

So it will take approximately 24·1 years for half of the mass to decay.

....

3. Problem:

In a town with a population of 40,000 people, a flu virus spread rapidly last winter. The percentage P of the population infected in t days after the initial outbreak satisfies the differential equation $\frac{dP}{dt} = kP$ where k is a constant.

a) If 100 people are infected initially, find, in terms of k, the percentage infected t days later.
b) Given that 500 people have flu after 7 days, how many more are likely to have contracted the virus after 10 days?

This example was taken from CSYS Mathematics 1998, Paper 1.

Solution:

a) Since $\frac{dP}{dt} = kP$ then:

$$\int \frac{1}{P} \, dP = \int k \, dt$$
$$\ln P = kt + C$$
$$P = e^{kt+C}$$
$$= Ae^{kt} \text{ where } A = e^C$$

Since 100 people were infected initially when $t = 0$, then:

P = the percentage of the population infected = $\frac{100}{40000} \times 100$ = 0·25

Substitute these values into the equation for P:
$P = Ae^{kt}$ becomes
$0 \cdot 25 = Ae^0$
$A = 0 \cdot 25$

So, in terms of k, the percentage infected t days later is given by: $P = 0 \cdot 25 e^{kt}$

b) Since 500 people have flu after 7 days, then when $t = 7$: $P = \frac{500}{40000} \times 100 = 1 \cdot 25$

Hence:

$$1 \cdot 25 = 0 \cdot 25 e^{7k}$$
$$e^{7k} = 5$$
$$7k = \ln 5$$
$$k = \frac{1}{7} \ln 5 \approx 0 \cdot 230 \text{ (3 s.f.)}$$

To calculate the total number of people with flu after 10 days, substitute $t = 10$ and $k = 0 \cdot 230$ into the formula for P.

$P = 0 \cdot 25 e^{0 \cdot 23(10)}$ (3 s.f.)

Therefore, there will be approximately 2·49% of 40000 = 996 people with flu. This is an increase of 496 people.

....

4. Problem:

The number of strands of bacteria $B(t)$ present in a culture after t days of growth is assumed to be increasing at a rate proportional to the number of strands present.

Write down the differential equation for B and solve it to find B in terms of t.

Given that the number of strands observed after one day is 502 and after 4 days is 1833, find the number of strands initially present.

This example was taken from CSYS Mathematics 1993, Paper 1.

Solution:

The differential equation for B is $\frac{dB}{dt} = kB$.

Separate the variables by arranging all B terms on the LHS and everything else on the RHS.

$\frac{dB}{B} = k \, dt$

Integrate both sides:

$\int \frac{dB}{B} = \int k \, dt$

$\ln |B| = kt + C$

Now start to rearrange for B by taking the exponential of both sides:

$e^{\ln|B|} = e^{kt+C}$

$B = e^{kt} e^C$

$B = Ae^{kt}$ where $A = e^C$

The general solution is therefore $B = Ae^{kt}$.

Given that when $t = 1$, $B_1 = 502$ we can then write: $502 = Ae^k$

Given that when $t = 4$, $B_4 = 1833$ we can then write: $1833 = Ae^{4k}$

To be able to continue, we need to rearrange each of the equations above for e^k and e^{4k}, respectively. We also need to recall the rule $(a^m)^n = a^{mn}$.

Rearranging gives:

$502 = Ae^k \quad \Rightarrow \quad e^k = \dfrac{502}{A}$

$1833 = Ae^{4k} \quad \Rightarrow \quad e^{4k} = \dfrac{1833}{A}$

Now applying the rule $(a^m)^n = a^{mn}$ to the LHS of the second equation above gives $e^{4k} = (e^k)^4$. We can now equate the two equations above by raising the first equation to the power of four and making it equal to the second equation, then solving for A.

$\left(e^k\right)^4 = e^{4k}$

$\left(\dfrac{502}{A}\right)^4 = \dfrac{1833}{A}$

$\dfrac{502^4}{A^4} = \dfrac{1833}{A}$

$\dfrac{502^4}{1833} = A^3$

$A \approx 326$

TOPIC 4. DIFFERENTIAL EQUATIONS

The general solution is therefore $B_t = 326e^{kt}$.

The number of strands of bacterial initially present occurs at time zero, i.e. when $t = 0$. Substitute for $t = 0$ into the particular solution to work out B.

$B_0 = 326e^{k(0)}$

$B_0 \approx 326$ strands

Growth and decay: Half-life of a radioactive element

Go online

The half-life of a radioactive element is the time required for half of the radioactive nuclei present in a sample to decay.

The following graph shows the percentage of 1000 atomic nuclei of a radioactive substance, whose half-life period (T) amounts to 20 seconds, that are not yet decayed at a given time.

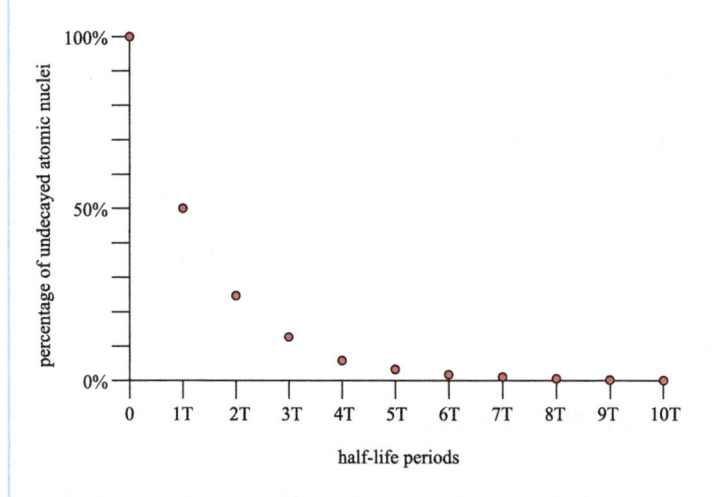

Q46:

a) Write down the differential equation that models the number N of radioactive nuclei present at time t.

b) Given that there are 1000 atomic nuclei at $t = 0$, find, in terms of t and k (the constant of proportionality), the number of nuclei still present at time t.

....

Q47: How many nuclei would you expect to be present after 20 seconds?

....

Q48: Estimate from the graph how long it will take (in seconds) for the radioactive element to decay to approximately 25% of its original mass. (Remember that T is 20 seconds.)

© HERIOT-WATT UNIVERSITY

Growth and decay exercise

Q49: A rumour spreads through a population in such a way that the number of people who hear the rumour n, grows in proportion to the number of people who have already heard the rumour. Three people started the rumour. After two hours have passed, 60 people have heard the rumour.

a) Write down the differential equation that models the number of people, n, who have heard the rumour after t hours.

b) Given that three people started the rumour, find, in terms of t and k, the number of people who know the rumour at time t.

c) After 2 hours, 60 people know the rumour; work out the value of k, the constant of proportionality. Give your answer to 3 decimal places.

d) Using this model, how many people would know the rumour after 3 hours? Give your answer to the nearest person.

....

Q50: If 10 grams of yoghurt is placed into 500 ml of warm milk and kept at a steady temperature of 45°C, then the amount of yoghurt doubles every 30 minutes.

a) Write down the differential equation that models the amount of yoghurt present, P, measured in grams, at time t, measured in hours.

b) Given there is 10 grams of yoghurt at $t = 0$, find, in terms of t and k, the amount of yoghurt present, P, at time t.

c) Given that the amount of yoghurt doubles after 30 minutes, work out the value of k, the constant of proportionality. Give your answer to 3 decimal places.

d) Using this model, how much yoghurt would be cultured after 4 hours? Give your answer to the nearest gram.

4.3.3 Further applications of differential equations

Newton's law of cooling

When a hot metal ingot is dropped into water, its temperature drops to that of the surrounding water.

When coffee is left standing in a mug, it cools until its temperature drops to that of the surrounding air.

These are examples that obey Newton's law of cooling.

The law states that the rate of cooling of a previously heated body is proportional to the *difference* between its temperature T after time t, and the temperature T_s of the surrounding environment.

This statement can be represented by a differential equation in the following way.

TOPIC 4. DIFFERENTIAL EQUATIONS

The rate of change of temperature T is $\frac{dT}{dt}$.

The rate of change of T is proportional to the *difference* between T and T_s.

Hence, $\frac{dT}{dt} = -k(T - T_s)$

Note that T_s is a constant, not a variable like T which is changing, i.e. the object is cooling to the surrounding temperature T_s.

Examples

1. Problem:

a) A hot liquid is cooling in a surrounding environment whose temperature remains at T_s. If T denotes the temperature of the liquid at time t, it follows from Newton's law of cooling that $\frac{dT}{dt} = -k(T - T_s)$ where k is a constant of proportionality.
If the temperature of the liquid is T_0 at time $t = 0$, show that
$T = T_s + (T_0 - T_s)e^{-kt}$

b) A cup of tea cools from 90°C to 60°C after 10 minutes in a room whose temperature was 20°C. Use Newton's law of cooling to calculate how much longer it would take the tea to cool to 40°C.

Solution:

a) Since $\frac{dT}{dt} = -k(T - T_s)$, separating the variables and integrating gives:

$$\int \frac{1}{T - T_s} dT = \int -k \, dt$$

$$\ln(T - T_s) = -kt + C$$

$$T - T_s = e^{-kt+C}$$

$$T - T_s = Ae^{-kt}, \text{ where } A = e^C$$

Using the initial condition at time $t = 0$, $T = T_0$, substitute into the general solution to work out the constant of integration, A:

$$T_0 - T_s = Ae^0$$

$$A = T_0 - T_s$$

Hence, $T - T_s = (T_0 - T_s)e^{-kt}$

$$T = T_s + (T_0 - T_s)e^{-kt}$$

b) For the cup of tea $T_0 = 90$ and $T_s = 20$:
Hence, $T = 20 + (90 - 20)e^{-kt}$

$$T = 20 + 70e^{-kt}$$

Also, $T = 60$ when $t = 10$, substitute into the line above as follows and solve for k:

$$60 = 20 + 70e^{-10k}$$

$$40 = 70e^{-10k}$$

$$e^{-10k} = \frac{40}{70}$$

$$-10k = \ln\left(\frac{4}{7}\right)$$

$$k = 0 \cdot 0560 \text{ (3 s.f.)}$$

© HERIOT-WATT UNIVERSITY

The particular solution obtained using this value of k enables us to calculate how long it will take for the tea to cool to 40°C from 90°C. This time we solve for t:

$$40 = 20 + 70e^{-0.0560t}$$

$$20 = 70e^{-0.0560t}$$

$$e^{-0.0560t} = \frac{20}{70}$$

$$-0.0560t = \ln\left(\frac{2}{7}\right)$$

$$t = -\frac{1}{0.0560}\ln\left(\frac{2}{7}\right) = 22.4 \text{ minutes}$$

It took 10 minutes for the tea to cool from 90°C to 60°C and 22.4 minutes for the tea to cool from 90°C to 40°C. It therefore took an extra 12.4 minutes for the tea to cool from 60°C to 40°C.

....

2. Problem:

Finally, here is an old CSYS question which would be considered to be rather difficult now!

An accident at a factory on a river results in the release of a polluting chemical. Immediately after the accident, the concentration of the chemical in the river becomes k g/m^3.

The river flows at a constant rate of w m^3/hour into a loch of volume V m^3.

Water flows over the dam at the other end of the loch at the same rate. The level, x g/m^3, of the pollutant in the loch t hours after the accident satisfies the differential equation $V\frac{dx}{dt} = w(k-x)$.

a) Find the general solution for x in terms of t.

b) In this particular case, the values of the constants are $V = 16000000$, $w = 8000$ and $k = 1000$.

   i. Before the accident, the level of chemical in the loch was zero. Fish in the loch will be poisoned if the level of pollutant in the loch reaches 10 g/m^3. How long do the authorities have to stop the leak before this level is reached?

   ii. In fact, when the level of pollutant in the loch has reached 5 g/m^3, the leak is located and plugged. The level in the river then drops to zero and the level in the loch falls according to the differential equation $V\frac{dx}{dt} = -wx$
   According to European Union standards, 1 g/m^3 is a safe level for the chemical. How much longer will it be before the level in the loch drops to this value?

This example comes from CSYS Mathematics 1997, Paper 1.

Solution:

a)

Since $V \frac{dx}{dt} = w(k-x)$

Separating the variables and integrating gives:

$$\int \frac{V}{k-x} dx = \int w \, dt$$

$$-V \ln(k-x) = wt + C$$

$$\ln(k-x) = -\frac{wt}{V} + C$$

$$k - x = e^{\left(-\frac{wt}{V} + C\right)}$$

$$x = k - A e^{\left(-\frac{wt}{V}\right)} \text{ where } A = e^C$$

b)

i) Since $V = 16000000$, $w = 8000$ and $k = 1000$ we have:

$$x = 1000 - A e^{\left(-\frac{8000t}{16000000}\right)}$$

$$= 1000 - A e^{\left(-\frac{t}{2000}\right)}$$

Since $x = 0$ when $t = 0$, $0 = 1000 - A e^{(0)} \Rightarrow A = 1000$

Hence: $x = 1000 \left(1 - e^{\left(-\frac{t}{2000}\right)}\right)$

To find how long it will take the level of pollutant in the loch to reach 10 g/m^3 substitute $x = 10$ into the formula:

$$10 = 1000 \left(1 - e^{\left(-\frac{t}{2000}\right)}\right)$$

$$1 - e^{\left(-\frac{t}{2000}\right)} = 0 \cdot 01$$

$$e^{\left(-\frac{t}{2000}\right)} = 1 - 0 \cdot 01 = 0 \cdot 99$$

$$-\frac{t}{2000} = \ln(0 \cdot 99)$$

$$t = -2000 \ln(0 \cdot 99) = 20 \cdot 1 \text{ hours}$$

ii) After the leak has been plugged the level of pollutant in the loch is now given by $V \frac{dx}{dt} = -wx$

Separating the variables and integrating gives:

$$\int \frac{1}{x} dx = -\int \frac{w}{V} dt$$

$$\ln(x) = -\frac{wt}{V} + C$$

$$x = e^{\left(-\frac{wt}{V} + C\right)}$$

$$= A e^{\left(-\frac{wt}{V}\right)} \text{ where } A = e^C$$

When the leak is plugged the level of pollutant in the loch is 5 g/m^3

Hence, $x = 5$ when $t = 0$ and so:

$5 = A e^{(0)}$

$A = 5$

Using the given values for the variables above then: $x = 5e^{\left(-\frac{t}{2000}\right)}$

Hence, calculating how long it takes for the level of pollution to reach 1 g/m³ by substituting $x = 1$ into the previous formula gives:

$$1 = 5e^{\left(-\frac{t}{2000}\right)}$$
$$e^{\left(-\frac{t}{2000}\right)} = 0\cdot 2$$
$$-\frac{t}{2000} = \ln\left(0\cdot 2\right)$$
$$t = -2000\ln\left(0\cdot 2\right)$$
$$= 3219 \text{ hours}$$

Further applications of differential equations exercise   Go online

Q51:
A hard-boiled egg is put in a sink of water at 16°C to cool. After 5 minutes, the egg's temperature is found to be 40°C, and then, after a further 15 minutes, it has cooled to 20°C.
Assuming that the water has not warmed significantly, use Newton's law of cooling to estimate the original temperature of the hard-boiled egg when it was put in the water.
(Give your answer to the nearest degree.)

....

Q52: The body of a murder victim was discovered in the early hours of the morning at 2:00 am. The police doctor arrived at 2:30 am and immediately took the temperature of the body, which was 34·8°C. One hour later, the temperature of the body was 34·1°C. The room temperature was a constant at 32·2°C. Normal body temperature is 37°C.

a) Let T be the temperature of the body, t be the passage of time, in hours, since 2:30 am, and k be the constant of proportionality. Formulate a differential equation model for the temperature of the body as a function of time.

b) Solve the differential equation to obtain the general solution in terms of T.

c) Given that the body had an initial temperature of 34·8°C and that 1 hour later it had dropped to 34·1°C, work out the value of k, the constant of proportionality, and A, the constant of integration. Give k to 3 decimal places.

d) Use the particular solution to estimate the time of death. Give your answer in 24 hour time.

4.4 Integrating factor

The previous section looked at solving first order linear differential equations. This section will continue to look at solving first order linear differential equations, but it will explore a different method of doing this. The method used here is the integrating factor. It should be noted that some of the differential equations presented in the previous section such as the equations for bacterial growth

TOPIC 4. DIFFERENTIAL EQUATIONS

and radioactive decay can also be solved using the method given in this section.

4.4.1 Method of integrating factor

Before starting this section we need to ensure that the first order linear differential equation is in the correct form. The differential equation must be in standard linear form:

> **Key point**
>
> Standard liner form of a first order linear differential equation is:
>
> $\frac{dy}{dx} + P(x)y = f(x)$
>
> Where:
>
> - the coefficient of $\frac{dy}{dx}$ is 1;
> - $P(x)$ is a function of x terms only;
> - y is linear, i.e. the power of y is 1;
> - $f(x)$ is a function of x terms only.

Examples

1. Problem:
Write the first order linear differential equation $4x\frac{dy}{dx} + 5y = -e^x$ in standard linear form and state $P(x)$ and $f(x)$.

Solution:
To write the differential equation in standard linear form it must look like $\frac{dy}{dx} + P(x)y = f(x)$, i.e. $\frac{dy}{dx}$ must have a coefficient of 1.

Given $4x\frac{dy}{dx} + 5y = -e^x$

Divide through by $4x$: $\frac{dy}{dx} + \frac{5y}{4x} = -\frac{e^x}{4x}$

This is now in standard linear form and so: $P(x) = \frac{5}{4x}$ and $f(x) = -\frac{e^x}{4x}$

....

2. Problem:
Write the first order linear differential equation $\frac{dy}{dx} = \frac{3y+x^2}{x}$ in standard linear form and state $P(x)$ and $f(x)$.

Solution:
To write the differential equation in standard linear form it must look like $\frac{dy}{dx} + P(x)y = f(x)$. We must therefore rearrange the differential equation so that the y term is on the LHS.

Given $\frac{dy}{dx} = \frac{3y+x^2}{x}$

Multiply through by x: $x\frac{dy}{dx} = 3y + x^2$

Subtract $3y$ from both sides: $x\frac{dy}{dx} - 3y = x^2$

© HERIOT-WATT UNIVERSITY

Divide through by x so that $\frac{dy}{dx}$ has a coefficient of 1:

$$\frac{dy}{dx} - \frac{3y}{x} = \frac{x^2}{x}$$

$$\frac{dy}{dx} - \frac{3y}{x} = x$$

This is now in standard linear form and so: $P(x) = -\frac{3}{x}$ and $f(x) = x$

Method of integrating factor: Identifying $P(x)$ and $f(x)$ exercise   Go online

For each of the following, rearrange the differential equation into standard linear form, if needed, then identify the functions for $P(x)$ and $f(x)$.

Q53: $\frac{dy}{dx} - 2y = e^{3x}$

Q54: $\frac{dy}{dx} = \frac{xy+4}{x^2}$

Q55: $x \frac{dy}{dx} = \frac{\cos x}{x} - 2y$

Q56: $\frac{dy}{dx} + \frac{1}{2}e^{-x} = 4y$

Q57: $2\frac{dy}{dx} - 2y = e^{\frac{x}{2}}$

Q58: $\frac{dy}{dx} + 3x^2 y = 6x^2$

The first order linear differential equation $\frac{dy}{dx} + P(x)y = f(x)$ can be solved by multiplying both sides of the equation by a suitable function called the **integrating factor** $I(x)$.

Key point

Integrating factor

$$I(x) = e^{\int P(x)\,dx}$$

Multiply both sides of $\frac{dy}{dx} + P(x)y = f(x)$ by $I(x)$: $I(x)\frac{dy}{dx} + I(x)P(x)y = I(x)f(x)$

Also note that $\dfrac{d}{dx}(I(x)) = \dfrac{d}{dx}\left(e^{\int P(x)\,dx}\right)$

Apply the Chain rule $= \dfrac{dy}{dx}\left(\int P(x)\,dx\right) e^{\int P(x)\,dx}$

$= P(x) e^{\int P(x)\,dx}$

$= P(x) I(x)$

TOPIC 4. DIFFERENTIAL EQUATIONS

To understand what comes next we have to note the important information from above.

The differential equation obtain by multiplying both sides by the integrating factor $I(x)$:
$I(x)\frac{dy}{dx} + I(x)P(x)y = I(x)f(x)$ (*)

We will compare this with a result later on.

The derivative of the integrating factor $\frac{d}{dx}I(x) = P(x)I(x)$ (+)

We will substitute this into a result later on.

To make a comparison with (*) we need to set up the next differential equation. This is $\frac{d}{dx}[I(x)y]$.
Now differentiate the equation using the product rule $\frac{dy}{dx} = u'v - uv'$

$\frac{d}{dx}[I(x)y] = \frac{d}{dx}I(x) \bullet y + I(x) \bullet \frac{dy}{dx}$

Notice the $\frac{d}{dx}I(x)$ term can be replaced by the RHS of (+): $\frac{d}{dx}[I(x)y] = P(x)I(x)y + I(x)\frac{dy}{dx}$

Re-order the RHS: $\frac{d}{dx}[I(x)y] = I(x)\frac{dy}{dx} + P(x)I(x)y$

The RHS of the differential equation above is the same as the LHS of (*). We can therefore replace it by $I(x)f(x)$: $\frac{d}{dx}[I(x)y] = I(x)f(x)$

This tells us that if we multiply the original first order linear differential equation by the integrating factor $I(x)$ it is the same as solving $\frac{d}{dx}[I(x)y]$.

Now integrating both sides of the new differential equation above gives:

$$\int \frac{d}{dx}[I(x)y]\,dx = \int I(x)f(x)\,dx$$

$$I(x)y = \int I(x)f(x)\,dx$$

Which we can now solve for y.

Hence, a first order differential equation can be solved using the following strategy.

> **Key point**
>
> **Strategy for solving first order linear differential equations**
>
> 1. Write the equation in standard linear form $\frac{dy}{dx} + P(x)y = f(x)$ and thus identify $P(x)$ and $f(x)$
> 2. Calculate the integrating factor $I(x) = e^{\int P(x)\,dx}$
> 3. Multiply the equation in standard linear form by $I(x)$ and obtain the equation $\frac{d}{dx}[I(x)y] = I(x)f(x)$
> 4. Integrate both sides to give $I(x)y = \int I(x)f(x)\,dx$
> 5. Rearrange to solve for y

Example

Problem:

Find the general solution of $\frac{dy}{dx} + 2y = e^{3x}$

Solution:

1. Note that the equation is given in standard linear form so:
   $P(x) = 2$ and $f(x) = e^{3x}$

2. Calculate the integrating factor:
   $$I(x) = e^{\int P(x)\,dx}$$
   $$= e^{\int 2\,dx}$$
   $$= e^{2x}$$
   Note that a constant of integration is not used in finding the integrating factor.

3. The equation can be written as $\frac{d}{dx}[I(x)y] = I(x)f(x)$
   With $I(x) = e^{2x}$ and $f(x) = e^{3x}$ the equation becomes:
   $$\frac{d}{dx}\left(e^{2x}y\right) = e^{2x}e^{3x}$$
   $$\frac{d}{dx}\left(e^{2x}y\right) = e^{5x}$$

4. Integrating both sides of this last equation gives:
   $$\int \frac{d}{dx}\left(e^{2x}y\right)\,dx = \int e^{5x}\,dx$$
   $$e^{2x}y = \frac{1}{5}e^{5x} + C$$

5. Rearranging to solve for y gives the general solution $y = \frac{1}{5}e^{3x} + Ce^{-2x}$

Top tip

Note that, in order to apply this method, it is first necessary to ensure that the equation is expressed in standard linear form.

$\frac{dy}{dx} + P(x)y = f(x)$

See the next example.

TOPIC 4. DIFFERENTIAL EQUATIONS

Example

Problem:

Find the general solution of $x\frac{dy}{dx} + y = \sin x$

Solution:

1. Rewrite the equation in standard linear form $\frac{dy}{dx} + P(x)y = f(x)$ so that the coefficient of $\frac{dy}{dx}$ is 1:

   $x\frac{dy}{dx} + y = \sin x$

   Divide both sides by x: $\frac{dy}{dx} + \frac{y}{x} = \frac{\sin x}{x}$

   Therefore, $P(x) = \frac{1}{x}$ and $f(x) = \frac{\sin x}{x}$

2. Hence, the integrating factor is:

   $$I(x) = e^{\int \frac{1}{x} dx}$$
   $$= e^{\ln x}$$
   $$= x$$

3. The equation can be written as $\frac{d}{dx}[I(x)y] = I(x)f(x)$

   Given $I(x) = x$ and $f(x) = \frac{\sin x}{x}$, then by substitution:

   $$\frac{d}{dx}(x \bullet y) = x \bullet \frac{\sin x}{x}$$
   $$\frac{d}{dx}(xy) = \sin x$$

4. Integrating gives:

   $$xy = \int \sin x \, dx$$
   $$xy = -\cos x + C$$

5. Rearranging gives the general solution $y = -\frac{\cos x}{x} + \frac{C}{x}$

Top tip

If an initial condition is known for a first order differential equation, then it should be possible to find a particular solution of the differential equation satisfying the initial condition.

The next example illustrates this point.

Example

Problem:

Solve the initial value problem $x\frac{dy}{dx} + 3y = 5x^2$ given that $y(1) = 0$

Recall that $y(1) = 0$ means that when $x = 1, y = 0$.

© HERIOT-WATT UNIVERSITY

Solution:

1. In standard linear form, the equation is $\frac{dy}{dx} + \frac{3y}{x} = 5x$ so: $P(x) = \frac{3}{x}$ and $f(x) = 5x$

2. Therefore, the integrating factor is:
$$I(x) = e^{\int \frac{3}{x}\,dx}$$
$$= e^{3\ln(x)}$$
$$= e^{\ln(x^3)}$$
$$= x^3$$

3. Multiplying both sides of $\frac{dy}{dx} + \frac{3y}{x} = 5x$ by $I(x)$ gives:
$$\frac{d}{dx}[I(x)\,y] = I(x)\,5x$$
$$\frac{d}{dx}(x^3 y) = x^3 5x$$
$$\frac{d}{dx}(x^3 y) = 5x^4$$

4. Integrating gives:
$$x^3 y = \int 5x^4\,dx$$
$$x^3 y = x^5 + C$$

5. Given that $y(1) = 0$, then it is possible to find C.
Substituting $x = 1$ and $y = 0$ into $x^3 y = x^5 + C$ gives:
$0 = 1 + C$ so $C = -1$
Therefore, $x^3 y = x^5 + C$ becomes: $x^3 y = x^5 - 1$

and rearranging to solve for y gives: $y = x^2 - x^{-3}$

Method of integrating factor extra help: Differential equations

Equations of this type can be rearranged in the form $\frac{dy}{dx} + P(x)\,y = f(x)$.

Once you have arranged it in this form, you may need to use any known process to integrate, i.e. algebraic manipulation to produce a standard integral, partial fractions, integration by parts etc.

The general solution will contain an arbitrary constant. If you are given an initial condition in order to determine the constant, then you will obtain a particular solution of the equation.

TOPIC 4. DIFFERENTIAL EQUATIONS

Find the general solution of $\frac{dy}{dx} + 2y = e^{-x}$ by answering the following set of multiple choice questions.

Q59: Compare $\frac{dy}{dx} + 2y = e^{-x}$ with $\frac{dy}{dx} + P(x)y = f(x)$ to determine $P(x)$.

a) $P(x) = 2$
b) $P(x) = y$
c) $P(x) = 2y$

....

Q60: Determine $\int P(x)\, dx$

a) $\int 2\, dx = 2x$
b) $\int 2\, dx = 2$

....

Q61: Determine the integrating factor $I(x)$, given that $I(x) = e^{\int P(x)\, dx}$.
Multiply both sides of the question by $I(x) = e^{2x}$

a) $e^{2x} + C$
b) $e^{2x + C}$
c) e^{2x}

....

Q62: The equation we will now solve takes the form $I(x)y = \int I(x) f(x)\, dx$.
Select the correct differential equation that is now to be solved.

a) $e^{2x} y = \int e^x\, dx$
b) $y = \int e^{3x}\, dx$
c) $y = \int e^{-x}\, dx$

....

Q63: We have identified the differential equation that we need to solve. Integrate the RHS and select the correct answer.

a) $e^{2x} y = x + C$
b) $e^{2x} y = e^x + C$
c) $e^{2x} y = x$
d) $e^{2x} y = e^x$

....

Q64: Express the answer as "$y = \ldots$" and simplify to give the final answer.

a) $y = e^{-x} + Ce^{-2x}$
b) $y = e^{-x} + C + e^{-2x}$
c) $y = e^x + C - e^{2x}$

© HERIOT-WATT UNIVERSITY

338 TOPIC 4. DIFFERENTIAL EQUATIONS

Find the particular integral of $\frac{dy}{dx} + y \tan x = \sin(2x)$ given the initial conditions that $y = 1$ when $x = 0$ by answering the following set of multiple choice questions.

Q65: Compare $\frac{dy}{dx} + y \tan x = \sin(2x)$ with $\frac{dy}{dx} + P(x)y = f(x)$ to determine $P(x)$.

a) $P(x) = y \tan x$
b) $P(x) = \tan x$
c) $P(x) = \ln(\sin x)$

....

Q66: Determine $\int P(x)\, dx$

a) $\ln(\sec x)$
b) $\sec^2 x$
c) $\ln(\sec^2 x)$

....

Q67: Determine the integrating factor $I(x)$, given that $I(x) = e^{\int P(x)\, dx}$.

a) $\sec x$
b) $e^{\sec x}$
c) $\ln|\sec x|$

....

Q68: The equation we will now solve takes the form $I(x)y = \int I(x)f(x)\, dx$. Select the correct differential equation that is now to be solved.

a) $\sec x \bullet y = \int \sec x \int \sin 2x\, dx$
b) $\sec x \bullet y = \int \tan x\, dx$
c) $\sec x \bullet y = 2\int \sin x\, dx$

....

Q69: From the previous question we have identified the differential equation that we need to solve. Integrate the RHS and select the correct answer.

a) $\sec x \bullet y = \ln|\sec x| + C$
b) $\sec x \bullet y = -2\cos x + C$
c) $\sec x \bullet y = \ln|\sec x|$
d) $\sec x \bullet y = -2\cos x$

....

Q70: Take the answer from the previous question and rearrange it for y to obtain the general solution in its simplest form. Select the correct answer.

a) $y = -2 + C\sin x$
b) $y = -2\cos^2 x + C\cos x$
c) $y = -2 + C\cos x$

....

© HERIOT-WATT UNIVERSITY

Q71: Use the initial condition that $y = 1$ when $x = 0$ to work out the value of the constant of integration, C.

a) $C = 0$
b) $C = 2$
c) $C = 3$

Method of integrating factor exercise

Go online

Q72: Solve the differential equation $\frac{dy}{dx} + \frac{2}{x}y = x$

a) Write down the integrating factor.
b) Given that the general solution is given by $y = g(x) + Ch(x)$, where C is the constant of integration, write down $g(x)$ and $h(x)$. Where necessary, write x with negative powers.
c) Write down the general solution to the differential equation.

....

Q73: Solve the differential equation $\frac{dy}{dx} - 2xy = 3x$

a) Write down the integrating factor.
b) Given that the general solution is given by $y = g(x) + Ch(x)$, where C is the constant of integration, write down $g(x)$ and $h(x)$. Where necessary write x with negative powers.
c) Write down the general solution to the differential equation.

....

Q74: Solve the differential equation $x^2 \frac{dy}{dx} - x^3 + xy = 0$ given that when $x = 1$, $y = 1$.

a) Write down the integrating factor.
b) Given that the general solution is given by $y = g(x) + Ch(x)$, where C is the constant of integration, write down $g(x)$ and $h(x)$. Where necessary write x with negative powers.
c) Given that when $x = 1$, $y = 1$, work out the value of C, the constant of integration, and write down the particular solution to the differential equation.

....

Q75: Solve the differential equation $\frac{dy}{dx} - 3y = x$ given that when $x = 0$, $y = 1$.

a) Write down the integrating factor.
b) Given that the general solution is given by $y = g(x) + Ch(x)$, where C is the constant of integration, write down $g(x)$ and $h(x)$. Where necessary write x with negative powers.
c) Given that when $x = 0$, $y = 1$, work out the value of C, the constant of integration, and write down the particular solution to the differential equation.

4.4.2 Integrating factor applications

As we have already seen, differential equations have important applications in engineering and science. This section looks at some further examples that involve solving first order linear differential equations using the integrating factor method.

Examples

1. Bacterial Growth

Problem:
Bacteria grow in a certain culture at a rate proportional to the amount of bacteria present. Given that there are 100 bacteria present initially, find an equation for bacterial growth and, by solving an appropriate differential equation, find the number of bacteria present at any subsequent time.

Solution:
This type of problem was discussed when using the method of separating variables to solve first order differential equations. However, an alternative method is to use an integrating factor.

Since bacteria grow at a rate proportional to the amount of bacteria present, the differential equation describing bacterial growth is $\frac{dB}{dt} = kB$ where $B(t)$ is the number of bacteria present at time t and k is the constant of proportionality.

In standard linear form, the equation is $\frac{dB}{dt} - kB = 0$ and therefore $P(t) = -k$ and $f(t) = 0$.

The integrating factor is $I(t) = e^{\int -k\, dt} = e^{-kt}$

The differential equation becomes:

$$\frac{d}{dt}[I(t)B] = I(t)f(t)$$

i.e. $\frac{d}{dt}(e^{-kt}B) = 0$

Integrating both sides gives $e^{-kt}B = C$ where C is a constant.

Rearranging to solve for B gives $B = Ce^{kt}$

Initially, there are 100 bacteria present in the culture, i.e. $B = 100$ when $t = 0$

Therefore, $100 = Ce^0$ and so $C = 100$

Hence, the number of bacteria present at time t is given by $B = 100e^{kt}$

....

TOPIC 4. DIFFERENTIAL EQUATIONS

2. A mixing problem

Problem:

A tank contains 50 kg of salt dissolved in 200 m³ of water.

Starting at time $t = 0$, water, which contains 2 kg of salt per m³, enters the tank at a rate of 4 m³/minute and the well-stirred solution leaves the tank at the same rate.

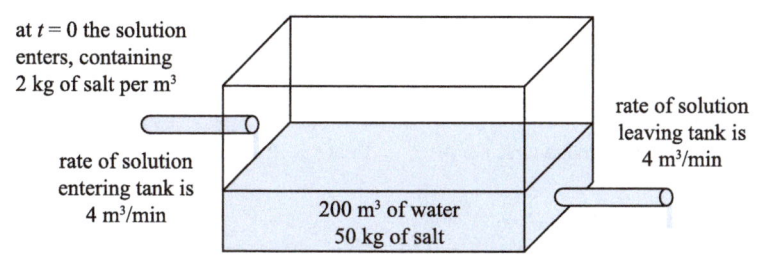

at $t = 0$ the solution enters, containing 2 kg of salt per m³

rate of solution entering tank is 4 m³/min

200 m³ of water
50 kg of salt

rate of solution leaving tank is 4 m³/min

Calculate how much salt there is in the tank after 30 minutes.

Solution: Let $M(t)$ represent the total mass of salt (in kg) in the tank at time t.

Then the rate of change of mass of salt is equal to $\frac{dM}{dt}$

However, it is also possible to calculate the rate of change of mass by calculating the *rate at which the total amount of salt in the tank changes* and so obtain a differential equation for M.

The rate at which salt enters the tank is $2 \times 4 = 8$ kg/min

The rate at which salt leaves the tank = the mass of salt in the tank × the proportion of the solution which leaves the tank per minute = $M \times \frac{4}{200} = 0 \cdot 02\, M$ kg/min

Thus, the rate of change of mass $= 8 - 0 \cdot 02\, M$ kg/min

Hence, $M(t)$ must satisfy the differential equation $\frac{dM}{dt} = 8 - 0 \cdot 02 M$

In standard linear form, the differential equation is $\frac{dM}{dT} + 0 \cdot 02 M = 8$

Using the integrating factor method, $P(t) = 0 \cdot 02$ and $f(t) = 8$

Integrating factor:

$I(t) = e^{\int P(t)\, dt}$
$= e^{\int 0 \cdot 02\, dt}$
$= e^{0 \cdot 02 t}$

© HERIOT-WATT UNIVERSITY

Therefore, the differential equation becomes:

$$I(t) M = \int I(t) f(t) \, dt$$

$$e^{0.02t} M = \int e^{0.02t} 8 \, dt$$

$$e^{0.02t} M = \frac{8}{0.02} e^{0.02t} + C$$

$$e^{0.02t} M = 400 e^{0.02t} + C$$

$$M = \frac{400 e^{0.02t}}{e^{0.02t}} + \frac{C}{e^{0.02t}}$$

$$M = 400 + C e^{-0.02t}$$

There is initially 50 kg of salt in the tank, i.e. $M(0) = 50$ at $t = 0$ and so:

$$50 = 400 + C e^{-0.02(0)}$$

$50 = 400 + C$ which means that $C = -350$

Hence, the equation becomes $M = 400 - 350 e^{-0.02t}$
When $t = 30$ then $M = 400 - 350 e^{-0.6} = 207 \cdot 9 kg$
After 30 minutes, there is 207·9 kg of salt in the tank.

....

3. An inductance-resistance circuit

Problem:

The following diagram represents an electrical circuit which contains a constant DC voltage source of 12 volts, a switch, a resistor of size 6 ohms and an inductor of size L henrys. The switch is initially open but is closed at time $t = 0$ and the current begins to flow at that time. Let $i(t)$ denote the current t seconds after the switch is closed. Then it is known that $i(t)$ satisfies $L \frac{di}{dt} + Ri = V$ when $i(0) = 0$

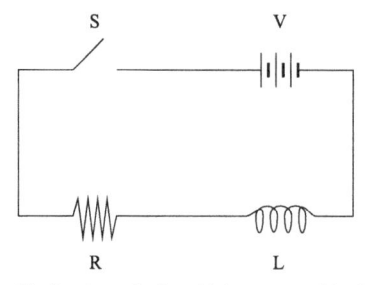

Find a formula for $i(t)$ in terms of L, R and V. (Note that in this question L, R and V are constants.)

Solution:

Rewrite the equation in standard linear form $\frac{di}{dt} + \frac{Ri}{L} = \frac{V}{L}$ and so $P(t) = \frac{R}{L}$ and $f(t) = \frac{V}{L}$

Calculate the integrating factor $I(t) = e^{\int \frac{R}{L} dt} = e^{\frac{R}{L} t}$

TOPIC 4. DIFFERENTIAL EQUATIONS

Therefore the differential equation becomes:

$$\frac{d}{dt}[I(t)\,i] = I(t)\,f(t)$$

$$\frac{d}{dt}\left(e^{\frac{R}{L}t}i\right) = \frac{V}{L}e^{\frac{R}{L}t}$$

Integrating both sides gives:

$$\int \frac{d}{dt}\left(e^{\frac{R}{L}t}i\right)\,dt = \int \frac{V}{L}e^{\frac{R}{L}t}\,dt$$

$$e^{\frac{R}{L}t}i = \frac{V}{L}\frac{L}{R}e^{\frac{R}{L}t} + C$$

$$= \frac{V}{R}e^{\frac{R}{L}t} + C$$

Substitute in the initial values and solve for y so when $t = 0$ then $i = 0$:

$$e^0 0 = \frac{V}{R}e^0 + C$$

$$0 = \frac{V}{R} + C$$

i.e. $C = -\frac{V}{R}$

Therefore:

$$e^{\frac{R}{L}t}i = \frac{V}{R}e^{\frac{R}{L}t} - \frac{V}{R}$$

$$i = \frac{V}{R} - \frac{V}{R}e^{-\frac{R}{L}t}$$

$$= \frac{V}{R}\left(1 - e^{-\frac{R}{L}t}\right)$$

Save the fish Go online

A rare species of fish is to be stored in a large tank containing 5000 litres of salt water. Initially, the concentration of salt in the water is 20 g/litre, but the fish will survive best in a concentration of 12 g/litre. Fresh water is poured into the tank and salt water is drained out at the same rate to reduce the concentration of salt in the water.

What flow of water is needed for the concentration of salt in the water to reach the correct level after 30 minutes? Give your answer to the nearest whole number of litres.

Integrating factor applications exercise Go online

Q76: In the first few weeks after birth, a baby gains weight at a rate proportion to its weight. A baby weighing 2·5 kg at birth weighs 3·7 kg after two weeks.

How much did the baby weigh after 5 days? Give your answer to 2 decimal places.

a) Use $W(t)$ to represent the weight of the baby at time t and k to represent the constant of proportionality to write down, in standard linear form, the differential equation that represents this situation.

© HERIOT-WATT UNIVERSITY

b) Write down the integrating factor for this differential equation.
c) Given that the general solution is given by $W = g(t) + Ch(t)$, where C is the constant of integration, write down $g(t)$ and $h(t)$.
d) Given that the baby weighs 2·5 kg at birth weight 3·7 kg after two weeks, work out the value of C, the constant of integration, and k, the constant of proportionality. Let time, t, be measured in days. (Give k to 3 decimal places.)
e) Write down the particular solution to the differential equation and work out the weight of the baby 5 days after birth. Give the weight of the baby to 2 decimal places.

....

Q77: A tank contains 100 kg of salt dissolved in 300 ml of water. Starting at time $t = 0$, water which contains 5 kg of salt per m^3 enters the tank at a rate of 3 m^3/min and the well-stirred solution leaves the tank at the same rate.

a) Let $M(t)$ represent the total amount of salt in the water at time t. Write down the differential equation that describes the rate of change of salt in the tank with respect to t.
b) Write down the integrating factor for this differential equation.
c) Given that the general solution is given by $W = g(t) + Ch(t)$, where C is the constant of integration, write down $g(t)$ and $h(t)$. Where necessary, write x with negative powers instead of as a fraction.
d) Using the initial condition that at time $t = 0$, there is 100kg of salt dissolved in the tank, work out the constant of integration C.
e) How much salt is in the tank after 1 hour? Give your answer to 1 decimal place.
f) How long does it take for there to be 1000 kg of salt in the water? Give your answer in minutes, to the nearest minute.
g) What happens to the level of salt in the water when $t \to \infty$?

4.5 Second order linear differential equations with constant coefficients

A second order linear differential equation is an equation which can be expressed in the standard form: $a \frac{d^2y}{dx^2} + b \frac{dy}{dx} + cy = f(x)$ where a, b and c are constants and $a \neq 0$.

This equation is second order because it contains a second derivative but no derivatives of higher order.

TOPIC 4. DIFFERENTIAL EQUATIONS

> **Key point**
>
> A second order linear differential equation is homogeneous when it can be expressed in the standard form:
>
> $$a\frac{d^2y}{dx^2} + b\frac{dy}{dx} + cy = 0$$
>
> where a, b and c are constants and $a \neq 0$

> **Key point**
>
> A second order linear differential equation is non-homogeneous when it can be expressed in the standard form:
>
> $$a\frac{d^2y}{dx^2} + b\frac{dy}{dx} + cy = f(x)$$
>
> where a, b and c are constants, $a \neq 0$ and $f(x) \neq 0$

If an equation is second order, we would expect it to be solved by carrying out two integrations and so we should also expect it to have a general solution containing two arbitrary constants.

In this section we shall see how to find such solutions for homogeneous and non-homogeneous second order linear differential equations with constant coefficients.

4.5.1 Homogeneous second order linear differential equations

It would be useful to be able to find the general solution of $a\frac{d^2y}{dx^2} + b\frac{dy}{dx} + cy = 0$, i.e. a solution containing two arbitrary constants.

Suppose that two independent solutions y_1 and y_2 are known (i.e. y_1 is not a multiple of y_2). Then, it is easy to see that:

1. $y_1 + y_2$ is also a solution;

2. if A is any constant, then Ay_1 is also a solution.

There are proofs in the proof section which justifies $y_1 + y_2$ as a solution and if A is any constant, then Ay_1 as a solution. This is not required to be reproduced for exam purposes, simply for your interest.

Proof 1: $y_1 + y_2$

For a homogenous second order linear differential equation, if y_1 and y_2 are two solutions to the equation, then so is $y = y_1 + y_2$.

We have:
$a\frac{d^2y_1}{dx^2} + b\frac{dy_1}{dx} + cy_1 = 0$ and $a\frac{d^2y_2}{dx^2} + b\frac{dy_2}{dx} + cy_2 = 0$

Addition gives:
$$a\frac{d^2y_1}{dx^2} + b\frac{dy_1}{dx} + cy_1 + a\frac{d^2y_2}{dx^2} + b\frac{dy_2}{dx} + cy_2 = 0$$
$$a\left(\frac{d^2y_1}{dx^2} + \frac{d^2y_2}{dx^2}\right) + b\left(\frac{dy_1}{dx} + \frac{dy_2}{dx}\right) + c(y_1 + y_2) = 0$$

By the additive law of differentiation, this gives:
$a\frac{d^2}{dx^2}(y_1 + y_2) + b\frac{d}{dx}(y_1 + y_2) + c(y_1 + y_2) = 0$

and so $y = y_1 + y_2$ is a solution to the differential equation.

Proof 2: If A is any constant

For a homogenous second order linear differential equation, if y_1 is a solution, then if A is any constant Ay_1 is also a solution.

We have:
$a\frac{d^2}{dx^2}(Ay_1) + b\frac{d}{dx}(Ay_1) + c(Ay_1) = 0$

Since a constant can be taken as a common factor to the front of the differentiation, then:
$$aA\frac{d^2y_1}{dx^2} + bA\frac{dy_1}{dx} + cAy_1 = 0$$
$$A\left(a\frac{d^2y_1}{dx^2} + b\frac{dy_1}{dx} + cy_1\right) = 0$$

y_1 is a solution and therefore so is Ay_1.

It follows that if y_1 and y_2 are two independent solutions of $a\frac{d^2y}{dx^2} + b\frac{dy}{dx} + cy = 0$, then $Ay_1 + By_2$ is also a solution where A and B are arbitrary constants,

$Ay_1 + By_2$ is the general solution of $a\frac{d^2y}{dx^2} + b\frac{dy}{dx} + cy = 0$

Thus, if two independent solutions y_1 and y_2 can be found, then it is possible to obtain the general solution.

The method which follows shows how to find these two distinct solutions.

The method used is to try and find appropriate values of m such that $y = e^{mx}$ is a solution.

There are proofs in the proof section which justifies the general solution of $a\frac{d^2y}{dx^2} + b\frac{dy}{dx} + cy = 0$ and $y = e^{mx}$ as a solution. This is not required to be reproduced for exam purposes, simply for your interest.

TOPIC 4. DIFFERENTIAL EQUATIONS

Proof 3: $y = Ay_1 + By_2$

For a homogenous second order linear differential equation if y_1 and y_2 are two solutions to the equation then so is $y = Ay_1 + By_2$.

We have:
$a\frac{d^2}{dx^2}(Ay_1) + b\frac{d}{dx}(Ay_1) + c(Ay_1) = 0$ and $a\frac{d^2}{dx^2}(By_2) + b\frac{d}{dx}(By_2) + c(By_2) = 0$

Addition gives:
$$a\frac{d^2}{dx^2}(Ay_1) + b\frac{d}{dx}(Ay_1) + c(Ay_1) + a\frac{d^2}{dx^2}(By_2) + b\frac{d}{dx}(By_2) + c(By_2) = 0$$
$$a\left(\frac{d^2}{dx^2}(Ay_1) + \frac{d^2}{dx^2}(By_2)\right) + b\left(\frac{d}{dx}(Ay_1) + \frac{d}{dx}(By_2)\right) + c(Ay_1 + By_2) = 0$$

Since we have established in Proof 2 that if y_1 and y_2 are two independent solutions then so are Ay_1 and By_2, we can use the additive law of differentiation to give:
$a\frac{d^2}{dx^2}(Ay_1 + By_2) + b\frac{d}{dx}(Ay_1 + By_2) + c(Ay_1 + By_2) = 0$

and so $y = Ay_1 + By_2$ is a solution to the differential equation.

Proof 4: A trial solution and the auxiliary equation

$y = Ae^{mx}$, for some A and m, is the solution to the first order linear differential equation $b\frac{dy}{dx} + cy = 0$. This was demonstrated in the 'Integrating factor' section.
It is therefore reasonable to consider it as a possible solution for $a\frac{d^2y}{dx^2} + b\frac{dy}{dx} + cy = 0$.

Now:
$y = Ae^{mx} \Rightarrow \frac{dy}{dx} = Ame^{mx} \Rightarrow \frac{d^2y}{dx^2} = Am^2e^{mx}$

If $y = Ae^{mx}$ is a solution, it will satisfy the equation.
Substituting into the original equation gives:
$aAm^2e^{mx} + bAme^{mx} + cAe^{mx} = 0$

Assuming $Ae^{mx} \neq 0$, by division, we get:
$am^2 + bm + c = 0$ (auxiliary equation A.E.)
The solutions to this quadratic, given by $m = \frac{-b \pm \sqrt{b^2 - 4ac}}{2a}$, provides two values of m which will make $y = Ae^{mx}$ a solution.

If we call the two solutions m_1 and m_2, then we have two solutions $y = Ae^{m_1x}$ and $y = Ae^{m_2x}$.

By proof 3, this gives the solution:
$y = Ae^{m_1x} + Be^{m_2x}$
The solution that we get from this auxiliary equation depends on the nature of the roots of this equation.

348 TOPIC 4. DIFFERENTIAL EQUATIONS

If $y = e^{mx}$, then $\frac{dy}{dx} = me^{mx}$ and $\frac{d^2y}{dx^2} = m^2 e^{mx}$

Substituting the above into the second order linear differential equation gives:

$$a \frac{d^2y}{dx^2} + b \frac{dy}{dx} + cy = 0$$
$$\Leftrightarrow am^2 e^{mx} + bm e^{mx} + c e^{mx} = 0$$
$$\Leftrightarrow e^{mx} \left(am^2 + bm + c\right) = 0$$

Key point

The second order, linear, homogeneous differential equation $a \frac{d^2y}{dx^2} + b \frac{dy}{dx} + cy = 0$ has the auxiliary equation

$$am^2 + bm + c = 0$$

Hence, $y = e^{mx}$ is a solution of the differential equation if, and only if, m satisfies the quadratic equation $am^2 + bm + c = 0$

This quadratic equation, which is termed the auxiliary equation (or characteristic equation) of $a \frac{d^2y}{dx^2} + b \frac{dy}{dx} + cy = 0$ has roots given by the quadratic formula $m = \frac{-b \pm \sqrt{b^2 - 4ac}}{2a}$. This gives two independent solutions which are required to find the general solution.

Three possible cases arise when solving the auxiliary equation:

1. $b^2 - 4ac > 0$ when the quadratic has two real and distinct roots;
2. $b^2 - 4ac = 0$ when the quadratic has one repeated real root (i.e. equal roots);
3. $b^2 - 4ac < 0$ when the quadratic has no real roots, i.e. it has complex roots.

4.5.1.1 Real and distinct roots

Key point

If the auxiliary equation gives two real, distinct roots m_1 and m_2, then $y_1 = e^{m_1 x}$ and $y_2 = e^{m_2 x}$ are independent solutions of $a \frac{d^2y}{dx^2} + b \frac{dy}{dx} + cy = 0$ and the general solution is:

$$y(x) = Ae^{m_1 x} + Be^{m_2 x}$$

Example

Problem:

Find the general solution for the differential equation $\frac{d^2y}{dx^2} - 4\frac{dy}{dx} + 3y = 0$

Solution:

The auxiliary equation takes the form $am^2 + bm + c = 0$.

To get the values of a, b and c we need to identify the coefficients of the differential equation.

© HERIOT-WATT UNIVERSITY

TOPIC 4. DIFFERENTIAL EQUATIONS 349

The differential equation is $\frac{d^2y}{dx^2} - 4\frac{dy}{dx} + 3y = 0$.

a is the coefficient of $\frac{d^2y}{dx^2}$, so $a = 1$

b is the coefficient of $\frac{dy}{dx}$, so $b = -4$

c is the coefficient of y, so $c = 3$

The auxiliary equation is therefore: $m^2 - 4m + 3 = 0$

Solve the auxiliary equation by factorising and setting each factor to zero:

$$m^2 - 4m + 3 = 0$$
$$(m-1)(m-3) = 0$$
$$(m-1) = 0 \text{ or } (m-3) = 0$$
$$m_1 = 1 \text{ or } m_2 = 3$$

The general solution takes the form $y(x) = Ae^{m_1 x} + Be^{m_2 x}$ so substituting for m_1 and m_2 gives:

$y(x) = Ae^x + Be^{3x}$

Real and distinct roots exercise

Go online

Q78: Find the general solution for the differential equation $\frac{d^2y}{dx^2} + \frac{dy}{dx} - 2y = 0$.

  a) The auxiliary equation is given by $am^2 + bm + c = 0$. Write down the auxiliary equation for the differential equation $\frac{d^2y}{dx^2} + \frac{dy}{dx} - 2y = 0$.

  b) Solve the auxiliary equation to determine the two roots m_1 and m_2, then write down the two solutions to the differential equation.

  c) Given that the general solution is given by $y(x) = Ae^{m_1 x} + Be^{m_2 x}$, write down the general solution to the differential equation $\frac{d^2y}{dx^2} + \frac{dy}{dx} - 2y = 0$.

....

Q79: Find the general solution for the differential equation $\frac{d^2y}{dx^2} + 2\frac{dy}{dx} - 15y = 0$.

....

Q80: Find the general solution for the differential equation $2\frac{d^2y}{dx^2} - \frac{dy}{dx} - 6y = 0$.

....

Q81: Find the general solution for the differential equation $6\frac{d^2y}{dx^2} - \frac{dy}{dx} - 2y = 0$.

© HERIOT-WATT UNIVERSITY

4.5.1.2 Equal roots

If $b^2 - 4ac = 0$, the auxiliary equation has a single repeated root given by $m = -\frac{b}{2a}$.
This is illustrated by using the quadratic formula and substituting for $b^2 - 4ac = 0$:

$$x = \frac{-b \pm \sqrt{b^2 - 4ac}}{2a}$$

$$x = \frac{-b \pm \sqrt{0}}{2a}$$

$$x = \frac{-b}{2a}$$

Hence, $a \frac{d^2y}{dx^2} + b \frac{dy}{dx} + cy = 0$ has the solution $y_1 = e^{mx}$

A straightforward calculation shows that, in this case, $y_2 = xe^{mx}$ is also a solution.

There is a proof in the proofs section which justifies $y_2 = xe^{mx}$ as a solution. This is not required to be reproduced for exam purposes, simply for your interest.

> **Proof 5:** $y_2 = xe^{mx}$
>
> For a homogeneous, second order, linear, differential equation with equal roots show that when $y_1 = e^{mx}$ is a solution, then $y_2 = xe^{mx}$ is also a solution.
>
> The differential equation $a \frac{d^2y}{dx^2} + b \frac{dy}{dx} + cy = 0$ has auxiliary equation $am^2 + bm + c = 0$
>
> The auxiliary equation has equal roots when $b^2 - 4ac = 0$ and from the quadratic formula $m = \frac{-b \pm \sqrt{b^2 - 4ac}}{2a}$ the repeated root is given by $m = -\frac{b}{2a}$
>
> $y_1 = e^{mx}$ is one solution of the homogeneous equation. In the case where m is a repeated root, try $y_2 = xe^{mx}$ as another solution.
>
> Then $\frac{dy_2}{dx} = mxe^{mx} + e^{mx}$ and $\frac{d^2y_2}{dx^2} = m^2xe^{mx} + 2me^{mx}$
>
> Hence $a \frac{d^2y_2}{dx^2} + b \frac{dy_2}{dx} + cy_2 = am^2xe^{mx} + 2ame^{mx} + bmxe^{mx} + be^{mx} + cxe^{mx}$
>
> $= (am^2 + bm + c) xe^{mx} + (2am + b) e^{mx}$
>
> $= 0 \left(\text{since } am^2 + bm + c = 0 \text{ and } m = -\frac{b}{2a} \right)$
>
> Therefore, if y is a repeated root of the auxiliary equation, the differential equation has solutions $y_1 = e^{mx}$ and $y_2 = xe^{mx}$ and the general solution is $y = Ae^{mx} + Bxe^{mx}$

Hence, in this situation, the equation has independent solutions e^{mx} and xe^{mx} and so has the general solution $y(x) = Ae^{mx} + Bxe^{mx}$

TOPIC 4. DIFFERENTIAL EQUATIONS

> **Key point**
>
> When the auxiliary equation, $am^2 + bm + c = 0$, of a second order linear differential equal has equal roots, i.e. $b^2 - 4ac = 0$, then the general solution takes the form:
>
> $$y(x) = Ae^{mx} + Bxe^{mx}$$
>
> In factorised form:
>
> $$y(x) = (A + Bx)e^{mx}$$

Example

Problem:

Find the general solution of the differential equation $\frac{d^2y}{dx^2} - 4\frac{dy}{dx} + 4y = 0$

Solution:

The auxiliary equation for $\frac{d^2y}{dx^2} - 4\frac{dy}{dx} + 4y = 0$ is $m^2 - 4m + 4 = 0$

Factorise this to give $(m - 2)(m - 2) = 0 \Rightarrow (m - 2) = 0$ (repeated root)

Therefore, the repeated root of the auxiliary equation is $m = 2$

Hence, the equation has independent solutions $y_1 = e^{2x}$ and $y_2 = xe^{2x}$ and therefore has general solution $y = Ae^{2x} + Bxe^{2x}$

In factorised form the solution is: $y = (A + Bx)e^{2x}$.

Note that the solution can be presented in either form. Both are correct.

Equal roots exercise   Go online

Q82: Find the general solution for the differential equation $\frac{d^2y}{dx^2} + 4\frac{dy}{dx} + 4y = 0$

a) The auxiliary equation is given by $am^2 + bm + c = 0$. Write down the auxiliary equation for $\frac{d^2y}{dx^2} + 4\frac{dy}{dx} + 4y = 0$

b) Solve the auxiliary equation to determine the two roots m_1 and m_2, then write down the two solutions to the differential equation.

c) Given that the general solution is given by $y(x) = Ae^{mx} + Bxe^{mx}$, write down the general solution to the differential equation.

....

Q83: Find the general solution for the differential equation $\frac{d^2y}{dx^2} - 6\frac{dy}{dx} + 9y = 0$

....

Q84: Find the general solution for the differential equation $9\frac{d^2y}{dx^2} - 6\frac{dy}{dx} + y = 0$.

....

© HERIOT-WATT UNIVERSITY

352 TOPIC 4. DIFFERENTIAL EQUATIONS

Q85: Find the general solution for the differential equation $4\frac{d^2y}{dx^2} + 12\frac{dy}{dx} + 9y = 0$

4.5.1.3 Complex roots

If $b^2 - 4ac < 0$, the auxiliary equation has complex roots. To understand what these are we need to introduce complex numbers and the letter i.

The letter i represents an imaginary number, where $i = \sqrt{(-1)}$. Sometimes it is easier to work with $i^2 = -1$. The introduction to the number i overcomes the problem of finding the square roots of a negative number e.g.

$$\begin{aligned}\sqrt{(-25)} &= \sqrt{25 \times (-1)} \\ &= \sqrt{25 \times i^2} \\ &= \sqrt{(5i)^2} \\ &= \pm 5i\end{aligned}$$

With the existence of the square roots of a negative number, it is possible to find the solutions of any quadratic equation of the form $ax^2 + bx + c = 0$ using the quadratic formula.

Example

Problem:

Solve the equation $x^2 - 6x + 25 = 0$

Solution:

$a = 1, b = -6, c = 25$

Use the quadratic formula: $x = \frac{-b \pm \sqrt{b^2 - 4ac}}{2a}$

$$\begin{aligned}x &= \frac{-(-6) \pm \sqrt{(-6)^2 - 4(1)(25)}}{2(1)} \\ &= \frac{6 \pm \sqrt{36 - 100}}{2} \\ &= \frac{6 \pm \sqrt{-64}}{2} \\ &= \frac{6 \pm \sqrt{64 \times (-1)}}{2} \\ &= \frac{6 \pm \sqrt{64i^2}}{2} \\ &= \frac{6 \pm 8i}{2} \\ &= 3 \pm 4i\end{aligned}$$

TOPIC 4. DIFFERENTIAL EQUATIONS

A complex number is a number of the form $a + bi$ where a and b are real numbers and $i = \sqrt{(-1)}$
The complex number may also be written as $a + bi$
It does not matter whether the i comes before or after the rest of a term. By putting it first confusion can often be avoided.
The set of complex numbers is written as $\mathbb{C} = \{a + ib : a, b \in \mathbb{R}\}$
Note that it is customary to denote a complex number by the letter z.
If $z = a + ib$ then:

- the number a is called the real part of z;
- the number b is called the imaginary part of z.

These are sometimes denoted $\mathrm{Re}\,(z)$ for a and $\mathrm{Im}\,(z)$ for b.

Given that $z = a + ib$ is the complex number, then we can also write:

- $\mathrm{Re}\,(a + ib) = a$ to mean the real part of the complex number is a;
- $\mathrm{Im}\,(a + ib) = b$ to mean the imaginary part of the complex number is b.

Now returning to solving the auxiliary equation, $am^2 + bm + c = 0$, for when $b^2 - 4ac < 0$

If we take the quadratic formula and algebraically determine the solutions the complex roots would be given by:

$m_1 = \frac{-b - \sqrt{b^2 - 4ac}}{2a}$ and $m_2 = \frac{-b + \sqrt{b^2 - 4ac}}{2a}$

which can be written as the complex conjugate pair $m_1 = p - iq$ and $m_2 = p + iq$.

This comes from separating the real and imaginary parts of m_1 and m_2 above. This is shown below:

$m_1 = \frac{-b}{2a} - \frac{\sqrt{b^2 - 4ac}}{2a}$ and $m_2 = \frac{-b}{2a} + \frac{\sqrt{b^2 - 4ac}}{2a}$

Given that $b^2 - 4ac < 0$, then to square root we need to multiply by -1 which is equivalent to i^2. We therefore have:

$m_1 = \frac{-b}{2a} - \frac{\sqrt{(b^2 - 4ac)i^2}}{2a}$ and $m_2 = \frac{-b}{2a} + \frac{\sqrt{(b^2 - 4ac)i^2}}{2a}$

By using the surd rule $\sqrt{ab} = \sqrt{a}\sqrt{b}$ we can simplify the numerator to the following:

$m_1 = \frac{-b}{2a} - \frac{\sqrt{(b^2 - 4ac)}}{2a}i$ and $m_2 = \frac{-b}{2a} + \frac{\sqrt{(b^2 - 4ac)}}{2a}i$

By letting $p = \frac{-b}{2a}$ and $q = \frac{\sqrt{(b^2 - 4ac)}}{2a}$ we obtain the complex conjugates $m_1 = p - iq$ and $m_2 = p + iq$.

The functions $e^{m_1 x}$ and $e^{m_2 x}$ now involve complex numbers and are not appropriate solutions of this differential equation. However, it is possible to deduce that, in this case, $a\frac{d^2y}{dx^2} + b\frac{dy}{dx} + cy = 0$ has independent real solutions $y_1 = e^{px}\cos(qx)$ and $y_2 = e^{px}\sin(qx)$

There is a proof in the proofs section which justifies $a\frac{d^2y}{dx^2} + b\frac{dy}{dx} + cy = 0$ having independent real solutions $y_1 = e^{px}\cos(qx)$ and $y_2 = e^{px}\sin(qx)$. This is not required to be reproduced for exam purposes, simply for your interest.

© HERIOT-WATT UNIVERSITY

Proof 6: $y_1 = e^{px}\cos(qx)$ and $y_1 = e^{px}\sin(qx)$

Show that the homogeneous, second order, linear, differential equation with complex roots has independent real solutions, $y_1 = e^{px}\cos(qx)$ and $y_2 = e^{px}\sin(qx)$.

The differential equation $a\frac{d^2y}{dx^2} + b\frac{dy}{dx} + cy = 0$ has auxiliary equation $am^2 + bm + c = 0$.

If $b^2 - 4ac < 0$ then the auxiliary equation has complex roots which can be written as $m = p \pm iq$.

Now if $m = p \pm iq$ is a solution to the auxiliary equation, then $y = e^{mx} = e^{(p+iq)x}$ is a solution of the differential equation. Unfortunately, y involves complex numbers but only require solutions of the equation involving only real numbers.
But
$y = e^{(p+iq)x}$
$= e^{px}e^{iqx}$
$= e^{px}\left(e^{ix}\right)^q$

From the topic "Sequences and series" we saw that $e^{ix} = \cos x + i\sin x$.
$y = e^{px}(\cos x + i\sin x)^q$

De Moivre's Theorem was discussed in the topic "Complex numbers". It states that $[r(\cos\theta + i\sin\theta)]^n = r^n[\cos(n\theta) + i\sin(n\theta)]$.
$y = e^{px}[\cos(qx) + i\sin(qx)]$

It can be checked that both the real and imaginary parts of y give rise to solutions of the differential equation and hence the independent solutions are $y_1 = e^{px}\cos(qx)$ and $y_2 = e^{px}\sin(qx)$ and we can write the general solution as:

$y = e^{px}(A\cos(qx) + B\sin(qx))$

The general solution is $y(x) = e^{px}(A\cos(qx) + B\sin(qx))$

Key point

When the auxiliary equation, $am^2 + bm + c = 0$, of a second order linear differential equal has non-real roots of the form $m = p \pm iq$, then the general solution takes the form:

$$y(x) = e^{px}\left(A\cos(qx) + B\sin(qx)\right)$$

Example

Problem:

Find the general solution for the differential equation $\frac{d^2y}{dx^2} + 2\frac{dy}{dx} + 5y = 0$

Solution:

The auxiliary equation for $\frac{d^2y}{dx^2} + 2\frac{dy}{dx} + 5y = 0$ is $m^2 + 2m + 5 = 0$

TOPIC 4. DIFFERENTIAL EQUATIONS

This equation has roots given by the following:

$$m = \frac{-2 \pm \sqrt{4 - 4(1)(5)}}{2(1)}$$

$$m = \frac{-2 \pm \sqrt{-16}}{2}$$

$$m = \frac{-2 \pm 4i}{2}$$

$$m = -1 \pm 2i$$

$m_1 = -1 - 2i$ or $m_2 = -1 + 2i$

Therefore, $p = -1$ and $q = 2$ and the differential equation has independent solutions:

$y_1 = e^{-x} \cos 2x$ and $y_2 = e^{-x} \sin 2x$

Hence, the general solution is $y = e^{-x}(A \cos(2x) + B \sin(2x))$

Complex roots exercise   Go online

Q86: Find the general solution for the differential equation $\frac{d^2y}{dx^2} - 6\frac{dy}{dx} + 13y = 0$

a) The auxiliary equation is given by $am^2 + bm + c = 0$. Write down the auxiliary equation for $\frac{d^2y}{dx^2} - 6\frac{dy}{dx} + 13y = 0$

b) Solve the auxiliary equation to determine the two roots m_1 and m_2, then write down the two solutions to the differential equation.

c) Write down the general solution of the differential equation in the form $y(x) = e^{px}(A \cos(qx) + B \sin(qx))$.

....

Q87: Find the general solution for the differential equation $\frac{d^2y}{dx^2} + 4\frac{dy}{dx} + 5y = 0$

....

Q88: Find the general solution for the differential equation $\frac{d^2y}{dx^2} + \frac{dy}{dx} + y = 0$

....

Q89: Find the general solution for the differential equation $\frac{d^2y}{dx^2} + 6y = 0$.

4.5.1.4 Homogeneous second order linear differential equations summary

Key point

Strategy for solving $a \frac{d^2y}{dx^2} + b \frac{dy}{dx} + cy = 0$

1. Write down the auxiliary equation $am^2 + bm + c = 0$
2. Solve the auxiliary equation by factorising or using the quadratic formula.
3. Write down the general solution, depending on the nature of the roots of the auxiliary equation, as follows:

    a) if roots m_1 and m_2 are real and distinct, then the general solution is $y = Ae^{m_1 x} + Be^{m_2 x}$
    b) if there is a repeated root m, then the general solution is $y = Ae^{mx} + Bxe^{mx}$
    c) if there are complex roots $p \pm iq$, then the general solution is $y = e^{px}(A\cos(qx) + B\sin(qx))$

4.5.2 Initial value problems

Given a second order differential equation and initial conditions which must be satisfied by the solution, then it is possible to find a particular solution of the equation satisfying the initial conditions.

As the general solution contains two arbitrary constants, it is appropriate to specify two initial conditions.

Examples

1. Problem:

Solve the initial value problem $\frac{d^2y}{dx^2} - 6\frac{dy}{dx} + 9y = 0$ where $y(0) = 2$ and $y'(0) = 9$

Solution:

The auxiliary equation for the differential equation is $m^2 - 6m + 9 = 0$

Factorising gives $(m-3)^2 = 0 \Rightarrow m = 3$ (repeated root)

The roots of the above auxiliary equation are equal and so the general solution of the differential equation takes the form $y(x) = (A + Bx)e^{mx} = (A + Bx)e^{3x}$

Find A and B to ensure that the initial conditions are satisfied.

Since $y(0) = 2$:

$(A + B \times 0)e^0 = 2 \Rightarrow A = 2$

Hence, $y(x) = (2 + Bx)e^{3x}$

Now find B to ensure that $y'(0) = 9$, i.e. when $x = 0$, $y' = 9$.

First calculate $y'(x)$ by using the product rule $\frac{dy}{dx} = u'v + uv'$:

Let $u = 2 + Bx \Rightarrow u' = B$

Let $v = e^{3x} \Rightarrow v' = 3e^{3x}$

Substituting into the product rule:

$$\frac{dy}{dx} = u'v + uv'$$
$$= Be^{3x} + (2 + Bx)\,3e^{3x}$$
$$= Be^{3x} + 3(2 + Bx)\,e^{3x}$$

Thus:
$$y'(0) = Be^0 + 3(2 + B(0))\,e^0$$
$$9 = B + 6$$
$$B = 3$$

Hence, the solution for the initial value problem is $y(x) = (2 + 3x)e^{3x}$

....

2. Simple harmonic motion

Problem:

A mass is attached to a vertical spring and hangs in equilibrium. The mass is then pulled a distance of 10 cm down from its equilibrium position and released from rest so that it oscillates at the end of the spring.

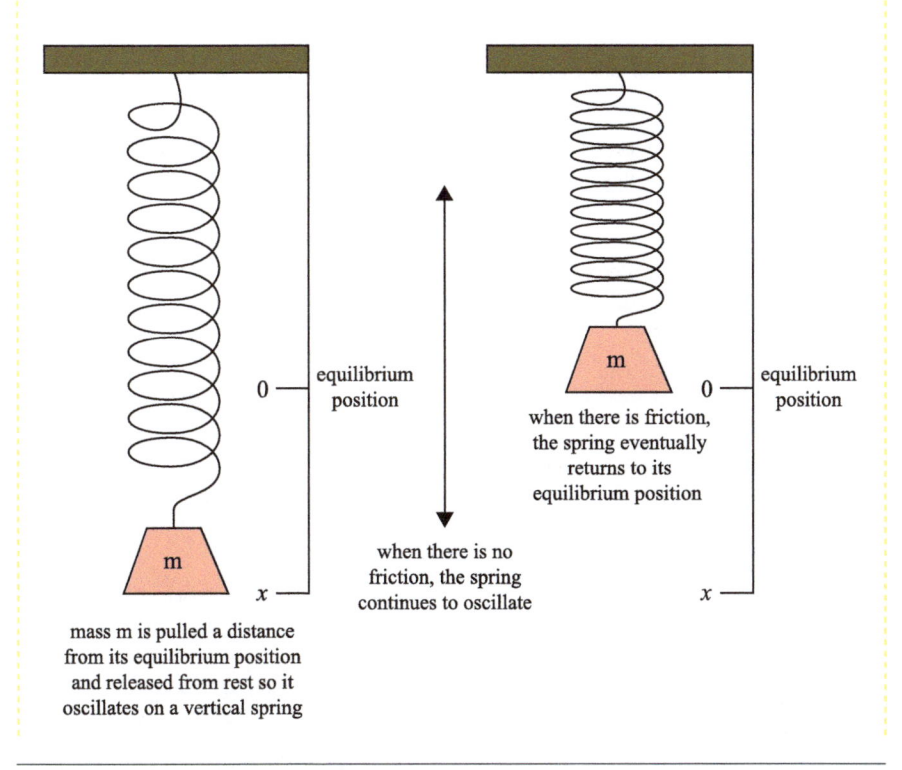

mass m is pulled a distance from its equilibrium position and released from rest so it oscillates on a vertical spring

equilibrium position

when there is no friction, the spring continues to oscillate

when there is friction, the spring eventually returns to its equilibrium position

equilibrium position

© HERIOT-WATT UNIVERSITY

If $x(t)$ denotes the distance of the mass below its equilibrium position at time t, it can be shown that $x(t)$ satisfies the differential equation $\frac{d^2x}{dt^2} + 9x = 0$

(The equation $\frac{d^2x}{dt^2} + w^2x = 0$ is the equation of simple harmonic motion.)

Find $x(t)$ and hence determine the position of the mass when $t = 2$

Solution:

The differential equation $\frac{d^2x}{dt^2} + 9x = 0$ has auxiliary equation $m^2 + 9 = 0$ which has solutions:

$m = \pm\sqrt{-9}$
$ = \pm\sqrt{9(-1)}$

Recall that $i^2 = -1$
$m = \pm\sqrt{9i^2}$
$ = \pm 3i$

These are complex roots of the form $p \pm iq$ where $p = 0$ and $q = 3$

Hence, the equation has independent solutions:

$y_1 = e^{0t}\cos(3t) = \cos(3t)$ and
$y_2 = e^{0t}\sin(3t) = \sin(3t)$

and so has general solution: $x = A\cos(3t) + B\sin(3t)$

Given that at $t = 0$:

- the spring is 10 cm from the equilibrium position, i.e. $x = 10$ at $t = 0$;
- the mass is at rest, i.e. $\frac{dx}{dt} = 0$ at $t = 0$

Choose A and B to obtain a particular solution which satisfies these initial conditions.

Since $x = 10$ when $t = 0$ then:
$10 = A\cos(0) + B\sin(0)$
$A = 10$

Hence, $x(t) = 10\cos(3t) + B\sin(3t)$

Since $\frac{dx}{dt} = 0$ when $t = 0$ then:
$x'(t) = -30\sin(3t) + 3B\cos(3t)$
$0 = -30\sin(0) + 3B\cos(0)$
$3B = 0$
$B = 0$

Thus, the position of the mass is given by $x = 10\cos(3t)$

Finally, when $t = 2$ the position of the mass is $x = 10\cos 6 \approx 9.6$ cm below the equilibrium position.

TOPIC 4. DIFFERENTIAL EQUATIONS

This process is summarised as follows.

> **Key point**
>
> To solve the homogeneous second order linear differential equation
>
> $$a \frac{d^2y}{dx^2} + b \frac{dy}{dx} + cy = 0$$
>
> given initial conditions:
>
> 1. Solve $a \frac{d^2y}{dx^2} + b \frac{dy}{dx} + cy = 0$ to find one of the following general solutions:
>   - $y(x) = Ae^{m_1 x} + Be^{m_2 x}$ (two real distinct roots)
>   - $y(x) = (A + Bx)e^{mx}$ (equal or repeated roots)
>   - $y(x) = e^{px}(A \cos(qx) + B \sin(qx))$ (complex roots of the form $p \pm iq$)
>
> 2. Substitute the first initial condition into the general solution to evaluate one of the arbitrary constants A or B.
>
> 3. The second initial condition will usually require the general solution to be differentiated. This initial condition is then substituted into the derivative to evaluate the second arbitrary constant.

Initial value problems exercise Go online

Q90: The equation of simple harmonic motion is $\frac{d^2x}{dt^2} = -w^2 x$, where x is the distance below the equilibrium of the spring at time t and w is a constant.

Find the particular solution when $w = 10$ given that, when $t = 0$, $x(0) = 5$ and $\frac{dx}{dt} = -10$

a) Work out the general solution, $x(t)$, for the differential equation.
b) Use the initial condition when $t = 0$, $x = 5$ to work out the arbitrary constant A.
c) Differentiate the general solution to use the second initial condition $\frac{dx}{dt} = -10$ at $t = 0$ and evaluate the arbitrary constant B.
d) Write down the particular solution to the differential equation.

....

Q91: Find the particular solution for $\frac{d^2y}{dx^2} - 5 \frac{dy}{dx} + 6y = 0$ given that $y(0) = 1$ and $y'(0) = 1$

....

Q92: A mass of 2 kg is attached to a vertical spring which is immersed in a sticky fluid. At time $t = 0$, the mass is lowered 0·2 m from its equilibrium position and given an initial velocity of 0·6 m/s in the upward direction, i.e. $\frac{dx}{dt} = -0 \cdot 6$.

If $x(t)$ denotes the distance of the mass below its equilibrium position at time t, it can be shown that $x(t)$ satisfies the differential equation $\frac{d^2x}{dt^2} + 20 \frac{dx}{dt} + 64x = 0$

Find $x(t)$. Round all coefficients to 2 decimal places.

© HERIOT-WATT UNIVERSITY

4.5.3 Non-homogeneous second order linear differential equations

Recall that:

> **Key point**
>
> A second order, linear differential equation is **non-homogeneous** when it can be expressed in the standard form
>
> $$a\frac{d^2y}{dx^2} + b\frac{dy}{dx} + cy = f(x)$$
>
> where a, b and c are constants, $a \neq 0$ and $f(x) \neq 0$

In this section we see how to find the general solution of the non-homogeneous (or inhomogeneous) differential equation where $f(x)$ is a simple function such as a polynomial, exponential or sine/cosine function.

As the equation is of second order, the general solution should contain two arbitrary constants. The general solution can be found in the following way.

1. Find a general solution, y_c of the corresponding homogeneous equation
   $a\frac{d^2y}{dx^2} + b\frac{dy}{dx} + cy = 0$
   y_c contains two arbitrary constants and is known as the **complementary function** for the non-homogeneous equation.

2. Find (by means of the method of undetermined coefficients, which is covered later) an 'obvious' special solution y_p of a $\frac{d^2y}{dx^2} + b\frac{dy}{dx} + cy = f(x)$
   y_p is a **particular integral** of the differential equation.

3. Write down the general solution of the non-homogeneous differential equation as $y_p + y_c$
   This is because y contains two arbitrary constants and

   $$a\frac{d^2y}{dx^2} + b\frac{dy}{dx} + cy = a\frac{d^2y_p}{dx^2} + b\frac{dy_p}{dx} + cy_p + a\frac{d^2y_c}{dx^2} + b\frac{dy_c}{dx} + cy_c$$
   $$= f(x) + 0$$
   $$= f(x)$$

   when substitution of the particular integral, y_p, and the complementary function, y_c are made in the differential equation.

Example

Problem:

Show that $y = e^{2x}$ is a particular integral of the differential equation $\frac{d^2y}{dx^2} + \frac{dy}{dx} - 2y = 4e^{2x}$ and hence find the general solution for this equation.

Solution:

To show that $y = e^{2x}$ is a particular solution of the differential equation we substitute it into

the LHS of the differential equation.
This means that we need to work out $\frac{dy}{dx}$ and $\frac{d^2y}{dx^2}$.

After substituting, if the LHS = RHS (= $4e^{2x}$), then $y = e^{2x}$ is a particular solution to the differential equation.

Differentiate: $y = e^{2x}$, $\frac{dy}{dx} = 2e^{2x}$ and $\frac{d^2y}{dx^2} = 4e^{2x}$

Substitute:

$$\frac{d^2y}{dx^2} + \frac{dy}{dx} - 2y = 4e^{2x} + 2e^{2x} - 2e^{2x}$$
$$= 4e^{2x}$$
$$= \text{RHS}$$

Since LHS = RHS, then $y_p = e^{2x}$ is a particular integral of the differential equation.

Remember that the general solution takes the form $y_p + y_c$

We already know $y_p = e^{2x}$, so we only need to work out y_c, the complementary function of the homogeneous differential equation $\frac{d^2y}{dx^2} + \frac{dy}{dx} - 2y = 0$

Auxiliary equation:

$$m^2 + m - 2 = 0$$
$$(m+2)(m-1) = 0$$
$$(m+2) = 0 \text{ or } (m-1) = 0$$
$$m = -2 \text{ or } m = 1$$

Complementary function: $y_c = Ae^{-2x} + Be^x$

General solution:

$$y(x) = y_P + y_c$$
$$y(x) = e^{2x} + Ae^{-2x} + Be^x$$

Key point

Strategy for solving $a\frac{d^2y}{dx^2} + b\frac{dy}{dx} + cy = f(x)$

1. Find the complementary function y_c, i.e. find the general solution of the corresponding homogeneous equation $a\frac{d^2y}{dx^2} + b\frac{dy}{dx} + cy = 0$.
2. Find a particular integral y_p.
3. Write down the required general solution $y = y_p + y_c$

The difficult step in the method is finding the particular solution y_p. This is explained in the next section.

Non-homogeneous second order linear differential equations exercise Go online

The general solution of non-homogenous second order linear differential equations takes the form $y(x) = y_p + y_c$

The following questions are second order linear non-homogeneous differential equations. For each question:

- Work out the auxiliary equation and solve it to find the complementary function, y_c
- Guess the form of the particular integral, y_p, and check it by substituting it into the LHS of the differential equation.
- Write down the general solution $y = y_c + y_p$

Q93: $2\frac{d^2y}{dx^2} - \frac{dy}{dx} - 3y = 3e^{2x}$

....

Q94: $\frac{d^2y}{dx^2} - 4\frac{dy}{dx} + 4y = 25e^{-3x}$

....

Q95: $\frac{d^2y}{dx^2} + 4\frac{dy}{dx} + 5y = 50e^{5x}$

....

Q96: $\frac{d^2y}{dx^2} + \frac{dy}{dx} - 12y = -12e^{-x}$

Particular integrals, y_p	Complementary functions, y_c
$t_p = e^{-x}$	$y_c = (A + Bx)e^{2x}$
$t_p = e^{2x}$	$y_c = Ae^{3x} + Be^{-4x}$
$t_p = e^{-3x}$	$y_c = Ae^{\frac{3}{2}x} + Be^{-x}$
$t_p = e^{5x}$	$y_c = e^{-2x}(A\cos x + B\sin x)$

4.5.3.1 Finding particular integrals

This section explores how to find particular integrals of the differential equation $a\frac{d^2y}{dx^2} + b\frac{dy}{dx} + cy = f(x)$ by using the method of undetermined coefficients.

This relies on the idea that, for simple functions, it is easy to guess the general form of a particular integral. Unknown coefficients in the general form can then be found by substituting into the differential equation.

The examples will focus on how this can be done in the cases where $f(x)$ is a polynomial, an

TOPIC 4. DIFFERENTIAL EQUATIONS

exponential or a sine/cosine function.

Polynomial functions

> **Key point**
>
> For the non-homogeneous second order, linear differential equation
>
> $$a \frac{d^2y}{dx^2} + b \frac{dy}{dx} + cy = f(x) \text{ where } (c \neq 0)$$
>
> if $f(x)$ is a polynomial of degree n, then the particular integral takes the form of a polynomial with the same degree:
>
> $$y_p = a_n x^n + a_{n-1} x^{n-1} + a_{n-2} x^{n-2} + \ldots + a_1 x^1 + a_0$$

Example

Problem:

Find a particular integral for $\frac{d^2y}{dx^2} - \frac{dy}{dx} + 2y = 2x$

Solution: In this example, the particular integral will be a polynomial of degree one and take the form $y_P = ax + b$ where a and b are the undetermined coefficients.

Find values for a and b to ensure that y_p is a particular integral by differentiating y_p and substituting into the LHS of the differential equation.

When $y_p = ax + b$ then $\frac{dy_p}{dx} = a$ and $\frac{d^2y_p}{dx^2} = 0$

Hence, y_p is a solution provided that:

$$\frac{d^2y_p}{dx^2} - \frac{dy_p}{dx} + 2y_p = 2x$$

i.e. $0 - a + 2(ax + b) = 2x$

Now collect together the x terms and the constant terms so that we can equate coefficients to work out the values of a and b: $2ax + (2b - a) = 2x$

Now equate the coefficients of the x terms on the LHS and the RHS: $2a = 2 \Rightarrow a = 1$

Now equate the constants on the LHS with the constants on the RHS and substitute for $a = 1$. Note that the constant on the RHS = 0: $2b - a = 0 \Rightarrow b = \frac{1}{2}$

Thus, the particular integral is $y_p = x + \frac{1}{2}$

> **Key point**
>
> - If $f(x)$ is a quadratic, e.g. $f(x) = 3x^2 - 4x + 1$ or $f(x) = x^2 - 5$, a particular integral would take the form $y_p = ax^2 + bx + c$
> - If $f(x)$ is a constant, e.g. $f(x) = 3$, a particular integral would be of the form $f(x) = a$

Examples

1. Problem:

Find the general solution of the differential equation $\frac{d^2y}{dx^2} + 2\frac{dy}{dx} - 3y = 6x - 1$

Solution:

The auxiliary equation is $m^2 + 2m - 3 = 0$
which factorises to $(m + 3)(m - 1) = 0$
which has the solution $m = -3$ or $m = 1$
so the complementary function is $y_c = Ae^{-3x} + Be^x$
Now we need to find a particular integral.
We choose one similar in appearance to $6x - 1$
Let $y_p = ax + b$ so $\frac{dy_p}{dx} = a$ and $\frac{d^2y_p}{dx^2} = 0$
Now substitute these values into the differential equation $\frac{d^2y}{dx^2} + 2\frac{dy}{dx} - 3y = 6x - 1$
$0 + 2(a) - 3(ax + b) = 6x - 1$
and simplify this to solve for a and b
$2a - 3ax - 3b = 6x - 1$
$-3ax + 2a - 3b = 6x - 1$
$-3ax + (2a - 3b) = 6x - 1$
By equating coefficients:
$x: \quad -3a = 6 \Rightarrow a = -2$
$x^0: \quad (2a - 3b) = -1$
$\quad\quad (2(-2) - 3b) = -1$
$\quad\quad\quad -3b = 3$
$\quad\quad\quad b = -1$
Hence the particular integral is $y_p = -2x - 1$
Now we have all that we need to find the general solution: $y = -2x - 1 + Ae^{-3x} + Be^x$

....

2. Problem:

Find the general solution of the differential equation $\frac{d^2y}{dx^2} - \frac{dy}{dx} - 2y = 6x^2 + 5$.

Solution:

Find the complementary function first by deriving the auxiliary equation and factorising to obtain the roots.

Auxiliary Equation:
$$m^2 - m - 2 = 0$$
$$(m + 1)(m - 2) = 0$$
$$(m + 1) = 0 \text{ or } (m - 2) = 0$$
$$m = -1 \text{ or } m = 2$$

Complementary function:
$$y_C = Ae^{-x} + Be^{2x}$$
Now find the particular integral.

The RHS of the differential equation is a quadratic and so the particular integral will also be a quadratic.

Let $y_P = ax^2 + bx + c$

$$\frac{dy_P}{dx} = 2ax + b$$

$$\frac{d^2y_P}{dx^2} = 2a$$

Substituting y_P, $\frac{dy_P}{dx}$ and $\frac{d^2y_P}{dx^2}$ into the LHS of the original differential equation:
$$2a - (2ax + b) - 2(ax^2 + bx + c) = 6x^2 + 5$$

Collect all the x^2 terms together, all the x terms together and all the coefficients together. Then equate coefficients.

$$2a - 2ax - b - 2ax^2 - 2bx - 2c = 6x^2 + 5$$
$$-2ax^2 - (2a + 2b)x + (2a - b - 2c) = 6x^2 + 5$$

Equate the coefficients of x^2:

$-2a = 6$

$a = -3$

Equate the coefficients of x and substitute for $a = -3$:

$-(2a + 2b) = 0$

$-(2(-3) + 2b) = 0$

$6 - 2b = 0$

$-2b = -6$

$b = 3$

Equate the constants and substitute for $a = -3$, $b = 3$:

$2a - b - 2c = 5$

$2(-3) - 3 - 2c = 5$

$-9 - 2c = 5$

$-2c = 4$

$c = -2$

Hence the particular integral is: $y_P = -3x^2 + 3x - 2$

The general solution is then given by:

$y = y_P + y_C$
$y = -3x^2 + 3x - 2 + Ae^{-x} + Be^{2x}$

Exponential functions

Key point

It is usually possible to find a particular integral for the equation $a\frac{d^2y}{dx^2} + b\frac{dy}{dx} + cy = e^{rx}$ of the form: $y_p = ke^{rx}$, where k is the undetermined coefficient.

Example

Problem:

Find a particular integral for $\frac{d^2y}{dx^2} + 2\frac{dy}{dx} - 3y = e^{2x}$

Solution:

Guess and try $y_p = k\,e^{2x}$

Then, $\frac{dy_p}{dx} = 2ke^{2x}$ and $\frac{d^2y_p}{dx^2} = 4ke^{2x}$

Hence, y_p is a particular integral provided:

$$4ke^{2x} + 2\left(2ke^{2x}\right) - 3ke^{2x} = e^{2x}$$
$$5ke^{2x} = e^{2x}$$
$$5k = 1$$
$$k = \frac{1}{5}$$

Thus, the function $y_p = \frac{1}{5}e^{2x}$ is a particular solution for the differential equation.

Key point

The above method for finding a particular solution of $a\frac{d^2y}{dx^2} + b\frac{dy}{dx} + cy = e^{rx}$ fails when r coincides with one of the roots of the auxiliary equation
$am^2 + bm + c = 0$

In this case, we can still find a particular integral of the form:

- $y_p = kxe^{rx}$ if r is a *single* root of the auxiliary equation;
- $y_p = kx^2e^{rx}$ if r is a *repeated* root of the auxiliary equation.

Examples

1. Choosing the particular integral

Problem:

By solving the auxiliary equation for $2\frac{d^2y}{dx^2} + \frac{dy}{dx} - 3y = -9e^{\frac{1}{2}x}$ determine the form of the particular integral.

Solution:
The auxiliary is given by:
$$2m^2 + m - 3 = 0$$
$$(2m+3)(m-1) = 0$$
$$(2m+3) = 0 \quad \text{or} \quad (m-1) = 0$$
$$m = -\frac{3}{2} \quad \text{or} \quad m = 1$$

The complementary function therefore takes the form $y_c = Ae^{-\frac{3}{2}x} + Be^x$.

Looking at each root, $m = -\frac{3}{2}$ and $m = 1$, and how these appear in the complementary function as $e^{-\frac{3}{2}x}$ and e^x we can see that neither coincide with $f(x) = e^{\frac{1}{2}x}$, which is the function on the RHS of the original differential equation. Since the roots of the auxiliary equation do not coincide with the RHS of the differential equation then the particular integral would take the form $y_P = Ce^{\frac{1}{2}x}$.

....

2. Choosing the particular integral
Problem:
By solving the auxiliary equation for $\frac{d^2y}{dx^2} - 3\frac{dy}{dx} + 2y = 3e^{2x}$ determine the form of the particular integral.

Solution:
The auxiliary is given by:
$$m^2 - 3m + 2 = 0$$
$$(m-1)(m-2) = 0$$
$$(m-1) = 0 \quad \text{or} \quad (m-2) = 0$$
$$m = 1 \quad \text{or} \quad m = 2$$

The complementary function therefore takes the form $y_c = Ae^x + Be^{2x}$.

Looking at each root, $m = 1$ and $m = 2$, and how these appear in the complementary function as e^x and e^{2x} we can see that the root $m = 2$ coincides with $f(x) = 3e^{2x}$, which is the function on the RHS of the original differential equation.

Since one root of the auxiliary equation coincides with the RHS of the differential equation then the particular integral would take the form $y_P = Cxe^{2x}$.

....

3. Coincidental roots
Problem:
Find the general solution for $\frac{d^2y}{dx^2} - 6\frac{dy}{dx} + 9y = 4e^{3x}$

Solution:
First we must work out the complementary function, y_c for the homogenous equation $\frac{d^2y}{dx^2} - 6\frac{dy}{dx} + 9y = 0$

Auxiliary equation:

$m^2 - 6m + 9 = 0$
$(m-3)(m-3) = 0$

$m - 3 = 0$ (repeated)
$m = 3$

Since $m = 3$ is a repeated root, the complementary function is $y_c = (A + Bx)e^{3x}$

Given the root of the complementary function, $m = 3$, coincides with $f(x) = 4e^{3x}$ and is a repeated root for the auxiliary equation, then the particular integral takes the form
$y_p = Cx^2 e^{3x}$

Using the method of undetermined coefficients, differentiate and substitute into the LHS of the differential equation to work out C.

$y_p = Cx^2 e^{3x}$

Using the product rule:

$$\frac{dy_p}{dx} = 2Cxe^{3x} + 3Cx^2 e^{3x}$$

$$\frac{d^2 y_p}{dx^2} = 2Ce^{3x} + 6Cxe^{3x} + 6Cxe^{3x} + 9Cx^2 e^{3x}$$

$$= 2Ce^{3x} + 12Cxe^{3x} + 9Cx^2 e^{3x}$$

Substituting:

$$\frac{d^2 y}{dx^2} - 6\frac{dy}{dx} + 9y = 4e^{3x}$$

$$\left(2Ce^{3x} + 12Cxe^{3x} + 9Cx^2 e^{3x}\right) - 6\left(2Cxe^{3x} + 3Cx^2 e^{3x}\right) + 9\left(Cx^2 e^{3x}\right) = 4e^{3x}$$

$$2Ce^{3x} + 12Cxe^{3x} + 9Cx^2 e^{3x} - 12Cxe^{3x} - 18Cx^2 e^{3x} + 9Cx^2 e^{3x} = 4e^{3x}$$

$$2Ce^{3x} = 4e^{3x}$$

Equate coefficients of e^{3x} so that: $2C = 4 \Rightarrow C = 2$

Therefore, the particular solution is: $y_p = 2x^2 e^{3x}$

The general solution of the differential equation then becomes:

$y(x) = y_p + y_c$
$= 2x^2 e^{3x} + (A + Bx)e^{3x}$

Particular integrals of exponential functions   Go online

For the following questions solve the auxiliary equation and determine the form of the particular integral.

Q97: $\frac{d^2 y}{dx^2} - \frac{dy}{dx} - 6y = 4e^{2x}$

....

Q98: $\frac{d^2 y}{dx^2} + \frac{dy}{dx} - 6y = 10e^{2x}$

....

Q99: $\frac{d^2 y}{dx^2} + 8\frac{dy}{dx} + 16y = 6e^{-4x}$

....

TOPIC 4. DIFFERENTIAL EQUATIONS

Q100: $2\frac{d^2y}{dx^2} - \frac{dy}{dx} - y = 9e^x$

....

Q101: $\frac{d^2y}{dx^2} - 6\frac{dy}{dx} + 9y = 8e^{3x}$

Sine and cosine functions

> **Key point**
>
> It is always possible to find a particular integral of the equations:
>
> $$a\frac{d^2y}{dx^2} + b\frac{dy}{dx} + cy = \sin(nx) \text{ where } (b \neq 0) \text{ or}$$
>
> $$a\frac{d^2y}{dx^2} + b\frac{dy}{dx} + cy = \cos(nx) \text{ where } (b \neq 0)$$
>
> of the form $y_p = p\sin(nx) + q\cos(nx)$ where p and q are the undetermined coefficients.

It would have been much more convenient if the particular integrals could have been, respectively, $y_p = p\sin(nx)$ and $y_p = q\cos(nx)$. Unfortunately, this simpler and more obvious approach does not work and the general form $y_p = p\sin(nx) + q\cos(nx)$ should always be used.

Example

Problem:

Find a particular solution for the differential equation $\frac{d^2y}{dx^2} + 2\frac{dy}{dx} + 2y = 5\sin x$

Solution:

Try a particular solution of the form $y_p = p\sin x + q\cos x$

Notice that the coefficient of x in y_p is 1 because the coefficient of x in $5\sin x$ is 1.

Hence, $\frac{dy_p}{dx} = p\cos x - q\sin x$ and $\frac{d^2y_p}{dx^2} = -p\sin x - q\cos x$

Hence y_p is a particular solution of $\frac{d^2y}{dx^2} + 2\frac{dy}{dx} + 2y = 5\sin x$ provided that:

$$-p\sin x - q\cos x + 2(p\cos x - q\sin x) + 2(p\sin x + q\cos x) = 5\sin x$$
$$-p\sin x - 2q\sin x + 2p\sin x - q\cos x + 2p\cos x + 2q\cos x = 5\sin x$$
$$p\sin x - 2q\sin x + 2p\cos x + q\cos x = 5\sin x$$
$$(p - 2q)\sin x + (2p + q)\cos x = 5\sin x$$

Now equate the coefficients of $\sin x$ on the LHS with the RHS: $p - 2q = 5$

Equate the coefficients of $\cos x$ on the LHS with the RHS: $2p + q = 0$

Using simultaneous equations, $p = 1$ and $q = -2$

The particular solution is therefore $y_p = \sin x - 2\cos x$

© HERIOT-WATT UNIVERSITY

> **Key point**
>
> For the differential equation $a \frac{d^2y}{dx^2} + b \frac{dy}{dx} + cy = f(x)$:
>
> If $f(x)$ is a sum of terms, i.e. $f(x) = p(x) + q(x)$ and
>
> $y = u(x)$ is a particular integral of $a \frac{d^2y}{dx^2} + b \frac{dy}{dx} + cy = p(x)$ and
>
> $y = v(x)$ is a particular integral of $a \frac{d^2y}{dx^2} + b \frac{dy}{dx} + cy = q(x)$
>
> then it can be shown that $y = u(x) + v(x)$ is a particular integral of
>
> $a \frac{d^2y}{dx^2} + b \frac{dy}{dx} + cy = p(x) + q(x)$

Examples

1. Problem:

State the form of the particular integral for the differential equation:

$\frac{d^2y}{dx^2} - 2\frac{dy}{dx} - 3y = 5e^{-x} + 3x$

Solution:

The function on the RHS is in the form $f(x) = p(x) + q(x)$, where $p(x) = 5e^{-x}$ and $q(x) = 3x$.

The particular integral, y_P, will therefore be made up of:

- an exponential function of one of the three forms:
    - $y_{p1} = Ce^{-x}$
    - $y_{p1} = Cxe^{-x}$
    - $y_{p1} = Cx^2e^{-x}$
- a polynomial of degree 1: $y_{p2} = Dx + E$

To determine the form of the exponential part of the particular integral we need to check for coincidental and repeated roots from the auxiliary equation.

Auxiliary equation:

$m^2 - 2m - 3 = 0$
$(m - 3)(m + 1) = 0$
$m - 3 = 0$ or $m + 1 = 0$
$m = 3$ or $m = -1$

The roots are real and distinct so the complementary function is $y_c = Ae^{3x} + Be^{-x}$

Since root $m = -1$ coincides with $p(x) = 5e^{-1x}$, then the general form of the particular integral corresponding to $p(x)$ is $y_{p1} = Cxe^{-x}$.

The general form of the particular integral is:

$y_p = y_{p1} + y_{p2}$
$= Cxe^{-x} + Dx + E$

....

2. Problem:

State the form of the particular integral for the differential equation: $\frac{d^2y}{dx^2} - 6\frac{dy}{dx} + 9y = 3x + 4\cos 2x$

Solution:

The function on the RHS is of the from $f(x) = p(x) + q(x)$, where $p(x) = 3x$ and $q(x) = 4\cos 2x$.

The particular integral, y_p, will therefore be made up of :

- a linear polynomial of the form: $y_{p1} = ax + b$;
- a trigonometric function of the form: $y_{p2} = c\cos 2x + d\sin 2x$

To check for coincidental roots we need to solve the auxiliary equation.

Auxiliary equation:
$$m^2 - 6m + 9 = 0$$
$$(m-3)(m-3) = 0$$
$(m-3) = 0 \quad \text{or} \quad (m-3) = 0$
$\qquad m = 3 \quad \text{or} \quad m = 3 \quad$ (repeated roots)

The root are real and repeated and so the complementary function is $y_c = (A + Bx)e^{3x}$.

The repeated roots, $m = 3$, do not coincide with the coefficient of x in y_{p2} and so the form of the particular integral is:

$f(x) = y_{p1} + y_{p2}$
$f(x) = ax + b + c\cos 2x + d\sin 2x$

Sum of terms   Go online

For the non-homogeneous second order linear differential equation

$a\frac{d^2y}{dx^2} + b\frac{dy}{dx} + cy = f(x)$

$f(x)$ can be made up of a sum of terms. The particular integral can therefore be made up of a sum of terms.

For the questions below, work out the particular integral for the non-homogeneous second order linear differential equations. Remember to work out the general solution to the homogeneous second order linear differential equation first.

Q102: $\frac{d^2y}{dx^2} - \frac{dy}{dx} - 2y = 3e^{2x} + 4x^2$

....

Q103: $\frac{d^2y}{dx^2} - 4\frac{dy}{dx} + 4y = 4e^{2x} + 2x - 1$

....

Q104: $\frac{d^2y}{dx^2} + \frac{dy}{dx} - 6y = 4e^{-3x} + 4\sin x$

....

Q105: $\frac{d^2y}{dx^2} - 5\frac{dy}{dx} + 4y = e^{-3x} - 5\cos(2x)$

Finding particular integrals exercise

Go online

Q106: Find the solution to the differential equation $\frac{d^2y}{dx^2} - 2\frac{dy}{dx} - 3y = -20\cos x$

....

Q107: Find the particular solution to the differential equation $\frac{d^2y}{dx^2} + 4y = 3x$ given that $y = 2$ when $x = 0$ and $y = \frac{\pi}{4}$ when $x = \frac{\pi}{4}$.

Further information

This is for the student to investigate further cases of coincidental and repeated roots for non-homogeneous second order linear differential equations.

We have previously considered one example of where the auxiliary equation of the differential equation $a\frac{d^2y}{dx^2} + b\frac{dy}{dx} + cy = f(x)$ has roots that coincide with $f(x)$. In the instance of this situation, the form of $f(x)$ needs to be considered before the particular integral can be found.

The following are the most important forms of $f(x)$ and their corresponding particular integrals that should be tried.

---

f(x) is a polynomial

Given $f(x) = a_n x^n + a_{n-1} x^{n-1} + \ldots + a_0$ then:

- $y_p = b_n x^n + b_{n-1} x^{n-1} + \ldots + b_0$ if $m = 0$ is *not* a root of $am^2 + bm + c = 0$
- $y_p = x(b_n x^n + b_{n-1} x^{n-1} + \ldots + b_0)$ if $m = 0$ is a root of $am^2 + bm + c = 0$
- $y_p = x^2(b_n x^n + b_{n-1} x^{n-1} + \ldots + b_0)$ if $m = 0$ is a repeated root of $am^2 + bm + c = 0$

---

f(x) is an exponential function

Given $f(x) = e^{mx}$ then:

- $y_p = Ae^{mx}$ if m is *not* a root of $am^2 + bm + c = 0$
- $y_p = Axe^{mx}$ if m is a root of $am^2 + bm + c = 0$
- $y_p = Ax^2 e^{mx}$ if m is a repeated root of $am^2 + bm + c = 0$

> **f(x) is the product of x and an exponential function**
>
> Given $f(x) = xe^{mx}$ then:
>
> - $y_p = (Ax + B)e^{mx}$ if m is *not* a root of $am^2 + bm + c = 0$
> - $y_p = (Ax^2 + Bx)e^{mx}$ if m is a root of $am^2 + bm + c = 0$
> - $y_p = (Ax^3 + Bx^2)e^{mx}$ if m is a repeated root of $am^2 + bm + c = 0$

> **f(x) is a sine or cosine function**
>
> Given $f(x) = C\cos(mx) + D\sin(mx)$ then:
>
> - $y_p = A\cos(mx) + B\sin(mx)$ if m is *not* a root of $am^2 + bm + c = 0$
> - $y_p = x(A\cos(mx) + B\sin(mx))$ if m is a root of $am^2 + bm + c = 0$

The particular integral is then determined by ensuring that it satisfies: $a\frac{d^2y}{dx^2} + b\frac{dy}{dx} + cy = f(x)$

4.6 Learning points

First order linear differential equations introduction

- A first order linear differential equation is an equation that can be expressed in the standard form: $\frac{dy}{dx} + P(x)y = f(x)$
- The equation is termed linear as it involves only first order terms in $\frac{dy}{dx}$ and y but not terms such as y^2, $y\frac{dy}{dx}$, y^3, $\cos y$ etc.

First order linear differential equations with separable variables

The first order linear differential equations with separable variables $\frac{dy}{dx} = f(x)g(y)$ can be solved in the following way:

1. Separate the variables $\frac{dy}{g(y)} = f(x)\,dx$
2. Integrate both sides $\int \frac{dy}{g(y)} = \int f(x)\,dx$
3. Rearrange for y

Growth and decay

- Exponential growth occurs when the population increases at a rate that is proportional to the size of the population, and is written as: $\frac{dP}{dt} = kP$
- This can be solved by the separable variables method to give the general solution: $P(t) = A\,e^{kt}$
- Exponential decay occurs when the population decreases at a rate that is proportional to the size of the population, and is written as: $\frac{dP}{dt} = -kP$
- This can be solved by the separable variables method to give the general solution: $P(t) = A\,e^{-kt}$

Further applications of differential equations

- Newton's law of cooling is given by: $\frac{dT}{dt} = -k(T - T_S)$
- T is the temperature after time t.
- T_s is the temperature of the surrounding environment.
- k is the constant of proportionality.
- This can be solved by the separable variables method.

Integrating factor

First order linear differential equations can be solved in the following way:

1. Write the equation in standard form: $\frac{dy}{dx} + P(x)y = f(x)$.
2. Calculate the integrating factor: $I(x) = e^{\int P(x)\,dx}$.

TOPIC 4. DIFFERENTIAL EQUATIONS

3. Multiply the equation in standard form by $I(x)$ and obtain the equation: $\frac{dy}{dx}[I(x)y] = I(x)f(x)$.
4. Integrate both sides of this last equation to give: $I(x)y = \int I(x)f(x)\,dx$.
5. Rearrange to solve for y.
6. If an initial condition is given for the first order linear differential equation, we should then be able to find a particular solution.

Second order linear differential equations with constant coefficients

A second order linear differential equation has the form $a\frac{d^2y}{dx^2} + b\frac{dy}{dx} + cy = f(x)$ where a, b and c are constants and $a \neq 0$

- It is homogeneous when $f(x) = 0$
- It is non-homogeneous when $f(x) \neq 0$

Homogeneous second order linear differential equations: Initial value problems

The homogeneous second order linear differential equation $a\frac{d^2y}{dx^2} + b\frac{dy}{dx} + cy = 0$ can be solved in the following way:

1. Write down the auxiliary equation $am^2 + bm + c = 0$.
2. Solve the auxiliary equation by factorising or using the quadratic formula.
3. Write down the general solution depending on the nature of the auxiliary equation as follows.

   a) If $b^2 - 4ac > 0$, then the general solution is $y = Ae^{m_1 x} + Be^{m_2 x}$
   b) If $b^2 - 4ac = 0$, then the general solution is $y = (A + Bx)e^{mx}$
   c) If $b^2 - 4ac < 0$, then the general solution is $y = e^{px}(A\cos(qx) + B\sin(qx))$

4. When we are given two initial conditions, we can find a particular solution of a second order linear differential equation satisfying these initial conditions.

Non-homogeneous second order linear differential equations: Particular integrals

The non-homogeneous, second order, linear differential equation $a\frac{d^2y}{dx^2} + b\frac{dy}{dx} + cy = f(x)$ can be solved in the following way:

1. Find the complementary function y_C, i.e. find the general solution of the corresponding homogeneous equation $a\frac{d^2y}{dx^2} + b\frac{dy}{dx} + cy = 0$
2. Find the particular integral y_p
3. Write down the required general solution $y(x) = y_p + y_c$

In certain cases, it is possible to find a particular integral for the differential equation $a\frac{d^2y}{dx^2} + b\frac{dy}{dx} + cy = f(x)$

1. If $f(x)$ is a polynomial of degree n, then the particular integral is of the form:
$y_P = a_n x^n + a_{n-1} x^{n-1} + a_{n-2} x^{n-2} + \ldots + a_1 x^1 + a_0$

© HERIOT-WATT UNIVERSITY

2. If $f(x) = e^{rx}$, then the particular integral takes one of three forms:

   a) $y_p = ke^{rx}$ (if r is distinct from the roots of the auxiliary equation).
   b) $y_p = kxe^{rx}$ (if r coincides with one of the roots of the auxiliary equation).
   c) $y_p = kx^2e^{rx}$ (if r coincides with the repeated root of the auxiliary equation).

3. If $f(x) = \sin(nx)$ or $f(x) = \cos(nx)$, then the particular integral is usually of the form $y_p = p\cos(nx) + q\sin(nx)$.

TOPIC 4. DIFFERENTIAL EQUATIONS

4.7 Proofs

Proof 1

For a homogenous second order linear differential equation, if y_1 and y_2 are two solutions to the equation, then so is $y = y_1 + y_2$.

We have:
$a\frac{d^2y_1}{dx^2} + b\frac{dy_1}{dx} + cy_1 = 0$ and $a\frac{d^2y_2}{dx^2} + b\frac{dy_2}{dx} + cy_2 = 0$

Addition gives:
$$a\frac{d^2y_1}{dx^2} + b\frac{dy_1}{dx} + cy_1 + a\frac{d^2y_2}{dx^2} + b\frac{dy_2}{dx} + cy_2 = 0$$
$$a\left(\frac{d^2y_1}{dx^2} + \frac{d^2y_2}{dx^2}\right) + b\left(\frac{dy_1}{dx} + \frac{dy_2}{dx}\right) + c(y_1 + y_2) = 0$$

By the additive law of differentiation, this gives:
$a\frac{d^2}{dx^2}(y_1 + y_2) + b\frac{d}{dx}(y_1 + y_2) + c(y_1 + y_2) = 0$

and so $y = y_1 + y_2$ is a solution to the differential equation.

Proof 2

For a homogenous second order linear differential equation, if y_1 is a solution, then if A is any constant Ay_1 is also a solution.

We have:
$a\frac{d^2}{dx^2}(Ay_1) + b\frac{d}{dx}(Ay_1) + c(Ay_1) = 0$

Since a constant can be taken as a common factor to the front of the differentiation, then:
$$aA\frac{d^2y_1}{dx^2} + bA\frac{dy_1}{dx} + cAy_1 = 0$$
$$A\left(a\frac{d^2y_1}{dx^2} + b\frac{dy_1}{dx} + cy_1\right) = 0$$

y_1 is a solution and therefore so is Ay_1.

Proof 3

For a homogenous second order linear differential equation if y_1 and y_2 are two solutions to the equation then so is $y = Ay_1 + By_2$.

We have:
$a\frac{d^2}{dx^2}(Ay_1) + b\frac{d}{dx}(Ay_1) + c(Ay_1) = 0$ and $a\frac{d^2}{dx^2}(By_2) + b\frac{d}{dx}(By_2) + c(By_2) = 0$

Addition gives:
$$a\frac{d^2}{dx^2}(Ay_1) + b\frac{d}{dx}(Ay_1) + c(Ay_1) + a\frac{d^2}{dx^2}(By_2) + b\frac{d}{dx}(By_2) + c(By_2) = 0$$
$$a\left(\frac{d^2}{dx^2}(Ay_1) + \frac{d^2}{dx^2}(By_2)\right) + b\left(\frac{d}{dx}(Ay_1) + \frac{d}{dx}(By_2)\right) + c(Ay_1 + By_2) = 0$$

© HERIOT-WATT UNIVERSITY

Since we have established in Proof 2 that if y_1 and y_2 are two independent solutions then so are Ay_1 and By_2, we can use the additive law of differentiation to give:

$$a\frac{d^2}{dx^2}(Ay_1 + By_2) + b\frac{d}{dx}(Ay_1 + By_2) + c(Ay_1 + By_2) = 0$$

and so $y = Ay_1 + By_2$ is a solution to the differential equation.

Proof 4: A trial solution and the auxiliary equation

$y = Ae^{mx}$, for some A and m, is the solution to the first order linear differential equation $b\frac{dy}{dx} + cy = 0$. This was demonstrated in the 'Integrating factor' section.
It is therefore reasonable to consider it as a possible solution for $a\frac{d^2y}{dx^2} + b\frac{dy}{dx} + cy = 0$.
Now:
$y = Ae^{mx} \Rightarrow \frac{dy}{dx} = Ame^{mx} \Rightarrow \frac{d^2y}{dx^2} = Am^2e^{mx}$

If $y = Ae^{mx}$ is a solution, it will satisfy the equation.
Substituting into the original equation gives:

$aAm^2e^{mx} + bAme^{mx} + cAe^{mx} = 0$

Assuming $Ae^{mx} \neq 0$, by division, we get:
$am^2 + bm + c = 0$ (auxiliary equation A.E.)

The solutions to this quadratic, given by $m = \frac{-b \pm \sqrt{b^2 - 4ac}}{2a}$, provides two values of m which will make $y = Ae^{mx}$ a solution.
If we call the two solutions m_1 and m_2, then we have two solutions $y = Ae^{m_1 x}$ and $y = Ae^{m_2 x}$.

By proof 3, this gives the solution: $y = Ae^{m_1 x} + Be^{m_2 x}$

The solution that we get from this auxiliary equation depends on the nature of the roots of this equation.

Proof 5

For a homogeneous, second order, linear, differential equation with equal roots show that when $y_1 = e^{mx}$ is a solution, then $y_2 = xe^{mx}$ is also a solution.

The differential equation $a\frac{d^2y}{dx^2} + b\frac{dy}{dx} + cy = 0$ has auxiliary equation $am^2 + bm + c = 0$
The auxiliary equation has equal roots when $b^2 - 4ac = 0$ and from the quadratic formula $m = \frac{-b \pm \sqrt{b^2 - 4ac}}{2a}$ the repeated root is given by $m = -\frac{b}{2a}$

$y_1 = e^{mx}$ is one solution of the homogeneous equation. In the case where m is a repeated root, try $y_2 = xe^{mx}$ as another solution.

Then $\frac{dy_2}{dx} = mxe^{mx} + e^{mx}$ and $\frac{d^2y_2}{dx^2} = m^2xe^{mx} + 2me^{mx}$

TOPIC 4. DIFFERENTIAL EQUATIONS

Hence $a \dfrac{d^2y_2}{dx^2} + b\dfrac{dy_2}{dx} + cy_2 = am^2xe^{mx} + 2ame^{mx} + bmxe^{mx} + be^{mx} + cxe^{mx}$

$= \left(am^2 + bm + c\right)xe^{mx} + (2am + b)e^{mx}$

$= 0 \left(\text{since } am^2 + bm + c = 0 \text{ and } m = -\dfrac{b}{2a}\right)$

Therefore, if y is a repeated root of the auxiliary equation, the differential equation has solutions $y_1 = e^{mx}$ and $y_2 = xe^{mx}$ and the general solution is $y = Ae^{mx} + Bxe^{mx}$

Proof 6

Show that the homogeneous, second order, linear, differential equation with complex roots has independent real solutions, $y_1 = e^{px}\cos(qx)$ and $y_2 = e^{px}\sin(qx)$.

The differential equation $a\dfrac{d^2y}{dx^2} + b\dfrac{dy}{dx} + cy = 0$ has auxiliary equation $am^2 + bm + c = 0$. If $b^2 - 4ac < 0$ then the auxiliary equation has complex roots which can be written as $m = p \pm iq$.

Now if $m = p \pm iq$ is a solution to the auxiliary equation, then $y = e^{mx} = e^{(p+iq)x}$ is a solution of the differential equation. Unfortunately, y involves complex numbers but only require solutions of the equation involving only real numbers.

But
$y = e^{(p+iq)x}$
$= e^{px}e^{iqx}$
$= e^{px}\left(e^{ix}\right)^q$

From the topic "Sequences and series" we saw that $e^{ix} = \cos x + i\sin x$.
$y = e^{px}(\cos x + i\sin x)^q$

De Moivre's Theorem was discussed in the topic "Complex numbers". It states that $[r(\cos\theta + i\sin\theta)]^n = r^n[\cos(n\theta) + i\sin(n\theta)]$.

$y = e^{px}[\cos(qx) + i\sin(qx)]$

It can be checked that both the real and imaginary parts of y give rise to solutions of the differential equation and hence the independent solutions are $y_1 = e^{px}\cos(qx)$ and $y_2 = e^{px}\sin(qx)$ and we can write the general solution as:
$y = e^{px}(A\cos(qx) + B\sin(qx))$

4.8 Extended information

On November 7, 1940, at approximately 11:00 AM, the first Tacoma Narrows suspension bridge collapsed due to wind-induced vibrations. Situated on the Tacoma Narrows in Puget Sound, near the city of Tacoma, Washington, the bridge had only been open for traffic a few months. The noticeable vertical undulations of the bridge had been witnessed from the early days of construction, and as a result, were being documented with photographs and film footage. On November 7, 1940, these oscillations became sufficiently large to snap a support cable at mid-span, producing an unbalanced loading condition that created severe torsional oscillations which eventually led to the bridge's collapse.

A video of the bridge collapsing can be viewed online:
https://commons.wikimedia.org/wiki/File:Tacoma_Narrows_Bridge_destruction.ogg

The standard textbook explanation for the collapse attributes the cause of the failure to the phenomenon of resonance. Like a mass hanging from a spring, a suspension bridge's deck hanging from its cables oscillates at a natural frequency. In order for a resonant phenomenon to exist, the driving force would have to be periodic, that is, varying regularly with respect to time. The mathematical model that most simply illustrates this type of behaviour is represented by the following differential equation $m \frac{d^2x}{dt^2} + b \frac{dx}{dt} + kx = F \cos(\alpha t)$ where the different variables represent the following:

- m = mass of the system;
- b = damping coefficient of the system;
- k = stiffness of the system;
- α = radian frequency of the input force;
- F = amplitude of the exciting force;
- x = characteristic (output) motion of the system.

S.O.S. Math - Differential Equations (http://www.sosmath.com/diffeq/diffeq.html) - this site is vast and details the solution of differential equations in great depth - much more than we need. However, you might find relevant information under "First Order Differential Equations/Linear Equations" and also "Second Order Differential Equations/Homogeneous Linear Equations" or "Homogeneous Equations with Constant Coefficients" or "Non-Homogeneous Linear Equations/Method of Undetermined Coefficients".

http://www.math.ualberta.ca/~ewoolgar/java/Hooke/Hooke.html - this site contains a Java applet which displays solutions of the differential equation $m \frac{d^2y}{dt^2} + c \frac{dy}{dt} + ky = 0$ which describes a damped harmonic oscillator with mass m, damping constant c, and spring constant k.

http://www-history.mcs.st-andrews.ac.uk/history/index.html - the MacTutor History of Mathematics archive provides a very comprehensive directory of biographies for hundreds of important mathematicians. This was the source for the following information on Euler and Hooke.

Leonhard Euler (1707 - 1783)

Leonhard Euler was born on the 15th April 1707 in Basel, Switzerland.

While living with his grandmother, he went to school in Basel but learned no mathematics at all there. His interest in mathematics was first sparked by his father, Paul, and indeed Euler read mathematics texts on his own and even took some private lessons.

He entered the University of Basel in 1720 at the age of 14 and in 1723 he completed his Masters degree having compared and contrasted the philosophical ideas of Descartes and Newton. He then went on to study mathematics and completed his studies in 1726. In 1727 he submitted an entry for the Grand Prize of the Paris Academy on the best arrangement of masts on a ship. His essay won him second place which was an excellent achievement for a young graduate. In later life he did go on to win this prize on two occasions, in 1738 and 1740.

He served as a medical lieutenant in the Russian navy from 1727 to 1730. However, when he became professor of physics at the St Petersburg Academy he was able to give up his Russian navy post.

On 7th January 1734 he married Katharina Gsell who was from a Swiss family and the daughter of a painter. They had 13 children altogether, although only five survived their infancy. It is said that Euler claimed to have made some of his greatest mathematical discoveries while holding a baby in his arms while others played around his feet.

Euler is best known for his analytical treatment of mathematics and his discussion of calculus concepts, but he is also credited for work in acoustics, mechanics, astronomy, and optics. Indeed his work in mathematics is so vast that he is considered the most prolific writer of mathematics of all time.

He discovered the procedure for solving linear differential equations with constant coefficients in 1739. This arose out of his work on various problems in dynamics that led to differential equations of this type.

We owe to Euler the notation $f(x)$ for a function, i for $\sqrt{-1}$, π for pi, σ for summation and many others.

In 1735 Euler had a severe fever and almost lost his life. In 1738 he started to have problems with his eyesight and by 1740 he had lost the use of an eye. By 1766 he became almost entirely blind after an illness. Incredibly after the age of 59, when he was now totally blind, he produced almost half of his total works. Upon losing the sight in his right eye he is quoted as saying *"Now I will have less distraction."*

Euler died in 1783. His last day has been described as follows.

On 18 September 1783 Euler spent the first half of the day as usual. He gave a mathematics lesson to one of his grandchildren, did some calculations with chalk on two boards on the motion of balloons; then discussed with Lexell and Fuss the recently discovered planet Uranus. About five o'clock in the afternoon he suffered a brain haemorrhage and uttered only *"I am dying"* before he lost consciousness. He died at about eleven o'clock in the evening.

After his death the St Petersburg Academy continued to publish Euler's unpublished work for nearly 50 more years.

Robert Hooke

Robert Hooke was born on the 18th July 1707 in Freshwater, Isle of Wight, and died on 3rd March 1703 in London, England.

He went to school in Westminster where he learned Latin and Greek and in 1653 he went to Christ College, Oxford.

In 1660 he discovered the law of elasticity, known as Hooke's law. He worked on optics, simple harmonic motion and stress in stretched springs. He applied these studies in his designs for the balance of springs in watches.

In 1662 he was appointed curator of experiments to the Royal Society of London and was elected a fellow the following year. In 1665 he became professor of geometry at Gresham College, London where he remained for 30 years. In addition to this he also held the post of City Surveyor and was a very competent architect. He was chief assistant to Sir Christopher Wren in his project to rebuild London after the great fire of 1666.

In 1665 he first achieved world wide fame after the publication of his book *Micrographia* ("Small

Drawings"). This book contained beautiful pictures of objects Hooke had studied through a microscope that he made himself, including the crystal structure of snowflakes. His studies of microscopic fossils led him to become one of the first proposers of a theory of evolution.

There is no portrait of Hooke that is known to exist. Perhaps because he has been described as a *'lean, bent and ugly man'* and so was maybe reluctant to sit and have his portrait painted.

Simple harmonic motion

One of the most useful applications of second order differential equations is in the study of vibrations or oscillations. This type of motion occurs in many situations but the classic example is the motion of a spring.

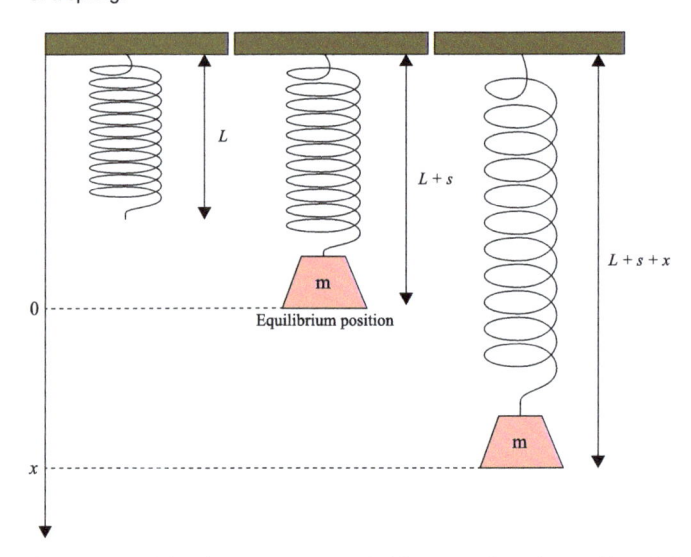

Consider a weight of mass m, suspended from a spring of natural length L. The weight will stretch the spring to a new length, $L + s$

Hooke's Law tells us that the tension in the spring acts back up towards the equilibrium position and is proportional to the amount the spring has stretched. Thus the upward force exerted by the spring is ks and since the mass is in equilibrium this is balanced by the force of gravity mg acting downwards on the mass. Thus we have:

$ks = mg$

Suppose that the spring is pulled downwards by an additional amount x (positive direction downwards) from this equilibrium position. Then the spring has been stretched by a total of $s + x$ and so the spring now exerts a force of $k(s + x)$ upwards. Thus the net downward force acting on the spring is:

$mg - k(s + x) = mg - ks - kx = -kx$

But, by Newton's Second Law: mass × acceleration = force

and so we obtain the differential equation:

TOPIC 4. DIFFERENTIAL EQUATIONS

$m \frac{d^2x}{dt^2} = -kx$

Dividing this equation by m and rearranging gives the equation:

$\frac{d^2x}{dt^2} + w^2 x = 0$ where $w^2 = \frac{k}{m}$

which is the equation of simple harmonic motion and has general solution:

$x = A\cos(wt) + B\sin(wt)$

This second order differential equation has auxiliary equation:

$m^2 + w^2 = 0$ which has solutions

$m = \pm \sqrt{(-w^2)}$
$= \pm wi$

Obviously, this has complex roots of the form:

$p \pm qi$ where $p = 0$ and $q = w$. Hence the general solution is:

$x = e^{0t}(A\cos(wt) + B\sin(wt))$
$= A\cos(wt) + B\sin(wt)$

This can be rewritten in the alternative form:

$y = R\cos(wt - \alpha)$ with $R = \sqrt{A^2 + B^2}$ and $\alpha = \tan^{-1}\left(\frac{B}{A}\right)$

This equation represents simple harmonic motion of amplitude R and period $\frac{2\pi}{w}$
The angle α is the phase angle.
The graph for this solution is as follows.

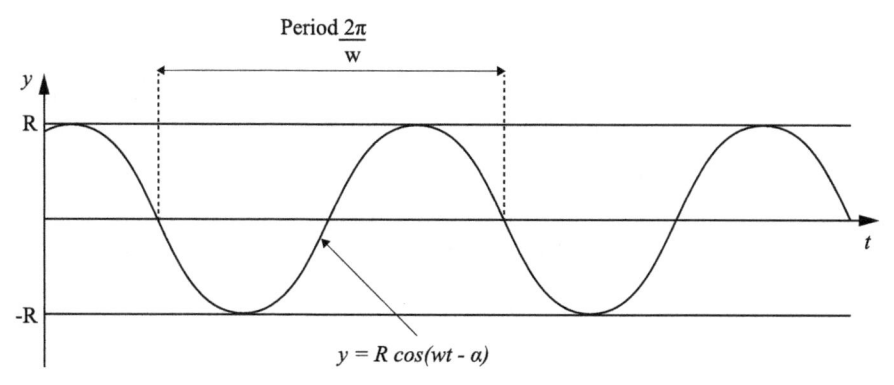

This graph represents the graph of simple harmonic motion of a frictionless spring. It is referred to as undamped vibration.

© HERIOT-WATT UNIVERSITY

Damped harmonic motion

If the motion of the spring is retarded by a friction force, which is proportional to the velocity, then the equation of motion becomes:

$m \frac{d^2x}{dt^2} = -kx - c \frac{dx}{dt}$

$c \frac{dx}{dt}$ is the friction force with c the constant of proportionality.

This equation can then be rewritten as:

$\frac{d^2x}{dt^2} + 2b \frac{dx}{dt} + w^2 x = 0$, where $2b = \frac{c}{m}$ and $w^2 = \frac{k}{m}$

The auxiliary equation is:

$m^2 + 2bm + w^2 = 0$ which has roots:

$m = -b \pm \sqrt{(b^2 - w^2)}$

If $b < w$ then the differential equation has complex roots:

$-b \pm qi$, $\left(q = \sqrt{w^2 - b^2}\right)$, and the general solution takes the form:

$x = e^{-bt}(A \cos(qt) + B \sin(qt))$
$\quad = e^{-bt} R (\cos(qt) - \alpha)$

As you can see, this is similar to the equation for undamped harmonic motion. However, due to the factor e^{-bt}, the amplitude diminishes exponentially as $t \to \infty$ because $e^{-bt} \to 0$ as $t \to \infty$

The graph for $x = Re^{-bt}(cos(wt) - \alpha)$ follows. This solution represents damped vibratory motion as the result of friction. Notice that the vibrations tend to die out as time progresses.

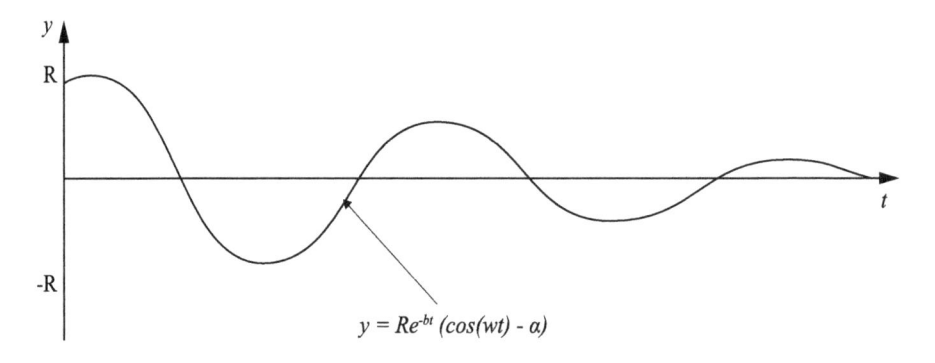

$y = Re^{-bt} (cos(wt) - \alpha)$

4.9 End of topic test

End of topic 4 test — Go online

First order linear differential equations with separable variables

Q108: Find the general solution of the differential equation $\frac{dy}{dx} = x^2 y$ giving y in terms of x.

Q109: Find the general solution of the differential equation $\frac{dy}{dx} = \frac{2y}{2x+1}$ giving y in terms of x.

Q110: Find the general solution of the differential equation $\frac{dy}{dx} = \frac{x+3}{y}$ giving y^2 in terms of x.

Q111: Find the general solution of the differential equation $\frac{dy}{dx} = \sqrt{xy}$ giving y in terms of x.

Q112: Find the general solution of the differential equation $\frac{dy}{dx} = 1 - 2x + y - 2xy$ giving y in terms of x.

Q113: Find the general solution of the differential equation $\frac{dy}{dx} = x\frac{dy}{dx} + y^2$ giving y in terms of x.

Q114: Find the particular solution of the differential equation $\frac{dy}{dx} = y \cos x$ with initial condition $y(0) = 1$.

Q115: Find the particular solution of the differential equation $\frac{dy}{dx} = \sqrt{1 - y^2}$ when $x = \frac{\pi}{6}$ and $y = 0$.

Q116: Find the particular solution of the differential equation $2(x-1)\frac{dy}{dx} - 2y(y+1) = 0$ when $y(2) = 1$ giving y in terms of x.

Growth and decay

Q117: A radioactive substance decreases at rate k proportional to the mass m (in grams) at time t (in years). The initial mass of radioactive material is m_0. Write down the differential equation for decay.

The half-life of the substance is known to be 10 years, i.e. $m = \frac{1}{2} m_0$ at $t = 10$. Calculate the constant of proportionality k and write down the particular solution for $m(t)$.

386 TOPIC 4. DIFFERENTIAL EQUATIONS

Q118: The rate at which alcohol is processed by the body, k, is proportional to the amount of alcohol still in the blood y (mg of alcohol per 100ml of blood) at time t (in minutes). Write down the differential equation for the amount of alcohol in the blood.

At time $t = 0$, the initial amount of alcohol in the blood is 70 mg; 40 minutes after drinking has stopped the blood alcohol level is 60 mg.

Calculate the constant of proportionality k and write down the particular solution for $y(t)$.

....

Q119: In a town with population 100,000 a cold virus spreads rapidly. The percentage of the population infected in t days after the initial outbreak satisfies the differential equation $\frac{dP}{dt} = kP$ where k is a constant.

a) If 100 people were infected initially, find in terms of k, the percentage infected t days later.

b) Given that 400 people have the cold 10 days later, how many more are likely to have contracted the virus after 14 days?

Further application of separable variable differential equations

Q120: When a valve is opened, the rate at which the water drains from a cask is proportional to the square root of the depth of the liquid inside. This can be represented by the differential equation $\frac{dh}{dt} = -2\sqrt{h}, h \geqslant 0$ where h is the depth (in centimetres) of liquid and t is the time (in minutes) elapsed since the value was opened.

a) Express h as a function of t.

b) Find the solution to the equation if initially the liquid is 100 cm deep in the cask.

c) How long would it take to drain a cask with liquid at a depth of 25 cm?

....

Q121: After 20 minutes in a room of temperature 17°C, a bowl of soup cools from 80°C to 60°C. Use Newton's Law of cooling, $\frac{dT}{dt} = -k(T - T_S)$, to estimate how much longer it will take to cool to 40°C.

(T is the temperature of the soup, T_S is the room temperature and k is the constant of proportionality)

Integrating factor

Q122: Find the general solution, in the form $y(x)$, of the first order linear differential equation $\frac{dy}{dx} + 5y = e^{3x}$.

....

Q123: Find the general solution, in the form $y(x)$, of the first order linear differential equation $\frac{dy}{dx} + \frac{2y}{x} = 6x^3$.

....

TOPIC 4. DIFFERENTIAL EQUATIONS

Q124: Find the general solution, in the form $y(x)$, of the first order linear differential equation $\frac{dy}{dx} + 5y = e^{-x}$.

....

Q125: Find the general solution, in the form $y(x)$, of the first order linear differential equation $\frac{dy}{dx} + 2xy = e^{-x^2} \cos x$.

....

Q126: Find the general solution, in the form $y(x)$, of the first order linear differential equation $x \frac{dy}{dx} - 3x^2 + 5y = 0$.

....

Q127: Find the general solution, in the form $y(x)$, of the first order linear differential equation $x \frac{dy}{dx} + y = x \sin x$.

....

Q128: Find the general solution, in the form $y(x)$, of the first order linear differential equation $\frac{dy}{dx} + y = \sin x$.

....

Q129: Solve the initial value problem $x \frac{dy}{dx} + 2y = \frac{\cos x}{x}$ given that $y(\pi) = 0$.

....

Q130: Solve the initial value problem $4 \frac{dy}{dx} + 6y = e^{-5x}$ given that $y(0) = 0$.

....

Q131: Find the particular solution of the initial value problem $x \frac{dy}{dx} + 2y = 6x^4$ given that $x = 1$ when $y = 5$.

....

Q132: Find the particular solution of the initial value problem $\frac{dy}{dx} + (\tan x) y = \cos^2 x$ given that $x = \frac{\pi}{4}$ when $y = \frac{3}{2}$.
(Hints: $\int \tan x = -\ln|\cos x| + C$)

Integrating factor applications

Q133: Bacteria grow in a certain culture at a rate proportional to the amount present.
Given that there are 150 bacteria present initially, find an equation for bacterial growth and by solving an appropriate differential equation find the number of bacteria present at time t.

....

Q134: A sick animal is fed glucose into the bloodstream at a constant rate. The rate of change of the overall concentration of glucose in the blood $G(t)$ satisfies the differential equation $\frac{dG}{dt} = \frac{1}{25} - 0 \cdot 005G$
The term $-(0 \cdot 005G)$ is included because it is assumed that the glucose is continually changing into other molecules at a rate proportional to its concentration. The concentration $G(t)$ is measured in milligrams per centilitre.

Given that the initial concentration of glucose in the bloodstream is 110 milligrams/centilitre, solve the equation for G(t).

....

Q135: An electric circuit contains a constant DC voltage source of V volts, a switch, a resistor of size $R = 4$ ohms, and an inductor of size $L = 10$ henrys. The circuit has no capacitance. The switch, initially open so that no current is flowing, is closed at time $t = 0$ so that current begins to flow at that time.
The current $I(t)$ satisfies the differential equation $10 \frac{dI}{dt} + 4I = V$
Find a formula for $I(t)$ in terms of V and t.

....

Q136: A tank contains 5000 litres of salt water.
40 litres of fresh water flows into the tank and 40 litres of salt water flows out of the tank each minute.
The mass of salt satisfies the differential equation $\frac{dM}{dt} = -0 \cdot 008M$
If the concentration of salt in the water was initially $s = 50$ g/litre calculate, how many minutes it takes for the amount of salt in the tank to reduce to 16 g/litre?

....

Q137: A tank contains 30 kg of salt dissolved in 500 m^3 of water. Starting at time $t = 0$, water containing 2·5 kg of salt per m^3 enters the tank at a rate of 8 m^3/minute. The well stirred solution also leaves the tank at a rate of 8 m^3/minute.
The mass of salt satisfies the differential equation $\frac{dM}{dt} = 20 - 0 \cdot 016M$
Hence, calculate the concentration of salt in the water after 40 minutes.

Homogeneous second order linear differential equations

Q138: Find the particular solution for the differential equation $\frac{d^2y}{dx^2} - 6\frac{dy}{dx} + 8y = 0$ given that $y = 9$ and $\frac{dy}{dx} = 20$ when $x = 0$

....

Q139: Find the particular solution for the differential equation $\frac{d^2y}{dx^2} + 4\frac{dy}{dx} + 8y = 0$ given that $y = 2$ and $\frac{dy}{dx} = 10$ when $x = 0$

....

Q140: Find the particular solution for the differential equation $\frac{d^2y}{dx^2} - 4\frac{dy}{dx} + 4y = 0$ given that $y = 7$ and $\frac{dy}{dx} = 5$ when $x = 0$

Q141: Find the particular solution of the initial value problem $\frac{d^2y}{dx^2} - 5\frac{dy}{dx} + 6y = 0$ given that $y(0) = 1$ and $y'(0) = 1$ in the form $y(x)$.

....

Q142: Find the particular solution of the initial value problem $\frac{d^2y}{dx^2} + 2\frac{dy}{dx} + y = 0$ given that $y(0) = 1$ and $y'(0) = 3$ in the form $y(x)$.

....

TOPIC 4. DIFFERENTIAL EQUATIONS

Q143: Find the particular solution of the initial value problem $\frac{d^2y}{dx^2} - 2\frac{dy}{dx} + 2y = 0$ given that $y(0) = 3$ and $y'(0) = 2$ in the form $y(x)$.

Q144: Find the particular solution of the initial value problem $\frac{d^2y}{dx^2} + 2\frac{dy}{dx} + y = 0$ given that $y(2) = 1$ and $y'(2) = 2$ in the form $y(x)$.

Q145: Find the particular solution for the differential equation $9\frac{d^2y}{dx^2} - 72\frac{dy}{dx} + 288y = 0$ given that $y = 3$ and $\frac{dy}{dx} = 20$ when $x = 0$

....

Q146: A mass is attached to a vertical spring and hangs in equilibrium. At time $t = 0$, the mass is lowered a distance of 10 cm down from its equilibrium position and given an initial velocity of 14 cm/second in the upward direction, i.e. $\frac{dx}{dt} = -14$, so that it oscillates at the end of the spring.
If $x(t)$ denotes the distance of the mass below its equilibrium position at time t, it can be shown that $x(t)$ satisfies the differential equation $\frac{d^2x}{dt^2} + x = 0$
Find $x(t)$ and hence determine the position of the mass when $t = 4$.

....

Q147: The motion of a certain mass on the end of a spring satisfies the differential equation $\frac{d^2x}{dt^2} + 2\frac{dx}{dt} + 37x = 0$ where $x(t)$ denotes the distance of the mass below its equilibrium position at time t.
At time $t = 0$, the mass is lowered a distance of 55 cm down from its equilibrium position and given an initial velocity of 8 cm/second in the upward direction, i.e. $\frac{dx}{dt} = -8$, so that it oscillates at the end of the spring.
Find $x(t)$ and hence determine the position of the mass when $t = 2$.

Q148: In the absence of any external electromotive force, the current i in a simple electrical circuit varies with time, t, according to the formula $L\frac{d^2i}{dt^2} + R\frac{di}{dt} + \frac{1}{C}i = 0$ (L, R, C are constants).
Find the general solution given that $L = 1$, $R = 10^3$ and $C = 5 \times 10^{-6}$.

....

Q149: The angular displacement, θ, of the bob of a simple pendulum satisfies the equation of motion $0 \cdot 1 \frac{d^2\theta}{dt^2} + 0 \cdot 2 \frac{d\theta}{dt} + \theta = 0$.
Find the particular solution if the bob is hit when in its rest position $\theta = 0$ such that it is given an angular speed $\frac{d\theta}{dt} = 0 \cdot 3$ rad s^{-1}.

....

Q150: A particle is moving along the x axis.
Its motion can be modelled by $\frac{d^2x}{dt^2} + 4\frac{dx}{dt} + 3x = 0$
At the start of the motion, $t = 0$, the displacement, x, is 10 units and the velocity is 0.
Find the formula for displacement and velocity.
(Recall that acceleration = $\frac{d^2x}{dt^2}$, velocity = $\frac{dx}{dt}$ and displacement = x)

Non-homogeneous second order linear differential equations

Finding particular integrals

Q151: Find the general solution to the differential equation $\frac{d^2y}{dx^2} - 2\frac{dy}{dx} - 24y = e^{-3x}$

....

Q152: Find the general solution to the differential equation $\frac{d^2y}{dx^2} + 2\frac{dy}{dx} + 17y = -520\cos x$

....

Q153: Find the general solution to the differential equation $\frac{d^2y}{dx^2} + 12\frac{dy}{dx} + 36y = 108x^2$

....

Q154: Find the general solution to the differential equation $\frac{d^2y}{dx^2} - 4\frac{dy}{dx} - 5y = 12e^{5x}$

....

Q155: Find the general solution to the differential equation $\frac{d^2y}{dx^2} + 5\frac{dy}{dx} + 4y = 34\sin x$

....

Q156: Find the general solution to the differential equation $\frac{d^2y}{dx^2} - 10\frac{dy}{dx} + 25y = e^{5x}$

Q157: If a potential V (volts) is applied to a simple circuit with inductance L (henry), resistance R (ohms) and capacitance C (farad), then q (coulomb) on the capacitor after time t seconds is given by $L\frac{d^2q}{dt^2} + R\frac{dq}{dt} + \frac{q}{C} = V$.
Find a formula for q where, $L = 5 \times 10^{-3}$, $R = 6$, $C = 10^{-3}$ and $V = 100$.

....

Q158: Find the solution of the differential equation $\frac{d^2y}{dx^2} + 6\frac{dy}{dx} + 5y = 4e^{-x}$ for which $y = \frac{dy}{dx} = 0$ at $x = 0$.

....

Q159: Find the solution of the differential equation $2\frac{d^2y}{dx^2} + \frac{dy}{dx} - y = x$ for which $y(0) = 0$ and $y'(0) = 1$.

....

Q160: Find the solution of the differential equation $\frac{d^2y}{dx^2} - \frac{dy}{dx} - 6y = 5 + 2x - 12x^2$ for which $y(0) = 12$ and $y'(0) = 10$.

Glossary

acceleration
> a vector quantity which is the rate of change of velocity with respect to time: if v is velocity then acceleration is $\frac{dv}{dt}$. If s is the displacement, then acceleration is $\frac{d^2s}{dt^2}$. An alternative notation uses x for displacement giving $\ddot{x} = \frac{d^2x}{dt^2}$

anti-differentiation
> is the reverse process of differentiation

arbitrary constant
> a constant that occurs in the general solution of a differential equation

auxiliary equation
> the second order linear homogeneous differential equation $a\frac{d^2y}{dx^2} + b\frac{dy}{dx} + cy = 0$ has the auxiliary equation $am^2 + bm + c = 0$; it is also called the characteristic equation

cartesian equation
> an equation which links the x-coordinate and the y-coordinate of the general point (x, y) on a curve

the chain rule
> Function notation: $h'(x) = g'(f(x)) \times f'(x)$, Leibniz notation: $\frac{dy}{dx} = \frac{dy}{du} \times \frac{du}{dx}$

characteristic equation
> the second order linear homogeneous differential equation $a\frac{d^2y}{dx^2} + b\frac{dy}{dx} + cy = 0$ has the characteristic equation $am^2 + bm + c = 0$; it is also called the auxiliary equation

closed interval
> $[a, b]$ the set of numbers $x \in \mathbb{R}$ where $a \leq x \leq b$, i.e. the end points are included

the closed interval [a, b]
> note that the notation $[a, b] = \{x \in \mathbb{R} : a \leq x \leq b\}$ means the closed interval with end points a and b

complementary function
> the non-homogeneous equation $a\frac{d^2y}{dx^2} + b\frac{dy}{dx} + cy = f(x)$ has the corresponding homogeneous equation $a\frac{d^2y}{dx^2} + b\frac{dy}{dx} + cy = 0$ which has the general solution y_c, containing two constants - y_c is the complementary function for the non-homogeneous equation

constant of integration
> in general if $\frac{d}{dx}(F(x)) = f(x)$ then $F(x)$ is called an anti-derivative; or integral, of $f(x)$, written as $\int f(x)\,dx = F(x) + C$ and C is the **constant of integration**

definite integral
> $\int_a^b f(x)\,dx$ is a definite integral because the limits of integration are known and $\int_a^b f(x)\,dx = [F(x)]_a^b = F(b) - F(a)$

degree
>for a polynomial, the degree is the value of the highest power

derived function
>the instantaneous speed, or rate of change of distance with respect to time, can be written as $f'(t)$ and is known as the derived function of $f(t)$

differential equation
>an equation involving an unknown function and its derivatives

displacement
>a vector quantity which is the shortest distance between the starting point and the current position of an object - it can be denoted by the letter s or x

distance
>a scalar quantity which is the total distance travelled from the starting point to the current position of an object - it can be denoted by the letter d

dividend
>the **dividend** in a long division calculation is the expression which is being divided; as a fraction it is the numerator

divisor
>the **divisor** is the expression which is doing the dividing; it is the expression outside the division sign; as a fraction it is the denominator

equation of simple harmonic motion
>the equation $\frac{d^2x}{dt^2} + w^2 x = 0$ is the equation of simple harmonic motion which has a general solution $x = A\cos(wt) + B\sin(wt)$

explicit function
>y can be written as an expression in which the only variable is x and we obtain only one value for y

exponential decay
>occurs when the population decreases at a rate proportional to the size of the population

exponential growth
>occurs when the population increases at a rate proportional to the size of the population

first order linear differential equation
>an equation that can be expressed in the standard form $\frac{dy}{dx} + P(x)y = f(x)$

general indefinite integral of f(x)
>$\int f(x)\,dx + C$, $F(x) + C$ is the **general indefinite integral of f(x)** with respect to x of the integral $\int f(x)\,dx$

general solution
>the general solution of a differential equation contains one or more arbitrary constants and gives infinitely many solutions that all satisfy the differential equation

homogeneous

a second order linear differential equation is homogeneous when it can be expressed in the standard form $a\frac{d^2y}{dx^2} + b\frac{dy}{dx} + cy = 0$ where a, b and c are constants, and $a \neq 0$

implicit function

if x is given any value then corresponding value(s) of y can be found and it is therefore not explicit

improper rational function

let $P(x)$ be a polynomial of degree n and $Q(x)$ be a polynomial of degree m

If $n \geq m$ then $\frac{P(x)}{Q(x)}$ is an **improper rational function**.

For example
$\frac{x^3 + 2x - 1}{x^2 - 4}$ or $\frac{x^2 + 2x - 1}{x^2 - 4}$

inhomogeneous

a second order linear differential equation is non-homogeneous when it can be expressed in the standard form $a\frac{d^2y}{dx^2} + b\frac{dy}{dx} + cy = f(x)$ where a, b and c are constants, and $a \neq 0$ and $f(x) \neq 0$; also called non-homogeneous

initial condition

for a differential equation, an initial condition is an additional condition which must be satisfied by the solution - this could be a coordinate on a curve, a velocity at $t = 0$, the amount of money in a bank account on 1st January 2000 etc.

integrand

$\int f(x)\,dx = F(x) + C$, $f(x)$ is the **integrand**

integrating factor

for the first order linear differential equation $\frac{dy}{dx} + P(x)y = f(x)$, the integrating factor is given by $I(x) = e^{\int P(x)\,dx}$

integration

the method used to find anti-derivatives

irreducible quadratic

a quadratic is irreducible when it has no real roots; it cannot be factorised

linear polynomial

a linear polynomial is one which has at most a degree of 1, graphically this would be represented by a straight line

monotonically

A monotonic function is one that is always either non-increasing, or non-decreasing. The sign of the derivative does not change e.g. for $f(x) = x^3$, $f'(x) = 3x^2$. The derivative is either zero or always positive and so the function is non-decreasing.

© HERIOT-WATT UNIVERSITY

non-homogeneous
   a second order linear differential equation is non-homogeneous when it can be expressed in the standard form $a \frac{d^2y}{dx^2} + b \frac{dy}{dx} + cy = f(x)$ where a, b and c are constants, and $a \neq 0$ and $f(x) \neq 0$; also called inhomogeneous

parameter
   a variable that is given a series of arbitrary values in order that a relationship between x and y may be established; the variable may denote time, angle or distance

parametric equations
   equations that are expressed in terms of an independent variable; they are of the form $x = f(t)$ and $y = g(t)$ where t is the independent variable for x and y and the variable t is called a parameter

partial fractions

   the process of taking a proper rational function and splitting it into separate terms each with a factor of the original denominator as its denominator is called expressing the function in **partial fractions**

particular integral
   for $\int f(x)\, dx = F(x) + C$ an anti-derivative given by a particular value of C is a **particular integral**

particular solution
   the particular solution of a differential equation is a solution which is often obtained from the general solution when values are evaluated for the arbitrary constants

polynomial of degree n
   if $P(x) = a_n x^n + a_{n-1} x^{n-1} + a_{n-2} x^{n-2} + \ldots + a_2 x^2 + a_1 x^1 + a_0$

   where $a_0, \ldots, a_n \in \mathbb{R}$
   then P is a **polynomial of degree** n.

product rule
   a method used to differentiate the product of two or more functions -

   it states that when $k(x) = f(x)\, g(x)$, then $k'(x) = f'(x)\, g(x) + f(x)\, g'(x)$

proper rational function
   let $P(x)$ be a polynomial of degree n and $Q(x)$ be a polynomial of degree m

   If $n < m$ then $\frac{P(x)}{Q(x)}$ is a **proper rational function**.

   For example,
   $\frac{x^2 + 2x - 1}{x^3 - 3x + 4}$

quotient
   the **quotient** is the answer to the division *but not including* the remainder

quotient rule

gives us a method that allows us to differentiate algebraic fractions - it states that when $k(x) = \frac{f(x)}{g(x)}$ then $k'(x) = \frac{f'(x)g(x) - f(x)g'(x)}{[g(x)]^2}$

rational function

if $P(x)$ and $Q(x)$ are polynomials then $\frac{P(x)}{Q(x)}$ is called a rational function

remainder

the **remainder** is what is left over after dividing

scalar

something that only has a magnitude

second order linear differential equation

an equation which can be expressed in the standard from $a\frac{d^2y}{dx^2} + b\frac{dy}{dx} + cy = f(x)$ where a, b and c are constants, and $a \neq 0$; this equation is second order because it contains a second derivative (but no derivatives of higher order)

speed

a scalar quantity which is the magnitude of velocity, i.e. the magnitude of the rate of change of distance with respect to time, that is to say, how fast the object is moving. If s is the displacement, then speed is the magnitude of $\frac{ds}{dt}$

vector

something that has a magnitude and a direction

velocity

a vector quantity which is the rate of change of displacement with respect to time: if s is the displacement then the velocity is $\frac{ds}{dt}$. An alternative notation uses x for displacement giving $\dot{x} = \frac{dx}{dt}$

Hints for activities

Topic 3: Integration

Challenge question 1

Hint 1: Complete the square in the denominator.

Hint 2: The substitution $u = x + 1$ may help.

Hint 3: Write the integrand as a multiple of $\frac{2u}{u^2 + a^2}$ plus a multiple of $\frac{1}{u^2 + a^2}$

Challenge question 2

Hint 1: Rewrite $\frac{1}{x(x^2 + x + 1)}$ in the form $\frac{A}{x} + \frac{Bx + C}{x^2 + x + 1}$

Hint 2: $\int \frac{x + 1}{x(x^2 + x + 1)} \, dx = \int \frac{x}{x^2 + x + 1} \, dx + \int \frac{1}{x^2 + x + 1} \, dx$

Hint 3: The integral $\int \frac{x}{x^2 + x + 1} \, dx$ may cause you consternation.
This may help: $\int \frac{x}{x^2 + x + 1} \, dx = \frac{1}{2} \int \frac{(2x + 1)}{x^2 + x + 1} \, dx - \frac{1}{2} \int \frac{1}{x^2 + x + 1} \, dx$

Hint 4: $\int \frac{1}{x^2 + x + 1} \, dx$ may also cause you to scratch your head. Remember that you can complete the square and then the integral will be a tan^{-1} function.

Challenge question 3

Hint 1: Use the same method as for $\int \ln x \, dx$
ie. Let $f = \sin^{-1} x$ and $g' = 1$

Hint 2: You might have difficulty with $\int \frac{x}{\sqrt{1-x^2}} \, dx$
Try this, $\int \frac{x}{\sqrt{1-x^2}} \, dx = \frac{1}{2} \int \frac{2x}{\sqrt{1-x^2}} \, dx$
It is now possible to solve this by substituting u for x^2

Topic 4: Differential equations

Save the fish

Hint 1:
Let S = total mass of salt in the tank and let w = flow of water in litres/minute. Then the differential equation is:
$\frac{dS}{dt} = -\frac{w}{5000} S$
In standard linear form this becomes:
$\frac{dS}{dt} + \frac{w}{5000} S = 0$ with $P(t) = \frac{w}{5000}$ and $f(t) = 0$

Hint 2:
This gives the differential equation:
$\frac{d}{dt} \left(e^{-\frac{wt}{5000}} S \right) = 0$

This can be integrated and solved for S to give $S = Ce^{-\frac{wt}{5000}}$ where C is an arbitrary constant.

Hint 3:
There is initially $S(0) = \frac{20 \times 5000}{1000} = 100$ kg of salt in the tank, therefore $C = 100$.
There has to be $S = \frac{12 \times 5000}{1000} = 60\ kg$ when the fish is put in for it to survive so the equation is $60 = 100e^{-\frac{wt}{5000}}$.

Answers to questions and activities

Topic 1: Partial fractions

Division by (x - a) exercise (page 5)

Q1: 8

Q2: 86

Q3: 55

Q4:
Remember we do not have an x term so $6x^3 + 7x^2 - 1$ must be interpreted as $6x^3 + 7x^2 + 0x - 1$ and the divisor is factorised to give $3\left(x - \frac{1}{3}\right)$.

```
 1/3 | 6   7   0   -1
     |     2   3    1
     | 6   9   3    0
```

So $f(x) = (x - \,^1/_3)(6x^2 + 9x + 3)$ but we have to take the common factor of 3 out of the quotient and put it back into the divisor.

$$\left(x - \frac{1}{3}\right)(6x^2 + 9x + 3) = \left(x - \frac{1}{3}\right) \times 3\left(2x^2 + 3x + 1\right)$$
$$= 3\left(x - \frac{1}{3}\right)\left(2x^2 + 3x + 1\right)$$

giving, $f(x) = (3x - 1)(2x^2 + 3x + 1)$.
So the quotient is $2x^2 + 3x + 1$ and the remainder $= 0$.

Factor theorem exercise (page 6)

Q5: Yes

Q6: Yes

Factorising polynomials exercise (page 8)

Q7:
Evaluate $f(1)$:
```
 1 | 1   0   -3    2
   |     1    1   -2
   | 1   1   -2    0
```
Since the remainder = 0, $(x - 1)$ is a factor and,
$$x^3 - 3x + 2 = (x - 1)\left(x^2 + x - 2\right)$$
$$= (x - 1)(x + 2)(x - 1)$$
$$= (x + 2)(x - 1)^2$$

© HERIOT-WATT UNIVERSITY

Q8:

Evaluate $f(1)$:

```
1 | 2   -3   -11    6
  |      2   -1   -12
  ─────────────────────
    2   -1   -12  | -6
```

Since the remainder $\neq 0$, $(x - 1)$ is not a factor.

Evaluate $f(-1)$:

```
-1 | 2   -3   -11    6
   |     -2    5     6
   ─────────────────────
     2   -5   -6   | 12
```

Since the remainder $\neq 0$, $(x + 1)$ is not a factor.

Evaluate $f(2)$:

```
2 | 2   -3   -11    6
  |      4    2   -18
  ─────────────────────
    2    1   -9   | -12
```

Since the remainder $\neq 0$, $(x - 2)$ is not a factor.

Evaluate $f(-2)$:

```
-2 | 2   -3   -11    6
   |     -4   14    -6
   ─────────────────────
     2   -7    3   | 0
```

Since the remainder = 0, $(x + 2)$ is a factor and,
$$2x^3 - 3x^2 - 11x + 6 = (x + 2)(2x^2 - 7x + 3)$$
$$= (x + 2)(2x - 1)(x - 3)$$

Linear factors exercise (page 14)

Q9:

Step 1: Factorise the denominator.
$$\frac{x+7}{x^2 - x - 2} = \frac{x+7}{(x+1)(x-2)}$$

Step 2: Identify the type of partial fraction. In this case it has linear factors.
$$\frac{x+7}{(x+1)(x-2)} = \frac{A}{(x+1)} + \frac{B}{(x-2)}$$

Step 3: Obtain the fractions with a common denominator.
$$\frac{x+7}{(x+1)(x-2)} = \frac{A(x-2)}{(x+1)(x-2)} + \frac{B(x+1)}{(x+1)(x-2)}$$

Step 4: Equate numerators since the denominators are equal.

$x + 7 = A(x - 2) + B(x + 1)$ (*)

Step 5: Select values of x and substitute into (*).

Let $x = 2$:
$$x + 7 = A(x - 2) + B(x + 1)$$
$$2 + 7 = A(2 - 2) + B(2 + 1)$$
$$9 = 3B$$
$$B = 3$$

Let $x = -1$:
$$x + 7 = A(x - 2) + B(x + 1)$$
$$-1 + 7 = A(-1 - 2) + B(-1 + 1)$$
$$6 = -3A$$
$$A = -2$$

Therefore: $\frac{x+7}{x^2 - x - 2} = \frac{3}{x-2} - \frac{2}{x+1}$

Q10:

Step 1: Factorise the denominator.
$$\frac{7x-4}{2x^2 - 3x - 2} = \frac{7x-4}{(2x+1)(x-2)}$$

Step 2: Identify the type of partial fraction. In this case it has linear factors.
$$\frac{7x-4}{(2x+1)(x-2)} = \frac{A}{(2x+1)} + \frac{B}{(x-2)}$$

Step 3: Obtain the fractions with a common denominator.
$$\frac{7x-4}{(2x+1)(x-2)} = \frac{A(x-2)}{(2x+1)(x-2)} + \frac{B(2x+1)}{(2x+1)(x-2)}$$

Step 4: Equate numerators since the denominators are equal.

$7x - 4 = A(x - 2) + B(2x + 1)$ (*)

Step 5: Select values of x and substitute into (*).

Let $x = 2$:
$$7x - 4 = A(x - 2) + B(2x + 1)$$
$$7(2) - 4 = A(2 - 2) + B(2(2) + 1)$$
$$10 = 5B$$
$$B = 2$$

ANSWERS: UNIT 1 TOPIC 1

Let $x = -\frac{1}{2}$:

$$7x - 4 = A(x - 2) + B(2x + 1)$$
$$7(-\frac{1}{2}) - 4 = A(-\frac{1}{2} - 2) + B(2(-\frac{1}{2}) + 1)$$
$$-\frac{7}{2} - 4 = -2\frac{1}{2}A$$
$$-\frac{15}{2} = -\frac{5}{2}A$$
$$A = \frac{15}{5}$$
$$A = 3$$

Therefore: $\frac{7x-4}{2x^2 - 3x - 2} = \frac{3}{2x+1} + \frac{2}{x-2}$

Q11:

Step 1: Factorise the denominator using synthetic division.

$\frac{8x}{x^3 + 3x^2 - x - 3} = \frac{8x}{(x+1)(x-1)(x+3)}$

Step 2: Identify the type of partial fraction. In this case it has linear factors.

$\frac{8x}{(x+1)(x-1)(x+3)} = \frac{A}{(x+1)} + \frac{B}{(x-1)} + \frac{C}{(x+3)}$

Step 3: Obtain the fractions with a common denominator.

$$\frac{8x}{(x+1)(x-1)(x+3)} = \frac{A(x-1)(x+3)}{(x+1)(x-1)(x+3)} + \frac{B(x+1)(x+3)}{(x+1)(x-1)(x+3)} + \frac{C(x+1)(x-1)}{(x+1)(x-1)(x+3)}$$

Step 4: Equate numerators since the denominators are equal.

$8x = A(x - 1)(x + 3) + B(x + 1)(x + 3) + C(x + 1)(x - 1)$ (*)

Step 5: Select values of x and substitute into (*).

Let $x = 1$:
$$8x = A(x-1)(x+3) + B(x+1)(x+3) + C(x+1)(x-1)$$
$$8(1) = A(1-1)(1+3) + B(1+1)(1+3) + C(1+1)(1-1)$$
$$8 = 8B$$
$$B = 1$$

Let $x = -1$:
$$8x = A(x-1)(x+3) + B(x+1)(x+3) + C(x+1)(x-1)$$
$$8(-1) = A(-1-1)(-1+3) + B(-1+1)(-1+3) + C(-1+1)(-1-1)$$
$$-8 = -4A$$
$$A = 2$$

Let $x = -3$:
$$8x = A(x-1)(x+3) + B(x+1)(x+3) + C(x+1)(x-1)$$
$$8(-3) = A(-3-1)(-3+3) + B(-3+1)(-3+3) + C(-3+1)(-3-1)$$
$$-24 = 8C$$
$$C = -3$$

Therefore: $\frac{8x}{2x^3 + 3x^2 - x - 3} = \frac{2}{x+1} + \frac{1}{x-1} - \frac{3}{x+3}$

© HERIOT-WATT UNIVERSITY

Q12:

Step 1: Factorise the denominator using synthetic division.

$$\frac{-11x - 34}{x^3 + 5x^2 + 2x - 8} = \frac{-11x - 34}{(x + 2)(x - 1)(x + 4)}$$

Step 2: Identify the type of partial fraction. In this case it has linear factors.

$$\frac{-11x - 34}{(x + 2)(x - 1)(x + 4)} = \frac{A}{(x + 2)} + \frac{B}{(x - 1)} + \frac{C}{(x + 4)}$$

Step 3: Obtain the fractions with a common denominator.

$$\frac{-11x-34}{(x+2)(x-1)(x+4)} = \frac{A(x-1)(x+4)}{(x+2)(x-1)(x+4)} + \frac{B(x+2)(x+4)}{(x+2)(x-1)(x+4)} + \frac{C(x+2)(x-1)}{(x+2)(x-1)(x+4)}$$

Step 4: Equate numerators since the denominators are equal.

$-11x - 34 = A(x - 1)(x + 4) + B(x + 2)(x + 4) + C(x + 2)(x - 1)$ (*)

Step 5: Select values of x and substitute into (*).

Let $x = 1$:
$-11x - 34 = A(x - 1)(x + 4) + B(x + 2)(x + 4) + C(x + 2)(x - 1)$
$-11(1) - 34 = A(1 - 1)(1 + 4) + B(1 + 2)(1 + 4) + C(1 + 2)(1 - 1)$
$\qquad -45 = 15B$
$\qquad B = -3$

Let $x = -2$:
$-11x - 34 = A(x - 1)(x + 4) + B(x + 2)(x + 4) + C(x + 2)(x - 1)$
$-11(-2) - 34 = A(-2 - 1)(-2 + 4) + B(-2 + 2)(-2 + 4) + C(-2 + 2)(-2 - 1)$
$\qquad -12 = -6A$
$\qquad A = 2$

Let $x = -4$:
$-11x - 34 = A(x - 1)(x + 4) + B(x + 2)(x + 4) + C(x + 2)(x - 1)$
$-11(-4) - 34 = A(-4 - 1)(-4 + 4) + B(-4 + 2)(-4 + 4) + C(-4 + 2)(-4 - 1)$
$\qquad 10 = 10C$
$\qquad C = 1$

Therefore: $\frac{-11x - 34}{x^3 + 5x^2 + 2x - 8} = \frac{2}{x + 2} - \frac{3}{x - 1} + \frac{1}{x + 4}$

Repeated factors exercise (page 20)

Q13:

Step 1: The denominator has already been factorised and can be recognised as a repeated linear factor.

$$\frac{2x + 2}{(x + 3)(x + 3)} = \frac{2x + 2}{(x + 3)^2}$$

Step 2: Identify the type of partial fraction. In this case it has repeated linear factors.

$$\frac{2x + 2}{(x + 3)^2} = \frac{A}{(x + 3)} + \frac{B}{(x + 3)^2}$$

Step 3: Obtain the fractions with a common denominator.

$$\frac{2x + 2}{(x + 3)^2} = \frac{A(x + 3)}{(x + 3)^2} + \frac{B}{(x + 3)^2}$$

Step 4: Equate numerators since the denominators are equal.

$2x + 2 = A(x + 3) + B$ (*)

ANSWERS: UNIT 1 TOPIC 1

Step 5: Select values of x and substitute into (*).
Let $x = -3$:
$$2x + 2 = A(x+3) + B$$
$$2(-3) + 2 = A(-3+3) + B$$
$$-6 + 2 = B$$
$$B = -4$$

Now substitute $B = -4$ back into (*) and select another value of x to make solving for A easy.
Let $x = 0$:
$$2x + 2 = A(x+3) + B$$
$$2(0) + 2 = A(0+3) - 4$$
$$2 = 3A - 4$$
$$6 = 3A$$
$$A = 2$$

We have $A = 2$ and $B = -4$.

Therefore: $\frac{2x+2}{(x+3)(x+3)} = \frac{2}{x+3} - \frac{4}{(x+3)^2}$

Q14:

Step 1: Use synthetic division to factorise the denominator.
$$\frac{x^2 - 7x + 2}{x^3 - 3x^2 + 3x - 1} = \frac{x^2 - 7x + 2}{(x-1)^3}$$

Step 2: Identify the type of partial fraction. In this case it has repeated linear factors.
$$\frac{x^2 - 7x + 2}{(x-1)^3} = \frac{A}{x-1} + \frac{B}{(x-1)^2} + \frac{C}{(x-1)^3}$$

Step 3: Obtain the fractions with a common denominator.
$$\frac{x^2 - 7x + 2}{(x-1)^3} = \frac{A(x-1)^2}{(x-1)^3} + \frac{B(x-1)}{(x-1)^3} + \frac{C}{(x-1)^3}$$

Step 4: Equate numerators since the denominators are equal.
$$x^2 - 7x + 2 = A(x-1)^2 + B(x-1) + C \quad (*)$$

Step 5: Select values of x and equate coefficients.

First we will select a value for x and substitute into (*).
Let $x = 1$:
$$x^2 - 7x + 2 = A(x-1)^2 + B(x-1) + C$$
$$1^2 - 7(1) + 2 = A(1-1)^2 + B(1-1) + C$$
$$-4 = C$$
$$C = -4$$

Now substitute $C = -4$ back into (*), expand the brackets and collect like terms.
$$x^2 - 7x + 2 = A(x-1)^2 + B(x-1) - 4$$
$$x^2 - 7x + 2 = A(x^2 - 2x + 1) + B(x-1) - 4$$
$$x^2 - 7x + 2 = Ax^2 - 2Ax + A + Bx - B - 4$$
$$x^2 - 7x + 2 = Ax^2 - 2Ax + Bx + A - B - 4$$
$$x^2 - 7x + 2 = Ax^2 + (-2A + B)x + A - B - 4$$

© HERIOT-WATT UNIVERSITY

Set the coefficient in front of x^2 on the LHS equal to the coefficient in front of x^2 on the RHS.
$x^2: \quad 1 = A$
$\qquad A = 1$
Now set the coefficient in front of x on the LHS equal to the coefficient in front of x on the RHS.
$x: \quad -7 = -2A + B$
$\qquad -7 = -2(1) + B$
$\qquad -5 = B$
$\qquad B = -5$
We have $A = 1, B = -5$ and $C = -4$.
Therefore: $\frac{x^2 - 7x + 2}{x^3 - 3x^2 + 3x - 1} = \frac{1}{x-1} - \frac{5}{(x-1)^2} - \frac{4}{(x-1)^3}$

Q15:
Step 1: Use synthetic division to factorise the denominator.
$\frac{-3x^2 + 6x + 20}{x^3 - x^2 - 8x + 12} = \frac{-3x^2 + 6x + 20}{(x+3)(x-2)^2}$
Step 2: Identify the type of partial fraction. In this case it has distinct linear factors and repeated linear factors.
$\frac{-3x^2 + 6x + 20}{(x+3)(x-2)^2} = \frac{A}{(x+3)} + \frac{B}{(x-2)} + \frac{C}{(x-2)^2}$
Step 3: Obtain the fractions with a common denominator.
$\frac{-3x^2 + 6x + 20}{(x+3)(x-2)^2} = \frac{A(x-2)^2}{(x+3)(x-2)^2} + \frac{B(x+3)(x-2)}{(x+3)(x-2)^2} + \frac{C(x+3)}{(x+3)(x-2)^2}$
Step 4: Equate numerators since the denominators are equal.
$-3x^2 + 6x + 20 = A(x-2)^2 + B(x+3)(x-2) + C(x+3)$ (*)
Step 5: Select values of x and substitute into (*).
Let $x = 2$:
$\qquad -3x^2 + 6x + 20 = A(x-2)^2 + B(x+3)(x-2) + C(x+3)$
$-3(2)^2 + 6(2) + 20 = A(2-2)^2 + B(2+3)(2-2) + C(2+3)$
$\qquad -12 + 12 + 20 = 5C$
$\qquad\qquad\qquad 20 = 5C$
$\qquad\qquad\qquad C = 4$
Let $x = -3$:
$\qquad -3x^2 + 6x + 20 = A(x-2)^2 + B(x+3)(x-2) + C(x+3)$
$-3(-3)^2 + 6(-3) + 20 = A(-3-2)^2 + B(-3+3)(-3-2) + C(-3+3)$
$\qquad -27 - 18 + 20 = 25A$
$\qquad\qquad\qquad -25 = 25A$
$\qquad\qquad\qquad A = -1$
Now substitute $A = -1$ and $C = 4$ back into (*) and select a value of x to simplify the calculation.
Let $x = 0$:
$\qquad -3x^2 + 6x + 20 = A(x-2)^2 + B(x+3)(x-2) + C(x+3)$
$-3(0)^2 + 6(0) + 20 = -1(0-2)^2 + B(0+3)(0-2) + 4(0+3)$
$\qquad\qquad\qquad 20 = -4 - 6B + 12$
$\qquad\qquad\qquad 12 = -6B$
$\qquad\qquad\qquad B = -2$

ANSWERS: UNIT 1 TOPIC 1

We have $A = -1$, $B = -2$ and $C = 4$.

Therefore: $\frac{-3x^2 + 6x + 20}{x^3 - x^2 - 8x + 12} = \frac{-1}{x+3} - \frac{2}{x-2} + \frac{4}{(x-2)^2}$

Q16:

Step 1: Use synthetic division to factorise the denominator.

$\frac{x^2 + 11x + 15}{x^3 + 3x^2 - 4} = \frac{x^2 + 11x + 15}{(x-1)(x+2)^2}$

Step 2: Identify the type of partial fraction. In this case it has distinct linear factors and repeated linear factors.

$\frac{x^2 + 11x + 15}{(x-1)(x+2)^2} = \frac{A}{(x-1)} + \frac{B}{(x+2)} + \frac{C}{(x+2)^2}$

Step 3: Obtain the fractions with a common denominator.

$\frac{x^2 + 11x + 15}{(x-1)(x+2)^2} = \frac{A(x+2)^2}{(x-1)(x+2)^2} + \frac{B(x-1)(x+2)}{(x-1)(x+2)^2} + \frac{C(x-1)}{(x-1)(x+2)^2}$

Step 4: Equate numerators since the denominators are equal.

$x^2 + 11x + 15 = A(x+2)^2 + B(x-1)(x+2) + C(x-1)$ (*)

Step 5: Select values of x and substitute into (*).

Let $x = -2$:
$$x^2 + 11x + 15 = A(x+2)^2 + B(x-1)(x+2) + C(x-1)$$
$$(-2)^2 + 11(-2) + 15 = A(-2+2)^2 + B(-2-1)(-2+2) + C(-2-1)$$
$$4 - 22 + 15 = -3C$$
$$-3 = -3C$$
$$C = 1$$

Let $x = 1$:
$$x^2 + 11x + 15 = A(x+2)^2 + B(x-1)(x+2) + C(x-1)$$
$$1^2 + 11(1) + 15 = A(1+2)^2 + B(1-1)(1+2) + C(1-1)$$
$$27 = 9A$$
$$A = 3$$

Now substitute $A = 3$ and $C = 1$ back into (*) and select a value of x to simplify the calculation.

Let $x = 0$:
$$x^2 + 11x + 15 = A(x+2)^2 + B(x-1)(x+2) + C(x-1)$$
$$0^2 + 11(0) + 15 = 3(0+2)^2 + B(0-1)(0+2) + 1(0-1)$$
$$15 = 12 - 2B - 1$$
$$4 = -2B$$
$$B = -2$$

We have $A = 3$, $B = -2$ and $C = 1$.

Therefore: $\frac{x^2 + 11x + 15}{x^3 + 3x^2 - 4} = \frac{3}{x-1} - \frac{2}{x+2} + \frac{1}{(x+2)^2}$

© HERIOT-WATT UNIVERSITY

Irreducible quadratic factors exercise (page 23)

Q17:

Step 1: Use synthetic division to factorise the denominator.
$$\frac{x^2 - 4x + 14}{x^3 - x^2 + 2x - 8} = \frac{x^2 - 4x + 14}{(x - 2)(x^2 + x + 4)}$$

Step 2: Identify the type of partial fraction. In this case it has a distinct linear factor and an irreducible quadratic factor.
$$\frac{x^2 - 4x + 14}{(x - 2)(x^2 + x + 4)} = \frac{A}{x - 2} + \frac{Bx+C}{x^2 + x + 4}$$

Step 3: Obtain the fractions with a common denominator.
$$\frac{x^2 - 4x + 14}{(x - 2)(x^2 + x + 4)} = \frac{A(x^2 + x + 4)}{(x - 2)(x^2 + x + 4)} + \frac{(Bx+C)(x - 2)}{(x - 2)(x^2 + x + 4)}$$

Step 4: Equate numerators since the denominators are equal.

$x^2 - 4x + 14 = A(x^2 + x + 4) + (Bx+C)(x - 2)$ (*)

Step 5: Select values of x and substitute into (*).

Let $x = 2$:
$$x^2 - 4x + 14 = A(x^2 + x + 4) + (Bx+C)(x - 2)$$
$$2^2 - 4(2) + 14 = A(2^2 + 2 + 4) + (B(2)+C)(2 - 2)$$
$$10 = 10A$$
$$A = 1$$

Now substitute $A = 1$ back into (*) and select another value of x to simplify the calculation.

Let $x = 0$:
$$x^2 - 4x + 14 = A(x^2 + x + 4) + (Bx+C)(x - 2)$$
$$0^2 - 4(0) + 14 = 1(0^2 + 0 + 4) + (B(0)+C)(0 - 2)$$
$$14 = 4 - 2C$$
$$10 = -2C$$
$$C = -5$$

Now substitute $A = 1$ and $C = -5$ back into (*) and select another value of x to simplify the calculation.

Let $x = 1$:
$$x^2 - 4x + 14 = A(x^2 + x + 4) + (Bx + C)(x - 2)$$
$$1^2 - 4(1) + 14 = 1(1^2 + 1 + 4) + (B(1) - 5)(1 - 2)$$
$$11 = 6 - B + 5$$
$$B = 0$$

We have $A = 1$, $B = 0$ and $C = -5$.

Therefore: $\frac{x^2 - 4x + 14}{x^3 - x^2 + 2x - 8} = \frac{1}{x - 2} - \frac{5}{x^2 + x + 4}$

Q18:

Step 1: Use synthetic division to factorise the denominator.

$$\frac{10x^2 + 16x + 9}{2x^3 + 7x^2 + 5x + 6} = \frac{10x^2 + 16x + 9}{(x + 3)(2x^2 + x + 2)}$$

Step 2: Identify the type of partial fraction. In this case it has a distinct linear factor and an irreducible quadratic factor.

$$\frac{10x^2 + 16x + 9}{(x + 3)(2x^2 + x + 2)} = \frac{A}{(x + 3)} + \frac{Bx + C}{(2x^2 + x + 2)}$$

Step 3: Obtain the fractions with a common denominator.

$$\frac{10x^2 + 16x + 9}{(x + 3)(2x^2 + x + 2)} = \frac{A(2x^2 + x + 2)}{(x + 3)(2x^2 + x + 2)} + \frac{(Bx + C)(x + 3)}{(x + 3)(2x^2 + x + 2)}$$

Step 4: Equate numerators since the denominators are equal.

$$10x^2 + 16x + 9 = A(2x^2 + x + 2) + (Bx + C)(x + 3) \quad (*)$$

Step 5: Select values of x and substitute into (*).

Let $x = -3$:
$$10x^2 + 16x + 9 = A(2x^2 + x + 2) + (Bx + C)(x + 3)$$
$$10(-3)^2 + 16(-3) + 9 = A\left(2(-3)^2 - 3 + 2\right) + (B(-3) + C)(-3 + 3)$$
$$51 = 17A$$
$$A = 3$$

Now substitute $A = 3$ back into (*) and select another value of x to simplify the calculation.

Let $x = 0$:
$$10x^2 + 16x + 9 = A(2x^2 + x + 2) + (Bx + C)(x + 3)$$
$$10(0)^2 + 16(0) + 9 = 3\left(2(0)^2 + 0 + 2\right) + (B(0) + C)(0 + 3)$$
$$9 = 6 + 3C$$
$$C = 1$$

Now substitute $A = 3$ and $C = 1$ back into (*) and select another value of x to simplify the calculation.

Let $x = 1$:
$$10x^2 + 16x + 9 = A(2x^2 + x + 2) + (Bx + C)(x + 3)$$
$$10(1)^2 + 16(1) + 9 = 3\left(2(1)^2 + 1 + 2\right) + (B(1) + 1)(1 + 3)$$
$$35 = 15 + 4B + 4$$
$$16 = 4B$$
$$B = 4$$

We have $A = 3$, $B = 4$ and $C = 1$.

Therefore: $\frac{10x^2 + 16x + 9}{2x^3 + 7x^2 + 5x + 6} = \frac{3}{x + 3} + \frac{4x + 1}{2x^2 + x + 2}$

Algebraic long division exercise (page 27)

Q19:

a) Set up the division correctly.

$$x^2 + 4 \overline{) 3x^3 \quad -2x^2 \quad +0x \quad +6}$$

b) Divide x^2 into $3x^3$ and type in the first term of the quotient.

$$\begin{array}{r} 3x \\ x^2 + 4 \overline{) 3x^3 \quad -2x^2 \quad +0x \quad +6} \end{array}$$

c) Multiply $3x$ by the divisor and write the answer below the dividend.

$$\begin{array}{r} 3x \\ x^2 + 4 \overline{) 3x^3 \quad -2x^2 \quad +0x \quad +6} \\ 3x^3 \quad +0x^2 \quad +12x \end{array}$$

d) Subtract to give a new last line in the dividend.

$$\begin{array}{r} 3x \\ x^2 + 4 \overline{) 3x^3 \quad -2x^2 \quad +0x \quad +6} \\ \underline{3x^3 \quad +0x^2 \quad +12x \downarrow} \\ -2x^2 \quad -12x \quad +6 \end{array}$$

e) Divide x^2 into $-2x^2$ and type in the second term of the quotient.

$$\begin{array}{r} 3x \quad -2 \\ x^2 + 4 \overline{) 3x^3 \quad -2x^2 \quad +0x \quad +6} \\ \underline{3x^3 \quad +0x^2 \quad +12x \downarrow} \\ -2x^2 \quad -12x \quad +6 \end{array}$$

f) Multiply -2 by the divisor and write the answer below new the dividend.

$$\begin{array}{r} 3x \quad -2 \\ x^2 + 4 \overline{) 3x^3 \quad -2x^2 \quad +0x \quad +6} \\ \underline{3x^3 \quad +0x^2 \quad +12x \downarrow} \\ -2x^2 \quad -12x \quad +6 \\ -2x^2 \quad +0x \quad -8 \end{array}$$

g) Subtract to give a new last line in the dividend.

$$\begin{array}{r} 3x \quad -2 \\ x^2 + 4 \overline{) 3x^3 \quad -2x^2 \quad +0x \quad +6} \\ \underline{3x^3 \quad +0x^2 \quad +12x \downarrow} \\ -2x^2 \quad -12x \quad +6 \\ \underline{-2x^2 \quad +0x \quad -8} \\ -12x \quad +14 \end{array}$$

h) Give the final solution.

$$\frac{3x^3 - 2x^2 + 6}{x^2 + 4} = 3x - 2 + \frac{14 - 12x}{x^2 + 4}$$

ANSWERS: UNIT 1 TOPIC 1

Q20:

$$\begin{array}{r} x^2 + 3x - 2 \\ x+5 \overline{\smash{)}\, x^3 + 8x^2 + 13x - 10} \\ \underline{x^3 + 5x^2 } \\ 3x^2 + 13x - 10 \\ \underline{3x^2 + 15x } \\ -2x - 10 \\ \underline{-2x - 10} \\ 0 \end{array}$$

Therefore: $\frac{x^3 + 8x^2 + 13x - 10}{x + 5} = x^2 + 3x - 2$

Q21:

$$\begin{array}{r} x^2 - 4x + 23 \\ x^2+4x \overline{\smash{)}\, x^4 + 0x^3 + 7x^2 + 13x - 4} \\ \underline{x^4 + 4x^3 } \\ -4x^3 + 7x^2 + 13x - 4 \\ \underline{-4x^3 - 16x^2 } \\ 23x^2 + 13x - 4 \\ \underline{23x^2 + 92x } \\ -79x - 4 \end{array}$$

Therefore: $\frac{x^4 + 7x^2 + 13x - 4}{x^2 + 4x} = x^2 - 4x + 23 - \frac{79x + 4}{x^2 + 4x}$

Q22:

$$\begin{array}{r} 1 \\ x^5+1 \overline{\smash{)}\, x^5 + 0x^4 - x^3 + 0x^2 + 0x + 0} \\ \underline{x^5 + 0x^4 + 0x^3 + 0x^2 + 0x + 1} \\ -x^3 + 0x^2 + 0x - 1 \end{array}$$

Therefore: $\frac{x^5 - x^3}{x^5 + 1} = 1 - \frac{x^3 + 1}{x^5 + 1}$

Summary of partial fraction forms (page 28)

Q23: 1-e, 2-d, 3-g, 4-h, 5-c, 6-a, 7-b, 8-f

Reduce improper rational functions by division exercise (page 32)

Q24:
If we expand the denominator it becomes $x^2 + 3x + 2$. It is then clear that we are working with an improper rational function as the degree of the numerator > degree of the denominator.

Step 1: Divide using algebraic long division.

$$\begin{array}{r} x - 3 \\ x^2 + 3x + 2 \overline{) x^3 + 0x^2 + 0x + 0} \\ \underline{x^3 + 3x^2 + 2x} \\ -3x^2 - 2x + 0 \\ \underline{-3x^2 - 9x - 6} \\ 7x + 6 \end{array}$$

This gives $\frac{x^3}{(x+1)(x+2)} = x - 3 + \frac{7x+6}{(x+1)(x+2)}$.

Step 2: Now express $\frac{7x+6}{(x+1)(x+2)}$ in partial fractions. In this case it has a distinct linear factors.

$\frac{7x+6}{(x+1)(x+2)} = \frac{A}{(x+1)} + \frac{B}{(x+2)}$

Step 3: Obtain the fractions with a common denominator.

$\frac{7x+6}{(x+1)(x+2)} = \frac{A(x+2)}{(x+1)(x+2)} + \frac{B(x+1)}{(x+1)(x+2)}$

Step 4: Equate numerators since the denominators are equal.

$7x + 6 = A(x+2) + B(x+1)$ (*)

Step 5: Select values of x and substitute into (*).

Let $x = -1$:
$7x + 6 = A(x+2) + B(x+1)$
$7(-1) + 6 = A(-1+2) + B(-1+1)$
$-1 = A$
$A = -1$

Let $x = -2$:
$7x + 6 = A(x+2) + B(x+1)$
$7(-2) + 6 = A(-2+2) + B(-2+1)$
$-8 = -B$
$B = 8$

Therefore: $\frac{7x+6}{(x+1)(x+2)} = \frac{-1}{(x+1)} + \frac{8}{(x+2)}$

Step 6: Substitute partial fractions back into the expression.

$$\frac{x^3}{(x+1)(x+2)} = x - 3 + \frac{7x+6}{(x+1)(x+2)}$$
$$= x - 3 + \left[\frac{-1}{(x+1)} + \frac{8}{(x+2)}\right]$$
$$= x - 3 - \frac{1}{(x+1)} + \frac{8}{(x+2)}$$

Q25:
It is clear that we are working with an improper rational function as the degree of the numerator > degree of the denominator.

ANSWERS: UNIT 1 TOPIC 1

Step 1: Divide using algebraic long division.

$$x^2 + 3x + 2 \overline{\smash{\big)}\, x^3 + 3x^2 + 4x + 3}$$

$$\phantom{x^2 + 3x + 2 \overline{\smash{\big)}\,}} \underline{x^3 + 3x^2 + 2x}$$

$$\phantom{x^2 + 3x + 2 \overline{\smash{\big)}\, x^3 + 3x^2 +\,}} 2x + 3$$

This gives $\frac{x^3 + 3x^2 + 4x + 3}{(x+1)(x+2)} = x + \frac{2x+3}{(x+1)(x+2)}$.

Step 2: Now express $\frac{2x+3}{(x+1)(x+2)}$ in partial fractions. In this case it has a distinct linear factors.

$$\frac{2x+3}{(x+1)(x+2)} = \frac{A}{(x+1)} + \frac{B}{(x+2)}$$

Step 3: Obtain the fractions with a common denominator.

$$\frac{2x+3}{(x+1)(x+2)} = \frac{A(x+2)}{(x+1)(x+2)} + \frac{B(x+1)}{(x+1)(x+2)}$$

Step 4: Equate numerators since the denominators are equal.

$$2x + 3 = A(x+2) + B(x+1) \; (*)$$

Step 5: Select values of x and substitute into (*).

Let $x = -1$:

$$2x + 3 = A(x+2) + B(x+1)$$
$$2(-1) + 3 = A(-1+2) + B(-1+1)$$
$$1 = A$$
$$A = 1$$

Let $x = -2$:

$$2x + 3 = A(x+2) + B(x+1)$$
$$2(-2) + 3 = A(-2+2) + B(-2+1)$$
$$-1 = -B$$
$$B = 1$$

Step 6: Substitute partial fractions back into the expression.

$$\frac{x^3 + 3x^2 + 4x + 3}{(x+1)(x+2)} = x + \frac{2x+3}{(x+1)(x+2)}$$

$$= x + \frac{1}{(x+1)} + \frac{1}{(x+2)}$$

End of topic 1 test (page 35)

Q26: $\frac{3}{x-1} + \frac{2}{x-2}$

Q27: $-\frac{1}{x+3} - \frac{5}{x+1}$

Q28: $-\frac{3}{x-2} - \frac{1}{x+4}$

Q29: $-\frac{1}{x+1} + \frac{3}{x+2} + \frac{2}{x-3}$

Q30: $\frac{9}{x-1} + \frac{13}{x-2} + \frac{1}{x+1}$

Q31: $\frac{2}{x-2} - \frac{3}{(x-2)^2}$

© HERIOT-WATT UNIVERSITY

Q32: $\frac{2}{x+3} + \frac{2}{(x+3)^2}$

Q33: $\frac{3}{x-1} - \frac{2}{x+2} - \frac{3}{(x+2)^2}$

Q34: $\frac{2}{x-2} + \frac{3}{(x-2)^2} + \frac{4}{(x-2)^3}$

Q35: $\frac{3}{x+1} + \frac{x+2}{x^2+4x+7}$

Q36: $-\frac{7}{2(x-1)} + \frac{7x+9}{2(x^2-2x+3)}$

Q37: $x - 3 + \frac{7x+6}{x^2+3x+2}$

Q38: $x - 3 - \frac{1}{x+1} + \frac{8}{x+2}$

Q39: $2x^2 + 4x + 6 - \frac{5x-8}{x^2-3x+2}$

Q40: $2x^2 + 4x + 6 - \frac{3}{x-1} - \frac{2}{x-2}$

Q41: $4x + \frac{4x-11}{x^2-4x+4}$

Q42: $4x + \frac{4}{x-2} - \frac{3}{(x-2)^2}$

Q43: $-3x + 5 - \frac{x^2-x+7}{x^3-2x^2+2x-5}$

Q44: $-3x + 5 - \frac{1}{x-1} - \frac{2}{x^2-3x+5}$

Topic 2: Differentiation

Differentiating sin x and cos x exercise (page 50)

Q1:

x	0	$\frac{\pi}{2}$	π	$\frac{3\pi}{2}$	2π
m_T	0	-1	0	1	0

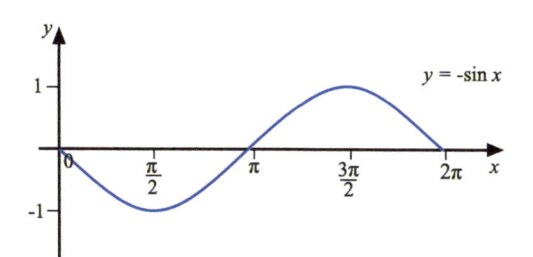

$y = -\sin x$

Q2: c) $-\sin x$

Differentiating using the chain rule exercise (page 54)

Q3:

Step 1:	Turn the square root into a power.	$h(x) = (x^2 + 6x)^{\frac{1}{2}}$
Step 2:	Bring down the power.	$\frac{1}{2}$
Step 3:	Write down the bracket.	$\frac{1}{2}(x^2 + 6x)$
Step 4:	Reduce the power by 1.	$\frac{1}{2}(x^2 + 6x)^{-\frac{1}{2}}$
Step 5:	Differentiate the bracket.	$\frac{1}{2}(x^2 + 6x)^{-\frac{1}{2}} \times (2x + 6)$
Step 6:	Simplify the answer.	$\frac{dy}{dx} = \frac{2x + 6}{2\sqrt{x^2 + 6x}}$
		$= \frac{2(x + 3)}{2\sqrt{x^2 + 6x}}$
		$= \frac{x + 3}{\sqrt{x^2 + 6x}}$

Q4:

Step 1:	Remember:	$\cos^3 x = (\cos x)^3$
Step 2:	Bring down the power.	3
Step 3:	Write down the bracket.	$3(\cos x)$
Step 4:	Reduce the power by 1.	$3(\cos x)^2$
Step 5:	Differentiate the bracket.	$3(\cos x)^2 \times (-\sin x)$
Step 6:	Simplify the answer.	$f'(x) = -3\cos^2 x \sin x$

Q5:

Step 1:	Bring down the power.	5
Step 2:	Write down the bracket.	$5(x^3 + 2x^2 - 3)$
Step 3:	Decrease the power by 1.	$5(x^3 + 2x^2 - 3)^4$
Step 4:	Differentiate the bracket.	$3x^2 - 4x$
Step 5:	Simplify the answer.	$f'(x) = 5(x^3 + 2x^2 - 3)^4(3x^2 - 4x)$ $f'(x) = 5x(x^3 + 2x^2 - 3)^4(3x - 4)$

Q6:

Step 1:	Re-write as a negative power.	$a(1 - 3x)^{-4}$
Step 2:	Bring down the power and decrease the power by 1.	$-16(1 - 3x)^{-5}$
Step 3:	Differentiate the bracket.	-3
Step 4:	Simplify the answer.	$k'(x) = -16(1 - 3x)^{-5} \times (-3)$ $k'(x) = \frac{48}{(1-3x)^5}$

Q7:

Step 1:	Differentiate sin	$\cos\left(5x - \frac{\pi}{3}\right)$
Step 2:	Differentiate the bracket.	5
Step 3:	Simplify the answer.	$h'(x) = 5\cos\left(5x - \frac{\pi}{3}\right)$

Determining the equation of a tangent to a curve exercise (page 57)

Q8:

Hints:

- Find $f'(x)$ and then find $f'(-1)$
- When $x = -1$, $y = (-1)^3 - 2 \times (-1)^2 + 3 \times (-1) - 6 = ?$
- Substitute the values into $y - b = m(x - a)$
- So the equation of the tangent is $y = ?$

Answer: $y = 10x - 2$

Leibniz notation exercise (page 63)

Q9: Find the derivative $\frac{dy}{dx} = 15x^4 + 6x$

Substitute $x = 2$ into $\frac{dy}{dx}$

$\frac{dy}{dx} = 15(2)^4 + 6(2) = 252$

ANSWERS: UNIT 1 TOPIC 2

Q10: Find the derivative:

$$\frac{dy}{dx} = 8x + 2 + 3x^{-2}$$
$$= 8x + 2 + \frac{3}{x^2}$$

Substitute $x = 1$ into $\frac{dy}{dx}$

$$\frac{dy}{dx} = 8(1) + 2 + \frac{3}{(1)^2}$$
$$\frac{dy}{dx} = 13$$

Q11: Find the derivative:

$\frac{dy}{d\theta} = 3\cos(3\theta)$

Substitute $\theta = \frac{\pi}{3}$ into $\frac{dy}{d\theta}$

$$\frac{dy}{d\theta} = 3\cos\left(3 \times \frac{\pi}{3}\right)$$
$$\frac{dy}{d\theta} = 3\cos(\pi) = -3$$

Q12: Find the derivative:

$$\frac{dS}{dt} = \frac{1}{2}(3t-4)^{-\frac{1}{2}} \times 3$$
$$\frac{dS}{dt} = \frac{3}{2\sqrt{(3t-4)}}$$

Closed intervals exercise (page 65)

Q13:
It is much easier to spot the maximum and minimum values if we make a sketch.
Step 1: Find the stationary points and their nature.
$\frac{dy}{dx} = 3x^2 - 12$
Stationary points occur when $\frac{dy}{dx} = 0$

$$3x^2 - 12 = 0$$
$$3(x^2 - 4) = 0$$
$$3(x-2)(x+2) = 0$$
$$x - 2 = 0 \text{ or } x + 2 = 0$$
$$x = 2 \text{ or } x = -2$$

When $x = 2$ then $y = 2^3 - 12 \times 2 = -16$
When $x = -2$ then $y = (-2)^3 - 12 \times (-2) = 16$
Thus the stationary points are (2,-16) and (-2,16)

© HERIOT-WATT UNIVERSITY

x	-3 →	-2	0 →	2	3 →
$(x - 2)$	-	-	-	0	+
$(x + 2)$	-	0	+	+	+
$3(x - 2)(x + 2)$	+	0	-	0	+
shape	↗	→	↘	→	↗

This gives us a maximum turning point at (-2,16) and a minimum turning point at (2,-16).

Step 2: Find the end points

When $x = -5$ then $y = -65$

When $x = 3$ then $y = -9$

Step 3: Make a sketch

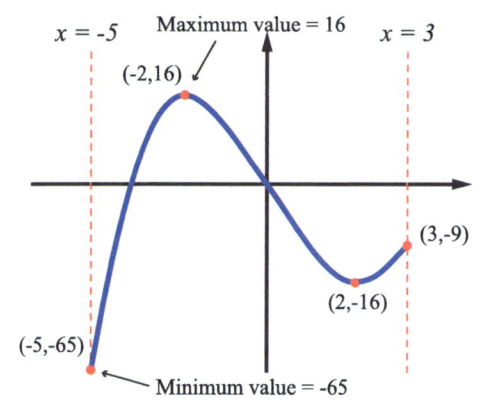

Hence the maximum value is 16 at the maximum turning point (-2,16) and the minimum value is -65 at the end point (-5,-65).

Answers from page 71.

Q14: b) $t = 4$

Q15: The gradient of the curve is steeper at $t = 4$, therefore the car is travelling faster at this point.

Q16: Since the distance-time graph given is a curve, the speed is not constant; since it is changing continuously we can only give an average.

Differentiability at a point: Continuous or discontinuous (page 78)

Graphs explained

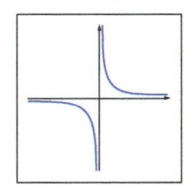

This graph is discontinuous at $x = 0$ and is not differentiable at this point. It does not have any corners so it is smooth.

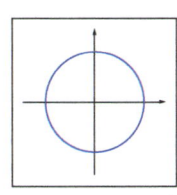

This graph is continuous since there are no "gaps" in the line. It is also smooth because a tangent can be drawn at each point along the curve.

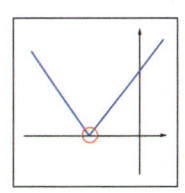

This graph is continuous since there are no "gaps" in the line. It is not smooth because there is a corner in the function.

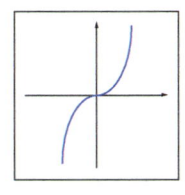

This graph is continuous since there are no "gaps" in the line. It is also smooth because there are no corners in the function.

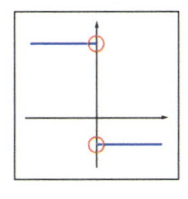

This graph is discontinuous at $x = 0$ because there is a "jump" in the function. It is smooth because there are no corners in the function.

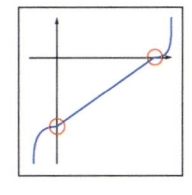

This graph is continuous since there are no "gaps" in the function. It is not smooth because there are corners in the function.

Q17:

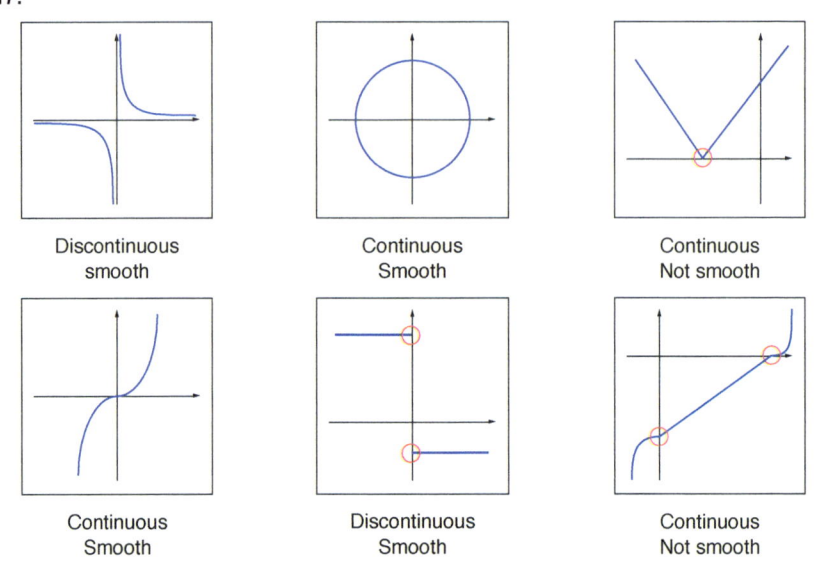

Discontinuous smooth	Continuous Smooth	Continuous Not smooth
Continuous Smooth	Discontinuous Smooth	Continuous Not smooth

Differentiability at a point exercise (page 79)

Q18: b) Discontinuous

Q19: $x = 0$

Q20: b) No

Q21: $x = 0$

Q22: Yes. This function is differentiable because it is a smooth curve with no corners, it is continuous and a tangent is defined everywhere.

Q23: No. This function is undefined for $x = 0$ so is not differentiable at this point. However it is a smooth and continuous curve so is differentiable everywhere except at $x = 0$.

Q24: No. This function is continuous but not smooth at $x = -1$ and $x = 2$. Therefore it is not differentiable at these points.

Q25: d) A, C, E, F

ANSWERS: UNIT 1 TOPIC 2

Q26:

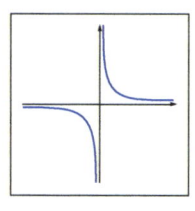

This function is not differentiable at the point $x = 0$ because there is a discontinuity at this point and the gradient is undefined here.

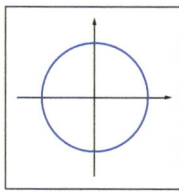

This graph is not differentiable at the points where curve meets the x-axis. At this point the gradient is undefined.

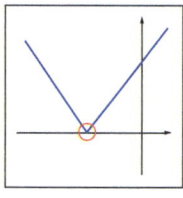

This function is not differentiable at the corner. It is not smooth even though it is continuous.

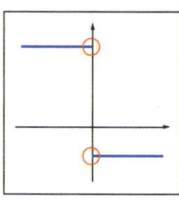

This graph has a discontinuity at $x = 0$. It is not differentiable at this point.

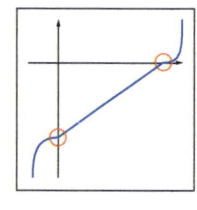

This graph is not smooth. It has corners at the highlighted points and is not differentiable at these points.

Differentiability over an interval exercise (page 82)

Q27: The function is not differentiable in the closed interval [-2,3].
The function is not smooth at $x = -1$ or $x = 2$

© HERIOT-WATT UNIVERSITY

Q28: It is differentiable over the interval [-5,-1] because it is smooth and continuous.
Any closed interval that contains 0 is not differentiable. This is because there is a discontinuity at $x = 0$.

Q29: The closed interval can start with any number other than -3 and go to any other number to the right of -3. For example [-2,5], [-2·9,100] etc.

Higher derivatives exercise (page 85)

Q30:
$$f'(x) = 16x^3 + 15x^2$$
$$f''(x) = 48x^2 + 30x$$
$$f^{(3)}(x) = 96x + 30$$
$$f^{(4)}(x) = 96$$
$$f^{(5)}(x) = 0 \text{ and } f^{(n)}(x) = 0 \text{ for } n = 5, 6, 7, \ldots$$

Q31:
$$f(x) = 3(2-x)^{-1}$$
$$f'(x) = -3(2-x)^{-2} \times (-1)$$
$$= 3(2-x)^{-2}$$
$$f''(x) = -6(2-x)^{-3} \times (-1)$$
$$= 6(2-x)^{-3}$$

Re-write the denominator as a negative power.
Remember the chain rule: $\frac{dy}{dx} = \frac{dy}{du} \times \frac{du}{dx}$
Multiply by the power and decrease the power by one.
Remember when differentiating a constant becomes zero.

Q32:
$$y' = -4\sin\left(2x - \frac{\pi}{3}\right) \times 2$$
$$= -8\sin\left(2x - \frac{\pi}{3}\right)$$
$$y'' = -8\cos\left(2x - \frac{\pi}{3}\right) \times 2$$
$$= -16\cos\left(2x - \frac{\pi}{3}\right)$$
$$y''' = 16\sin\left(2x - \frac{\pi}{3}\right) \times 2$$
$$= 32\sin\left(2x - \frac{\pi}{3}\right)$$

Q33: 10

Differentiate x^{10} until you get zero.

When differentiating, the degree of polynomial is reduced by one each time if the derivative exists.

We do not need to do the exact differentiation. The only term that we need to differentiate and wait to become zero is x^{10}. We will differentiate this term only and ignore the coefficient in the front.

ANSWERS: UNIT 1 TOPIC 2 421

$f'(x) = x^9$
$f''(x) = x^8$
$f'''(x) = x^7$
$f^{(4)}(x) = x^6$
$f^{(5)}(x) = x^5$
$f^{(6)}(x) = x^4$
$f^{(7)}(x) = x^3$
$f^{(8)}(x) = x^2$
$f^{(9)}(x) = x$
$f^{(9)}(x) = x$
$f^{(10)}(x) = 1$

We differentiate 10 times before it becomes zero.

Discontinuity exercise (page 87)

Q34:
$f'(x) = 7(x+1)^{\frac{2}{5}}$
$f''(x) = \frac{14}{5}(x+1)^{-\frac{3}{5}}$

Q35:

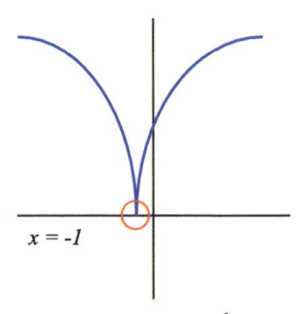

$f'(x) = 7(x+1)^{\frac{2}{5}}$ 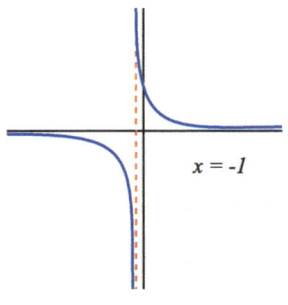 $f''(x) = (x+1)^{-\frac{3}{5}}$

$f'(x)$ is not differentiable at $x = -1$
$f''(x)$ is discontinuous at $x = -1$ as can be seen from the graphs above.

Q36: $f'(x) = 8(2x+1)^{\frac{1}{3}}$
$f''(x) = \frac{16}{3}(2x+1)^{-\frac{2}{3}}$

Q37: $f'(x)$ is undefined for $x = -\frac{1}{2}$
$f''(x)$ is undefined for $x = -\frac{1}{2}$

© HERIOT-WATT UNIVERSITY

Q38:
a)

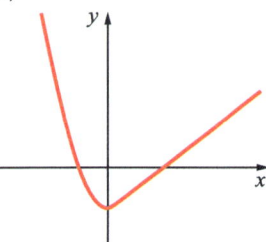

b)
$$f'(x) = \begin{cases} 2x, & \text{for } x < 0 \\ 1, & \text{for } x \geq 0 \end{cases}$$

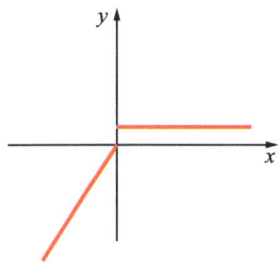

c)
$f'(x)$ is not differentiable for $x = 0$.

The product rule exercise (page 91)

Q39:
By definition, the product rule is: $k'(x) = f'(x) g(x) + f(x) g'(x)$
Let $f(x) = x^2 + 7$ and $g(x) = x^3 - 6x + 9$
Then $f'(x) = 2x$ and $g'(x) = 3x^2 - 6$
From the product rule we have:
$k'(x) = f'(x) g(x) + f(x) g'(x)$
$k'(x) = 2x(x^3 - 6x + 9) + (x^2 + 7)(3x^2 - 6)$
Multiplying this out and simplifying gives us:
$k'(x) = 2x^4 - 12x^2 + 18x + 3x^4 - 6x^2 + 21x^2 - 42$
$k'(x) = 5x^4 + 3x^2 + 18x - 42$

ANSWERS: UNIT 1 TOPIC 2

Q40:

By definition, the product rule is: $k'(x) = f'(x)g(x) + f(x)g'(x)$

Let $f(x) = \sin x$ and $g(x) = \cos x$

Then $f'(x) = \cos x$ and $g'(x) = -\sin x$

From the product rule we have:

$k'(x) = f'(x)g(x) + f(x)g'(x)$
$k'(x) = \cos x \cos x + \sin x (-\sin x)$

Simplifying gives us:

$k'(x) = \cos^2 x - \sin^2 x$ remember from Higher that $\cos 2x = \cos^2 x - \sin^2 x$
$k'(x) = \cos 2x$

Q41:

By definition, the product rule is: $k'(x) = f'(x)g(x) + f(x)g'(x)$

Let $f(x) = x$ and $g(x) = (2-3x^2)^5$

Then $f'(x) = 1$ and $g'(x) = -30x(2-3x^2)^4$

From the product rule we have:

$k'(x) = f'(x)g(x) + f(x)g'(x)$
$k'(x) = 1(2-3x^2)^5 + x \times (-30x)(2-3x^2)^4$
$k'(x) = (2-3x^2)^4 [2 - 3x^2 - 30x^2]$
$k'(x) = (2-3x^2)^4 (2 - 33x^2)$

Q42:

By definition, the product rule is: $k'(x) = f'(x)g(x) + f(x)g'(x)$

Let $f(x) = x$ and $g(x) = \sin^2 x$

Then $f'(x) = 1$ and $g'(x) = 2\sin x \cos x$

From the product rule we have:

$k'(x) = f'(x)g(x) + f(x)g'(x)$
$k'(x) = 1 \times \sin^2 x + x \times 2\sin x \cos x$
$k'(x) = \sin^2 x + x \sin 2x$
$k'\left(\frac{\pi}{4}\right) = \sin^2\left(\frac{\pi}{4}\right) + \left(\frac{\pi}{4}\right) \sin 2\left(\frac{\pi}{4}\right)$
$k'\left(\frac{\pi}{4}\right) = \frac{1}{2}$

Q43:

By definition, the product rule is: $k'(x) = f'(x)g(x) + f(x)g'(x)$

Let $f(x) = x^2$ and $g(x) = (5x^2 - 8)^{\frac{1}{2}}$

Then $f'(x) = 2x$ and $g'(x) = 5x(5x^2 - 8)^{-\frac{1}{2}}$

From the product rule we have:

© HERIOT-WATT UNIVERSITY

$k'(x) = f'(x)g(x) + f(x)g'(x)$

$k'(x) = 2x \times (5x^2 - 8)^{\frac{1}{2}} + x^2 \times 5x(5x^2 - 8)^{-\frac{1}{2}}$

$k'(x) = x(5x^2 - 8)^{-\frac{1}{2}} (2(5x^2 - 8) + 5x^2)$

$k'(x) = x(5x^2 - 8)^{-\frac{1}{2}} (15x^2 - 16)$

$k'(x) = \dfrac{x(15x^2 - 16)}{(5x^2 - 8)^{\frac{1}{2}}}$

Q44:

By definition, the product rule is: $k'(x) = f'(x)g(x) + f(x)g'(x)$

Let $f(x) = \sin 4x$ and $g(x) = \cos(2x + 3)$

Then $f'(x) = 4\cos 4x$ and $g'(x) = -2\sin(2x + 3)$

From the product rule we have:

$k'(x) = f'(x)g(x) + f(x)g'(x)$

$k'(x) = 4\cos 4x \times \cos(2x + 3) + \sin 4x \times (-2)\sin(2x + 3)$

$k'(x) = 4\cos 4x \cos(2x + 3) - 2\sin 4x \sin(2x + 3)$

Q45:

By definition, the product rule is: $k'(x) = f'(x)g(x)h(x) + f(x)g'(x)h(x) + f(x)g(x)h'(x)$

Let $f(x) = x$, $g(x) = (x + 3)^2$ and $h(x) = \sin x$

Then $f'(x) = 1$, $g'(x) = 2x + 6$ and $h'(x) = \cos x$

From the product rule we have:

$k'(x) = f'(x)g(x)h(x) + f(x)g'(x)h(x) + f(x)g(x)h'(x)$

$k'(x) = 1 \times (x+3)^2 \times \sin x + x \times (2x+6) \times \sin x + x \times (x+3)^2 \times \cos x$

$k'(x) = (x+3)\{(x+3)\sin x + 2x\sin x + (2x+6)\cos x\}$

Q46:

By definition, the product rule is: $k'(x) = f'(x)g(x)h(x) + f(x)g'(x)h(x) + f(x)g(x)h'(x)$

Let $f(x) = x^2$, $g(x) = \sin 2x$ and $h(x) = \cos 3x$

Then $f'(x) = 2x$, $g'(x) = 2\cos 2x$ and $h'(x) = -3\sin 3x$

From the product rule we have:

$k'(x) = f'(x)g(x)h(x) + f(x)g'(x)h(x) + f(x)g(x)h'(x)$

$k'(x) = 2x\sin 2x\cos 3x + x^2 \times 2\cos 2x\cos 3x + x^2\sin 2x(-3\sin 3x)$

$k'(x) = 2x\sin 2x\cos 3x + 2x^2\cos 2x\cos 3x - 3x^2\sin 2x\sin 3x$

The quotient rule exercise (page 95)

Q47:

By definition from the quotient rule:

$\frac{d}{dx}\left(\frac{f}{g}\right) = \dfrac{f'(x)g(x) - f(x)g'(x)}{[g(x)]^2}$

ANSWERS: UNIT 1 TOPIC 2

Let $f(x) = 4x^2$ and $g(x) = (3x-1)^{\frac{1}{2}}$

Then $f'(x) = 8x$ and $g'(x) = \frac{3}{2}(3x-1)^{-\frac{1}{2}}$

From the quotient rule we have:

$$k'(x) = \frac{f'(x)g(x) - f(x)g'(x)}{[g(x)]^2}$$

$$= \frac{8x(3x-1)^{\frac{1}{2}} - 4x^2 \frac{3}{2}(3x-1)^{-\frac{1}{2}}}{3x-1}$$

Take out the common factor of $2x(3x-1)^{-\frac{1}{2}}$, or alternatively written as $\frac{2x}{(3x-1)^{\frac{1}{2}}}$.

But first we will re-write the fraction in a way that is easier to take out he common factor.

$$k'(x) = \frac{8x\frac{(3x-1)}{(3x-1)^{\frac{1}{2}}} - 6x^2 \frac{1}{(3x-1)^{\frac{1}{2}}}}{3x-1}$$

Now taking out the common factor $\frac{2x}{(3x-1)^{\frac{1}{2}}}$ we get:

$$k'(x) = \frac{\frac{2x}{(3x-1)^{\frac{1}{2}}}\{4(3x-1) - 3x\}}{3x-1}$$

Re-writing so that the $(3x-1)^{\frac{1}{2}}$ is put onto the denominator and tidying the bracket.

$$k'(x) = \frac{2x\{9x-4\}}{(3x-1)^{\frac{3}{2}}}$$

Q48:

By definition from the quotient rule:

$$\frac{d}{dx}\left(\frac{f}{g}\right) = \frac{f'(x)g(x) - f(x)g'(x)}{[g(x)]^2}$$

Let $f(x) = \cos^2 x$ and $g(x) = \sin x$

Then $f'(x) = -2\cos x \sin x$ and $g'(x) = \cos x$

From the quotient rule we have:

$$k'(x) = \frac{f'(x)g(x) - f(x)g'(x)}{[g(x)]^2}$$

$$= \frac{-2\cos x \sin x \sin x - \cos^2 x \cos x}{(\sin x)^2}$$

Separating into two fractions and simplifying we have:

$$k'(x) = \frac{-2\cos x (\sin x)^2}{(\sin x)^2} - \frac{(\cos x)^3}{(\sin x)^2}$$

$$= -2\cos x - \cos x \times \frac{(\cos x)^2}{(\sin x)^2}$$

$$= -2\cos x - \cos x \times \frac{1}{\tan^2 x}$$

$$= -2\cos x - \frac{\cos x}{\tan^2 x}$$

Remember from Higher that $\tan x = \frac{\sin x}{\cos x}$

© HERIOT-WATT UNIVERSITY

Q49:

By definition from the quotient rule:

$\frac{d}{dx}\left(\frac{f}{g}\right) = \frac{f'(x)g(x) - f(x)g'(x)}{[g(x)]^2}$

Let $f(x) = 5x$ and $g(x) = 2x - 3$
Then $f'(x) = 5$ and $g'(x) = 2$
From the quotient rule we have:

$k'(x) = \frac{f'(x)g(x) - f(x)g'(x)}{[g(x)]^2}$

$k'(x) = \frac{5(2x-3) - 5x \times 2}{(2x-3)^2}$

$k'(x) = \frac{-15}{(2x-3)^2}$

Q50:

By definition from the quotient rule:

$\frac{d}{dx}\left(\frac{f}{g}\right) = \frac{f'(x)g(x) - f(x)g'(x)}{[g(x)]^2}$

Let $f(x) = 6x^3 + 2x - 3$ and $g(x) = x^2 - 4x + 1$
Then $f'(x) = 18x^2 + 2$ and $g'(x) = 2x - 4$
From the quotient rule we have:

$k'(x) = \frac{f'(x)g(x) - f(x)g'(x)}{[g(x)]^2}$

$= \frac{(18x^2 + 2)(x^2 - 4x + 1) - (6x^3 + 2x - 3)(2x - 4)}{(x^2 - 4x + 1)^2}$

$= \frac{18x^4 - 72x^3 + 18x^2 + 2x^2 - 8x + 2 - (12x^4 - 24x^3 + 4x^2 - 8x - 6x + 12)}{(x^2 - 4x + 1)^2}$

$= \frac{6x^4 - 48x^3 + 12x^2 - 22x - 10}{(x^2 - 4x + 1)^2}$

Q51:

By definition from the quotient rule:

$\frac{d}{dx}\left(\frac{f}{g}\right) = \frac{f'(x)g(x) - f(x)g'(x)}{[g(x)]^2}$

Let $f(x) = \sin^3 x$ and $g(x) = x^2$
Then $f'(x) = 3\sin^2 x \cos x$ and $g'(x) = 2x$
From the quotient rule we have:

$k'(x) = \frac{f'(x)g(x) - f(x)g'(x)}{[g(x)]^2}$

$= \frac{3\sin^2 x \cos x (x^2) - \sin^3 x (2x)}{(x^2)^2}$

$= \frac{3\sin^2 x \cos x (x^2) - \sin^3 x (2x)}{x^4}$

ANSWERS: UNIT 1 TOPIC 2

Q52:
By definition from the quotient rule:
$$\frac{d}{dx}\left(\frac{f}{g}\right) = \frac{f'(x)g(x) - f(x)g'(x)}{[g(x)]^2}$$
Let $f = 1 + 5x$ and $g = (2x - 5)^{\frac{1}{2}}$
Then $f' = 5$ and $g' = (2x - 5)^{-\frac{1}{2}}$
Giving,
$$\left(\frac{f}{g}\right)' = \frac{5(2x-5)^{\frac{1}{2}} - (1+5x)(2x-5)^{-\frac{1}{2}}}{\left((2x-5)^{\frac{1}{2}}\right)^2}$$
Simplifying the denominator:
$$\left(\frac{f}{g}\right)' = \frac{5(2x-5)^{\frac{1}{2}} - (1+5x)(2x-5)^{-\frac{1}{2}}}{2x-5}$$
Taking out a common factor of $(2x-5)^{-\frac{1}{2}}$ or written as $\frac{1}{(2x-5)^{\frac{1}{2}}}$

Before we do this, we will re-write the numerator in a way that shows clearly how this factor is taken out:
$$\left(\frac{f}{g}\right)' = \frac{\frac{5(2x-5)}{(2x-5)^{\frac{1}{2}}} - (1+5x)\frac{1}{(2x-5)^{\frac{1}{2}}}}{2x-5}$$
Now take out the common factor $\frac{1}{(2x-5)^{\frac{1}{2}}}$
$$\left(\frac{f}{g}\right)' = \frac{\frac{1}{(2x-5)^{\frac{1}{2}}}[5(2x-5) - (1+5x)]}{2x-5}$$
Simplifying by taking $(2x-5)^{\frac{1}{2}}$ to the denominator and tidying the bracket to get:
$$\left(\frac{f}{g}\right)' = \frac{[5(2x-5) - (1+5x)]}{(2x-5)^{\frac{3}{2}}}$$
$$\left(\frac{f}{g}\right)' = \frac{5x - 26}{(2x-5)^{\frac{3}{2}}}$$

Differentiate cot(x), sec(x), cosec(x) and tan(x) exercise (page 97)

Q53:
By definition $\cot x = \frac{1}{\tan x}$ so $k(x) = \frac{\cos x}{\sin x}$
By definition from the quotient rule: $\frac{d}{dx}\left(\frac{f}{g}\right) = \frac{f'(x)g(x) - f(x)g'(x)}{[g(x)]^2}$
Let $f(x) = \cos x$ and $g(x) = \sin x$
Then $f'(x) = -\sin x$ and $g'(x) = \cos x$
From the quotient rule we have: $k'(x) = \frac{f'(x)g(x) - f(x)g'(x)}{(g(x))^2}$
$k'(x) = \frac{-\sin x \sin x - \cos x \cos x}{\sin^2 x}$

© HERIOT-WATT UNIVERSITY

Simplifying we have:

$$k'(x) = \frac{-(\sin x \sin x + \cos x \cos x)}{\sin^2 x} \qquad (\sin^2 x + \cos^2 x = 1)$$

$$= \frac{-1}{\sin^2 x} \qquad \left(\csc x = \frac{1}{\sin x}\right)$$

$$= -\csc^2 x$$

Q54:

By definition $\csc x = \frac{1}{\sin x}$ therefore $k(x) = \frac{1}{\sin x}$

Let $f(x) = 1$ and $g(x) = \sin x$

Then $f\,'(x) = 0$ and $g\,'(x) = \cos x$

From the quotient rule we have: $k'(x) = \frac{f'(x)g(x) - f(x)g'(x)}{(g(x))^2}$

$$k'(x) = \frac{0 \sin x - \cos x\,(1)}{\sin^2 x}$$

Simplifying we have:

$$k'(x) = \frac{-\cos x}{\sin^2 x}$$

$$= -\frac{1}{\sin x} \times \frac{\cos x}{\sin x} \qquad \csc x = \frac{1}{\sin x} \text{ and } \cot x = \frac{1}{\tan x}$$

$$= -\csc x \cot x$$

Differentiate exp(x) exercise (page 101)

Q55: Use the chain rule with $u = 5x$ and $y = e^u$

Hence,

$$\frac{dy}{dx} = \frac{dy}{du} \times \frac{du}{dx}$$
$$= e^u \times 5$$
$$= 5e^{5x}$$

Q56: Use the chain rule with $u = \cot x$ then $y = e^u$

Hence,

$$\frac{dy}{dx} = \frac{dy}{du} \times \frac{du}{dx}$$
$$= e^u \times -\csc^2 x$$
$$= -\csc^2 x\, e^{\cot x}$$

Q57: Use the chain rule with $u = 6x - 1$ and $y = e^u$

Hence,

$$\frac{dy}{dx} = \frac{dy}{du} \times \frac{du}{dx}$$
$$= e^u \times 6$$
$$= 6e^{(6x-1)}$$

ANSWERS: UNIT 1 TOPIC 2

Q58:
Use the product rule where,
$f = x^2$ and $g = e^{\cos x}$
$f' = 2x$ and $g' = -\sin x e^{\cos x}$
Hence,
$$\begin{aligned} y' &= f'g + fg' \\ &= 2x e^{\cos x} - x^2 \sin x e^{\cos x} \\ &= x e^{\cos x}(2 - x \sin x) \end{aligned}$$

Q59:
Use the chain rule with $u = \tan 3x$ and $y = e^u$
Hence,
$$\begin{aligned} \frac{dy}{dx} &= \frac{dy}{du} \times \frac{du}{dx} \\ &= e^u \times \sec^2 3x \times 3 \\ &= 3\sec^2 3x e^{\tan 3x} \end{aligned}$$

Q60:
Use the product rule where,
$f = 3 - 4x$ and $g = e^{4x^3}$
$f' = -4$ and $g' = 12x^2 e^{4x^3}$
Hence,
$$\begin{aligned} y' &= f'g + fg' \\ &= -4e^{4x^3} + 12x^2 e^{4x^3}(3 - 4x) \\ &= -4e^{4x^3}\left(12x^3 - 9x^2 + 1\right) \end{aligned}$$

Q61:
Product rule: $k' = f'g + fg'$
Let $f = \cos^2 x$ and $g = e^{\tan x}$
Then $f' = -2\cos x \sin x = -\sin(2x)$ and $g' = \sec^2 x\, e^{\tan x}$
So:
$$\begin{aligned} k' &= -\sin 2x e^{\tan x} + \cos^2 x \sec^2 x e^{\tan x} \\ &= e^{\tan x}(1 - \sin 2x) \end{aligned}$$
Double angle formula: $\sin 2x = 2 \sin x \cos x$
$\sec x = \frac{1}{\cos x}$

Q62:
Use the product rule, $k' = f'g + fg'$, where,
$f = 5x$ and $g = e^x$
$f' = 5$ and $g' = e^x$

© HERIOT-WATT UNIVERSITY

Hence,
$$y' = f'g + fg'$$
$$= 5e^x + 5xe^x$$
$$= 5e^x(1+x)$$
$$\frac{dy}{dx} = 5e^x(1+x)$$

For stationary points $\frac{dy}{dx} = 0$

$$5e^x(1+x) = 0$$
$$(x+1) = 0$$
$$x = -1$$

Nature table to identify stationary point:

x	$\rightarrow$	-1	$\rightarrow$
$\frac{dy}{dx}$	-	0	+
slope	↗	→	↘

There is a maximum turning point at $\left(-1, -\frac{5}{e^x}\right)$

Differentiating ln(x) exercise (page 105)

Q63:
$$\frac{dy}{dx} = 5 \times \frac{1}{5x-2}$$
$$= \frac{5}{5x-2}$$

Q64:
Apply logarithmic rules first: $\log_a x^n = n \log_a x$
$$y = \ln\left|\sqrt{x^2+3}\right|$$
$$= \ln\left|x^2+3\right|^{\frac{1}{2}}$$
$$= \frac{1}{2}\ln\left|x^2+3\right|$$

Apply $\frac{dy}{dx}\ln|f(x)| = f'(x) \times \frac{1}{f(x)}$

$$\frac{dy}{dx} = \frac{1}{2}\frac{d}{dx}\left|\ln\left|x^2+3\right|\right|$$
$$= \frac{1}{2}\left|2x \times \frac{1}{x^2+3}\right|$$
$$= \frac{1}{2}\left|\frac{2x}{x^2+3}\right|$$
$$= \frac{x}{x^2+3}$$

//ANSWERS: UNIT 1 TOPIC 2// 431

Q65:

Apply the quotient rule: $y' = \frac{u'v - uv'}{v^2}$

$u = x^3$ and $v = \ln x$

$u' = 3x^2$ and $v' = \frac{1}{x}$

Then apply: $\frac{dy}{dx} \ln |f(x)| = f'(x) \times \frac{1}{f(x)}$

$$\frac{dy}{dx} = \frac{3x^2(\ln x) - x^3\left(\frac{1}{x}\right)}{(\ln x)^2}$$

$$= \frac{3x^2 \ln x - x^2}{(\ln x)^2}$$

Q66:

Apply: $\frac{dy}{dx} \ln |f(x)| = f'(x) \times \frac{1}{f(x)}$

$$\frac{dy}{dx} = \sec^2 x \times \frac{1}{\tan x}$$

$$= \frac{1}{\cos^2 x} \times \frac{\cos x}{\sin x}$$

$$= \frac{1}{\cos x \, \sin x}$$

$$= \sec x \, \csc x$$

Reminder: $\frac{d}{dx}(\tan x) = \sec^2 x$, $\sec x = \frac{1}{\cos x}$ and $\csc x = \frac{1}{\cos x}$

Logarithmic differentiation exercise (page 107)

Q67:

Apply the natural log and its rules:

$\ln |y| = \ln \left|(2x+5)^3 \left(x^2+1\right)^{-2}\right|$

$\ln |y| = 3\ln |2x+5| - 2\ln |x^2+1|$

Differentiating:

$$\frac{1}{y}\frac{dy}{dx} = \frac{3}{2x+5} \times 2 - \frac{2}{x^2+1} \times 2x$$

$$\frac{dy}{dx} = y\left(\frac{6}{2x+5} - \frac{4x}{x^2+1}\right)$$

$$\frac{dy}{dx} = \frac{(2x+5)^3}{(x^2+1)^2}\left(\frac{6}{2x+5} - \frac{4x}{x^2+1}\right)$$

$$\frac{dy}{dx} = \cdots$$

© HERIOT-WATT UNIVERSITY

$$\frac{1}{y}\frac{dy}{dx} = \cdots$$

$$\frac{dy}{dx} = \frac{(2x+5)^3}{(x^2+1)^2}\left(\frac{6(x^2+1) - 4x(2x+5)}{(2x+5)(x^2+1)}\right)$$

$$\frac{dy}{dx} = -\frac{2(2x+5)^3}{(x^2+1)^2}\left(\frac{x^2+10x-3}{(2x+5)(x^2+1)}\right)$$

$$\frac{dy}{dx} = -\frac{2(2x+5)^2(x^2+10x-3)}{(x^2+1)^3}$$

Q68:
Apply the natural log and its rules:

$$\ln|y| = \ln\left|\frac{\sqrt{5-2x}}{(x+1)^2}\right|$$

$$\ln|y| = \frac{1}{2}\ln|5-2x| - 2\ln|x+1|$$

Differentiating:

$$\frac{1}{y}\frac{dy}{dx} = \frac{-2}{2(5-2x)} - \frac{2}{x+1}$$

$$\frac{dy}{dx} = y\left(\frac{-1}{(5-2x)} - \frac{2}{x+1}\right)$$

$$\frac{dy}{dx} = -\frac{\sqrt{5-2x}}{(x+1)^2}\left(\frac{1}{5-2x} + \frac{2}{x+1}\right)$$

$$\frac{dy}{dx} = -\frac{\sqrt{5-2x}}{(x+1)^2}\left(\frac{x+1+2(5-2x)}{(5-2x)(x+1)}\right)$$

$$\frac{dy}{dx} = \frac{3x-11}{(x+1)^3(5-2x)^{\frac{1}{2}}}$$

Q69:
Apply $\ln(x)$ to both sides: $\ln|y| = \ln|x^x|$
Apply logarithmic rules: $\ln|y| = x\ln|x|$
Apply Implicit differentiation (and the product rule): $\frac{1}{y}\frac{dy}{dx} = \ln|x| + x \times \frac{1}{x}$
Simplify and rearrange for $\frac{dy}{dx}$: $\frac{dy}{dx} = y(\ln|x| + 1)$
Replace y: $\frac{dy}{dx} = x^x(\ln|x| + 1)$

Q70:
Apply $\ln(x)$ to both sides: $\ln|y| = \ln\left|\ln x^{\ln x}\right|$
Apply logarithmic rules: $\ln|y| = \ln x \ln|\ln x|$
Apply Implicit differentiation (and the product rule): $\frac{1}{y}\frac{dy}{dx} = \frac{1}{x}\ln|\ln x| + \ln x \times \frac{1}{x} \times \frac{1}{\ln x}$
Simplify and rearrange for $\frac{dy}{dx}$: $\frac{dy}{dx} = y\left(\frac{1}{x}\ln|\ln x| + \frac{1}{x}\right)$
Replace y:

$$\frac{dy}{dx} = \ln x^{\ln x}\left(\frac{1}{x}\ln|\ln x| + \frac{1}{x}\right)$$

$$= \frac{\ln x^{\ln x}}{x}(\ln|\ln x| + 1)$$

ANSWERS: UNIT 1 TOPIC 2   433

Q71:

Apply $\ln(x)$ to both sides: $\ln|y| = \ln\left|\dfrac{(2x+1)^{\frac{1}{2}}(3x-1)^{\frac{2}{3}}}{(4x+3)^{\frac{3}{4}}}\right|$

Apply logarithmic rules:

$\ln|y| = \ln\left|(2x+1)^{\frac{1}{2}}\right| + \ln\left|(3x-1)^{\frac{2}{3}}\right| - \ln\left|(4x+3)^{\frac{3}{4}}\right|$

$\ln|y| = \dfrac{1}{2}\ln|2x+1| + \dfrac{2}{3}\ln|3x-1| - \dfrac{3}{4}\ln|4x+3|$

Apply implicit differentiation: $\dfrac{1}{y}\dfrac{dy}{dx} = \dfrac{1}{2} \times 2 \times \dfrac{1}{2x+1} + \dfrac{2}{3} \times 3 \times \dfrac{1}{3x-1} - \dfrac{3}{4} \times 4 \times \dfrac{1}{4x+3}$

Simplify and rearrange for $\dfrac{dy}{dx}$: $\dfrac{dy}{dx} = y\left(\dfrac{1}{2x+1} + \dfrac{2}{3x-1} - \dfrac{3}{4x+3}\right)$

Replace y: $\dfrac{dy}{dx} = \dfrac{(2x+1)^{\frac{1}{2}}(3x-1)^{\frac{2}{3}}}{(4x+3)^{\frac{3}{4}}}\left(\dfrac{1}{2x+1} + \dfrac{2}{3x-1} - \dfrac{3}{4x+3}\right)$

Q72: Apply the natural log and its rules:

$\ln|y| = \ln\left|\dfrac{(2x+3)^2}{\sqrt{x+1}}\right|$

$\ln|y| = 2\ln|2x+3| - \dfrac{1}{2}\ln|x+1|$

Differentiating:

$\dfrac{1}{y}\dfrac{dy}{dx} = \dfrac{4}{2x+3} - \dfrac{1}{2(x+1)}$

$\dfrac{dy}{dx} = y\left(\dfrac{4}{2x+3} - \dfrac{1}{2(x+1)}\right)$

$\dfrac{dy}{dx} = \dfrac{(2x+3)^2}{\sqrt{x+1}}\left(\dfrac{4}{2x+3} - \dfrac{1}{2(x+1)}\right)$

$\dfrac{dy}{dx} = \dfrac{(2x+3)^2}{\sqrt{x+1}}\left(\dfrac{8(x+1)-(2x+3)}{2(2x+3)(x+1)}\right)$

$\dfrac{dy}{dx} = \dfrac{(2x+3)(6x+5)}{2\sqrt{(x+1)^3}}$

Q73:

Apply the natural log and its rules:

$\ln|y| = \ln\left|\dfrac{2^x}{2x+1}\right|$

$\ln|y| = x\ln|2| - \ln|2x+1|$

Differentiating:

$\dfrac{1}{y}\dfrac{dy}{dx} = \ln|2| - \dfrac{2}{(2x+1)}$

$\dfrac{dy}{dx} = y\left(\ln|2| - \dfrac{2}{(2x+1)}\right)$

$\dfrac{dy}{dx} = \dfrac{2^x}{2x+1}\left(\ln|2| - \dfrac{2}{(2x+1)}\right)$

© HERIOT-WATT UNIVERSITY

Q74:

Apply the natural log and its rules:

$$\ln|y| = \ln\left|\sqrt{\frac{3+x}{3-x}}\right|$$

$$\ln|y| = \frac{1}{2}\ln|3+x| - \frac{1}{2}\ln|3-x|$$

Differentiating:

$$\frac{1}{y}\frac{dy}{dx} = \frac{1}{2(3+x)} + \frac{1}{2(3-x)}$$

$$\frac{dy}{dx} = y\left(\frac{1}{2(3+x)} + \frac{1}{2(3-x)}\right)$$

$$\frac{dy}{dx} = \sqrt{\frac{3+x}{3-x}}\left(\frac{1}{2(3+x)} + \frac{1}{2(3-x)}\right)$$

$$\frac{dy}{dx} = \sqrt{\frac{3+x}{3-x}}\left(\frac{(3-x)+(3+x)}{2(3+x)(3-x)}\right)$$

$$\frac{dy}{dx} = \sqrt{\frac{3+x}{3-x}}\left(\frac{6}{2(3+x)(3-x)}\right)$$

$$\frac{dy}{dx} = \left(\frac{6}{2(3+x)^{\frac{1}{2}}(3-x)^{\frac{3}{2}}}\right)$$

$$\frac{dy}{dx} = \left(\frac{3}{(3+x)^{\frac{1}{2}}(3-x)^{\frac{3}{2}}}\right)$$

Q75:

Apply the natural log and its rules:

$$\ln|y| = \ln|(1+x)(2+3x)(x-5)|$$

$$\ln|y| = \ln|(1+x)(2+3x)(x-5)|$$

Differentiating:

$$\frac{1}{y}\frac{dy}{dx} = \frac{1}{(1+x)} + \frac{3}{(2+3x)} + \frac{1}{(x-5)}$$

$$\frac{dy}{dx} = (1+x)(2+3x)(x-5)\left(\frac{1}{(1+x)} + \frac{3}{(2+3x)} + \frac{1}{(x-5)}\right)$$

$$\frac{dy}{dx} = (1+x)(2+3x)(x-5)\left(\frac{(2+3x)(x-5)+3(1+x)(x-5)+(1+x)(2+3x)}{(1+x)(2+3x)(x-5)}\right)$$

$$\frac{dy}{dx} = (1+x)(2+3x)(x-5)\left(\frac{9x^2-20x-23}{(1+x)(2+3x)(x-5)}\right)$$

$$\frac{dy}{dx} = 9x^2 - 20x - 23$$

ANSWERS: UNIT 1 TOPIC 2   435

Q76:
Apply the natural log and its rules:
$$\ln|y| = \ln\left|\frac{x^2\sqrt{7x-3}}{1-x}\right|$$
$$\ln|y| = 2\ln|x| + \frac{1}{2}\ln|7x-3| - \ln|1-x|$$
Differentiating:
$$\frac{1}{y}\frac{dy}{dx} = \frac{2}{x} + \frac{7}{2(7x-3)} + \frac{1}{1-x}$$
$$\frac{dy}{dx} = y\left(\frac{2}{x} + \frac{7}{2(7x-3)} + \frac{1}{1-x}\right)$$
$$\frac{dy}{dx} = \frac{x^2\sqrt{7x-3}}{1-x}\left(\frac{2}{x} + \frac{7}{2(7x-3)} + \frac{1}{1-x}\right)$$
$$\frac{dy}{dx} = \frac{x^2\sqrt{7x-3}}{1-x}\left(\frac{4(7x-3)(1-x) + 7x(1-x) + 2x(7x-3)}{2x(7x-3)(1-x)}\right)$$
$$\frac{dy}{dx} = \frac{x^2\sqrt{7x-3}}{1-x}\left(\frac{-21x^2 + 41x - 12}{2x(7x-3)(1-x)}\right)$$
$$\frac{dy}{dx} = \frac{-x(21x^2 - 41x + 12)}{2(7x-3)^{\frac{1}{2}}(1-x)^2}$$

Explicit and implicit functions exercise (page 109)

Q77: b) Implicit

Q78: a) Explicit

Q79: a) Explicit

Q80: b) Implicit

Q81: b) Implicit

Q82: a) Explicit

Implicit differentiation exercise (page 114)

Q83:
$$\frac{d}{dx}(y^2 - x^2) = \frac{d}{dx}(12)$$
$$\frac{d}{dx}(y^2) - \frac{d}{dx}(x^2) = \frac{d}{dx}(12)$$
$$2y\frac{dy}{dx} - 2x = 0$$
$$\frac{dy}{dx} = \frac{2x}{2y}$$
$$\frac{dy}{dx} = \frac{x}{y}$$

© HERIOT-WATT UNIVERSITY

Q84:

$$\frac{d}{dx}\left(x^2 + xy + y^2\right) = \frac{d}{dx}(7)$$

$$\frac{d}{dx}(x^2) + \frac{d}{dx}(xy) + \frac{d}{dx}(y^2) = \frac{d}{dx}(7) \quad \text{Note: apply the product rule when differentiating xy}$$

$$2x + \left(y + x\frac{dy}{dx}\right) + 2y\frac{dy}{dx} = 0$$

$$(x + 2y)\frac{dy}{dx} = -(2x + y)$$

$$\frac{dy}{dx} = \frac{-(2x + y)}{x + 2y}$$

Q85:

$$\frac{d}{dx}(3y) = \frac{d}{dx}(2x^3 + \sin y)$$

$$\frac{d}{dx}(3y) = \frac{d}{dx}(2x^3) + \frac{d}{dx}(\sin y)$$

$$3\frac{dy}{dx} = 6x^2 + \cos y \frac{dy}{dx}$$

$$(3 - \cos y)\frac{dy}{dx} = 6x^2$$

$$\frac{dy}{dx} = \frac{6x^2}{3 - \cos y}$$

Q86:

$$\frac{d}{dx}(x \tan y) = \frac{d}{dx}(e^x)$$

$$\tan y + x\sec^2 y \frac{dy}{dx} = e^x$$

$$x\sec^2 y \frac{dy}{dx} = e^x - \tan y$$

$$\frac{dy}{dx} = \frac{e^x - \tan y}{x\sec^2 y}$$

Remember $\sec y = \frac{1}{\cos y}$ so we can simplify this.

$$\frac{dy}{dx} = \frac{1}{x}\cos^2 y \left[ex - \tan y\right]$$

$$= \frac{1}{x}\left[\cos^2 y \, e^x - \cos y \sin y\right]$$

Second derivative exercise (page 120)

Q87:

$\frac{dy}{dx} = -6x^2\sqrt{y}$

$\frac{d^2y}{dx^2} = 18x^4 - 12x\sqrt{y}$

ANSWERS: UNIT 1 TOPIC 2

Q88:
First derivative:
$$\tfrac{d}{dx}\left(2y^3\right) + \tfrac{d}{dx}\left(xy\right) = \tfrac{d}{dx}\left(3\right)$$
$$6y^2\tfrac{dy}{dx} + y + x\tfrac{dy}{dx} = 0$$
$$\tfrac{dy}{dx}\left(6y^2 + x\right) = -y$$
$$\tfrac{dy}{dx} = -\tfrac{y}{6y^2 + x}$$
$$\tfrac{dy}{dx} = -y\left(6y^2 + x\right)^{-1}$$

Second derivative: $\tfrac{d^2y}{dx^2} = -\tfrac{dy}{dx}\left(6y^2 + x\right)^{-1} - y(-1)\left(6y^2 + x\right)^{-2}\left(12y\tfrac{dy}{dx} + 1\right)$

Substitute for $\tfrac{dy}{dx} = -y\left(6y^2 + x\right)^{-1}$

$$\tfrac{d^2y}{dx^2} = \tfrac{y}{6y^2 + x}\left(6y^2 + x\right)^{-1} + \tfrac{y}{\left(6y^2 + x\right)^2}\left(-12y\tfrac{y}{6y^2 + x} + 1\right)$$

$$\tfrac{d^2y}{dx^2} = \tfrac{y}{\left(6y^2 + x\right)^2} - \tfrac{12y^3}{\left(6y^2 + x\right)^3} + \tfrac{y}{\left(6y^2 + x\right)^2}$$

$$\tfrac{d^2y}{dx^2} = \tfrac{2y}{\left(6y^2 + x\right)^2}\left(1 - \tfrac{6y^2}{6y^2 + x}\right)$$

Q89:
First derivative: $\tfrac{d}{dx}\left(5y^2\right) + \tfrac{d}{dx}\left(x^3\right) = \tfrac{d}{dx}\left(4\right)$

Apply the chain rule to $\tfrac{d}{dx}\left(5y^2\right)$:

$$10y\tfrac{dy}{dx} + 3x^2 = 0$$
$$\tfrac{dy}{dx} = -\tfrac{3x^2}{10y}$$

Second derivative: $\tfrac{d}{dx}\left(\tfrac{dy}{dx}\right) = \tfrac{d}{dx}\left(-\tfrac{3x^2}{10y}\right)$

Apply the quotient rule to the RHS:

$$\tfrac{d^2y}{dx^2} = -\tfrac{6x(10y) - 3x^2\left(10\tfrac{dy}{dx}\right)}{100y^2}$$

Simplify as much as possible:

$$\tfrac{d^2y}{dx^2} = -\tfrac{60xy - 30x^2\left(\tfrac{dy}{dx}\right)}{100y^2}$$

$$\tfrac{d^2y}{dx^2} = -\tfrac{6xy - 3x^2\left(\tfrac{dy}{dx}\right)}{10y^2}$$

Substitute for $\tfrac{dy}{dx}$

$$\tfrac{d^2y}{dx^2} = -\tfrac{6xy - 3x^2\left(-\tfrac{3x^2}{10y}\right)}{10y^2}$$

Multiply the numerator and denominator by $10y$: $\tfrac{d^2y}{dx^2} = -\tfrac{60xy^2 + 9x^4}{10y^2}$

© HERIOT-WATT UNIVERSITY

Q90:

First derivative: $\frac{d}{dx}(y^2) - \frac{d}{dx}(x^2y) = \frac{d}{dx}(x)$

Apply the chain rule to $\frac{d}{dx}(y^2)$ and the product rule to $\frac{d}{dx}(x^2y)$: $2y\frac{dy}{dx} - \left(2xy + x^2\frac{dy}{dx}\right) = 1$

Rearrange for $\frac{dy}{dx}$

$$\frac{dy}{dx}(2y - x^2) - 2xy = 1$$

$$\frac{dy}{dx} = \frac{1 + 2xy}{2y - x^2}$$

Second derivative:

Applying the quotient rule and remembering to apply the product rule to $2xy$:

$$\frac{d^2y}{dx^2} = \frac{\left(2y + 2x\frac{dy}{dx}\right)(2y - x^2) - (1 + 2xy)\left(2\frac{dy}{dx} - 2x\right)}{(2y - x^2)^2}$$

Substitute for $\frac{dy}{dx}$

$$\frac{d^2y}{dx^2} = \frac{\left(2y + 2x\left(\frac{1+2xy}{2y-x^2}\right)\right)(2y - x^2) - (1 + 2xy)\left(2\left(\frac{1+2xy}{2y-x^2}\right) - 2x\right)}{(2y - x^2)^2}$$

To simplify multiply numerator and denominator by $(2y - x^2)$:

$$\frac{d^2y}{dx^2} = \frac{\left(2y + 2x\left(\frac{1+2xy}{2y-x^2}\right)\right)(2y - x^2)^2 - (2y - x^2)(1 + 2xy)\left(2\left(\frac{1+2xy}{2y-x^2}\right) - 2x\right)}{(2y - x^2)^3}$$

$$\frac{d^2y}{dx^2} = \frac{2y(2y - x^2)^2 + 2x(1 + 2xy)(2y - x^2) - 2(1 + 2xy)^2 - 2x(2y - x^2)(1 + 2xy)}{(2y - x^2)^3}$$

Take out a factor of $(2y - x^2)$:

$$\frac{d^2y}{dx^2} = \frac{(2y - x^2)\left[2y(2y - x^2) + 2x(1 + 2xy) - 2x(1 + 2xy)\right] - 2(1 + 2xy)^2}{(2y - x^2)^3}$$

$$\frac{d^2y}{dx^2} = \frac{(2y - x^2)\left[4y^2 - 2x^2y + 2x + 4x^2y - 2x - 2x^2y\right] - 2(1 + 2xy)^2}{(2y - x^2)^3}$$

$$\frac{d^2y}{dx^2} = \frac{(2y - x^2)\left[4y^2\right] - 2\left(1 + 4xy + 4x^2y^2\right)}{(2y - x^2)^3}$$

$$\frac{d^2y}{dx^2} = \frac{8y^3 - 4x^2y^2 - 2 - 8xy - 8x^2y^2}{(2y - x^2)^3}$$

$$\frac{d^2y}{dx^2} = \frac{8y^3 - 12x^2y^2 - 8xy - 2}{(2y - x^2)^3}$$

ANSWERS: UNIT 1 TOPIC 2

Q91:

First derivative: $\frac{d}{dx}(2y^2) - \frac{d}{dx}(xy) = \frac{d}{dx}(3x^2) + \frac{d}{dx}(2)$

Apply the chain rule to $\frac{d}{dx}(2y^2)$ and the product rule to $\frac{d}{dx}(xy)$: $4y\frac{dy}{dx} - \left(y + x\frac{dy}{dx}\right) = 6x$

Factorise for $\frac{dy}{dx}$:

$$\frac{dy}{dx}(4y - x) = 6x + y$$

$$\frac{dy}{dx} = \frac{6x + y}{4y - x}$$

Second derivative:

Apply the quotient rule:

$$\frac{d}{dx}\left(\frac{dy}{dx}\right) = \frac{d}{dx}\left(\frac{6x + y}{4y - x}\right)$$

$$\frac{d^2y}{dx^2} = \frac{\left(6 + \frac{dy}{dx}\right)(4y - x) - (6x + y)\left(\frac{dy}{dx} - 1\right)}{(4y - x)^2}$$

Substitute for $\frac{dy}{dx}$:

$$\frac{d^2y}{dx^2} = \frac{\left(6 + \left(\frac{6x+y}{4y-x}\right)\right)(4y-x) - (6x+y)\left(\left(\frac{6x+y}{4y-x}\right) - 1\right)}{(4y-x)^2}$$

Simplify as much as possible:

$$\frac{d^2y}{dx^2} = \frac{(6(4y-x) + 6x+y) - \left(\left(\frac{(6x+y)^2}{4y-x}\right) - (6x+y)\right)}{(4y-x)^2}$$

Multiply the numerator and the denominator by $(4y - x)$:

$$\frac{d^2y}{dx^2} = \frac{6(4y - x)^2 + (6x + y)(4y - x) - \left((6x + y)^2 - (6x + y)(4y - x)\right)}{(4y - x)^3}$$

$$\frac{d^2y}{dx^2} = \frac{6(4y - x)^2 + (6x + y)(4y - x) - (6x + y)^2 + (6x + y)(4y - x)}{(4y - x)^3}$$

Take out a factor of $(4y - x)$:

$$\frac{d^2y}{dx^2} = \frac{(4y - x)[24y - 6x + 6x + y + 6x + y] - (6x + y)^2}{(4y - x)^3}$$

$$\frac{d^2y}{dx^2} = \frac{(4y - x)[26y + 6x] - (36x^2 + 12xy + y^2)}{(4y - x)^3}$$

$$\frac{d^2y}{dx^2} = \frac{104y^2 + 24xy - 26xy - 6x^2 - (36x^2 + 12xy + y^2)}{(4y - x)^3}$$

$$\frac{d^2y}{dx^2} = \frac{103y^2 - 14xy - 42x^2}{(4y - x)^3}$$

Q92:

$\frac{dy}{dx} = \frac{y - 3x^2}{2y - x}$

At (1,1) $\frac{dy}{dx} = -2$

$\frac{d^2y}{dx^2} = \left(\frac{dy}{dx} - 6x\right)(2y - x)^{-1} - (y - 3x^2)(2y - x)^{-2}\left(2\frac{dy}{dx} - 1\right)$

© HERIOT-WATT UNIVERSITY

At (1,0) $\frac{dy}{dx} = \frac{-3}{-1} = 3$

At (1,0) $\frac{d^2y}{dx^2} = \frac{3-6}{-1} - \frac{(-3)(2\times 3-1)}{(-1)^2} = 3 + 15 = 18$

Applying implicit differentiation to problems exercise (page 123)

Q93:

Use implicit differentiation:

$\frac{d}{dx}(xy) - \frac{d}{dx}(3y^2) = \frac{d}{dx}(-1)$

Apply the product rule to $\frac{d}{dx}(xy)$ and the chain rule to $\frac{d}{dx}(3y^2)$:

$\left(y + x\frac{dy}{dx}\right) - 6y\frac{dy}{dx} = 0$

Rearrange for $\frac{dy}{dx}$:

$\frac{dy}{dx}(x - 6y) = -y$

$\frac{dy}{dx} = -\frac{y}{x - 6y}$

Calculate the gradient by substituting for (2,1):

$\frac{dy}{dx} = -\frac{1}{2 - 6(1)}$

$\frac{dy}{dx} = \frac{1}{4}$

Substitute (2,1) and $m = \frac{1}{4}$ into the straight line formula:

$y - b = m(x - a)$

$y - 1 = \frac{1}{4}(x - 2)$

$4y - 4 = x - 2$

$4y = x + 2$

Q94:

Use implicit differentiation:

$\frac{d}{dx}(y^3) - \frac{d}{dx}(2xy) = \frac{d}{dx}(x^2) + \frac{d}{dx}(4y)$

Apply the chain rule to $\frac{d}{dx}(y^3)$ and $\frac{d}{dx}(4y)$.

Apply the product rule to $\frac{d}{dx}(2xy)$

$3y^2\frac{dy}{dx} - \left(2y + 2x\frac{dy}{dx}\right) = 2x + 4\frac{dy}{dx}$

Rearrange for $\frac{dy}{dx}$:

$\frac{dy}{dx}(3y^2 - 2x - 4) = 2x + 2y$

$\frac{dy}{dx} = \frac{2x + 2y}{3y^2 - 2x - 4}$

ANSWERS: UNIT 1 TOPIC 2

For the gradient substitute for (3,-1):
$$\frac{dy}{dx} = \frac{2(3)+2(-1)}{3(-1)^2 - 2(3) - 4}$$
$$\frac{dy}{dx} = -\frac{4}{7}$$

To find the equation of the tangent substitute $m = -\frac{4}{7}$ and (3,-1) into the equation of a straight line:
$$y - b = m(x - a)$$
$$y + 1 = -\frac{4}{7}(x - 3)$$
$$7y + 7 = -4x + 12$$
$$7y + 4x = 5$$

Q95:

Differentiate to obtain an expression for the gradient $\frac{dy}{dx}$.

Apply the product rule $y' = u'v + uv'$ to the LHS.

$$\frac{dy}{dx}\left(y(x+y)^2\right) = \frac{dy}{dx}\left(3(x^3 - 5)\right)$$
$$(x+y)^2 \frac{dy}{dx} + 2y(x+y)\left(1 + \frac{dy}{dx}\right) = 9x^2$$
$$(x^2 + 2xy + y^2)\frac{dy}{dx} + 2y\left(x + x\frac{dy}{dx} + y + y\frac{dy}{dx}\right) = 9x^2$$
$$(x^2 + 2xy + y^2)\frac{dy}{dx} + 2xy + 2xy\frac{dy}{dx} + 2y^2 + 2y^2\frac{dy}{dx} = 9x^2$$
$$(x^2 + 4xy + 3y^2)\frac{dy}{dx} = 9x^2 - 2xy - 2y^2$$
$$\frac{dy}{dx} = \frac{9x^2 - 2xy - 2y^2}{x^2 + 4xy + 3y^2}$$

Evaluate the derivative at (2,1):
$$\frac{dy}{dx} = \frac{9(2)^2 - 2(2)(1) - 2(1)^2}{(2)^2 + 4(2)(1) + 3(1)^2}$$
$$= \frac{36 - 4 - 2}{4 + 8 + 3}$$
$$= \frac{30}{15}$$
$$= 2$$

Substitute the gradient $m = 2$ and the coordinate (2,1) into the straight line equation:
$$y - b = m(x - a)$$
$$y - 1 = 2(x - 2)$$
$$y - 1 = 2x - 4$$
$$2x - y - 3 = 0$$

The equation of the tangent is given by $2x - y - 3 = 0$

Q96:

Using implicit differentiation:

$\frac{d}{dx}(3y^4) - \frac{d}{dx}(2x^2y) - \frac{d}{dx}(y) + \frac{d}{dx}(5x^2) = 0$

© HERIOT-WATT UNIVERSITY

Apply the chain rule to $\frac{d}{dx}\left(3y^4\right)$ and $\frac{d}{dx}\left(y\right)$

Apply the product rule to $\frac{d}{dx}\left(2x^2y\right)$: $12y^3\frac{dy}{dx} - \left(4xy + 2x^2\frac{dy}{dx}\right) - \frac{dy}{dx} + 10x = 0$

Rearrange for $\frac{dy}{dx}$ by factorising:

$$\frac{dy}{dx}\left(12y^3 - 2x^2 - 1\right) = -10x + 4xy$$

$$\frac{dy}{dx} = \frac{-10x + 4xy}{12y^3 - 2x^2 - 1}$$

For Stationary Points $\frac{dy}{dx} = 0$:

$$\frac{-10x + 4xy}{12y^3 - 2x^2 - 1} = 0$$

$$-10x + 4xy = 0$$

$$x(-10 + 4y) = 0$$

$$x = 0$$

Stationary point occurs when $x = 0$.

Q97:

Using implicit differentiation: $\frac{d}{dx}\left(xy^2\right) + \frac{d}{dx}\left(5x^2y\right) = \frac{d}{dx}\left(10\right)$

Apply the product rule to LHS as well as the chain rule to y^2 and y.

$$\left(y^2 + x \times 2y\frac{dy}{dx}\right) + \left(10xy + 5x^2\frac{dy}{dx}\right) = 0$$

Rearrange for $\frac{dy}{dx}$ by factorising:

$$\frac{dy}{dx}\left(2xy + 5x^2\right) = -10xy - y^2$$

$$\frac{dy}{dx} = \frac{-10xy - y^2}{2xy + 5x^2}$$

Horizontal tangents occur when $\frac{dy}{dx} = 0$:

$$\frac{-10xy - y^2}{2xy + 5x^2} = 0$$

$$-10xy - y^2 = 0$$

$$x = \frac{y^2}{-10y}$$

$$-10x = y$$

Substitute this into the original curve:

$$x(-10x)^2 + 5x^2(-10x) = 10$$

$$100x^3 - 50x^3 = 10$$

$$50x^3 = 10$$

$$x^3 = \frac{10}{50}$$

$$x^3 = \frac{1}{5}$$

$$x = \frac{1}{\sqrt[3]{5}}$$

ANSWERS: UNIT 1 TOPIC 2

Q98:

The tangent parallel to the x-axis will have a gradient of 0, i.e. $\frac{dy}{dx} = 0$
The tangent parallel to the y-axis will have an undefined gradient.
First we must differentiate to obtain an expression for $\frac{dy}{dx}$
Apply the product rule to the $\frac{d}{dx}(xy)$

$$\frac{dy}{dx}(x^2 - xy + y^2) = \frac{dy}{dx}(3)$$

$$2x - \left(y + x\frac{dy}{dx}\right) + 2y\frac{dy}{dx} = 0$$

$$2x - y - x\frac{dy}{dx} + 2y\frac{dy}{dx} = 0$$

$$(2y - x)\frac{dy}{dx} = y - 2x$$

$$\frac{dy}{dx} = \frac{y - 2x}{2y - x}$$

For the tangent parallel to the x-axis set $\frac{dy}{dx} = 0$ and solve:

$$\frac{y - 2x}{2y - x} = 0$$

$$y - 2x = 0$$

$$y = 2x$$

Substitute $y = 2x$ into the original function $x^2 - xy + y^2 = 3$:

$$x^2 - x(2x) + (2x)^2 = 3$$

$$x^2 - 2x^2 + 4x^2 = 3$$

$$3x^2 = 3$$

$$x^2 = 1$$

$$x = \pm 1$$

For $x = 1 \Rightarrow$ (1,2)
For $x = -1 \Rightarrow$ (-1,2)

For the tangent parallel to the y-axis, set $\frac{dy}{dx}$ is undefined. This happens when the denominator is equal to zero so set the denominator of $\frac{dy}{dx} = 0$ and solve:

$$2y - x = 0$$

$$x = 2y$$

Substitute $x = 2y$ into the original function $x^2 - xy + y^2 = 3$:

$$(2y)^2 - (2y)y + y^2 = 3$$

$$4y^2 - 2y^2 + y^2 = 3$$

$$3y^2 = 3$$

$$y^2 = 1$$

$$y = \pm 1$$

For $y = 1 \Rightarrow$ (2,1)
For $y = -1 \Rightarrow$ (-2,-1)

© HERIOT-WATT UNIVERSITY

Q99:

Differentiate to obtain an expression for $\frac{dy}{dx}$ and apply the product rule to the $\frac{d}{dx}(xy)$ $y' = u'v + uv'$:

$$\frac{d}{dx}\left(2x^2 - xy + 3y^2\right) = \frac{d}{dx}(46)$$

$$4x - \left(y + x\frac{dy}{dx}\right) + 6y\frac{dy}{dx} = 0$$

$$4x - y - x\frac{dy}{dx} + 6y\frac{dy}{dx} = 0$$

$$(6y - x)\frac{dy}{dx} = y - 4x$$

$$\frac{dy}{dx} = \frac{y - 4x}{6y - x}$$

Stationary points occur when $\frac{dy}{dx} = 0$:

$$\frac{y - 4x}{6y - x} = 0$$

$$y - 4x = 0$$

$$y = 4x$$

Substitute $y = 4x$ into $2x^2 - xy + 3y^2 = 46$:

$$2x^2 - x(4x) + 3(4x)^2 = 46$$

$$2x^2 - 4x^2 + 48x^2 = 46$$

$$46x^2 = 46$$

$$x^2 = 1$$

$$x = \pm 1$$

The x-coordinates of the stationary points are $x = 1$ and $x = -1$.

Astroid (page 124)

Q100:

a)

$$\frac{dy}{dx}(x^{\frac{2}{3}} + y^{\frac{2}{3}}) = \frac{dy}{dx}(2^{\frac{2}{3}})$$

$$\frac{2}{3}x^{-\frac{1}{3}} + \frac{2}{3}y^{-\frac{1}{3}}\frac{dy}{dx} = 0$$

$$x^{-\frac{1}{3}} + y^{-\frac{1}{3}}\frac{dy}{dx} = 0$$

$$\frac{dy}{dx} = -\frac{x^{-\frac{1}{3}}}{y^{-\frac{1}{3}}}$$

$$= -\left(\frac{y}{x}\right)^{\frac{1}{3}}$$

b)

At $\left(\frac{1}{\sqrt{2}}, \frac{1}{\sqrt{2}}\right)$ $\frac{dy}{dx} = -1^{\frac{1}{3}} = -1$

ANSWERS: UNIT 1 TOPIC 2

Q101: (0,2), (2,0), (0,-2), (-2,0)

Differentiating inverse functions exercise (page 129)

Q102:
Let $y = f^{-1}(x)$ so $f(y) = 3e^{2y} + 4$.
Then $f'(y) = 6e^{2y}$.
Using the result: $\frac{d}{dx}\left(f^{-1}(x)\right) = \frac{1}{f'(y)}$
We have: $\frac{d}{dx}\left(f^{-1}(x)\right) = \frac{1}{6e^{2y}}$
Since y is a function of x, we have: $x = 3e^{2y} + 4$
We can rearrange this for e^{2y}: $e^{2y} = \frac{x-4}{3}$
Now substitute for this in the derivative: $\frac{dy}{dx}\left(f^{-1}(x)\right) = \frac{1}{6\left(\frac{x-4}{3}\right)}$
Which simplifies to: $\frac{d}{dx}\left(f^{-1}(x)\right) = \frac{1}{2x-8}$

Q103:

a) Let $y = f^{-1}(x)$ so $f(y) = y^3 + 4y + 2$.
Then $f'(y) = 3y^2 + 4$.
Using the result: $\frac{d}{dx}\left(f^{-1}(x)\right) = \frac{1}{f'(y)}$
We have: $\frac{d}{dx}\left(f^{-1}(x)\right) = \frac{1}{3y^2+4}$

b) The gradient of the tangent is given by $m = \frac{1}{3y^2+4}$
Substituting in the y point of contact, $y = 1$, gives:
$$m = \frac{1}{3(1)^2 + 4}$$
$$= \frac{1}{7}$$
The equation of a straight line is given by: $y - b = m(x - a)$
Substituting for the point (7,1) and $m = \frac{1}{7}$ we have:
$$y - 1 = \frac{1}{7}(x - 7)$$
$$7y - x = 0$$

Q104:

Steps:

- What is the expression for $f'(y)$? $f'(y) = \frac{5}{4}y^{\frac{1}{4}}$
- What is the derivative of the inverse in terms of y?
$\frac{d}{dx}\left(f^{-1}(x)\right) = \frac{4}{5y^{\frac{1}{4}}}$

Answer: $\frac{d}{dx}\left(f^{-1}(x)\right) = \frac{4}{5x^{\frac{1}{5}}}$

© HERIOT-WATT UNIVERSITY

Q105:
Steps:
- What is the expression for $f'(y)$? $f'(y) = -\frac{8}{y^3}$
- What is the derivative of the inverse in terms of y? $\frac{d}{dx}\left(f^{-1}(x)\right) = -\frac{y^3}{8}$

Answer: $\frac{d}{dx}\left(f^{-1}(x)\right) = -\frac{1}{x^{\frac{3}{2}}}$

Q106:
Steps:
- What is the expression for $f'(y)$? $f'(y) = 2y + 4$

Answer: $\frac{d}{dx}\left(f^{-1}(x)\right) = \frac{1}{2y+4}$

Q107:
Steps:
- What is the expression for $f'(y)$? $f'(y) = 15e^{3y}$
- What is the derivative of the inverse in terms of y? $\frac{d}{dx}\left(f^{-1}(x)\right) = \frac{1}{15e^{3y}}$

Answer: $\frac{d}{dx}\left(f^{-1}(x)\right) = \frac{1}{3x+18}$

Q108:
a) *Steps:*
- What is the expression for $f'(y)$? $f'(y) = 3y^2 - 6$

Answer: $\frac{d}{dx}\left(f^{-1}(x)\right) = \frac{1}{3y^2-6}$

b) *Steps:*
- What is the gradient of the tangent? $m = -\frac{1}{6}$

Answer: $6y + x - 2 = 0$

Q109:
a) *Steps:*
- What is the expression for $f'(y)$? $f'(y) = 3y^2 + 10y - 1$

Answer: $\frac{d}{dx}\left(f^{-1}(x)\right) = \frac{1}{3y^2+10y-1}$

b) *Steps:*
- What is the gradient of the tangent? $m = \frac{1}{124}$

Answer: $124y - x - 619 = 0$

Q110:
a) *Steps:*
- What is the expression for $f'(y)$? $f'(y) = 6y^2 - 5$

Answer: $\frac{d}{dx}\left(f^{-1}(x)\right) = \frac{1}{6y^2-5}$

… ANSWERS: UNIT 1 TOPIC 2

b) **Steps:**
- What is the gradient of the tangent? $m = -\frac{1}{5}$

Answer: $5y + x - 55 = 0$

Differentiating inverse trigonometric functions exercise (page 135)

Q111:
Using the standard result of $\frac{d}{dx}\left(\sin^{-1}x\right) = \frac{1}{\sqrt{1-x^2}}$
Now let $u = 3x$ so that $y = \sin^{-1}u$ and use the chain rule $\frac{dy}{dx} = \frac{dy}{du} \times \frac{du}{dx}$
$\frac{dy}{du} = \frac{1}{\sqrt{1-u^2}}$ and $\frac{du}{dx} = 3$
So $\frac{dy}{dx} = \frac{1}{\sqrt{1-u^2}} \times 3$
Substituting $u = 3x$ back in we have:
$$\frac{dy}{dx} = \frac{3}{\sqrt{1-(3x)^2}}$$
$$\frac{dy}{dx} = \frac{3}{\sqrt{1-9x^2}}$$

Q112: Use the chain rule to perform this differentiation.
Let $u = \frac{x}{3}$ and $y = \sin^{-1}u$
So $\frac{du}{dx} = \frac{1}{3}$ and $\frac{dy}{du} = \frac{1}{\sqrt{1-u^2}}$
Using the chain rule: $\frac{dy}{dx} = \frac{dy}{du} \times \frac{du}{dx}$
Substituting in: $\frac{dy}{dx} = \frac{1}{\sqrt{1-u^2}} \times \frac{1}{3}$
Replacing $u = \frac{x}{3}$, gives: $\frac{dy}{dx} = \frac{1}{3\sqrt{1-\frac{x^2}{9}}} = \frac{1}{3\sqrt{\frac{9-x^2}{9}}}$
Rearranging: $\frac{dy}{dx} = \frac{1}{\frac{3}{3}\sqrt{9-x^2}}$
$\frac{dy}{dx} = \frac{1}{\sqrt{9-x^2}}$

Q113:
Use the chain rule to perform this differentiation.
Let $u = e^x$ and $y = \cos^{-1}u$.
So $\frac{du}{dx} = e^x$ and $\frac{dy}{du} = -\frac{1}{\sqrt{1-u^2}}$
Using the chain rule: $\frac{dy}{dx} = \frac{dy}{du} \times \frac{du}{dx}$
Substituting in: $\frac{dy}{dx} = -\frac{1}{\sqrt{1-u^2}} \times e^x$
Replacing $u = e^x$:
$\frac{dy}{dx} = -\frac{e^x}{\sqrt{1-e^{2x}}}$

Q114:
Using the standard result of $\frac{d}{dx}\left(\cos^{-1}x\right) = -\frac{1}{\sqrt{1-x^2}}$

© HERIOT-WATT UNIVERSITY

Now let $u = 4x$ so that $y = \cos^{-1} u$ and use the chain rule $\frac{dy}{dx} = \frac{dy}{du} \times \frac{du}{dx}$
$\frac{dy}{du} = -\frac{1}{\sqrt{1-u^2}}$ and $\frac{du}{dx} = 4$
So $\frac{dy}{dx} = -\frac{1}{\sqrt{1-u^2}} \times 4$
Substituting $u = 4x$ back in we have:
$$\frac{dy}{dx} = -\frac{4}{\sqrt{1-(4x)^2}}$$
$$\frac{dy}{dx} = -\frac{4}{\sqrt{1-16x^2}}$$

Q115:
Using the standard result of $\frac{d}{dx}\left(\tan^{-1} x\right) = \frac{1}{1+x^2}$
Now let $u = 5x$ so that $y = \tan^{-1} u$ and use the chain rule $\frac{dy}{dx} = \frac{dy}{du} \times \frac{du}{dx}$
$\frac{dy}{du} = \frac{1}{1+u^2}$ and $\frac{du}{dx} = 5$
So $\frac{dy}{dx} = \frac{1}{1+u^2} \times 5$
Substituting $u = 5x$ back in we have:
$$\frac{dy}{dx} = \frac{5}{\left(1+(5x)^2\right)}$$
$$\frac{dy}{dx} = \frac{5}{(1+25x^2)}$$

Q116:
Using the standard result of $\frac{d}{dx}\left(\sin^{-1} x\right) = \frac{1}{\sqrt{1-x^2}}$
Now let $u = \frac{x}{6}$ so that $y = \sin^{-1} u$ and use the chain rule $\frac{dy}{dx} = \frac{dy}{du} \times \frac{du}{dx}$
$\frac{dy}{du} = \frac{1}{\sqrt{1-u^2}}$ and $\frac{du}{dx} = \frac{1}{6}$
So $\frac{dy}{dx} = \frac{1}{\sqrt{1-u^2}} \times \frac{1}{6}$
Substituting $u = \frac{x}{6}$ back in we have:
$$\frac{dy}{dx} = \frac{1}{6\sqrt{1-\left(\frac{x}{6}\right)^2}}$$
$$\frac{dy}{dx} = \frac{1}{6\sqrt{1-\frac{x^2}{36}}}$$
To simplify get a common denominator for $1 - \frac{x^2}{36}$
This becomes: $\frac{36-x^2}{36}$
We have:
$$\frac{dy}{dx} = \frac{1}{6\sqrt{\frac{36-x^2}{36}}}$$
Take out the common denominator of 36 outside the square root.
$$\frac{dy}{dx} = \frac{1}{\frac{6}{6}\sqrt{36-x^2}}$$
$$\frac{dy}{dx} = \frac{1}{\sqrt{36-x^2}}$$

Q117:

Using the standard result of $\frac{d}{dx}\left(\cos^{-1}x\right) = -\frac{1}{\sqrt{1-x^2}}$

Now let $u = \frac{x}{3}$ so that $y = \cos^{-1}u$ and use the chain rule $\frac{dy}{dx} = \frac{dy}{du} \times \frac{du}{dx}$

$\frac{dy}{du} = -\frac{1}{\sqrt{1-u^2}}$ and $\frac{du}{dx} = \frac{1}{3}$

So $\frac{dy}{dx} = -\frac{1}{\sqrt{1-u^2}} \times \frac{1}{3}$

Substituting $u = \frac{x}{3}$ back in we have:

$$\frac{dy}{dx} = -\frac{1}{3\sqrt{1-\left(\frac{x}{3}\right)^2}}$$

$$\frac{dy}{dx} = -\frac{1}{3\sqrt{1-\frac{x^2}{9}}}$$

To simplify get a common denominator for $1 - \frac{x^2}{9}$

This becomes: $\frac{9-x^2}{9}$

We have:

$$\frac{dy}{dx} = -\frac{1}{3\sqrt{\frac{9-x^2}{9}}}$$

Take out the common denominator of 9 outside the square root.

$$\frac{dy}{dx} = -\frac{1}{\frac{3}{9}\sqrt{9-x^2}}$$

$$\frac{dy}{dx} = -\frac{1}{\sqrt{9-x^2}}$$

Q118:

Using the standard result of $\frac{d}{dx}\left(\tan^{-1}x\right) = \frac{1}{1+x^2}$

Now let $u = \frac{x}{2}$ so that $y = \tan^{-1}u$ and use the chain rule $\frac{dy}{dx} = \frac{dy}{du} \times \frac{du}{dx}$

$\frac{dy}{du} = \frac{1}{1+u^2}$ and $\frac{du}{dx} = \frac{1}{2}$

So $\frac{dy}{dx} = \frac{1}{1+u^2} \times \frac{1}{2}$

Substituting $u = \frac{x}{2}$ back in we have:

$$\frac{dy}{dx} = -\frac{1}{2\left(1+\left(\frac{x}{2}\right)^2\right)}$$

$$\frac{dy}{dx} = -\frac{1}{2\left(1+\frac{x^2}{4}\right)}$$

$$\frac{dy}{dx} = -\frac{2}{x^2+4}$$

Q119:

Using the standard result of $\frac{d}{dx}\left(\sin^{-1}x\right) = \frac{1}{\sqrt{1-x^2}}$

Now let $u = \frac{3}{x}$ so that $y = \sin^{-1}u$ and use the chain rule $\frac{dy}{dx} = \frac{dy}{du} \times \frac{du}{dx}$

$\frac{dy}{du} = \frac{1}{\sqrt{1-u^2}}$ and $\frac{du}{dx} = -\frac{3}{x^2}$

So $\frac{dy}{dx} = \frac{1}{\sqrt{1-u^2}} \times -\frac{3}{x^2}$

Substituting $u = \frac{3}{x}$ back in we have:

$$\frac{dy}{dx} = -\frac{3}{x^2\sqrt{1-\left(\frac{3}{x}\right)^2}}$$

$$\frac{dy}{dx} = -\frac{3}{x^2\sqrt{1-\frac{9}{x^2}}}$$

To simplify get a common denominator for $1 - \frac{9}{x^2}$

This becomes: $\frac{x^2-9}{x^2}$

We have:

$$\frac{dy}{dx} = -\frac{3}{x^2\sqrt{\frac{x^2-9}{x^2}}}$$

Take out the common denominator of x^2 outside the square root.

$$\frac{dy}{dx} = -\frac{3}{\frac{x^2}{x}\sqrt{x^2-9}}$$

$$\frac{dy}{dx} = -\frac{3}{x\sqrt{x^2-9}}$$

Q120:

Using the standard result of $\frac{d}{dx}\left(\cos^{-1}x\right) = -\frac{1}{\sqrt{1-x^2}}$

Now let $u = \frac{4}{x}$ so that $y = \cos^{-1}u$ and use the chain rule $\frac{dy}{dx} = \frac{dy}{du} \times \frac{du}{dx}$

$\frac{dy}{du} = -\frac{1}{\sqrt{1-u^2}}$ and $\frac{du}{dx} = -\frac{4}{x^2}$

So $\frac{dy}{dx} = -\frac{1}{\sqrt{1-u^2}} \times -\frac{4}{x^2}$

Substituting $u = \frac{4}{x}$ back in we have:

$$\frac{dy}{dx} = \frac{4}{x^2\sqrt{1-\left(\frac{4}{x}\right)^2}}$$

$$\frac{dy}{dx} = \frac{4}{x^2\sqrt{1-\frac{16}{x^2}}}$$

To simplify get a common denominator for $1 - \frac{16}{x^2}$

This becomes: $\frac{x^2-16}{x^2}$

We have:

$$\frac{dy}{dx} = \frac{4}{x^2\sqrt{\frac{x^2-16}{x^2}}}$$

Take out the common denominator of x^2 outside the square root.

$$\frac{dy}{dx} = \frac{4}{\frac{x^2}{x}\sqrt{x^2-16}}$$

$$\frac{dy}{dx} = \frac{4}{x\sqrt{x^2-16}}$$

Q121:

Using the standard result of $\frac{d}{dx}\left(\tan^{-1}x\right) = \frac{1}{1+x^2}$

Now let $u = \frac{5}{x}$ so that $y = \tan^{-1}u$ and use the chain rule $\frac{dy}{dx} = \frac{dy}{du} \times \frac{du}{dx}$

$\frac{dy}{du} = \frac{1}{1+u^2}$ and $\frac{du}{dx} = -\frac{5}{x^2}$

So $\frac{dy}{dx} = \frac{1}{1+u^2} \times -\frac{5}{x^2}$

Substituting $u = \frac{5}{x}$ back in we have:

$\frac{dy}{dx} = -\dfrac{5}{x^2\left(1+\left(\frac{5}{x}\right)^2\right)}$

$\frac{dy}{dx} = -\dfrac{5}{x^2\left(1+\frac{25}{x^2}\right)}$

$\frac{dy}{dx} = -\dfrac{5}{x^2+25}$

Q122:

Using the standard result of $\frac{d}{dx}\left(\sin^{-1}x\right) = \frac{1}{\sqrt{1-x^2}}$

Now let $u = \frac{3x}{4}$ so that $y = \sin^{-1}u$ and use the chain rule $\frac{dy}{dx} = \frac{dy}{du} \times \frac{du}{dx}$

$\frac{dy}{du} = \frac{1}{\sqrt{1-u^2}}$ and $\frac{du}{dx} = \frac{3}{4}$

So $\frac{dy}{dx} = \frac{1}{\sqrt{1-u^2}} \times \frac{3}{4}$

Substituting $u = \frac{3x}{4}$ back in we have:

$\frac{dy}{dx} = \dfrac{3}{4\sqrt{1-\left(\frac{3x}{4}\right)^2}}$

$\frac{dy}{dx} = \dfrac{3}{4\sqrt{1-\frac{9x^2}{16}}}$

To simplify get a common denominator for $1 - \frac{9x^2}{16}$

This becomes: $\frac{16-9x^2}{16}$

We have:

$\frac{dy}{dx} = \dfrac{3}{4\sqrt{\frac{16-9x^2}{16}}}$

Take out the common denominator of 16 outside the square root.

$\frac{dy}{dx} = \dfrac{3}{\frac{4}{4}\sqrt{16-9x^2}}$

$\frac{dy}{dx} = \dfrac{3}{\sqrt{16-9x^2}}$

Q123:

Using the standard result of $\frac{d}{dx}\left(\cos^{-1}x\right) = -\frac{1}{\sqrt{1-x^2}}$

Now let $u = \frac{2x}{3}$ so that $y = \cos^{-1}u$ and use the chain rule $\frac{dy}{dx} = \frac{dy}{du} \times \frac{du}{dx}$

$\frac{dy}{du} = -\frac{1}{\sqrt{1-u^2}}$ and $\frac{du}{dx} = \frac{2}{3}$

So $\frac{dy}{dx} = -\frac{1}{\sqrt{1-u^2}} \times \frac{2}{3}$

© HERIOT-WATT UNIVERSITY

Substituting $u = \frac{2x}{3}$ back in we have:
$$\frac{dy}{dx} = -\frac{2}{3\sqrt{1 - \left(\frac{2x}{3}\right)^2}}$$
$$\frac{dy}{dx} = -\frac{2}{3\sqrt{1 - \frac{4x^2}{9}}}$$
To simplify get a common denominator for $1 - \frac{4x^2}{9}$
This becomes: $\frac{9-4x^2}{9}$
We have:
$$\frac{dy}{dx} = -\frac{2}{3\sqrt{\frac{9-4x^2}{9}}}$$
Take out the common denominator of 9 outside the square root.
$$\frac{dy}{dx} = -\frac{2}{\frac{9}{9}\sqrt{9-4x^2}}$$
$$\frac{dy}{dx} = -\frac{2}{\sqrt{9-4x^2}}$$

Q124:
Using the standard result of $\frac{d}{dx}\left(\tan^{-1} x\right) = \frac{1}{1+x^2}$
Now let $u = \frac{5x}{2}$ so that $y = \tan^{-1} u$ and use the chain rule $\frac{dy}{dx} = \frac{dy}{du} \times \frac{du}{dx}$
$\frac{dy}{du} = \frac{1}{1+u^2}$ and $\frac{du}{dx} = \frac{5}{2}$
So $\frac{dy}{dx} = \frac{1}{1+u^2} \times \frac{5}{2}$
Substituting $u = \frac{5x}{2}$ back in we have:
$$\frac{dy}{dx} = -\frac{5}{2\left(1 + \left(\frac{5x}{2}\right)^2\right)}$$
$$\frac{dy}{dx} = -\frac{5}{2\left(1 + \frac{25x^2}{4}\right)}$$
$$\frac{dy}{dx} = -\frac{5}{x^2 + 25}$$

Q125:
Using the standard result of $\frac{d}{dx}\left(\sin^{-1} x\right) = \frac{1}{\sqrt{1-x^2}}$
Now let $u = \frac{5}{2x}$ so that $y = \sin^{-1} u$ and use the chain rule $\frac{dy}{dx} = \frac{dy}{du} \times \frac{du}{dx}$
$\frac{dy}{du} = \frac{1}{\sqrt{1-u^2}}$ and $\frac{du}{dx} = -\frac{5}{2x^2}$
So $\frac{dy}{dx} = \frac{1}{\sqrt{1-u^2}} \times -\frac{5}{2x^2}$
Substituting $u = \frac{5}{2x}$ back in we have:
$$\frac{dy}{dx} = -\frac{5}{2x^2\sqrt{1 - \left(\frac{5}{2x}\right)^2}}$$
$$\frac{dy}{dx} = -\frac{5}{2x^2\sqrt{1 - \frac{25}{4x^2}}}$$

ANSWERS: UNIT 1 TOPIC 2

To simplify get a common denominator for $1 - \frac{25}{4x^2}$

This becomes: $\frac{4x^2 - 25}{4x^2}$

We have:

$\frac{dy}{dx} = -\frac{5}{2x^2 \sqrt{\frac{4x^2-25}{4x^2}}}$

Take out the common denominator of $4x^2$ outside the square root.

$\frac{dy}{dx} = -\frac{5}{\frac{4x^2}{2x}\sqrt{4x^2-25}}$

$\frac{dy}{dx} = -\frac{5}{2x\sqrt{4x^2-2}}$

Q126:

Using the standard result of $\frac{d}{dx}\left(\cos^{-1} x\right) = -\frac{1}{\sqrt{1-x^2}}$

Now let $u = \frac{3}{2x}$ so that $y = \cos^{-1} u$ and use the chain rule $\frac{dy}{dx} = \frac{dy}{du} \times \frac{du}{dx}$

$\frac{dy}{du} = -\frac{1}{\sqrt{1-u^2}}$ and $\frac{du}{dx} = -\frac{3}{2x^2}$

So $\frac{dy}{dx} = -\frac{1}{\sqrt{1-u^2}} \times -\frac{3}{2x^2}$

Substituting $u = \frac{3}{2x}$ back in we have:

$\frac{dy}{dx} = \frac{3}{2x^2 \sqrt{1-\left(\frac{3}{2x}\right)^2}}$

$\frac{dy}{dx} = \frac{3}{2x^2 \sqrt{1-\frac{9}{4x^2}}}$

To simplify get a common denominator for $1 - \frac{9}{4x^2}$

This becomes: $\frac{4x^2-9}{4x^2}$

We have:

$\frac{dy}{dx} = \frac{3}{2x^2 \sqrt{\frac{4x^2-9}{4x^2}}}$

Take out the common denominator of $4x^2$ outside the square root.

$\frac{dy}{dx} = \frac{3}{\frac{4x^2}{2x}\sqrt{4x^2-9}}$

$\frac{dy}{dx} = \frac{3}{2x\sqrt{4x^2-9}}$

Q127:

Using the standard result of $\frac{d}{dx}\left(\tan^{-1} x\right) = \frac{1}{1+x^2}$

Now let $u = \frac{2}{7x}$ so that $y = \tan^{-1} u$ and use the chain rule $\frac{dy}{dx} = \frac{dy}{du} \times \frac{du}{dx}$

$\frac{dy}{du} = \frac{1}{1+u^2}$ and $\frac{du}{dx} = -\frac{2}{7x^2}$

So $\frac{dy}{dx} = \frac{1}{1+u^2} \times -\frac{2}{7x^2}$

Substituting $u = \frac{2}{7x}$ back in we have:

© Heriot-Watt University

$$\frac{dy}{dx} = -\frac{2}{7x^2\left(1+\left(\frac{2}{7x}\right)^2\right)}$$

$$\frac{dy}{dx} = -\frac{2}{7x^2\left(1+\frac{4}{49x^2}\right)}$$

$$\frac{dy}{dx} = -\frac{14}{49x^2+4}$$

Q128:
Using the standard result of
Now let $u = 4x - 1$ so that $y = \sin^{-1}u$ and use the chain rule $\frac{dy}{dx} = \frac{dy}{du} \times \frac{du}{dx}$
$\frac{dy}{du} = \frac{1}{\sqrt{1-u^2}}$ and $\frac{du}{dx} = 4$
So $\frac{dy}{dx} = \frac{1}{\sqrt{1-u^2}} \times 4$
Substituting $u = 4x - 1$ back in we have:

$$\frac{dy}{dx} = \frac{4}{\sqrt{1-(4x-1)^2}}$$

$$\frac{dy}{dx} = \frac{4}{\sqrt{1-(16x^2-8x+1)}}$$

$$\frac{dy}{dx} = \frac{4}{\sqrt{8x-16x^2}}$$

Q129: $\frac{d}{dx}\left(\cos^{-1}x\right) = -\frac{1}{\sqrt{1-x^2}}$
Using the standard result of
Now let $u = 3\sin x$ so that $y = \cos^{-1}u$ and use the chain rule $\frac{dy}{dx} = \frac{dy}{du} \times \frac{du}{dx}$
$\frac{dy}{du} = -\frac{1}{\sqrt{1-u^2}}$ and $\frac{du}{dx} = 3\cos x$
So $\frac{dy}{dx} = -\frac{1}{\sqrt{1-u^2}} \times 3\cos x$
Substituting $u = 3\sin x$ back in we have:

$$\frac{dy}{dx} = -\frac{3\cos x}{\sqrt{1-(3\sin x)^2}}$$

$$\frac{dy}{dx} = -\frac{3\cos x}{\sqrt{1-9\sin^2 x}}$$

Q130:
Using the standard result of
Now let $u = \ln|2x|$ so that $y = \tan^{-1}u$ and use the chain rule $\frac{dy}{dx} = \frac{dy}{du} \times \frac{du}{dx}$
$\frac{dy}{du} = \frac{1}{1+u^2}$ and $\frac{du}{dx} = \frac{1}{x}$
So $\frac{dy}{dx} = \frac{1}{1+u^2} \times \frac{1}{x}$
Substituting $u = \ln|2x|$ back in we have:

$$\frac{dy}{dx} = \frac{1}{x\left(1+(\ln|2x|)^2\right)}$$

$$\frac{dy}{dx} = \frac{1}{x(1+\ln|2x|)}$$

ANSWERS: UNIT 1 TOPIC 2

Differentiating inverse trigonometric functions implicitly exercise (page 138)

Q131: When $y = \cos^{-1}(3x)$ then $\cos y = 3x$
Using implicit differentiation gives:
$$\frac{d}{dy}(\cos(y))\frac{dy}{dx} = \frac{d}{dx}(3x)$$
$$-\sin(y)\frac{dy}{dx} = 3$$
$$\frac{dy}{dx} = -\frac{3}{\sin(y)}$$
$\cos(y) = 3x$, therefore $\cos^2(y) = 9x^2$ and so:
$$\sin^2(y) = 1 - \cos^2(y)$$
$$= 1 - 9x^2$$
$$\sin(y) = \sqrt{1 - 9x^2}$$
Therefore, when $y = \cos^{-1}(3x)$, then:
$$\frac{dy}{dx} = -\frac{3}{\sin(y)}$$
$$= -\frac{3}{\sqrt{1 - 9x^2}}$$

Q132:
When $y = \sin^{-1}\left(\frac{3x^2}{2}\right)$, then $\sin(y) = \frac{3x^2}{2}$
Using implicit differentiation gives:
$$\frac{d}{dy}(\sin(y))\frac{dy}{dx} = \frac{d}{dx}\left(\frac{3x^2}{2}\right)$$
$$\cos(y)\frac{dy}{dx} = 3x$$
$$\frac{dy}{dx} = \frac{3x}{\cos(y)}$$
Now, since $\sin(y) = \frac{3x^2}{2}$, then:
$$\sin^2(y) = \frac{9x^4}{4}$$
$$\cos^2(y) = 1 - \sin^2(y)$$
$$= 1 - \frac{9x^4}{4}$$
$$= \frac{4 - 9x^4}{4}$$
$$\cos(y) = \sqrt{\frac{4 - 9x^4}{4}} = \frac{\sqrt{4 - 9x^4}}{2}$$
Therefore, when $y = \sin^{-1}\left(\frac{3x^2}{2}\right)$, then: $\frac{dy}{dx} = \frac{3x}{\cos(y)} = \frac{6x}{\sqrt{4 - 9x^4}}$

© HERIOT-WATT UNIVERSITY

Q133:

When $y = \tan^{-1}\left(\frac{x}{6}\right)$ then $\tan y = \frac{x}{6}$

Differentiating this equation implicitly gives,

$$\frac{d}{dx}(\tan y) = \frac{d}{dx}\left(\frac{x}{6}\right)$$

$$\sec^2 y \frac{dy}{dx} = \frac{1}{6}$$

$$\frac{dy}{dx} = \frac{1}{6\sec^2 y}$$

Since $1 + \tan^2 y = \sec^2 y$

Then,

$$\frac{1}{6\sec^2 y} = \frac{1}{6(1 + \tan^2 y)}$$

$$\frac{dy}{dx} = \frac{1}{6\left(1 + \left(\frac{x}{6}\right)^2\right)}$$

$$= \frac{1}{6\left(1 + \frac{x^2}{36}\right)}$$

$$= \frac{6}{(36 + x^2)}$$

Q134:

When $y = \sin^{-1}\left(\frac{5}{x}\right)$ then $\sin y = \frac{5}{x}$

Differentiating implicitly gives,

$$\frac{d}{dx}(\sin y) = \frac{d}{dx}\left(\frac{5}{x}\right)$$

$$\cos y \frac{dy}{dx} = -\frac{5}{x^2}$$

$$\frac{dy}{dx} = -\frac{5}{x^2 \cos y}$$

Since $x^2 \cos y = x^2\sqrt{1 - \sin^2 y}$

Then,

$$\frac{dy}{dx} = -\frac{5}{x^2\sqrt{1 - \sin^2 y}}$$

$$= -\frac{5}{x^2\sqrt{1 - \left(\frac{5}{x}\right)^2}}$$

$$= -\frac{5}{x^2\sqrt{1 - \frac{25}{x^2}}}$$

$$= -\frac{5}{x\sqrt{x^2 - 25}}$$

Q135:

When $y = \cos^{-1}\left(\sqrt{2-5x}\right)$ then $\cos y = \sqrt{2-5x}$
Differentiating implicitly gives,
$$\frac{d}{dx}(\cos y) = \frac{d}{dx}\left(\sqrt{2-5x}\right)$$
$$-\sin y \frac{dy}{dx} = -\frac{5}{2}(2-5x)^{-\frac{1}{2}}$$
$$\frac{dy}{dx} = \frac{5}{2\left(\sqrt{2-5x}\right)\sin y}$$
Since $\sin y = \sqrt{1-\cos^2 y}$
Then,
$$\frac{dy}{dx} = \frac{5}{2\left(\sqrt{2-5x}\right)\sqrt{1-\cos^2 y}}$$
$$= \frac{5}{2\left(\sqrt{2-5x}\right)\sqrt{1-\left(\sqrt{2-5x}\right)^2}}$$
$$= \frac{5}{2\left(\sqrt{2-5x}\right)\sqrt{1-(2-5x)}}$$
$$= \frac{5}{2\sqrt{2-5x}\sqrt{5x-1}}$$

Q136:

When $y = \tan^{-1}\left(\sqrt{x-1}\right)$ then $\tan y = \sqrt{x-1}$
Differentiating this equation implicitly gives,
$$\frac{d}{dx}(\tan y) = \frac{d}{dx}\left(\sqrt{x-1}\right)$$
$$\sec^2 y \frac{dy}{dx} = \frac{1}{2(x-1)^{\frac{1}{2}}}$$
$$\frac{dy}{dx} = \frac{1}{2(x-1)^{\frac{1}{2}}\sec^2 y}$$
Since $1 + \tan^2 y = \sec^2 y$
Then,
$$\frac{dy}{dx} = \frac{1}{2(x-1)^{\frac{1}{2}}(1+\tan^2 y)}$$
$$= \frac{1}{2(x-1)^{\frac{1}{2}}\left(1+\left(\sqrt{x-1}\right)^2\right)}$$
$$= \frac{1}{2(x-1)^{\frac{1}{2}}(x)}$$
$$= \frac{1}{2x(x-1)^{\frac{1}{2}}}$$

© HERIOT-WATT UNIVERSITY

Differentiation using the product, quotient and chain rules exercise (page 140)

Q137:
Use the product rule: $k' = f'g + fg'$
Let $f = x^3$ and $g = \tan 2x$
Then $f' = 3x^2$ and $g' = 2\sec^2 2x$
Substituting into the product rule: $k' = 3x^2 \tan 2x + x^3 2\sec^2 2x$
Tidying this up: $k' = 3x^2\tan 2x + 2x^3 \sec^2 2x$

Q138:
Applying the product rule:
Let $y = fg$ then $f(x) = \sqrt{3x}$ and $g(x) = \exp(-2x)$
So, $f'(x) = \frac{3}{2\sqrt{3x}}$ and $g'(x) = -2e^{-2x}$
Then,
$y' = f'g + fg'$
$y' = \frac{3}{2\sqrt{3x}} \times e^{-2x} + \sqrt{3x} \times \left(-2e^{-2x}\right)$
Gives,
$f'(x) = \frac{3}{2e^{2x}\sqrt{3x}} - \frac{2\sqrt{3x}}{e^{2x}}$
$= \frac{3}{2e^{2x}\sqrt{3x}} - \frac{2\sqrt{3x} \times 2\sqrt{3x}}{2e^{2x}\sqrt{3x}}$
$= \frac{3 - 12x}{2e^{2x}\sqrt{3x}}$

Q139:
Apply the product rule:
Let $y = fg$ then $f(x) = \sin^2 x$ and $g(x) = e^{\csc x}$
So, $f'(x) = 2\sin x \cos x$ and $g'(x) = -\csc x \cot x e^{\csc x}$
Then,
$y' = f'g + fg'$
$y' = 2\sin x \cos x \times e^{\csc x} + \sin^2 x \times \left(-\csc x \cot x e^{\csc x}\right)$
Remember that:
$\sin 2x = 2\sin x \cos x$

$\csc x = \frac{1}{\sin x}$

$\cot x = \frac{\cos x}{\sin x}$
Therefore,
$f'(x) = \sin 2x e^{\csc x} - \frac{\sin^2 x}{\sin x} \times \frac{\cos x}{\sin x} \times e^{\csc x}$
$f'(x) = \sin 2x e^{\csc x} - \cos x e^{\csc x}$

ANSWERS: UNIT 1 TOPIC 2

Q140:
Apply the quotient rule:
Let $y = \frac{f}{g}$ then $f(x) = \cos^{-1}(7x)$ and $g(x) = 2x - 1$
So, $f'(x) = -\frac{7}{\sqrt{1-49x^2}}$ and $g'(x) = 2$
Then,
$$y' = \frac{f'g - fg'}{g^2}$$
$$y' = \frac{-\frac{7}{\sqrt{1-49x^2}} \times (2x-1) - \left(\cos^{-1}(7x)\right) \times 2}{(2x-1)^2}$$
$$y' = \frac{-14x + 7 - 2\cos^{-1}(7x)\sqrt{1-49x^2}}{(2x-1)^2\sqrt{1-49x^2}}$$

Q141:
Apply the quotient rule:
Let $y = \frac{f}{g}$, then $f(x) = x^5$ and $g(x) = 1 - \tan(2x)$
So, $f'(x) = 5x^4$ and $g'(x) = -2\sec^2 2x$
Then,
$$y' = \frac{f'g - fg'}{g^2}$$
$$y' = \frac{5x^4 \times (1 - \tan(2x)) - (x^5) \times (-2\sec^2 2x)}{(1 - \tan(2x))^2}$$
$$y' = \frac{5x^4 - 5x^4 \tan(2x) - 2x^5 \sec^2 2x}{(1 - \tan(2x))^2}$$

Q142:
Apply the quotient rule:
Let $y = \frac{f}{g}$ then $f(x) = 6x^3$ and $g(x) = \ln|\cos x|$
So, $f'(x) = 18x^2$ and $g'(x) = -\frac{\sin x}{\cos x} = -\tan x$
Then,
$$y' = \frac{f'g - fg'}{g^2}$$
$$y' = \frac{18x^2 \times (\ln|\cos x|) - (6x^3) \times (-\tan x)}{(\ln|\cos x|)^2}$$
$$y' = \frac{18x^2 \ln|\cos x| + 6x^3 \tan x}{(\ln|\cos x|)^2}$$

Q143:
Apply the chain rule:
Let $y = f(g(h(x)))$ where $f(x) = \tan^{-1}(x)$, $g(x) = \cot x$ and $h(x) = 5x$
So, $f'(x) = \frac{1}{1+x^2}$, $g'(x) = -\csc^2 x$ and $h'(x) = 5$
Then,

© HERIOT-WATT UNIVERSITY

$y' = f'(g(h(x))) \times g'(h(x)) \times h'(x)$

$y' = \dfrac{1}{1 + (\cot 5x)^2} \times -\csc^2 5x \times 5$

$y' = -\dfrac{5\csc^2 5x}{1 + \cot^2 5x}$

Parametric curves exercise (page 146)

Q144:

a)
$x^2 + y^2 = (2\cos t)^2 + (2\sin t)^2$
$ = 4\cos^2 t + 4\sin^2 t$
$ = 4\left(\cos^2 t + \sin^2 t\right)$
$x^2 + y^2 = 4$

This is a circle with centre at the origin and radius 2.

b)
The circle starts at (2,0) and travels anti-clockwise round to the start again.

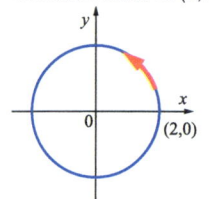

Q145:

a)
$x = 2t \Rightarrow t = \tfrac{x}{2}$

$y = 4t^2 \Rightarrow y = 4\left(\dfrac{x}{2}\right)^2$

$ y = 4\left(\dfrac{x^2}{4}\right)$

$ y = x^2$

b)

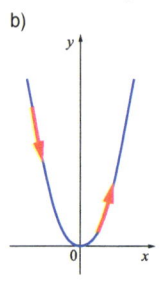

ANSWERS: UNIT 1 TOPIC 2

Q146:

a)

$x = 2t - 3 \Rightarrow 2t = x + 3$

$y = 4t - 3 \Rightarrow 4t = y + 3$

So,

$2x + 6 = y + 3 \Rightarrow y = 2x + 3$

b)

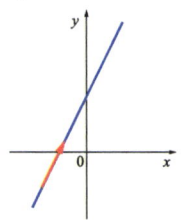

Q147:

a)

$$\begin{aligned} x^2 + y^2 &= (3\cos t)^2 + (-3\sin t)^2 \\ &= 9\cos^2 t + 4\sin^2 t \\ &= 9\left(\cos^2 t + \sin^2 t\right) \end{aligned}$$

$x^2 + y^2 = 9$

This is circle with centre at the origin and radius 3.

b)

The circle starts at (3,0) and travels clockwise round to the start again.

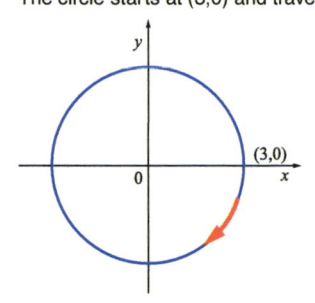

Q148:

a)

$$\begin{aligned} x^2 + y^2 &= (\cos(\pi - t))^2 + (\sin(\pi - t))^2 \\ &= \cos^2(\pi - t) + \sin^2(\pi - t) \end{aligned}$$

$x^2 + y^2 = 1$

Notice that since $0 \leqslant x \leqslant \pi$ then $\cos(\pi - t) \geqslant 0$, that is $y \geqslant 0$

© HERIOT-WATT UNIVERSITY

b)
The curve starts at (-1,0) and finishes at (1,0).

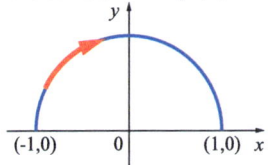

Q149:

a)
$x = t$ and $y = \sqrt{1 - t^2}$ then $y = \sqrt{1 - x^2}$
Since $0 \leqslant t \leqslant 1$ then $\sqrt{1 - x^2} \geqslant 0$, therefore $y \geqslant 0$

b)
The curve starts at (0,1) and ends at (1,0).

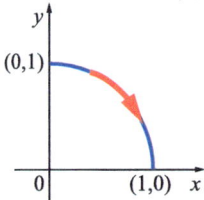

Q150:

a)
$x = 2t$ and $y = 3 - 3t$
$x = 2t \Rightarrow t = \frac{x}{2}$
Substitute this into $y = 3 - 3t$
$$y = 3 - 3\left(\frac{x}{2}\right)$$
$$y = 3 - \frac{3x}{2}$$
$$2y = 6 - 3x$$
$2y + 3x = 6$
Since $0 \leqslant t \leqslant 1$ then $x = 2t$ and $y = 3 - 3t$ are both $\geqslant 0$

b)
The line starts at (0,3) and ends at (2,0).

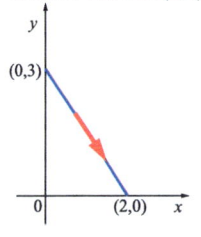

ANSWERS: UNIT 1 TOPIC 2

Q151:

a)

$x + y = \cos^2 t + \sin^2 t$
$x + y = 1$
Since $0 \leqslant t \leqslant \frac{\pi}{2}$ then $\cos^2 t \geqslant 0$ and $\sin^2 t \geqslant 0$

b)
The line starts at (1,0) and ends at (0,1).

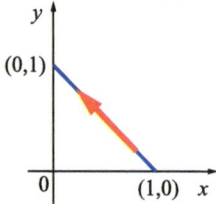

First order parametric differentiation exercise (page 148)

Q152:

$$x = \ln |t^2 - 3|$$
$$\Rightarrow \frac{dx}{dt} = \frac{2t}{t^2 - 3}$$

and

$$y = \ln |t^2|$$
$$\Rightarrow \frac{dy}{dt} = \frac{2t}{t^2}$$
$$= \frac{2}{t}$$

$$\frac{dy}{dx} = \frac{\frac{dy}{dt}}{\frac{dx}{dt}}$$
$$\frac{dy}{dx} = \frac{\frac{2}{t}}{\frac{2t}{t^2 - 3}}$$
$$\frac{dy}{dx} = \frac{2}{t} \times \frac{t^2 - 3}{2t}$$
$$\frac{dy}{dx} = \frac{t^2 - 3}{t^2}$$

Q153:

$$x = \csc \theta$$
$$\Rightarrow \frac{dx}{d\theta} = -\csc \theta \cot \theta$$

and

$$y = \sec \theta$$
$$\Rightarrow \frac{dy}{d\theta} = \sec \theta \tan \theta$$

© HERIOT-WATT UNIVERSITY

$$\frac{dy}{dx} = \frac{\frac{dy}{d\theta}}{\frac{dx}{d\theta}}$$
$$\frac{dy}{dx} = \frac{\sec\theta\tan\theta}{-\csc\theta\cot\theta}$$
$$\frac{dy}{dx} = \sec\theta\tan\theta \times \frac{1}{-\csc\theta\cot\theta}$$
$$\frac{dy}{dx} = -\frac{\sin\theta}{\cos\theta\cos\theta} \times \frac{\sin\theta\sin\theta}{\cos\theta}$$
$$\frac{dy}{dx} = -\frac{\sin^3\theta}{\cos^3\theta}$$
$$\frac{dy}{dx} = -\tan^3\theta$$

Q154:

$$x = t^3 + t^2 \qquad\qquad y = t^2 + t$$
$$\Rightarrow \frac{dx}{dt} = 3t^2 + 2t \quad \text{and} \quad \Rightarrow \frac{dy}{dt} = 2t + 1$$

$$\frac{dy}{dx} = \frac{\left(\frac{dy}{dt}\right)}{\left(\frac{dx}{dt}\right)}$$
$$= \frac{2t+1}{3t^2+2t}$$

Q155:

$$x = 4\cos^2\theta \qquad\qquad y = 4\sin^3\theta$$
$$\Rightarrow \frac{dx}{d\theta} = -8\cos\theta\sin\theta \quad \text{and} \quad \Rightarrow \frac{dy}{d\theta} = 12\sin^2\theta\cos\theta$$

$$\frac{dy}{dx} = \frac{\left(\frac{dy}{d\theta}\right)}{\left(\frac{dx}{d\theta}\right)}$$
$$= -\frac{12\sin^2\theta\cos\theta}{8\cos\theta\sin\theta}$$
$$= -\frac{3\,\sin\theta\sin\theta\,\cos\theta}{2\cos\theta\sin\theta}$$
$$= -\frac{3}{2}\sin\theta$$

Q156:

$$x = 4 - 3t \qquad\qquad y = \frac{3}{t}$$
$$\Rightarrow \frac{dx}{dt} = -3 \quad \text{and} \quad \Rightarrow \frac{dy}{dt} = -\frac{3}{t^2}$$

$$\frac{dy}{dx} = \frac{\left(\frac{dy}{dt}\right)}{\left(\frac{dx}{dt}\right)}$$
$$= \frac{-3}{-3t^2}$$
$$= \frac{1}{t^2}$$

From $x = 4 - 3t$, we rearrange to make t the subject:
$x = 4 - 3t$
$3t = 4 - x$
$t = \dfrac{4 - x}{3}$

Substitute $t = \dfrac{(4-x)}{3}$ into $\dfrac{dy}{dx}$ to get $\dfrac{dy}{dx}$ in terms of x only:

$$\dfrac{dy}{dx} = \dfrac{1}{\left(\frac{4-x}{3}\right)^2}$$
$$= \dfrac{9}{(4-x)^2}$$

Motions in a plane exercise (page 151)

Q157:
At $t = 0$:
$x = 4(0) - 1 \Rightarrow x = -1$
$y = (0)^2 + 3(0) \Rightarrow y = 0$
$Distance = \sqrt{x^2 + y^2}$
$= \sqrt{(-1)^2 + 0^2}$
$= 1$

The position of the particle (-1,0) is 1 metre from the origin.

Q158: b) 4 ms^{-1}

Q159: d) $(2t + 3)$ ms^{-1}

Q160: a) 4 ms^{-1}

Q161: c) 7 ms^{-1}

Q162: b) $8 \cdot 06$ ms^{-1}

Q163: The instantaneous direction of motion of a particle is given by:

$\dfrac{dy}{dx} = \dfrac{\left(\frac{dy}{dt}\right)}{\left(\frac{dx}{dt}\right)}$

$\dfrac{dx}{dt} = 4 \quad \dfrac{dy}{dt} = 2t + 3$

$\dfrac{dy}{dx} = \dfrac{\left(\frac{dy}{dt}\right)}{\left(\frac{dx}{dt}\right)}$
$= \dfrac{1}{4}(2t + 3)$

© HERIOT-WATT UNIVERSITY

Q164:

Evaluate $\frac{dy}{dx}$ at $t = 2$.

$$\frac{dy}{dx} = \frac{1}{4}(2t + 3)$$
$$= \frac{1}{4}(2(2) + 3)$$
$$= \frac{7}{4}$$

Set $m = \frac{7}{4}$ and evaluate θ using $m = \tan \theta$

$$\tan \theta = \frac{7}{4}$$
$$\theta = \tan^{-1}\left(\frac{7}{4}\right)$$
$$\theta = 60 \cdot 3°$$

Q165:

$x = 5 \cos t$ $\qquad$ $y = 3 \sin t$
$\frac{dx}{dt} = -5 \sin t$ $\qquad$ $\frac{dy}{dt} = 3 \cos t$

$$\text{speed} = \sqrt{\left(\frac{dx}{dt}\right)^2 + \left(\frac{dy}{dt}\right)^2}$$
$$= \sqrt{(-5\sin t)^2 + (3\cos t)^2}$$
$$= \sqrt{25\sin^2 t + 9\cos^2 t}$$

When $t = \frac{\pi}{4}$:

$$\text{speed} = \sqrt{25\sin^2\left(\frac{\pi}{4}\right) + 9\cos^2\left(\frac{\pi}{4}\right)}$$
$$= \sqrt{25\left(\frac{1}{\sqrt{2}}\right)^2 + 9\left(\frac{1}{\sqrt{2}}\right)^2}$$
$$= \sqrt{\frac{25}{2} + \frac{9}{2}}$$
$$= \sqrt{17}$$
$$= 4 \cdot 12 \ ms^{-1}$$

Q166:

Instantaneous direction of the particle is given by:

$\frac{dx}{dt} = -5 \sin t$ and $\frac{dy}{dt} = 3 \cos t$

$$\frac{dy}{dx} = \frac{\left(\frac{dy}{dt}\right)}{\left(\frac{dx}{dt}\right)}$$
$$= \frac{3 \cos t}{-5 \sin t}$$
$$= -\frac{3}{5} \cot t$$

Remember: $\cot t = \frac{1}{\tan t} = \frac{\cos t}{\sin t}$

At $t = \frac{\pi}{4}$:

$$\begin{aligned}\frac{dy}{dx} &= -\frac{3}{5}\cot\left(\frac{\pi}{4}\right) \\ &= -\frac{3\cos\left(\frac{\pi}{4}\right)}{5\sin\left(\frac{\pi}{4}\right)} \\ &= -\frac{3\times\frac{1}{\sqrt{2}}}{5\times\frac{1}{\sqrt{2}}} \\ &= -\frac{3}{5}\end{aligned}$$

Evaluate $m = \tan\theta$ at $\frac{dy}{dx} = -\frac{3}{5}$:

$$\begin{aligned}\tan\theta &= -\frac{3}{5} \\ \theta &= \tan^{-1}\left(\frac{3}{5}\right) \\ \theta &= 31° \\ \theta &= 180° - 31° \\ &= 149°\end{aligned}$$

Tangents to parametric curves exercise (page 153)

Q167:

$\frac{dy}{dx} = \frac{\frac{dy}{dt}}{\frac{dx}{dt}}$

$\frac{dy}{dt} = 2\cos 2t$ and $\frac{dx}{dt} = -2\sin 2t$

$\frac{dy}{dx} = \frac{2\cos 2t}{-2\sin 2t}$

$\frac{dy}{dx} = -\cot 2t$

Gradient when $t = \frac{\pi}{3}$:

$\frac{dy}{dx} = -\cot 2t \quad \Rightarrow \quad \frac{dy}{dx} = -\cot\left(\frac{2\pi}{3}\right)$

$\qquad\qquad\qquad\qquad\quad \frac{dy}{dx} = \frac{1}{2}$

When $t = \frac{\pi}{3}$ then $x = \cos 2t \Rightarrow x = \cos\left(\frac{2\pi}{3}\right) \Rightarrow x = -\frac{1}{2}$

When $t = \frac{\pi}{3}$ then $y = \sin 2t \Rightarrow y = \sin\left(\frac{2\pi}{3}\right) \Rightarrow y = \frac{\sqrt{3}}{2}$

Coordinate $\left(-\frac{1}{2}, \frac{\sqrt{3}}{2}\right)$
The straight line is given by:
$$y - b = m(x - a)$$
$$y - \frac{\sqrt{3}}{2} = \frac{1}{2}\left(x + \frac{1}{2}\right)$$
$$y - \frac{\sqrt{3}}{2} = \frac{1}{2}x + \frac{1}{4}$$
$$4y - 2\sqrt{3} = 2x + 1$$
$$4y - 2x = 1 + 2\sqrt{3}$$

Q168:

$$\frac{dy}{dx} = \frac{\frac{dy}{d\theta}}{\frac{dx}{d\theta}}$$
$$\frac{dy}{d\theta} = -\sin\theta \text{ and } \frac{dx}{d\theta} = -\csc\theta\cot\theta$$
$$\frac{dy}{dx} = \frac{-\sin\theta}{-\csc\theta\cot\theta}$$
$$\frac{dy}{dx} = \sin\theta \times \frac{\cos\theta}{\sin\theta} \times \frac{1}{\sin\theta}$$
$$\frac{dy}{dx} = \frac{\cos\theta}{\sin\theta}$$

Gradient when $\theta = \frac{\pi}{2}$:
$$\frac{dy}{dx} = \frac{\cos\theta}{\sin\theta} \quad \Rightarrow \quad \frac{dy}{dx} = \frac{\cos\left(\frac{\pi}{2}\right)}{\sin\left(\frac{\pi}{2}\right)}$$
$$\frac{dy}{dx} = 0$$

When $\theta = \frac{\pi}{2}$ then $x = \csc\theta \Rightarrow x = \csc\left(\frac{\pi}{2}\right) \Rightarrow x = 1$
When $\theta = \frac{\pi}{2}$ then $y = \cos\theta \Rightarrow y = \cos\left(\frac{\pi}{2}\right) \Rightarrow y = 0$

Coordinate (1,0)
The straight line is given by:
$$y - b = m(x - a)$$
$$y - 0 = 0(x - 1)$$
$$y = 0$$

Q169:

$$\frac{dy}{dx} = \frac{\frac{dy}{d\theta}}{\frac{dx}{d\theta}}$$
$$\frac{dy}{dt} = 3t^2 - 3t \text{ and } \frac{dx}{dt} = \frac{1}{2}t^{-\frac{1}{2}}$$
$$\frac{dy}{dx} = \frac{3t^2 - 3t}{\frac{1}{2}t^{-\frac{1}{2}}}$$
$$\frac{dy}{dx} = 6t^{\frac{1}{2}}\left(t^2 - t\right)$$
$$\frac{dy}{dx} = 6t^{\frac{5}{2}} - 6t^{\frac{3}{2}}$$

Gradient when $t = 4$:

$$\frac{dy}{dx} = 6t^{\frac{5}{2}} - 6t^{\frac{3}{2}} \Rightarrow \frac{dy}{dx} = 6(4)^{\frac{5}{2}} - 6(4)^{\frac{3}{2}}$$
$$\frac{dy}{dx} = 144$$

When $t = 4$ then $x = \sqrt{t} \Rightarrow x = \sqrt{4} \Rightarrow x = 2$
When $t = 4$ then $y = t^3 - \frac{3}{2}t^2 \Rightarrow y = (4) - \frac{3}{2}(4)^2 \Rightarrow y = -20$
Coordinate (2,-20)
The straight line is given by:
$$y - b = m(x - a)$$
$$y + 20 = 144(x - 2)$$
$$y = 144x - 308$$

Second order parametric differentiation exercise (page 155)

Q170:

For the first derivative:
$\frac{dx}{dt} = \cos t$ and $\frac{dy}{dt} = -\sin t$
So,
$$\frac{dy}{dx} = \frac{\frac{dy}{dt}}{\frac{dx}{dt}}$$
$$\frac{dy}{dx} = -\frac{\sin t}{\cos t}$$
$$\frac{dy}{dx} = -\tan t$$

For the second derivative:
$$\frac{d^2y}{dx^2} = \frac{\frac{d}{dt}\left(\frac{dy}{dx}\right)}{\frac{dx}{dt}}$$
$$\frac{d^2y}{dx^2} = \frac{-\sec^2 t}{\cos t}$$
$$\frac{d^2y}{dx^2} = -\frac{1}{\cos^2 t} \times \frac{1}{\cos t}$$
$$\frac{d^2y}{dx^2} = -\sec^3 t$$

Substituting into the formula in the question:
$$\cos t \frac{d^2y}{dx^2} + \left(\frac{dy}{dx}\right)^2 = k$$
$$\cos t \left(-\sec^3 t\right) + \left(-\tan t\right)^2 = k$$
$$-\cos t \times \frac{1}{\cos^3 t} + \tan^2 t = k$$
$$-\sec^2 t + \tan^2 t = k$$
(since $1 + \tan^2 t = \sec^2 t$)
$$k = -1$$

Q171:

For the first derivative:

$\frac{dx}{dt} = 4t$ and $\frac{dy}{dt} = 10t - 7t^2$

So,

$$\frac{dy}{dx} = \frac{\frac{dy}{dt}}{\frac{dx}{dt}}$$

$$\frac{dy}{dx} = \frac{10t - 7t^2}{4t}$$

$$\frac{dy}{dx} = \frac{10 - 7t}{4}$$

For the second derivative:

$$\frac{d^2y}{dx^2} = \frac{\frac{d}{dt}\left(\frac{dy}{dx}\right)}{\frac{dx}{dt}}$$

$$\frac{d^2y}{dx^2} = \frac{\frac{7}{4}}{4t}$$

$$\frac{d^2y}{dx^2} = \frac{7}{16t}$$

Substituting into the formula in the question:

$$\frac{d^2y}{dx^2} = a\frac{1}{t}$$

$$a = \frac{7}{16}$$

Q172:

For the first derivative:

$\frac{dx}{dt} = 3 + 2t^2$ and $\frac{dy}{dt} = t - 4$

So,

$$\frac{dy}{dx} = \frac{\frac{dy}{dt}}{\frac{dx}{dt}}$$

$$\frac{dy}{dx} = \frac{t - 4}{3 + 2t^2}$$

For the second derivative:

$$\frac{d^2y}{dx^2} = \frac{\frac{d}{dt}\left(\frac{dy}{dx}\right)}{\frac{dx}{dt}}$$

$$\frac{d^2y}{dx^2} = \frac{\frac{1 \times (4t) - (t-4) \times 4t}{(3+2t^2)^2}}{3+2t^2}$$

$$\frac{d^2y}{dx^2} = \frac{4t(1-t-4)}{(3+2t^2)^3}$$

$$\frac{d^2y}{dx^2} = \frac{-4t(t+3)}{(3+2t^2)^3}$$

Explicitly defined rate related problems exercise (page 156)

Q173:

$\frac{dL}{dt} = \frac{dL}{dh} \times \frac{dh}{dt}$

ANSWERS: UNIT 1 TOPIC 2

$\frac{dL}{dh} = 15h^2 - 2$ and $\frac{dh}{dt} = 3t - 5$

$$\frac{dL}{dt} = (15h^2 - 2)(3t - 5)$$

$$\frac{dL}{dt} = 45h^2 t - 75h^2 - 6t + 10$$

Q174:

$\frac{dK}{dt} = \frac{dK}{d\theta} \times \frac{d\theta}{dt}$

$\frac{dK}{d\theta} = 2\sec 2\theta \tan 2\theta + 6\theta$ and $\frac{d\theta}{dt} = t - 4t^3$

$\frac{dK}{dt} = (2\sec 2\theta \tan 2\theta + 6\theta)(t - 4t^3)$

Q175:

$$\frac{dS}{dt} = \frac{dS}{dh} \times \frac{dh}{dt}$$

$$\frac{dS}{dt} = \pi r \times 5$$

$$= 5\pi r$$

Therefore, when $r = 2$, $\frac{dS}{dt} = 10\pi$

Q176:

$$\frac{dB}{dx} = \frac{dB}{dt} \times \frac{dt}{dx}$$

$$= (2t + 12\cos 3t) \times \frac{1}{2}$$

$$= t + 6\cos 3t$$

When $t = \pi$

$$\frac{dB}{dx} = \pi + 6\cos 3\pi$$

$$= \pi - 6$$

Implicitly defined rate related problems exercise (page 158)

Q177:

Differentiate $x^2 - 3xy + 2y^2 = 4$ with respect to t:

$$\frac{d}{dt}(x^2) - \frac{d}{dt}(3xy) + \frac{d}{dt}(2y^2) = \frac{d}{dt}(4)$$

$$2x\frac{dx}{dt} - \left(3y\frac{dx}{dt} + 3x\frac{dy}{dt}\right) + 4y\frac{dy}{dt} = 0$$

$$2x\frac{dx}{dt} - 3y\frac{dx}{dt} - 3x\frac{dy}{dt} + 4y\frac{dy}{dt} = 0$$

$$(2x - 3y)\frac{dx}{dt} + (4y - 3x)\frac{dy}{dt} = 0$$

$$\frac{dx}{dt} = \frac{-(4y - 3x)}{(2x - 3y)}\frac{dy}{dt}$$

Calculate the rate of change of the x-coordinate with respect to time at the point (2,3) when the rate of change of the y-coordinate is 3 units per second.

$\frac{dx}{dt} = 3 \cdot 6$ units per second

© HERIOT-WATT UNIVERSITY

Q178:

$$\frac{d}{dt}(x^2) - \frac{d}{dt}(5xy) - \frac{d}{dt}(3y^2) = \frac{d}{dt}(-7)$$

$$2x\frac{dx}{dt} - \left(5y\frac{dx}{dt} + 5x\frac{dy}{dt}\right) - 6y\frac{dy}{dt} = 0$$

$$2x\frac{dx}{dt} - 5y\frac{dx}{dt} - 5x\frac{dy}{dt} - 6y\frac{dy}{dt} = 0$$

$$(2x - 5y)\frac{dx}{dt} = (5x + 6y)\frac{dy}{dt}$$

$$\frac{dx}{dt} = \frac{(5x + 6y)}{(2x - 5y)}\frac{dy}{dt}$$

Substitute (1,1) and $\frac{dy}{dt} = 2$:

$$\frac{dx}{dt} = \frac{(5(1) + 6(1))}{(2(1) - 5(1))}(2)$$

$$= \frac{22}{-3}$$

Q179:

$\frac{dy}{dt} = 2$ and $\frac{dx}{dt} = -\frac{22}{3}$

$$\frac{dy}{dx} = \frac{2}{\left(-\frac{22}{3}\right)}$$

$$= 2 \times \left(-\frac{3}{22}\right)$$

$$= -\frac{3}{11}$$

Substitute into $m = \tan \theta$

$$\tan \theta = -\frac{3}{11}$$

$$\tan^{-1}\left(\frac{3}{11}\right) = 15 \cdot 3°$$

So:

$$\theta = 180° - 15 \cdot 3°$$

$$= 164 \cdot 7°$$

ANSWERS: UNIT 1 TOPIC 2

Related rate problems exercise (page 162)

Q180:
Radius is decreasing: $\frac{dr}{dt} = -0 \cdot 5 \; cms^{-1}$

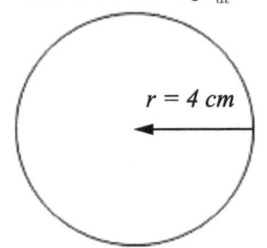

(Diagram of a circle with $r = 4\;cm$ and an inward arrow)

The variables in the problem are:
- $V = \frac{4}{3}\pi r^3$
- r = the radius of the sphere

We need to calculate $\frac{dV}{dt}$:

$\frac{dV}{dt} = \frac{dV}{dr} \times \frac{dr}{dt}$

Substituting

$\frac{dV}{dr} = 4\pi r^2$ and $\frac{dr}{dt} = -0 \cdot 5$ and $r = 4$ cm

So,
$$\frac{dV}{dt} = 4\pi (4)^2 \times (-0 \cdot 5)$$
$$\frac{dV}{dt} = -32\pi cm^3 s^{-1}$$

Q181:
The surface area of the cube is given by: $S = 6x^2$

So,
$$\frac{dS}{dt} = \frac{dS}{dx} \times \frac{dx}{dt}$$
$$\frac{dS}{dt} = 12x \frac{dx}{dt}$$

Q182:
The velocity is given by: $v = 2s^3 + 5s$

The acceleration is given by: $a = \frac{dv}{dt}$

So,
$$\frac{dv}{dt} = \frac{dv}{ds} \times \frac{ds}{dt}$$
$$\frac{dv}{dt} = (6s^2 + 5) \times \frac{ds}{dt}$$

Now $\frac{ds}{dt}$ by definition is v

We have:

$\frac{dv}{dt} = (6s^2 + 5)(2s^3 + 5s)$

© HERIOT-WATT UNIVERSITY

Q183:

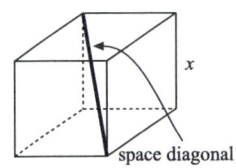

We need an expression for the space diagonal.

This is a cube where all the sides are x cm. We can find out the length of the space diagonal by first finding the face diagonal using Pythagoras' Theorem:

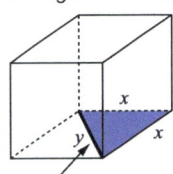

$$y^2 = x^2 + x^2$$
$$y^2 = 2x^2$$

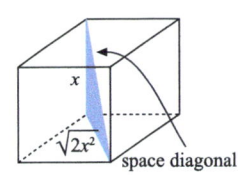

$$D^2 = 2x^2 + x^2$$
$$D^2 = 3x^2$$
$$D = \sqrt{3x^2}$$

The space diagonal is given by: $D = \sqrt{3x^2}$

So the rate of change of the space diagonal with respect to time, t, where x is a function of time is:

$\frac{dD}{dt} = \frac{dD}{dx} \times \frac{dx}{dt}$

Now $D = (3x^2)^{\frac{1}{2}}$ then,

$\frac{dD}{dx} = \frac{1}{2}(3x^2)^{-\frac{1}{2}} 6x \Rightarrow \frac{dD}{dx} = 3x(3x^2)^{-\frac{1}{2}}$

$\Rightarrow \frac{dD}{dx} = \frac{3}{\sqrt{3}}$

When $\frac{dD}{dx} = \frac{3}{\sqrt{3}}$, $\frac{dx}{dt} = -2\ cms^{-1}$

Given $\frac{dD}{dt} = \frac{dD}{dx} \times \frac{dx}{dt}$

Then,

$\frac{dD}{dt} = \frac{3}{\sqrt{3}} \times -2$

$\frac{dD}{dt} = -\frac{6}{\sqrt{3}} \times \frac{\sqrt{3}}{\sqrt{3}}$

$\frac{dD}{dt} = -\frac{6\sqrt{3}}{3}$

$\frac{dD}{dt} = -2\sqrt{3}\ cms^{-1}$

ANSWERS: UNIT 1 TOPIC 2

Displacement, velocity and acceleration exercise (page 167)

Q184:

a) $v = \frac{ds}{dt} = 6t^2$ and $a = \frac{dv}{dt} = 12t$

b) When $t = 10$: $v = 6 \times 10^2 = 600$ m/s and $a = 12 \times 10 = 120$ m/s^2

Q185:

a) $v = \frac{ds}{dt} = 9 - 10t$
When $t = 1 \cdot 5$ then $v = 9 - 10(1 \cdot 5) = -6$ m/s so the speed is 6 m/s

b) The maximum height is reached when $v = 0$ m/s since this will be when it changes direction from travelling upwards to travelling back down.
Solve $v = 9 - 10t = 0 \rightarrow t = 0 \cdot 9$ s
$s(0 \cdot 9) = 2 + 9(0 \cdot 9) - 5(0 \cdot 9)^2 = 6 \cdot 05$ m

c) The stone hits the ground when $s = 0$ m
Solve $2 + 9t - 5t^2 = 0 \rightarrow t = 2$ s
$v(2) = 9 - 10(2) = -11$ m/s so the speed is 11 m/s

© HERIOT-WATT UNIVERSITY

End of topic 2 test (page 184)

Q186: $\frac{dy}{dx} = \frac{3x}{y}$

Q187: $\frac{dy}{dx} = \frac{-(y^2 + 3x^2y)}{(2xy + x^3)}$

Q188: $x - y - 1 = 0$

Q189: $-\frac{2}{3}$

Q190: $\frac{d^2y}{dx^2} = \frac{2y-1}{x^2}$

Q191: $\frac{dy}{dx} = 4^x \ln|4|$

Q192:
$$\frac{dy}{dx} = (\cos x)^{5x}\left(5\ln|\cos x| - \frac{5x\sin x}{\cos x}\right)$$
$$= (\cos x)^{5x}(5\ln|\cos x| - 5x\tan x)$$

Q193: $x = \frac{9}{10}, x = 0, x = 3$

Q194:
You divide by $\sin^2 \theta$.
$\frac{dy}{dx} = -\frac{1}{1+x^2}$

Q195: $\frac{dy}{dx} = -\frac{1}{\sqrt{1-x^2}}$

Q196:
When $y = 5$, $\frac{d}{dx}(f^{-1}(x)) = \frac{1}{6}$

Q197: $3x^3(4\ln x + 1)$

Q198: $(x^3 + 3x^2 - 2x - 1)e^x$

Q199: $(3x^2 - 2)\sin x + (x^3 - 2x)\cos x$

Q200: $e^x(\sin x + \cos x)$

Q201: $\cos x \ln x + \frac{\sin x}{x}$

Q202: $\frac{-5}{(4x+1)^2}$

Q203: $-\frac{\ln x}{(4x^2)}$

Q204: $-\frac{\sin x(1 + 4x^2) + 8x\cos x}{(1 + 4x^2)^2}$

Q205: $-\sin x \exp(\cos x)$

Q206: $\frac{-3x^2}{\sqrt{1-x^6}}$

Q207: $\frac{1}{\sqrt{1-x^2}}e^{\sin^{-1}x}$

ANSWERS: UNIT 1 TOPIC 2   477

Q208: $(3x^2 - 2)e^{x^3 - 2x}$

Q209: cot x

Q210: $\frac{3x^2 - 2}{x^3 - 2x}$

Q211: $xe^{-x^3}(2 - 3x^3)$

Q212: $\frac{2\cos x + 1}{(2 + \cos x)^2}$

Q213: $-2x \sin x^2 \sin 3x + 3 \cos x^2 \cos 3x$

Q214: $\frac{1 - \ln(x + 4)}{(x + 4)^2}$

Q215: $2\sec^2(2x) \exp(\tan 2x)$

Q216: $3x^2 \ln|\cos x| - x^3 \tan x$

Q217: $\frac{1}{\sqrt{x(1-x)}}$

Q218: $6(2x - 1)^2(5x^2 - x + 5)$

Q219: $f'(x) = \frac{-6}{(x-2)^2}$ which is always negative.

Q220: $f'(x) = (x - 1)(x + 1)e^x$
Since $e^x > 0$ for all $x \in \mathbb{R}$
$f'(x) < 0$ for $-1 < x < 1$ hence $f(x)$ is decreasing.

Q221: $\frac{3\pi}{4}$

Q222:

$$\frac{dy}{dx} = \frac{1}{x^2}(-\cos(kx) - xk\sin(kx))$$

$$\frac{d^2y}{dx^2} = \frac{1}{x^4}\left(2x^2k\sin(kx) + 2x\cos(kx) - x^3k^2\cos(kx)\right)$$

Substitute into the equation:

$$\frac{d^2y}{dx^2} + \frac{2}{x}\frac{dy}{dx} + k^2y$$

$$=\frac{1}{x^4}\left(2x^2k\sin(kx) + 2x\cos(kx) - x^3k^2\cos(kx)\right) + \frac{2}{x}\left(\frac{1}{x^2}(-\cos(kx) - xk\sin(kx))\right) + k^2\left(\frac{\cos kx}{x}\right)$$

$$=\frac{2k\sin(kx)}{x^2} + \frac{2\cos(kx)}{x^3} - \frac{k^2\cos(kx)}{x} - \frac{2\cos(kx)}{x^3} - \frac{2k\sin(kx)}{x^2} + \frac{k^2\cos kx}{x}$$

$$=0$$

Q223: $\frac{dy}{dx} = \frac{3}{4}t$

Q224: $\frac{dy}{dx} = -\frac{4\cos(t)}{3\sin(t)}$

Q225:

Steps:

- What are the coordinates of x and y when $t = 4$? (44,2)

© HERIOT-WATT UNIVERSITY

- What is the gradient of the curve when $t = 4$? $\frac{dy}{dx} = \frac{1}{92}$
- Find the equation of the tangent, give your answer in standard from i.e. $Ax + By + C = 0$

Answer: $x - 92y + 140 = 0$

Q226:

$$\frac{dy}{dx} = -2\cos 4t$$

$$\frac{d^2y}{dx^2} = -2$$

Q227:

a) 6 m/s

b) $\frac{3}{2}$

Q228:

Steps:

- What is the rate of change of the volume? $0 \cdot 02$ m^3 s^{-1}
- What is $\frac{dV}{dr}$? $\frac{dV}{dr} = 4\pi r^2$

Answer: $\frac{dr}{dt} = -\frac{1}{2\pi}$ ms^{-1}

Q229:

Steps:

- What is the rate of change for x? $\frac{dx}{dt} = 2$ cm s^{-1}
- What is an expression for y in terms of x? $y = \frac{50}{x}$
- What is an expression for the perimeter in terms of x? $P = 2x + \frac{100}{x}$

Answer: $\frac{dP}{dt} = -4$ cm s^{-1}; The perimeter is decreasing.

Q230:

a) If $v(t)$ is the velocity, what is the acceleration in relation to this?
Using the quotient rule to differentiate:

$$a = \frac{dv}{dt}$$

$$a = \frac{d}{dt}\frac{150t}{12+25t}$$

$$a = \frac{150(12+25t) - 150t(25)}{(12+25t)^2}$$

Therefore, the unsimplified numerator of the expression for acceleration is:
$150(12 + 25t) - 150t(25)$

b) When $t = 5$:

$$a = \frac{1800}{18769}$$

$$a = 0 \cdot 1 \text{ m/s}^2$$

ANSWERS: UNIT 1 TOPIC 2   479

Q231:

a) If $v(t)$ is the velocity, what is the acceleration in relation to this? $a = \frac{dv}{dt} \Rightarrow a = 6t - 4$
$a(0) = -4$ m/s^2

b) If $v(t)$ is the velocity, what is the displacement in relation to this?
$s = \int v \, dt \Rightarrow s(t) = t^3 - 2t^2 + 5t + c$
But $s(0) = 0 \Rightarrow s(t) = t^3 - 2t^2 + 5t$

Q232:

a) If $a(t)$ is the acceleration, what is the velocity in relation to this? $v = \int a \, dt \Rightarrow v = \frac{3t^2}{2} + t + c$
But, $v(0) = 0 \Rightarrow v = \frac{3t^2}{2} + t$
So $v(2) = 6 + 2 = 8$ m/s^2

b) If $v(t)$ is the velocity, what is the displacement in relation to this? $s = \int v \, dt \Rightarrow s = \frac{t^3}{2} + \frac{t^2}{2} + c$
But, $s(0) = 0 \Rightarrow s = \frac{t^3}{2} + \frac{t^2}{2}$
$s(4) = 32 + 8 = 40$ m

Q233:

a) If $s(t)$ is the displacement, what is the velocity in relation to this? $v = \frac{ds}{dt} \Rightarrow v = 12t^2 - 2t + 5$
$v(1) = 12 - 2 + 5 = 15$ m/s

b) If $v(t)$ is the velocity, what is the acceleration in relation to this? $a = \frac{dv}{dt} \Rightarrow a = 24t - 2$
$a(3) = 72 - 2 = 70$ m/s^2

Q234:

a) If $s(t)$ is the displacement, what is the velocity in relation to this? $v = \frac{ds}{dt} \Rightarrow v = -4 + 12t$
$v(1) = -4 + 12 = 8$ m/s

b) What is the velocity at the greatest height?
It reaches its greatest height when $v(t) = 0$
$0 = -4 + 12t \Rightarrow t = \frac{1}{3}$
The ball reaches its greatest height at $\frac{1}{3}$ seconds.
When $t = \frac{1}{3}$ seconds, then $s\left(\frac{1}{3}\right) = 2 - 4\left(\frac{1}{3}\right) + 6\left(\frac{1}{3}\right)^2 = \frac{4}{3}$
Therefore, the greatest height that the ball reaches is $\frac{4}{3}$ m.

c) If $v(t)$ is the velocity, what is the acceleration in relation to this?
$a = \frac{dv}{dt} \Rightarrow a = 12$ m/s^2

Q235:

a) If $s(t)$ is the displacement, what is the velocity in relation to this? $v = \frac{ds}{dt} \Rightarrow 6\cos(3t)$
$v(1) = 6\cos(3) = -5 \cdot 9$ m/s

b) Solve $v(t) = 0$
$6\cos 3t = 0 \Rightarrow 3t = \frac{\pi}{2} \Rightarrow t = \frac{\pi}{6}$ s

c) If $v(t)$ is the velocity, what is the acceleration in relation to this? $a = \frac{dv}{dt} \Rightarrow a = -18\sin(3t)$
This is at a maximum when $a(t) = 18$ m/s^2 so solve for this.

© HERIOT-WATT UNIVERSITY

$$-18\sin 3t = 18$$
$$\sin 3t = -1$$
$$3t = \frac{3\pi}{2}$$
$$t = \frac{\pi}{2}$$

The acceleration is first at a maximum when $t = \frac{\pi}{2}$ s

Q236:

a) If x represents one number and y the other, then $x + y = 18$ as stated in the question.
We can rearrange this, so that y is written in terms of x: $y = 18 - x$
So the sum of the product and the difference becomes:
$xy + (x - y) = x(18 - x) + (x - (18 - x))$
$xy + (x - y) = -x^2 + 20x - 18$
Note that we could have re-written it in terms of y. However, since the question asks us to maximise x first, it was written in terms of x.

b) To maximise the product and difference, we need to differentiate the expression and identify the nature of the stationary point, if it exists.
$\frac{d}{dx}\left(-x^2 + 20x - 18\right) = -2x + 20$
For stationary points, $\frac{d}{dx} = 0$
$-2x + 20 = 0$
$x = 10$
Using the second derivative will determine if it is a maximum or minimum.
$\frac{d^2}{dx^2}(-2x + 20) = -2 < 0$ therefore, a maximum.
When $x = 10$, the sum is maximised

c) To find the value of y that maximises the sum, substitute $x = 10$ back into $y = 18 - x$:
$y = 18 - 10 = 8$

Q237:

a) $(x, \sqrt{x})$

b) To find the distance between two points, we use the distance formula.
The distance between the points $(x, \sqrt{x})$ and $(4, 0)$ is:
$(x - 4)^2 + (\sqrt{x} - 0)^2 = (x - 4)^2 + x$
or
$(4 - x)^2 + (0 - \sqrt{x})^2 = (4 - x)^2 + x$

c) To find the minimum distance between these points, we need to differentiate this expression.
$$\frac{d}{dx}\left((x - 4)^2 + x\right) = 2(x - 4) + 1$$
$$= 2x - 7$$
For stationary points, $\frac{d}{dx} = 0$
$2x - 7 = 0$
$$x = \frac{7}{2}$$
Using the second derivative will determine if it is a maximum or minimum.
$\frac{d^2}{dx^2}(2x - 7) = 2 > 0$ therefore, it is a minimum.
The closest distance is when $x = 3 \cdot 5$
If a nature table is preferred, this can be used determine the nature of the stationary points.

… ANSWERS: UNIT 1 TOPIC 2

x	$\rightarrow$	3·5	$\rightarrow$
$\frac{dy}{dx}$	-	0	+
Slope	↘	$\rightarrow$	↗

When $x = 3 \cdot 5$, then $y = \sqrt{3 \cdot 5} = 1 \cdot 9$
The closest point to (4, 0) is (3·5, 1·9)

Q238:

a) Using the quotient rule, the gradient of the tangent is:
$\frac{d}{dx} \frac{6}{x^2+3} = \frac{(0)(x^2+3)-6(2x)}{(x^2+3)^2} = \frac{-12x}{(x^2+3)^2}$

b) The ideal minimum slope of the tangent would occur at a maximum or minimum since this is where the gradient would be zero. We will check this first to see if it exists here.
For stationary points, $\frac{d}{dx} = 0$ so $0 = \frac{-12x}{(x^2+3)^2} \Rightarrow x = 0$
It does not matter if this is a maximum or minimum. The tangent line will still have a gradient of zero.
When $x = 0$, then $y = \frac{6}{0^2+3} \Rightarrow y = 2$
The point of contact is (0,2)

Q239:

a) The standard derivative for $\tan^{-1}\left(\frac{a}{x}\right) = \frac{-a}{x^2+a^2}$
So $\frac{d}{dx}\left(\tan^{-1}\left(\frac{30}{x}\right) - \tan^{-1}\left(\frac{10}{x}\right)\right) = \frac{-30}{x^2+900} + \frac{10}{x^2+100}$

b) To maximise the viewing angle, use the differentiated expression above and equate to zero to find stationary points.

$$\frac{-30}{x^2+900} + \frac{10}{x^2+100} = 0$$
$$10\left(x^2+900\right) = 30(x^2+100)$$
$$20x^2 = 6000$$
$$x^2 = 300$$

Since we are dealing with real lengths, then we only consider positive values for x.
To determine if this is a maximum, we will use a table of values.

x	$\rightarrow$	$\sqrt{300}$	$\rightarrow$
$\frac{d}{dx}$	+	0	-
Slope	↗	$\rightarrow$	↘

So $x^2 = 300$ is a maximum.

c) The maximum viewing angle is found by substituting $x^2 = 300$ into:
$\tan^{-1}\left(\frac{30}{x}\right) - \tan^{-1}\left(\frac{10}{x}\right)$
$\theta = \tan^{-1}\left(\frac{30}{\sqrt{300}}\right) - \tan^{-1}\left(\frac{10}{\sqrt{300}}\right)$
$\theta = \frac{\pi}{6}$ radians

© HERIOT-WATT UNIVERSITY

Topic 3: Integration

Completing the square exercise (page 201)

Q1: $q = -9$

Q2: $q = -10$

Q3: $q = 23$

Q4:
$$\begin{aligned} y &= x^2 + 5x - 1 \\ &= \left(x^2 + 5x\right) - 1 \\ &= \left(x + \frac{5}{2}\right)^2 - \left(\frac{5}{2}\right)^2 - 1 \\ &= \left(x + \frac{5}{2}\right)^2 - \left(\frac{25}{4}\right) - 1 \\ y &= \left(x + \frac{5}{2}\right)^2 - \frac{29}{4} \end{aligned}$$
or
$$y = (x + 2 \cdot 5)^2 - 7 \cdot 25$$

Integrating expressions exercise (page 205)

Q5:
$$\begin{aligned} \int 3x^2 - 6x - 5 \, dx &= \int 3x^2 \, dx - \int 6x \, dx - \int 5 \, dx \\ &= \frac{3x^3}{3} - \frac{6x^2}{2} - 5x + C \\ &= x^3 - 3x^2 - 5x + C \end{aligned}$$

Q6:
$$\begin{aligned} \int 3x^{\frac{1}{2}} + x^{-4} dx &= \int 3x^{\frac{1}{2}} dx + \int x^{-4} dx \\ &= \frac{3x^{\frac{3}{2}}}{\frac{3}{2}} + \frac{x^{-3}}{-3} + C \\ &= 3x^{\frac{3}{2}} \times \frac{2}{3} - \frac{x^{-3}}{3} + C \\ &= 2x^{\frac{3}{2}} - \frac{1}{3x^3} + C \\ &= 2\sqrt{x^3} - \frac{1}{3x^3} + C \end{aligned}$$

Q7:

$$\int \frac{1}{\sqrt[3]{r}} - \frac{1}{2r^3} dr = \int \frac{1}{\sqrt[3]{r}} dr - \int \frac{1}{2r^3} dr$$
$$= \int r^{-\frac{1}{3}} dr - \int \frac{1}{2} r^{-3} dr$$
$$= \frac{r^{\frac{2}{3}}}{\frac{2}{3}} - \frac{1}{2} \times \frac{r^{-2}}{-2} + C$$
$$= \frac{3}{2} r^{\frac{2}{3}} + \frac{r^{-2}}{4} + C$$
$$= \frac{3\sqrt[3]{r^2}}{2} + \frac{1}{4r^2} + C$$

Q8:
Step 1: Pull apart the fraction to simplify

$$\frac{p^5 + 1}{\sqrt{p^3}} = \frac{p^5}{\sqrt{p^3}} + \frac{1}{\sqrt{p^3}}$$
$$= \frac{p^5}{p^{\frac{3}{2}}} + \frac{1}{p^{\frac{3}{2}}}$$
$$= p^{\frac{7}{2}} + p^{-\frac{3}{2}}$$

Step 2: Integrate

$$\int p^{\frac{7}{2}} + p^{-\frac{3}{2}} dp = \frac{p^{\frac{9}{2}}}{\frac{9}{2}} + \frac{p^{-\frac{1}{2}}}{-\frac{1}{2}} + C$$

Step 3: Simplify

$$= \frac{2p^{\frac{9}{2}}}{9} - \frac{2p^{-\frac{1}{2}}}{1} + C$$
$$= \frac{2\sqrt{p^9}}{9} - \frac{2}{\sqrt{p}} + C$$

Integration of sin and cos exercise (page 207)

Q9: $\int \frac{6}{\sqrt{u}} - \sin u \, du = 12\sqrt{u} + \cos u + C$

Q10: $\int \frac{\cos x}{4} + \sqrt{7} \sin x - \pi dx = \frac{\sin x}{4} - \sqrt{7} \cos x - \pi x + C$

Q11: $\int \sin 2x \, dx = -\frac{1}{2} \cos 2 + C$

Q12: $\int \cos(7x + 6) dx = \frac{1}{7} \sin(7x + 6) + C$

Q13: $\int \sin(4x + 5) dx = -\frac{1}{4} \cos(4x + 5) + C$

Q14: $\int 6 \cos\left(4x + \frac{\pi}{4}\right) dx = \frac{3}{2} \sin\left(4x + \frac{\pi}{4}\right) + C$

Q15: $\int 5 \sin\left(\frac{x}{5} - \frac{\pi}{3}\right) dx = -25 \cos\left(\frac{x}{5} - \frac{\pi}{3}\right) + C$

Q16:

a) $\int \sin^2 x \, dx = \frac{1}{2} - \frac{1}{2} \cos 2x + C$

b) $\int \sin^2 x \, dx = \frac{1}{2} x - \frac{1}{4} \sin 2x + C$

Integrating composite functions exercise (page 209)

Q17: $f(x) = \frac{(8x+7)^{\frac{5}{4}}}{10} + C$

Q18: $f(x) = -(8-3x)^{\frac{1}{3}} + C$

Q19:

Steps:

- $\sqrt{2x+7} = (2x+7)^n \ n = ? \ 1/2$

Answer: $f(x) = \frac{(2x+7)^{\frac{3}{2}}}{3} + C$

Q20:

Steps:

- $\frac{1}{(4x+1)^2} = (4x+1)^m \ m = ? \ -2$

Answer: $f(x) = -\frac{1}{4(4x+1)} + C$

Q21:

Steps:

- $\frac{1}{(\sqrt{4x+5})^3} = (4x+5)^k \ k = ? \ -\frac{3}{2}$

Answer: $f(x) = -\frac{1}{2\sqrt{4x+5}} + C$

Q22:

Steps:

- Given that $\frac{ds}{dt} = (1-8t)^{-7}$ then $s = \int \frac{ds}{dt} dt = \int (1-8t)^{-7} dt$

Answer: $s(t) = \frac{1}{48(1-8t)^6} + C$

Finding the area under a curve exercise (page 211)

Q23:

$$\int_1^3 (3x^2 - 12x + 9) dx = \left[\frac{3x^3}{3} - \frac{12}{2}x^2 + 9x \right]_1^3$$
$$= \left[x^3 - 6x^2 + 9x \right]_1^3$$
$$= \left(3^3 - 6(3)^2 + 9(3) \right) - \left(1^3 - 6(1)^2 + 9(1) \right)$$
$$= 0 - 4$$
$$= -4$$

Note that the answer is **negative** because the area being integrated is **below** the x-axis.

ANSWERS: UNIT 1 TOPIC 3

Q24:

$$\int_3^4 (3x^2 - 12x + 9)\,dx = \left[\frac{3x^3}{3} - \frac{12}{2}x^2 + 9x\right]_3^4$$
$$= \left[x^3 - 6x^2 + 9x\right]_3^4$$
$$= \left(4^3 - 6(4)^2 + 9(4)\right) - \left(3^3 - 6(3)^2 + 9(3)\right)$$
$$= 4$$

Note that the answer is **positive** because the area being integrated is **above** the x-axis.

Q25:

To calculate the total shaded area in the above diagram the areas above and below the axis have to be calculated separately. We ignore the negative sign for the area below the axis because an area cannot be negative.

Total shaded area $= 4 + 4$
$ = 8$

Finding the area between two curves exercise (page 213)

Q26:

Steps:

- What is the definite integral needed to find the shaded area? $\int_{-2}^{3} 6 + x - x^2\,dx$
- What is $\int 6 + x - x^2\,dx$? $f(x) = 6x + \frac{x^2}{2} - \frac{x^3}{3} + C$

Answer: 20·83 units^2

Q27:

Steps:

- To find the points of intersection solve $x^2 - 3x - 4 = 4 + 3x - x^2$. $A = -1$ and $B = 4$
- What is the definite integral needed to find the shaded area? $\int_A^B 8 + 6x - 2x^2\,dx$
- What is $\int 8 + 6x - 2x^2\,dx$? $f(x) = 8x + 3x^2 - \frac{2x^3}{3} + C$

Answer: 41·67 units^2

Q28:

Steps:

- Sketch the curve and shade the area to be calculated.

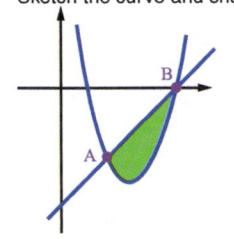

- Solve $x - 4 = x^2 - 5x + 4$ to find the points of intersection. $A = 2$ and $B = 4$
- What is the definite integral needed to find the shaded area? $\int_A^B (-x)^2 + 6x - 8\, dx$
- What is $\int (-x)^2 + 6x - 8\, dx$? $f(x) = -\frac{x^3}{3} + 3x^2 - 8x + C$

Answer: 1·33 units^2

Q29:

Steps:

- The graphs intersect at two points where $x = a$ and where $x = b$ with $a < b$. To find the points of intersection solve x^2 - 20x + 17 = -1 - x^2. $a = 1$ and $b = 9$
- What is the definite integral needed to find the shaded area? $\int_a^b -18 + 20x - 2x^2\, dx$
- What is $\int -18 + 20x - 2x^2\, dx$? $f(x) = -18x + 10x^2 - \frac{2x^3}{3} + C$

Answer: 170·67

Q30:

Steps:

- To find the points of intersection solve $x^3 - 3x - 2 = x - 2$. At A, $x = -2$; At B $x = 0$; At C $x = 2$
- What is the definite integral needed to find the shaded area between A and B? $\int_{-2}^0 x^3 - 4x\, dx$
- What is the area between A and B? 4 units^2
- What is the definite integral needed to find the shaded area between B and C? $\int_0^2 4x - x^3\, dx$
- What is the area between B and C? 4 units^2

Answer: 8 units^2

Q31:

Steps:

- The line and the curve intersect when $2x^2 - 7x + 7 = x + 7$ when $x = 0$ and when $x = $? 4

Answer: $\int_0^4 8x - 2x^2\, dx$

Q32:

Steps:

- Solve $y = 16 - \frac{x^2}{4} = 0$ to find A and B.

Answer: A(-8,0) and B(8,0)

Q33:

Steps:

- What is the area of the rectangle? 560 feet^2
- What is $\int 16 - \frac{x^2}{4}\, dx$? $f(x) = 16x - \frac{x^3}{12} + C$
- What is the area under the parabola? 170·67 feet^2
- What is the area to be repainted? 389·33 feet^2

Answer: £1557

ANSWERS: UNIT 1 TOPIC 3

Integrate using standard results exercise (page 220)

Q34:

Standard Result: $\int d(ax+b)^n dx = \frac{d(ax+b)^{n-1}}{a(n-1)} + C$

$\int 3(2x+5)^4 dx = \frac{3}{10}(2x+5)^5 + C$

Q35:

Standard Result: $\int d\cos(ax+b) dx = \frac{d}{a}\sin(ax+b) + C$

$\int 2\cos(3x-1) dx = \frac{2}{3}\sin(3x-1) + C$

Q36:

Standard Result: $\int d\sin(ax+b) dx = -\frac{d}{a}\cos(ax+b) + C$

$\int 5\sin(2x+3) dx = -\frac{5}{2}\cos(2x+3) + C$

Q37:

Standard Result: $\int d\sec^2(ax+b) dx = \frac{d}{a}\tan(ax+b) + C$

$\int 2\sec^2(3x-7) dx = \frac{2}{3}\tan(3x-7) + C$

Q38:

Standard Result: $\int de^{(ax+b)} dx = \frac{d}{a}e^{(ax+b)} + C$

$\int -7e(2x-4) dx = -\frac{7}{2}e(2x-4) + C$

Note that if it is easier to read as $\int -7\exp(2x-4) dx = -\frac{7}{2}\exp(2x-4) + C$ then retain this form.

Q39:

Standard Result: $\int \frac{d}{ax+b} dx = \frac{d}{a}\ln|ax+b| + C$

$\int \frac{-5}{3x+2} dx = -\frac{5}{3}\ln(3x+2) + C$

Q40:

Standard Result: $\int \frac{d}{ax+b} dx = \frac{d}{a}\ln|ax+b| + C$

$$\int \frac{x}{x-6} dx = \int \frac{(x-6)+6}{x-6} dx$$
$$= \int \frac{x-6}{x-6} + \frac{1}{x-6} dx$$
$$= \int 1\, dx + \int \frac{1}{x-6} dx$$
$$= x + 6\ln|x-6| + C$$

Q41:

Standard Result: $\int \frac{d}{ax+b} dx = \frac{d}{a}\ln|ax+b| + C$

$$\int \frac{r+3}{r+7} dr = \int \frac{(r+7)-7+3}{r+7} dr$$
$$= \int \frac{(r+7)-4}{r+7} dr$$
$$= \cdots$$

© HERIOT-WATT UNIVERSITY

$$\int \frac{r+3}{r+7} dr = \cdots$$
$$= \int \frac{r+7}{r+7} - \frac{4}{r+7} dr$$
$$= \int 1\, dr - 4\int \frac{1}{r+7} dr$$
$$= r - 4\ln|r+7| + C$$

Q42:

Standard Result: $\int \frac{d}{ax+b} dx = \frac{d}{a} \ln|ax+b| + C$

$$\int \frac{2x}{6x+3} dx = \int \frac{\frac{1}{3}(6x+3) - 1}{6x+3} dx$$
$$= \frac{1}{3} \int \frac{(6x+3)}{6x+3} dx - \int \frac{1}{6x+3} dx$$
$$= \frac{1}{3} \int 1\, dx - \int \frac{1}{6x+3} dx$$
$$= \frac{1}{3}x - \frac{1}{6}\ln|6x+3| + C$$

Integration by substitution (substitution given) exercise 1 (page 222)

Q43:

Let $u = 2 + 3x^2$ then $\frac{du}{dx} = 6x$ and $\frac{dx}{du} = \frac{1}{6x}$

$$\int 5x\sqrt{2+3x^2}\, dx = \int 5x\sqrt{2+3x^2} \frac{dx}{du} du$$
$$= \int 5x\sqrt{u} \frac{1}{6x} du$$
$$= \int \frac{5}{6} u^{\frac{1}{2}} du$$
$$= \frac{5}{6} \int u^{\frac{1}{2}} du$$
$$= \frac{5}{6} \times \frac{2}{3} u^{\frac{3}{2}} + C$$
$$= \frac{5}{9}(2+3x)^{\frac{3}{2}} + C$$
$$= \frac{5}{9}\sqrt{(2+3x)^3} + C$$

ANSWERS: UNIT 1 TOPIC 3

Q44:

Let $u = \sin(x)$ then $\frac{du}{dx} = \cos(x)$ and $\frac{dx}{du} = \frac{1}{\cos(x)}$

$$\begin{aligned}
\int \sin^3(x)\cos(x)\,dx &= \int \sin^3(x)\cos(x)\,\frac{dx}{du}\,du \\
&= \int u^3 \cos(x)\,\frac{1}{\cos(x)}\,du \\
&= \int u^3\,du \\
&= \frac{1}{4}u^4 + C \\
&= \frac{1}{4}\sin^4(x) + C
\end{aligned}$$

Q45:

Let $u = 3x^2 + 7$ then $\frac{du}{dx} = 6x$ and $\frac{dx}{du} = \frac{1}{6x}$

$$\begin{aligned}
\int \frac{5x}{3x^2 + 7}\,dx &= \int \frac{5x}{3x^2 + 7}\,\frac{dx}{du}\,du \\
&= \int \frac{5x}{u} \times \frac{1}{6x}\,du \\
&= \int \frac{5}{6} \times \frac{1}{u}\,du \\
&= \frac{5}{6} \int \frac{1}{u}\,du \\
&= \frac{5}{6} \ln|u| + C \\
&= \frac{5}{6} \ln|3x^2 + 7| + C
\end{aligned}$$

© HERIOT-WATT UNIVERSITY

Integration by substitution (substitution given) exercise 2 (page 225)

Q46:

Let $u = x - 5$ then $\frac{du}{dx} = 1$ and $\frac{dx}{du} = 1$

Also, since $u = x - 5$ then $x = u + 5$

$$\int 4x(x-5)^3 dx = \int 4x(x-5)^3 \frac{dx}{du} du$$
$$= \int 4(u+5)u^3 du$$
$$= \int 4u^4 + 20u^3 du$$
$$= \frac{4}{5}u^5 + 5u^4 + C$$
$$= \frac{4}{5}u^5 + \frac{25}{5}u^4 + C$$
$$= \frac{1}{5}u^4(4u+25) + C$$
$$= \frac{1}{5}(x-5)^4(4(x-5)+25) + C$$
$$= \frac{1}{5}(x-5)^4(4x+5) + C$$

Q47:

Let $u = 2x + 1$ then $\frac{du}{dx} = 2$ and $\frac{dx}{du} = \frac{1}{2}$

Also, since $u = 2x + 1$ then $x = \frac{1}{2}(u-1)$

$$\int 5x(2x+1)^{-4} dx = \int 5x(2x+1)^{-4} \frac{dx}{du} du$$
$$= \int 5 \times \frac{1}{2}(u-1)u^{-4} \times \frac{1}{2} du$$
$$= \frac{5}{4} \int (u-1)u^{-4} du$$
$$= \frac{5}{4} \int u^{-3} - u^{-4} du$$
$$= \frac{5}{4}\left(\frac{u^{-2}}{-2} - \frac{u^{-3}}{-3}\right) + C$$
$$= \frac{5}{4}u^{-2}\left(-\frac{3}{6} + \frac{2u^{-1}}{6}\right) + C$$
$$= \frac{5}{24u^2}\left(-3 + \frac{2}{u}\right) + C$$
$$= \frac{5}{24(2x+1)^2}\left(\frac{2}{2x+1} - 3\right) + C$$

ANSWERS: UNIT 1 TOPIC 3

Q48:

Let $u = x + 6$ then $\frac{du}{dx} = 1$ and $\frac{dx}{du} = 1$
Also, since $u = x + 6$ then $x = u - 6$

$$\int x(x+6)^{-2} dx = \int x(x+6)^{-2} \frac{dx}{du} du$$

$$= \int (u-6)u^{-2} du$$

$$= \int u^{-1} - 6u^{-2} du$$

$$= \int \frac{1}{u} - 6u^{-2} du$$

$$= \ln|u| - \frac{6u^{-1}}{-1} + C$$

$$= \ln|u| + \frac{6}{u} + C$$

$$= \ln|x+6| + \frac{6}{x+6} + C$$

Definite integrals with substitution exercise (page 227)

Q49:

Let $u = 1 + 2\sqrt{x}$ then $\frac{du}{dx} = x^{-\frac{1}{2}}$ and $\frac{dx}{du} = x^{\frac{1}{2}}$
Change the limits:
When $x = 0$, $u = 1 + 2\sqrt{0} = 1$
When $x = \frac{1}{4}$, $u = 1 + 2\sqrt{\frac{1}{4}} = 2$
Therefore, we have

$$\int_0^{\frac{1}{4}} \frac{(1+2\sqrt{x})^3}{\sqrt{x}} dx = \int_1^2 \frac{(1+2\sqrt{x})^3}{\sqrt{x}} \frac{dx}{du} du$$

$$= \int_1^2 \frac{u^3}{x^{\frac{1}{2}}} x^{\frac{1}{2}} du$$

$$= \int_1^2 u^3 du$$

$$= \left[\frac{1}{4}u^4\right]_1^2$$

$$= \frac{1}{4}(2)^4 - \frac{1}{4}(1)^4$$

$$= 4 - \frac{1}{4}$$

$$= 3\frac{3}{4}$$

© HERIOT-WATT UNIVERSITY

Q50:
Let $u = 2x^2 - 3x$ then $\frac{du}{dx} = 4x - 3$ and $\frac{dx}{du} = \frac{1}{4x-3}$
Change the limits:
When $x = 0$, $u = 2(0)^2 - 3(0) = 0$
When $x = 2$, $u = 2(2)^2 - 3(2) = 2$
Therefore, we have
$$\int_0^2 (4x-3)e(2x^2-3x)dx = \int_0^2 (4x-3)e(2x^2-3x)\frac{dx}{du}du$$
$$= \int_0^2 (4x-3)e(u)\frac{1}{4x-3}du$$
$$= \int_0^2 e(u)du$$
$$= [\exp(u)]_0^2$$
$$= e^2 - e^0$$
$$= e^2 - 1$$

Q51:
Let $u = \tan(x)$ then $\frac{du}{dx} = \sec^2(x)$ and $\frac{dx}{du} = \frac{1}{\sec^2(x)}$
Change the limits:
When $x = 0$, $u = \tan(0) = 0$
When $x = \frac{\pi}{4}$, $u = \tan\left(\frac{\pi}{4}\right) = 1$
Therefore, we have
$$\int_0^{\frac{\pi}{4}} \sec^2(x)\tan^5(x)dx = \int_0^1 \sec^2(x)\tan^5(x)\frac{dx}{du}du$$
$$= \int_0^1 \sec^2(x)u^5 \frac{1}{\sec^2(x)}du$$
$$= \int_0^1 u^5 du$$
$$= \left[\frac{u^6}{6}\right]_0^1$$
$$= \frac{1^6}{6} - \frac{0^6}{6}$$
$$= \frac{1}{6}$$

Q52:
Let $u = \cos(x)$ then $\frac{du}{dx} = -\sin(x)$ and $\frac{dx}{du} = -\frac{1}{\sin(x)}$
Change the limits:
When $x = 0$, $u = \cos(0) = 1$
When $x = \frac{\pi}{3}$, $u = \cos\left(\frac{\pi}{3}\right) = \frac{1}{2}$

Therefore, using the identity $\tan(x) = \frac{\sin(x)}{\cos(x)}$, we have

$$\int_0^{\frac{\pi}{3}} \tan(x)dx = \int_1^{\frac{1}{2}} \frac{\sin(x)}{\cos(x)} \frac{dx}{du} du$$

$$= \int_1^{\frac{1}{2}} \frac{\sin(x)}{u} \frac{-1}{\sin(x)} du$$

$$= -\int_1^{\frac{1}{2}} \frac{1}{u} du$$

$$= -[\ln|u|]_1^{\frac{1}{2}}$$

$$= -\left(\ln\left|\frac{1}{2}\right| - \ln|1|\right)$$

Using the logarithmic rule: $\log_a \left|\frac{x}{y}\right| = \log_a |x| - \log_a |y|$

$$= -(\ln|1| - \ln|2| - \ln|1|)$$
$$= -(-\ln|2|)$$
$$= \ln|2|$$

Definite integrals with substitution and trigonometric identity exercise (page 230)

Q53:
Let $x = \frac{6}{\sqrt{5}} \sin(u)$ then $\frac{dx}{du} = \frac{6}{\sqrt{5}} \cos(u)$ and $x^2 = \frac{36}{5} \sin^2(u)$
The limits must also be changed:
When $x = 0$,
$\frac{6}{\sqrt{5}} \sin(u) = 0$
$\sin(u) = 0$ and so $u = 0$
$u = \sin^{-1}(0)$
$u = 0$
When $x = \frac{6}{\sqrt{5}}$,
$\frac{6}{\sqrt{5}} \sin(u) = \frac{6}{\sqrt{5}}$
$\sin(u) = 1$ and so $u = \frac{\pi}{2}$
$u = \sin^{-1}(1)$
$u = \frac{\pi}{2}$

$$\int_0^{\frac{6}{\sqrt{5}}} \frac{1}{\sqrt{36-5x^2}} dx = \int_0^{\frac{\pi}{2}} \frac{1}{\sqrt{36-5x^2}} \frac{dx}{du} du$$

$$= \int_0^{\frac{\pi}{2}} \frac{1}{\sqrt{36 - 5\left(\frac{36}{5}\sin^2(u)\right)}} \frac{6}{\sqrt{5}} \cos(u) du$$

$$= \frac{6}{\sqrt{5}} \int_0^{\frac{\pi}{2}} \frac{1}{\sqrt{36 - 36\sin^2(u)}} \cos(u) du$$

$$= \frac{6}{\sqrt{5}} \int_0^{\frac{\pi}{2}} \frac{1}{6\sqrt{1-\sin^2(u)}} \cos(u) du \quad \text{Recall that } 1 - \sin^2(x) = \cos^2(x)$$

$$= \frac{6}{\sqrt{5}} \int_0^{\frac{\pi}{2}} \frac{1}{6\sqrt{\cos^2(u)}} \cos(u) du$$

$$= \frac{1}{\sqrt{5}} \int_0^{\frac{\pi}{2}} \frac{1}{\cos(u)} \cos(u) du$$

$$= \frac{1}{\sqrt{5}} \int_0^{\frac{\pi}{2}} 1 du$$

$$= \frac{1}{\sqrt{5}} [u]_0^{\frac{\pi}{2}}$$

$$= \frac{1}{\sqrt{5}} \left(\frac{\pi}{2} - 0\right)$$

$$= \frac{\pi}{2\sqrt{5}}$$

Q54:

Hints:

- Use $1 + \tan^2(x) = \sec^2(x)$

Answer:

Let $x = \frac{3}{2}\tan(u)$ then $\frac{dx}{du} = \frac{3}{2}\sec^2(u)$ and $x^2 = \frac{9}{4}\tan^2(u)$

The limits must also be changed:

When $x = 0$,
$\frac{3}{2}\tan(u) = 0$

$\tan(u) = 0$ and so $u = 0$
$u = \tan^{-1}(0)$
$u = 0$

When $x = \frac{3}{2}$,
$\frac{3}{2}\sin(u) = \frac{3}{2}$

$\tan(u) = 1$ and so $u = \frac{\pi}{4}$
$u = \tan^{-1}(1)$
$u = \frac{\pi}{4}$

ANSWERS: UNIT 1 TOPIC 3

$$\int_0^{\frac{3}{2}} \frac{1}{9+4x^2}dx = \int_0^{\frac{\pi}{4}} \frac{1}{9+4x^2}\frac{dx}{du}du$$

$$= \int_0^{\frac{\pi}{4}} \frac{1}{9+4\left(\frac{9}{4}\tan^2(u)\right)} \frac{3}{2}\sec^2(u)du$$

$$= \frac{3}{2} \int_0^{\frac{\pi}{4}} \frac{1}{9+9\tan^2(u)}\sec^2(u)du$$

$$= \frac{3}{2} \int_0^{\frac{\pi}{4}} \frac{1}{9(1+\tan^2(u))}\sec^2(u)du \quad \text{Recall: } 1+\tan^2(x) = \sec^2(x)$$

$$= \frac{3}{2} \int_0^{\frac{\pi}{4}} \frac{1}{9\sec^2(u)}\sec^2(u)du$$

$$= \frac{1}{6} \int_0^{\frac{\pi}{4}} \frac{1}{\sec^2(u)}\sec^2(u)du$$

$$= \frac{1}{6} \int_0^{\frac{\pi}{4}} 1\,du$$

$$= \frac{1}{6}[u]_0^{\frac{\pi}{4}}$$

$$= \frac{1}{6}\left(\frac{\pi}{4} - 0\right)$$

$$= \frac{\pi}{24}$$

Q55:
Hints:

- Use $\cos(2x) = 2\cos^2(x) - 1$ so $\cos^2(x) = \frac{1}{2}(\cos(2x) + 1)$

Answer:

Let $x = \frac{1}{2}\sin(t)$ then $\frac{dx}{dt} = \frac{1}{2}\cos(t)$ and $x^2 = \frac{1}{4}\sin^2(t)$

The limits must also be changed:

When $x = 0$,
$$\frac{1}{2}\sin(t) = 0$$
$$\sin(t) = 0 \quad \text{and so } t = 0$$
$$t = \sin^{-1}(0)$$
$$t = 0$$

When $x = \frac{1}{4}$,
$$\frac{1}{2}\sin(t) = \frac{1}{4}$$
$$\sin(t) = \frac{1}{2}$$
$$t = \sin^{-1}\left(\frac{1}{2}\right) \quad \text{and so } t = \frac{\pi}{6}$$
$$t = \frac{\pi}{6}$$

© HERIOT-WATT UNIVERSITY

$$\int_0^{\frac{1}{4}} \sqrt{1-4x^2}\,dx = \int_0^{\frac{\pi}{6}} \sqrt{1-4x^2}\,\frac{dx}{dt}\,dt$$

$$= \int_0^{\frac{\pi}{6}} \sqrt{1-4\left(\frac{1}{4}\sin^2(t)\right)}\,\frac{1}{2}\cos(t)\,dt$$

$$= \frac{1}{2}\int_0^{\frac{\pi}{6}} \sqrt{1-\sin^2(t)}\,\cos(t)\,dt \quad \text{Recall: } 1-\sin^2(t) = \cos^2(t)$$

$$= \frac{1}{2}\int_0^{\frac{\pi}{6}} \sqrt{\cos^2(t)}\,\cos(t)\,dt$$

$$= \frac{1}{2}\int_0^{\frac{\pi}{6}} \cos^2(t)\,dt \quad \text{Recall: } \cos^2(x) = \frac{1}{2}(1+\cos(2x))$$

$$= \frac{1}{2}\int_0^{\frac{\pi}{6}} \frac{1}{2}(1+\cos(2t))\,dt$$

$$= \frac{1}{4}\int_0^{\frac{\pi}{6}} (1+\cos(2t))\,dt$$

$$= \frac{1}{4}\left[t + \frac{1}{2}\sin(2t)\right]_0^{\frac{\pi}{6}}$$

$$= \frac{1}{4}\left[\left(\frac{\pi}{6} + \frac{1}{2}\sin\left(\frac{\pi}{3}\right)\right) - \left(0 + \frac{1}{2}\sin(0)\right)\right]$$

$$= \frac{1}{4}\left[\left(\frac{\pi}{6} + \frac{1}{2}\left(\frac{\sqrt{3}}{2}\right)\right) - 0\right]$$

$$= \frac{\pi}{24} + \frac{\sqrt{3}}{16}$$

Integration by substitution (substitution not given) exercise (page 233)

Q56: b) $\int f'(x)[f(x)]^n\,dx$

Q57: e) $\int \frac{f'(x)}{f(x)}\,dx$

Q58: a) $\int f(ax+b)\,dx$

Q59: d) $\int f'(x)e^{f(x)}\,dx$

Q60: c) $\int (ax+b)(cx+d)^n\,dx$

Q61:
From $\int f(ax+b)\,dx$, we are going to use the substitution $u = 4x - 5$.
This gives $\frac{du}{dx} = 4$ and $\frac{dx}{du} = \frac{1}{4}$.

$$\begin{aligned}
\int 2\sec^2(4x-5)\,dx &= 2\int \sec^2(4x-5)\frac{dx}{du}\,du \\
&= 2\int \sec^2(4x-5)\left(\frac{1}{4}\right)du \\
&= \frac{1}{2}\int \sec^2(u)\,du \\
&= \frac{1}{2}\tan(u) + C \\
&= \frac{1}{2}\tan(4x-5) + C
\end{aligned}$$

Q62:
From $\int f'(x)[f(x)]^n\,dx$, we are going to use the substitution $u = 4x^3 + 3x^2 + 5$.
This gives $\frac{du}{dx} = 12x^2 + 6x$ and $\frac{dx}{du} = \frac{1}{12x^2+6x}$.

$$\begin{aligned}
\int (2x^2+x)(4x^3+3x^2+5)^5\,dx &= \int (2x^2+x)(4x^3+3x^2+5)^5 \frac{dx}{du}\,du \\
&= \int (2x^2+x)u^5 \left(\frac{1}{12x^2+6x}\right)du \\
&= \int \frac{2x^2+x}{6(2x^2+x)} u^5\,du \\
&= \frac{1}{6}\int u^5\,du \\
&= \frac{1}{36}u^6 + C \\
&= \frac{1}{36}(4x^3+3x^2+5)^6 + C
\end{aligned}$$

Q63:
From $\int (ax+b)(cx+d)^n\,dx$, we are going to use the substitution $u = 2x - 3$.
This gives $x = \frac{1}{2}(u+3)$, $\frac{du}{dx} = 2$ and $\frac{dx}{du} = \frac{1}{2}$.

$$\begin{aligned}
\int (x+4)(2x-3)^4\,dx &= \int (x+4)(2x-3)^4 \frac{dx}{du}\,du \\
&= \int \left(\frac{1}{2}(u+3)+4\right)u^4 \left(\frac{1}{2}\right)du \\
&= \frac{1}{4}\int u^5 + 11u^4\,du \\
&= \frac{1}{4}\left(\frac{1}{6}u^6 + \frac{11}{5}u^5\right) + C \\
&= \frac{1}{120}u^5(5u+66) + C \\
&= \frac{1}{120}(2x-3)^5(10x+51) + C
\end{aligned}$$

© HERIOT-WATT UNIVERSITY

Q64:

From $\int f'(x) e^{f(x)} dx$, we are going to use the substitution $u = x^2 - 3$.

This gives $\frac{du}{dx} = 2x$ and $\frac{dx}{du} = \frac{1}{2x}$.

$$\int 4xe^{(x^2-3)} dx = \int 4xe^{(x^2-3)} \frac{dx}{du} du$$
$$= \int 4xe^u \left(\frac{1}{2x}\right) du$$
$$= 2\int e^u du$$
$$= 2e^u + C$$
$$= 2e^{(x^2-3)} + C$$

Q65:

From $\int \frac{f'(x)}{f(x)} dx$, we are going to use the substitution $u = 2x^2 - 8x + 3$.

This gives $\frac{du}{dx} = 4x - 8$ and $\frac{dx}{du} = \frac{1}{4x-8}$.

$$\int \frac{2x-4}{2x^2 - 8x + 3} dx = \int \frac{2x-4}{2x^2 - 8x + 3} \frac{dx}{du} du$$
$$= \int \frac{2x-4}{u} \left(\frac{1}{4x-8}\right) du$$
$$= \frac{1}{2} \int \frac{2x-4}{2(2x-4)} \frac{1}{u} du$$
$$= \frac{1}{2} \int \frac{1}{u} du$$
$$= \frac{1}{2} \ln|u| + C$$
$$= \frac{1}{2} \ln|2x^2 - 8x + 3| + C$$

Integration involving inverse trigonometric functions exercise (page 237)

Q66:

$$\int \frac{4}{\sqrt{64 - x^2}} dx = 4\int \frac{1}{\sqrt{64 - x^2}} dx$$
$$= 4\int \frac{1}{\sqrt{8^2 - x^2}} dx$$
$$= 4\sin^{-1}\left(\frac{x}{8}\right) + C$$

ANSWERS: UNIT 1 TOPIC 3 499

Q67:

$$\int \frac{6}{36 + 4x^2} \, dx = 6 \int \frac{1}{4(9 + x^2)} \, dx$$
$$= \frac{6}{4} \int \frac{1}{(3^2 + x^2)} \, dx$$
$$= \frac{6}{4} \times \frac{1}{3} \tan^{-1}\left(\frac{x}{3}\right) + C$$
$$= \frac{1}{2} \tan^{-1}\left(\frac{x}{3}\right) + C$$

Q68:

$$\int_0^{\frac{3}{4}} \frac{dx}{\sqrt{9 - 4x^2}} = \int_0^{\frac{3}{4}} \frac{dx}{\sqrt{4\left(\frac{9}{4} - x^2\right)}}$$
$$= \int_0^{\frac{3}{4}} \frac{dx}{2\sqrt{\left(\left(\frac{3}{2}\right)^2 - x^2\right)}}$$
$$= \frac{1}{2} \int_0^{\frac{3}{4}} \frac{dx}{\sqrt{\left(\left(\frac{3}{2}\right)^2 - x^2\right)}}$$
$$= \left[\frac{1}{2} \sin^{-1}\left(\frac{x}{\frac{3}{2}}\right)\right]_0^{\frac{3}{4}}$$
$$= \frac{1}{2} \left[\sin^{-1}\left(\frac{2x}{3}\right)\right]_0^{\frac{3}{4}}$$
$$= \frac{1}{2} \left(\sin^{-1}\left(\frac{1}{2}\right) - \sin^{-1}(0)\right)$$
$$= \frac{1}{2} \left(\frac{\pi}{6}\right)$$
$$= \frac{\pi}{12}$$

© HERIOT-WATT UNIVERSITY

Q69:

$$\int_0^3 \frac{16}{3x^2+9}dx = \int_0^3 \frac{16}{3(x^2+3)}dx$$
$$=\frac{16}{3}\int_0^3 \frac{1}{x^2+\left(\sqrt{3}\right)^2}dx$$
$$=\frac{16}{3}\left[\frac{1}{\sqrt{3}}\tan^{-1}\left(\frac{x}{\sqrt{3}}\right)\right]_0^3$$
$$=\frac{16}{3\sqrt{3}}\left[\tan^{-1}\left(\frac{x}{\sqrt{3}}\right)\right]_0^3$$
$$=\frac{16}{3\sqrt{3}}\left(\tan^{-1}\left(\frac{3}{\sqrt{3}}\right)-\tan^{-1}(0)\right)$$
$$=\frac{16}{3\sqrt{3}}\left(\tan^{-1}\left(\sqrt{3}\right)\right)$$
$$=\frac{16}{3\sqrt{3}}\left(\frac{\pi}{3}\right)$$
$$=\frac{16\sqrt{3}}{9}\left(\frac{\pi}{3}\right)$$
$$=\frac{16\sqrt{3}\pi}{27}$$

Integration involving inverse tangent and logarithms exercise (page 240)

Q70:

$$\int \frac{x-1}{x^2+4}dx = \int \frac{x}{x^2+4}dx - \int \frac{1}{x^2+4}dx$$
$$= \frac{1}{2}\int \frac{2x}{x^2+4}dx - \int \frac{1}{x^2+2^2}dx$$
$$= \frac{1}{2}\ln\left|x^2+4\right| - \frac{1}{2}\tan^{-1}\left(\frac{x}{2}\right) + C$$

Q71:

$$\int \frac{x+3}{x^2+16}dx = \int \frac{x}{x^2+16}dx + \int \frac{3}{x^2+16}dx$$
$$= \frac{1}{2}\int \frac{2x}{x^2+16}dx + 3\int \frac{1}{x^2+4^2}dx$$
$$= \frac{1}{2}\ln\left|x^2+16\right| + \frac{3}{4}\tan^{-1}\left(\frac{x}{4}\right) + C$$

Q72:

$$\int \frac{x-7}{x^2+1}dx = \int \frac{x}{x^2+1}dx - \int \frac{7}{x^2+1}dx$$
$$= \frac{1}{2}\int \frac{2x}{x^2+1}dx - 7\int \frac{1}{x^2+1^2}dx$$
$$= \frac{1}{2}\ln\left|x^2+1\right| - 7\tan^{-1}(x) + C$$

Q73:

$$\int \frac{x-1}{4x^2+36} dx = \int \frac{x}{4x^2+36} dx - \int \frac{1}{4x^2+36} dx$$
$$= \frac{1}{8} \int \frac{8x}{4x^2+36} dx - \frac{1}{4} \int \frac{1}{x^2+9} dx$$
$$= \frac{1}{8} \int \frac{8x}{4x^2+36} dx - \frac{1}{4} \int \frac{1}{x^2+3^2} dx$$
$$= \frac{1}{8} \ln|4x^2+36| - \frac{1}{4} \cdot \frac{1}{3} \tan^{-1}\left(\frac{x}{3}\right) + C$$
$$= \frac{1}{8} \ln|4x^2+36| - \frac{1}{12} \tan^{-1}\left(\frac{x}{3}\right) + C$$

Q74:

$$\int \frac{3x+2}{5x^2+4} dx = \int \frac{3x}{5x^2+4} dx + \int \frac{2}{5x^2+4} dx$$
$$= \frac{3}{10} \int \frac{10x}{5x^2+4} dx + \int \frac{2}{5x^2+4} dx$$
$$= \frac{3}{10} \int \frac{10x}{5x^2+4} dx + \frac{2}{5} \int \frac{1}{x^2+\frac{4}{5}} dx$$
$$= \frac{3}{10} \int \frac{10x}{5x^2+4} dx + \frac{2}{5} \int \frac{1}{x^2+\left(\frac{2}{\sqrt{5}}\right)^2} dx$$
$$= \frac{3}{10} \ln|5x^2+4| + \frac{2}{5} \times \frac{\sqrt{5}}{2} \tan^{-1}\left(\frac{\sqrt{5}x}{2}\right) + C$$
$$= \frac{3}{10} \ln|5x^2+4| + \frac{\sqrt{5}}{5} \tan^{-1}\left(\frac{\sqrt{5}x}{2}\right) + C$$

Integration involving inverse tangent and completing the square exercise (page 242)

Q75:

Step 1: First we complete the square for the denominator.
$$x^2 - 6x + 12 = (x^2 - 6x) + 12$$
$$= (x-3)^2 - (-3)^2 + 12$$
$$= (x-3)^2 - 9 + 12$$
$$= (x-3)^2 + 3$$

Therefore the integral then becomes: $\int_3^4 \frac{1}{x^2-6x+12} dx = \int_3^4 \frac{1}{(x-3)^2+3} dx$

Step 2: Now apply the method of substitution, remembering to change the limits to the new variable.
Let $u = x - 3$ then $\frac{du}{dx} = 1$ and $\frac{dx}{du} = 1$
When $x = 3$ then $u = 0$
When $x = 4$ then $u = 1$

Therefore the integral then becomes:

$$\int_3^4 \frac{1}{x^2 - 6x + 12} dx = \int_3^4 \frac{1}{(x-3)^2 + 3} dx$$
$$= \int_0^1 \frac{1}{u^2 + 3} \frac{dx}{du} du$$
$$= \int_0^1 \frac{1}{u^2 + 3} \times 1 \, du$$
$$= \int_0^1 \frac{du}{u^2 + 3}$$

Step 3: Integrate with respect to u using the standard results and evaluate for the limits.

$$\int_3^4 \frac{1}{x^2 - 6x + 12} dx = \int_0^1 \frac{du}{u^2 + 3}$$
$$= \int_0^1 \frac{du}{u^2 + (\sqrt{3})^2}$$
$$= \left[\frac{1}{\sqrt{3}} \tan^{-1} \left(\frac{u}{\sqrt{3}} \right) \right]_0^1$$
$$= \left(\frac{1}{\sqrt{3}} \tan^{-1} \left(\frac{1}{\sqrt{3}} \right) \right) - \left(\frac{1}{\sqrt{3}} \tan^{-1} \left(\frac{0}{\sqrt{3}} \right) \right)$$
$$= \frac{1}{\sqrt{3}} \times \frac{\pi}{6}$$
$$= \frac{\sqrt{3}\pi}{18}$$

Q76:

Step 1: First we complete the square for the denominator.

$$\begin{aligned} x^2 + 8x + 19 &= (x^2 + 8x) + 19 \\ &= (x + 4)^2 - 4^2 + 19 \\ &= (x + 4)^2 - 16 + 19 \\ &= (x + 4)^2 + 3 \end{aligned}$$

Therefore the integral then becomes: $\int_{-4}^{-3} \frac{7}{x^2+8x+19} dx = \int_{-4}^{-3} \frac{7}{(x+4)^2+3} dx$

Step 2: Now apply the method of substitution, remembering to change the limits to the new variable.

Let $u = x + 4$ then $\frac{du}{dx} = 1$ and $\frac{dx}{du} = 1$
When $x = -4$ then $u = 0$
When $x = -3$ then $u = 1$

Therefore the integral then becomes:

$$\int_{-4}^{-3} \frac{7}{x^2 + 8x + 19} dx = \int_{-4}^{-3} \frac{7}{(x+4)^2 + 3} dx$$
$$= \int_0^1 \frac{7}{(x+4)^2 + 3} \frac{dx}{du} du$$
$$= \int_0^1 \frac{7}{u^2 + 3} \times 1 \, du$$
$$= 7 \int_0^1 \frac{du}{u^2 + 3}$$

ANSWERS: UNIT 1 TOPIC 3

Step 3: Integrate with respect to u using the standard results and evaluate for the limits.

$$\int_{-4}^{-3} \frac{7}{x^2+8x+19} dx = 7 \int_0^1 \frac{du}{u^2+3}$$
$$= 7 \int_0^1 \frac{du}{u^2+\left(\sqrt{3}\right)^2}$$
$$= \frac{7}{\sqrt{3}} \left[\tan^{-1}\left(\frac{u}{\sqrt{3}}\right) \right]_0^1$$
$$= \frac{7}{\sqrt{3}} \left(\left(\tan^{-1}\left(\frac{1}{\sqrt{3}}\right) \right) - \left(\tan^{-1}\left(\frac{0}{\sqrt{3}}\right) \right) \right)$$
$$= \frac{7}{\sqrt{3}} \times \frac{\pi}{6}$$
$$= \frac{7\sqrt{3}\pi}{18}$$

Challenge question 1 (page 242)

Expected answer

$\int \frac{x-2}{x^2+2x+5} dx$

For $x^2 + 2x + 5$, $b^2 - 4ac < 0$ so the first step is to complete the square on the denominator.

$$x^2 + 2x + 5 = (x+1)^2 - 1^2 + 5$$
$$= (x+1)^2 - 1 + 5$$
$$= (x+1)^2 + 4$$

The integrand now becomes:

$$\int \frac{x-2}{x^2+2x+5} dx = \int \frac{x-2}{(x+1)^2+4} dx$$
$$= \int \frac{x}{(x+1)^2+4} - \frac{2}{(x+1)^2+4} dx$$
$$= \int \frac{x}{(x+1)^2+4} dx - \int \frac{2}{(x+1)^2+4} dx$$

For both integrands we will use substitution.

Let $u = x + 1$, then $\frac{du}{dx} = 1$ and $\frac{dx}{du} = 1$

Also, since $u = x + 1$ then $x = u - 1$

$$\int \frac{x-2}{x^2+2x+5} dx = \int \frac{x}{(x+1)^2+4} \frac{dx}{du} du - \int \frac{2}{(x+1)^2+4} \frac{dx}{du} du$$
$$= \int \frac{u-1}{u^2+4} \times 1 \times du - \int \frac{2}{u^2+4} \times 1 \times du$$
$$= \int \frac{u}{u^2+4} du - \int \frac{1}{u^2+4} du - \int \frac{2}{u^2+4} du$$

The first integrand will integrate to a logarithmic function. The numerator must therefore look like the derivative of the denominator i.e. $2u$, so multiply the integrand by $\frac{2}{2}$ and take the fraction $\frac{1}{2}$ to the front of the integrand.

© HERIOT-WATT UNIVERSITY

Then use the standard result $\int \frac{f'(x)}{f(x)} dx = \ln |f(x)| + C$ to integrate.

To integrate the second and third integral use the standard result $\int \frac{dx}{a^2+x^2} dx = \frac{1}{a} tan^{-1} \left(\frac{x}{a}\right) + C$

$$\int \frac{x-2}{x^2+2x+5} dx = \frac{1}{2} \int \frac{2u}{u^2+4} du - \int \frac{1}{u^2+4} du - 2 \int \frac{1}{u^2+4} du$$

$$= \frac{1}{2} \int \frac{2u}{u^2+4} du - \int \frac{1}{u^2+2^2} du - 2 \int \frac{1}{u^2+2^2} du$$

$$= \frac{1}{2} ln |u^2+4| - \frac{1}{2} tan^{-1}\left(\frac{u}{2}\right) - 2 \times \frac{1}{2} tan^{-1}\left(\frac{u}{2}\right) + C$$

$$= \frac{1}{2} ln |u^2+4| - \frac{1}{2} tan^{-1}\left(\frac{u}{2}\right) - tan^{-1}\left(\frac{u}{2}\right) + C$$

$$= \frac{1}{2} ln |u^2+4| - \frac{3}{2} tan^{-1}\left(\frac{u}{2}\right) + C$$

Substitute back for the original variable u:

$$\int \frac{x-2}{x^2+2x+5} dx = \frac{1}{2} ln |u^2+4| - \frac{3}{2} tan^{-1}\left(\frac{u}{2}\right) + C$$

$$= \frac{1}{2} ln \left|(x+1)^2+4\right| - \frac{3}{2} tan^{-1}\left(\frac{x+1}{2}\right) + C$$

$$= \frac{1}{2} ln |x^2+2x+5| - \frac{3}{2} tan^{-1}\left(\frac{x+1}{2}\right) + C$$

Distinct linear factors exercise (page 244)

Q77:

Step 1: Rewrite $\frac{9}{2x^3+3x^2-3x-2}$ as partial fractions.

To be able to split the rational function into the sum of partial fractions first factorise the denominator. (Synthetic division could be used.)

$2x^3 + 3x^2 - 3x - 2 = (x-1)(x+2)(2x+1)$

$\frac{9}{(x-1)(x+2)(2x+1)} = \frac{A}{(x-1)} + \frac{B}{(x+2)} + \frac{C}{(2x+1)}$

Step 2: Obtain a common denominator and equation the numerators.

$A(x+2)(2x+1) + B(x-1)(2x+1) + C(x-1)(x+2) = 9$

Step 3: Find the values of A, B and C.

Let $x = 1$ then $A(3)(3) = 9$
$A = 1$

Let $x = -2$ then $B(-3)(-3) = 9$
$B = 1$

Let $x = 0$ then $A(2)(1) + B(-1)(1) + C(-1)(2) = 9$
We know that $A = 1$ and $B = 1$ which gives $(1)(2)(1) + (1)(-1)(1) + C(-1)(2) = 9$
$C = -4$

Step 4: Substitute the values of A, B and C into the equation.

$\frac{9}{(x-1)(x+2)(2x+1)} = \frac{1}{(x-1)} + \frac{1}{(x+2)} + \frac{4}{(2x+1)}$

Step 5: Now integrate.

$\int \frac{9}{2x^3+3x^2-3x-2} dx = \ln|x-1| + \ln|x+2| - 2\ln|2x+1| + C$

ANSWERS: UNIT 1 TOPIC 3 505

Q78:

Step 1: Rewrite $\frac{-x^2+x+22}{x^3+3x^2-4x-12}$ as partial fractions.
To be able to split the rational function into the sum of partial fractions first factorise the denominator. (Synthetic division could be used.)
$x^3 + 3x^2 - 4x - 12 = (x - 2)(x + 2)(x + 3)$
$\frac{-x^2+x+22}{(x-2)(x+2)(x+3)} = \frac{A}{(x-2)} + \frac{B}{(x+2)} + \frac{C}{(x+3)}$

Step 2: Obtain a common denominator and equation the numerators.
$A(x+2)(x+3) + B(x-2)(x+3) + C(x-2)(x+2) = -x^2 + x + 22$

Step 3: Find the values of A, B and C.
Let $x = 2$ then $A(4)(5) = -4 + 2 + 22$
$A = 1$
Let $x = -2$ then $B(-4)(1) = -4 - 2 + 22$
$B = -4$
Let $x = 0$ then $A(2)(3) + B(-2)(3) + C(-2)(2) = 22$
We know that $A = 1$ and $B = -4$ which gives $(1)(2)(3) + (-4)(-2)(3) + C(-2)(2) = 22$
$C = 2$

Step 4: Substitute the values of A, B and C into the equation.
$\frac{-x^2+x+22}{(x-2)(x+2)(x+3)} = \frac{1}{(x-2)} + \frac{4}{(x+2)} + \frac{2}{(x+3)}$

Step 5: Now integrate.
$\int \frac{-x^2 + x + 22}{x^3 + 3x^2 - 4x - 12} dx = \ln|x - 2| - 4\ln|x + 2| + 2\ln|x + 3| + C$

Repeated linear factors exercise (page 247)

Q79:

Step 1: Rewrite the indefinite integral as the sum of partial fractions.
$\frac{x^2 - 4}{(x + 1)^3} = \frac{A}{(x + 1)} + \frac{B}{(x + 1)^2} + \frac{C}{(x + 1)^3}$

Step 2: Obtain a common denominator and equate numerators.
$x^2 - 4 = A(x + 1)^2 + B(x + 1) + C$

Step 3: Find the values of A, B and C.
Let $x = -1$ then $-3 = 0 \times A + 0 \times B + C$
$C = -3$
Let $x = 0$ then $-4 = A + B - 3$
$A + B = -1$
Let $x = 1$ then $-3 = 4A + 2B - 3$
$4A + 2B = 0$
Using simultaneous equations gives us $2A = 2$
$A = 1$
$A + B = -1$ and $A = 1$
$B = -2$

© HERIOT-WATT UNIVERSITY

Step 4: Substitute the values of A, B and C into the original equation.
$$\frac{x^2-4}{(x+1)^3} = \frac{1}{x+1} - \frac{2}{(x+1)^2} - \frac{3}{(x+1)^3}$$
Step 5: Integrate, begin by rewriting into a form that can be integrated.
$$\int \frac{x^2-4}{(x+1)^3}\,dx = \int \frac{1}{x+1} - 2(x+1)^{-2} - 3(x+1)^{-3}\,dx$$
$$= \ln|x+1| - \frac{2}{-1}(x+1)^{-1} - \frac{3}{-2}(x+1)^{-2} + C$$
$$= \ln|x+1| + \frac{2}{x+1} + \frac{3}{2(x+1)^2} + C$$

Q80:
Step 1: Factorise the denominator. You may choose to use synthetic division.
$$x^3 - 6x^2 + 12x - 8 = (x-2)^3$$
Step 2: Rewrite the integral as the sum of partial fractions.
$$\frac{2x^2-7x+5}{(x-2)^3} = \frac{A}{(x-2)} + \frac{B}{(x-2)^2} + \frac{C}{(x-2)^3}$$
Step 3: Obtain a common denominator and equate numerators.
$$2x^2 - 7x + 5 = A(x-2)^2 + B(x-2) + C$$
Step 4: Find the values of A, B and C.
Let $x = 2$ then $8 - 14 + 5 = 0 \times A + 0 \times B + C$
$C = -1$
Let $x = 0$ then $5 = 4A - 2B - 1$
$2A - B = 3$
Let $x = 1$ then $0 = A - B - 1$
$A - B = 1$
Using simultaneous equations gives $A = 2$
$A - B = 1$ and $A = 2$
$B = 1$
Step 5: Substitute the values of A, B and C into the original equation.
$$\frac{2x^2-7x+5}{x^3-6x^2+12x-8} = \frac{2}{x-2} + \frac{1}{(x-2)^2} - \frac{1}{(x-2)^3}$$
Step 6: Integrate, begin by rewriting into a form that can be integrated.
$$\int_3^4 \frac{2x^2-7x+5}{(x-2)^3}\,dx = \int_3^4 \frac{2}{x-2} + (x-2)^{-2} - (x-2)^{-3}\,dx$$
$$= \left[2\ln|x-2| - (x-2)^{-1} - \frac{1}{-2}(x-2)^{-2}\right]_3^4$$
$$= \left[2\ln|x-2| - \frac{1}{x-2} + \frac{1}{2(x-2)^2}\right]_3^4$$
$$= \left(2\ln|2| - \frac{1}{2} + \frac{1}{8}\right) - \left(2\ln|1| - 1 + \frac{1}{2}\right)$$
$$= 2\ln|2| + \frac{1}{8}$$

Irreducible quadratic factors exercise (page 250)

Q81:
Step 1: Factorise the denominator. We could use synthetic division.
For synthetic division let $x = 3$.

$$\begin{array}{c|cccc} 3 & 1 & -3 & 2 & -6 \\ & 0 & 3 & 0 & 6 \\ \hline & 1 & 0 & 2 & 0 \end{array}$$

Since the remainder is zero then $(x - 3)$ is a factor.
$x^3 - 3x^2 + 2x - 6 = (x - 3)(x^2 + 2)$

Step 2: Write the integrand as the sum of partial fractions.

$$\frac{4x^2 - 9x + 13}{x^3 - 3x^2 + 2x - 6} = \frac{4x^2 - 9x + 13}{(x - 3)(x^2 + 2)}$$
$$= \frac{A}{x - 3} + \frac{Bx + C}{x^2 + 2}$$

Step 3: Obtain a common denominator and equate the numerators.

$$\frac{4x^2 - 9x + 13}{x^3 - 3x^2 + 2x - 6} = \frac{A(x^2 + 2)}{(x - 3)(x^2 + 2)} + \frac{(Bx + C)(x - 3)}{(x - 3)(x^2 + 2)}$$
$$4x^2 - 9x + 13 = A(x^2 + 2) + (Bx + C)(x - 3)$$

Step 4: Find the values of A, B and C.
Let $x = 3$ then
$$4(3)^2 - 9(3) + 13 = A(3^2 + 2) + (B(3) + C)(3 - 3)$$
$$22 = 11A$$
$$A = 2$$

Given $A = 2$ let $x = 0$ then,
$$4(0)^2 - 9(0) + 13 = 2(0^2 + 2) + (B(0) + C)(0 - 3)$$
$$13 = 4 - 3C$$
$$9 = -3C$$
$$C = -3$$

Given $A = 2$ and $C = -3$ let $x = 1$ then,
$$4(1)^2 - 9(1) + 13 = 2(1^2 + 2) + (B(1) - 3)(1 - 3)$$
$$8 = 6 - 2B + 6$$
$$-4 = -2B$$
$$B = 2$$

Step 5: Substitute $A = 2$, $B = 2$ and $C = -3$ back into the sum of partial fractions

$$\frac{4x^2 - 9x + 13}{x^3 - 3x^2 + 2x - 6} = \frac{2}{x - 3} + \frac{2x - 3}{x^2 + 2}$$

Step 6: Integrate, begin by rewriting into a form that can be integrated.

$$\int \frac{4x^2 - 9x + 13}{x^3 - 3x^2 + 2x - 6} dx = \int \frac{2}{x-3} + \frac{2x-3}{x^2+2} dx$$

$$= \int \frac{2}{x-3} dx + \int \frac{2x}{x^2+2} dx - \int \frac{3}{x^2+2} dx$$

$$= 2 \int \frac{1}{x-3} dx + \int \frac{2x}{x^2+2} dx - 3 \int \frac{1}{x^2 + \left(\sqrt{2}\right)^2} dx$$

$$= 2ln|x-3| + ln|x^2+2| - 3 \times \frac{1}{\sqrt{2}} tan^{-1}\left(\frac{x}{\sqrt{2}}\right) + C$$

$$= 2ln|x-3| + ln|x^2+2| - \frac{3}{\sqrt{2}} tan^{-1}\left(\frac{x}{\sqrt{2}}\right) + C$$

Q82:

Step 1: Write the integrand as the sum of partial fractions.
$\frac{6x^2+x+11}{(x+1)(x^2+3)} = \frac{A}{x+1} + \frac{Bx+C}{x^2+3}$

Step 2: Obtain a common denominator and equate the numerators.

$$\frac{6x^2 + x + 11}{(x+1)(x^2+3)} = \frac{A(x^2+3)}{(x+1)(x^2+3)} + \frac{(Bx+C)(x+1)}{(x+1)(x^2+3)}$$

$$6x^2 + x + 11 = A(x^2+3) + (Bx+C)(x+1)$$

Step 3: Find the values of A, B and C.

Let $x = -1$ then,

$$6(-1)^2 + (-1) + 11 = A\left((-1)^2+3\right) + (B(-1)+C)(-1+1)$$

$$16 = 4A$$

$$A = 4$$

Given $A = 4$ let $x = 0$ then,

$$6(0)^2 + 0 + 11 = 4(0^2+3) + (B(0)+C)(0+1)$$

$$11 = 12 + C$$

$$C = -1$$

Given $A = 4$ and $C = -1$ let $x = 1$ then,

$$6(1)^2 + 1 + 11 = 4(1^2+3) + (B(1)-1)(1+1)$$

$$18 = 16 + 2B - 2$$

$$4 = 2B$$

$$B = 2$$

Step 4: Substitute $A = 4$, $B = 2$ and $C = -1$ back into the sum of partial fractions.

$\frac{6x^2+x+11}{(x+1)(x^2+3)} = \frac{4}{x+1} + \frac{2x-1}{x^2+3}$

Step 5: Integrate, begin by rewriting into a form that can be integrated.

$$\int_0^1 \frac{6x^2 + x + 11}{(x+1)(x^2+3)} dx = \int_0^1 \frac{4}{x+1} + \frac{2x-1}{x^2+3} dx$$

$$= 4\int_0^1 \frac{1}{x+1} dx + \int_0^1 \frac{2x}{x^2+3} - \int_0^1 \frac{1}{x^2 + (\sqrt{3})^2} dx$$

$$= 4\left[ln|x+1|\right]_0^1 + \left[ln|x^2+3|\right]_0^1 + \left[\frac{1}{\sqrt{3}} tan^{-1}\left(\frac{x}{\sqrt{3}}\right)\right]_0^1$$

$$= 4\left[ln|x+1|\right]_0^1 + \left[ln|x^2+3|\right]_0^1 + \frac{1}{\sqrt{3}}\left[tan^{-1}\left(\frac{x}{\sqrt{3}}\right)\right]_0^1$$

$$= 4(ln|2| - ln|1|) + (ln|4| - ln|3|) + \frac{1}{\sqrt{3}}\left(tan^{-1}\left(\frac{1}{\sqrt{3}}\right) - tan^{-1}(0)\right)$$

$$= 4ln|2| + ln|2^2| - ln|3| + \frac{1}{\sqrt{3}}\left(\frac{\pi}{6}\right)$$

$$= 4ln|2| + 2ln|2| - ln|3| + \frac{\pi}{6\sqrt{3}}$$

$$= 6ln|2| - ln|3| + \frac{\sqrt{3}\pi}{18}$$

Recall the logarithmic rules:

- $\ln x^a = a \ln x$
- $\ln 1 = 0$

Challenge question 2 (page 250)

Expected answer

Step 1: Write the integrand as the sum of partial fractions, noting that $x^2 + x + 1$ i.e. $b^2 - 4ac < 0$.

$\frac{1}{x(x^2+x+1)} = \frac{A}{x} + \frac{Bx+C}{x^2+x+1}$

Step 2: Obtain a common denominator and equate the numerators.

$$\frac{1}{x(x^2+x+1)} = \frac{A(x^2+x+1)}{x(x^2+x+1)} + \frac{(Bx+C)x}{x(x^2+x+1)}$$

$$1 = A(x^2+x+1) + (Bx+C)x$$

$$1 = A(x^2+x+1) + Bx^2 + Cx$$

Step 3: Find the values of A, B and C.

Let $x = 0$ then,

$1 = A(0^2 + 0 + 1) + B(0)^2 + C(0)$

$1 = A$

$A = 1$

Given $A = 1$ let $x = 1$ then,

$1 = (1)(1^2 + 1 + 1) + B(1)^2 + C(1)$

$1 = 3 + B + C$

$-2 = B + C$ (1)

Given $A = 1$ let $x = -1$ then,

$1 = (1)\left((-1)^2 + (-1) + 1\right) + B(-1)^2 + C(-1)$

$1 = 1 + B - C$

$0 = B - C$

$B = C$

Substitute $B = C$ into (1):

$-2 = C + C$

$-2 = 2C$

$C = -1$

Since $B = C$, then $B = -1$.

Step 4: Substitute $A = 1$, $B = -1$ and $C = -1$ back into the sum of partial fractions.

$\frac{1}{x(x^2+x+1)} = \frac{1}{x} + \frac{-x-1}{x^2+x+1}$

Step 5: Integrate, begin by rewriting into a form that can be integrated.

$$\int \frac{1}{x(x^2+x+1)} dx = \int \frac{1}{x} + \frac{-x-1}{x^2+x+1} dx$$

$$= \int \frac{1}{x} dx - \int \frac{x+1}{x^2+x+1} dx$$

The numerator of the second integrand must look like the derivative of the denominator i.e. $\frac{d}{dx}\{x^2+x+1\} = 2x+1$

$$\int \frac{1}{x(x^2+x+1)} dx = \int \frac{1}{x} dx - \int \frac{\frac{1}{2}(2x+1) + \frac{1}{2}}{x^2+x+1} dx$$

$$= \int \frac{1}{x} dx - \frac{1}{2}\int \frac{2x+1}{x^2+x+1} dx - \frac{1}{2}\int \frac{1}{x^2+x+1} dx$$

Before integrating we must complete the square for the denominator of the third integrand.

$x^2 + x + 1 = \left(x + \frac{1}{2}\right)^2 - \left(\frac{1}{2}\right)^2 + 1$

$= \left(x + \frac{1}{2}\right)^2 - \frac{1}{4} + 1$

$= \left(x + \frac{1}{2}\right)^2 + \frac{3}{4}$

ANSWERS: UNIT 1 TOPIC 3   511

To get rid of the fraction inside the bracket write $x + \frac{1}{2}$ as one fraction and take $\frac{1}{2}$ out as a common factor.

$$\begin{aligned}
x^2 + x + 1 &= \left(x + \frac{1}{2}\right)^2 + \frac{3}{4} \\
&= \left(\frac{2x+1}{2}\right)^2 + \frac{3}{4} \\
&= \left(\frac{1}{2}(2x+1)\right)^2 + \frac{3}{4} \\
&= \frac{1}{4}(2x+1)^2 + \frac{3}{4} \\
&= \frac{1}{4}\left((2x+1)^2 + 3\right)
\end{aligned}$$

We therefore have,

$$\begin{aligned}
\int \frac{1}{x(x^2+x+1)} dx &= \int \frac{1}{x} dx - \frac{1}{2}\int \frac{2x+1}{x^2+x+1} dx - \frac{1}{2}\int \frac{1}{\frac{1}{4}\left((2x+1)^2+3\right)} dx \\
&= \int \frac{1}{x} dx - \frac{1}{2}\int \frac{2x+1}{x^2+x+1} dx - 2\int \frac{1}{(2x+1)^2+(\sqrt{3})^2} dx \\
&= \ln|x| - \frac{1}{2}\ln|x^2+x+1| - 2\int \frac{1}{(2x+1)^2+(\sqrt{3})^2} dx \quad (2)
\end{aligned}$$

Use substitution to integrate the third integrand.

$\int \frac{1}{(2x+1)^2+(\sqrt{3})^2} dx$

Let $u = 2x+1$, then $\frac{du}{dx} = 2$ and $\frac{dx}{du} = \frac{1}{2}$ giving,

$$\begin{aligned}
\int \frac{1}{(2x+1)^2+(\sqrt{3})^2} dx &= \int \frac{1}{(2x+1)^2+(\sqrt{3})^2} \frac{dx}{du} du \\
&= \int \frac{1}{u^2+(\sqrt{3})^2} \times \frac{1}{2} du \\
&= \frac{1}{2}\int \frac{1}{u^2+(\sqrt{3})^2} du
\end{aligned}$$

Using the standard result $\int \frac{dx}{a^2+x^2} dx = \frac{1}{a}\tan^{-1}\left(\frac{x}{a}\right) + C$ we get:

$$\begin{aligned}
\int \frac{1}{(2x+1)^2+(\sqrt{3})^2} dx &= \frac{1}{2} \times \frac{1}{\sqrt{3}}\tan^{-1}\left(\frac{u}{\sqrt{3}}\right) + C \\
&= \frac{1}{2\sqrt{3}}\tan^{-1}\left(\frac{2x+1}{\sqrt{3}}\right) + C
\end{aligned}$$

Substitute this result back into (2) to obtain the solution.

$$\begin{aligned}
\int \frac{1}{x(x^2+x+1)} dx &= \ln|x| - \frac{1}{2}\ln|x^2+x+1| - 2 \times \frac{1}{2\sqrt{3}}\tan^{-1}\left(\frac{2x+1}{\sqrt{3}}\right) + C \\
&= \ln|x| - \frac{1}{2}\ln|x^2+x+1| - \frac{1}{\sqrt{3}}\tan^{-1}\left(\frac{2x+1}{\sqrt{3}}\right) + C
\end{aligned}$$

© HERIOT-WATT UNIVERSITY

Improper rational functions exercise (page 252)

Q83:

$\int \frac{2x^3 + 5x^2 - 9x - 18}{x^2 + x - 6} dx$

The integrand is an improper ration function. The degree of the numerator is larger than the degree of the denominator. We must therefore use *algebraic long division*.

Step 1: Algebraic long division

```
                    2x  +  3
           ─────────────────────────
x² + x - 6 │ 2x³ + 5x²  - 9x  - 18
             2x³ + 2x²  -12x   ↓
             ───────────────
                   3x²  + 3x  - 18
                   3x²  + 3x  - 18
                   ───────────────
                                 0
```

We therefore have,

$\frac{2x^3+5x^2-9x-18}{x^2+x-6} = 2x+3$

Step 2: Integrate

Since the solution has a remainder of zero we simply integrate the quotient function as follows:

$$\int \frac{2x^3 + 5x^2 - 9x - 18}{x^2 + x - 6} dx = \int 2x + 3 dx$$
$$= \frac{2x^2}{2} + 3x + C$$
$$= x^2 + 3x + C$$

Q84:

$\int \frac{4x^3+6x^2-17x-36}{x^3+x^2-8x-12} dx$

The integrand is an improper ration function. The degree of the numerator is equal to the degree of the denominator. We must therefore use algebraic long division.

Step 1: Algebraic long division

```
                           4
                ─────────────────────────
x³ + x² - 8x - 12 │ 4x³ + 6x²  - 17x  - 36
                    4x³ + 4x²  - 32x  - 48
                    ─────────────────────
                          2x²  + 15x  + 12
```

We therefore have,

$\frac{4x^3+6x^2-17x-36}{x^3+x^2-8x-12} = 4 + \frac{2x^2+15x+12}{x^3+x^2-8x-12}$

Step 2: Now consider the proper rational function and factorise the denominator. We can use synthetic division to do this.

Let $x = 3$

$$\begin{array}{c|rrrr} 3 & 1 & 1 & -8 & -12 \\ & 0 & 3 & 12 & 12 \\ \hline & 1 & 4 & 4 & 0 \end{array}$$

Since the remainder is zero $x = 3$ is a root and so $(x - 3)$ is a factor.
We therefore have,
$$x^3 + x^2 - 8x - 12 = (x - 3)(x^2 + 4x + 4)$$
$$= (x - 3)(x + 2)^2$$
Which gives us,
$$\frac{2x^2+15x+12}{x^3+x^2-8x-12} = \frac{2x^2+15x+12}{(x-3)(x+2)^2}$$

Step 3: Write the as the sum of partial fractions.
$$\frac{2x^2+15x+12}{(x-3)(x+2)^2} = \frac{A}{x-3} + \frac{B}{x+2} + \frac{C}{(x+2)^2}$$

Step 4: Obtain a common denominator and equate the numerators.
$$\frac{2x^2 + 15x + 12}{(x - 3)(x + 2)^2} = \frac{A(x + 2)^2}{(x - 3)(x + 2)^2} + \frac{B(x - 3)(x + 2)}{(x - 3)(x + 2)^2} + \frac{C(x - 3)}{(x - 3)(x + 2)^2}$$
$$2x^2 + 15x + 12 = A(x + 2)^2 + B(x - 3)(x + 2) + C(x - 3)$$

Step 5: Find the values of A, B and C.

Let $x = 3$ then,
$$2(3)^2 + 15(3) + 12 = A(3 + 2)^2 + B(3 - 3)(3 + 2) + C(3 - 3)$$
$$75 = 25A$$
$$A = 3$$

Given $A = 3$ let $x = -2$ then,
$$2(-2)^2 + 15(-2) + 12 = 3(-2 + 2)^2 + B(-2 - 3)(-2 + 2) + C(-2 - 3)$$
$$-10 = -5C$$
$$C = 2$$

Given $A = 3$ and $C = 2$ let $x = 0$ then,
$$2(0)^2 + 15(0) + 12 = 3(0 + 2)^2 + B(0 - 3)(0 + 2) + (2)(0 - 3)$$
$$12 = 12 - 6B - 6$$
$$6 = -6B$$
$$B = -1$$

Step 6: Substitute $A = 3$, $B = -1$ and $C = 2$ back into the sum of partial fractions.
$$\frac{2x^2+15x+12}{(x-3)(x+2)^2} = \frac{3}{x-3} - \frac{1}{x+2} + \frac{2}{(x+2)^2}$$

© HERIOT-WATT UNIVERSITY

Step 7: Integrate, begin by rewriting into a form that can be integrated.

$$\int \frac{4x^3 + 6x^2 - 17x - 36}{x^3 + x^2 - 8x - 12} dx = \int 4 + \frac{2x^2 + 15x + 12}{(x-3)(x+2)^2} dx$$

$$= \int 4 + \frac{3}{x-3} - \frac{1}{x+2} + \frac{2}{(x+2)^2} dx$$

$$= \int 4 dx + 3 \int \frac{1}{x-3} dx - \int \frac{1}{x+2} dx + 2 \int (x+2)^{-2} dx$$

$$= 4x + 3ln\,|x-3| - ln\,|x+2| + \frac{2}{(-1)}(x+2)^{-1} + C$$

$$= 4x + 3ln\,|x-3| - ln\,|x+2| + \frac{2}{(x+2)} + C$$

Q85:

$\int \frac{x^3 + 4x^2 + 3x + 2}{x^3 - x^2 + 4x - 4} dx$

The integrand is an improper ration function. The degree of the numerator is equal to the degree of the denominator. We must therefore use algebraic long division.

Step 1: Algebraic long division

$$\begin{array}{r} 1 \\ x^3 - x^2 + 4x - 4 \overline{) x^3 + 4x^2 + 3x + 2} \\ \underline{x^3 - x^2 + 4x - 4} \\ 5x^2 - x + 6 \end{array}$$

We therefore have,

$\frac{x^3+4x^2+3x+2}{x^3-x^2+4x-4} = 1 + \frac{5x^2-x+6}{x^3-x^2+4x-4}$

Step 2: Now consider the proper rational function and factorise the denominator. We can use synthetic division to do this.

Let $x = 1$,

$$\begin{array}{c|cccc} 1 & 1 & -1 & 4 & -4 \\ & 0 & 1 & 0 & 4 \\ \hline & 1 & 0 & 4 & 0 \end{array}$$

Since the remainder is zero $x = 1$ is a root and so $(x - 1)$ is a factor.

We therefore have,

$x^3 - x^2 + 4x - 4 = (x-1)(x^2 + 4)$

Which gives us,

$\frac{5x^2-x+6}{x^3-x^2+4x-4} = \frac{5x^2-x+6}{(x-1)(x^2+4)}$

Step 3: Write the as the sum of partial fractions.

$\frac{5x^2-x+6}{(x-1)(x^2+4)} = \frac{A}{x-1} + \frac{Bx+C}{x^2+4}$

Step 4: Obtain a common denominator and equate the numerators.
$$\frac{5x^2 - x + 6}{(x-1)(x^2+4)} = \frac{A(x^2+4)}{(x-1)(x^2+4)} + \frac{(Bx+C)(x-1)}{(x-1)(x^2+4)}$$
$$5x^2 - x + 6 = A(x^2+4) + (Bx+C)(x-1)$$

Step 5: Find the values of A, B and C.

Let $x = 1$ then,
$$5(1)^2 - 1 + 6 = A\left((1)^2 + 4\right) + (B(1) + C)(1-1)$$
$$10 = 5A$$
$$A = 2$$

Given $A = 2$ let $x = 0$ then,
$$5(0)^2 - 0 + 6 = 2(0^2 + 4) + (B(0) + C)(0 - 1)$$
$$6 = 8 - C$$
$$-2 = -C$$
$$C = 2$$

Given $A = 2$ and $C = 2$ let $x = -1$ then,
$$5(-1)^2 - (-1) + 6 = 2\left((-1)^2 + 4\right) + (B(-1) + 2)(-1 - 1)$$
$$12 = 10 + 2B - 4$$
$$6 = 2B$$
$$B = 3$$

Step 6: Substitute $A = 2$, $B = 3$ and $C = 2$ back into the sum of partial fractions
$$\frac{5x^2 - x + 6}{(x-1)(x^2+4)} = \frac{2}{x-1} + \frac{3x+2}{x^2+4}$$
$$= \frac{2}{x-1} + \frac{3x}{x^2+4} + \frac{2}{x^2+4}$$

Step 7: Integrate, begin by rewriting into a form that can be integrated.
$$\int \frac{x^3 + 4x^2 + 3x + 2}{x^3 - x^2 + 4x - 4} dx = \int 1 + \frac{2x^2 + 15x + 12}{(x-3)(x+2)^2} dx$$
$$= \int 1 + \frac{2}{x-1} + \frac{3x}{x^2+4} + \frac{2}{x^2+4} dx$$
$$= \int 1 dx + 2\int \frac{1}{x-1} dx + 3\int \frac{x}{x^2+4} dx + 2\int \frac{1}{x^2+4} dx$$
$$= \int 1 dx + 2\int \frac{1}{x-1} dx + \frac{3}{2}\int \frac{2x}{x^2+4} dx + 2\int \frac{1}{x^2+2^2} dx$$
$$= x + 2ln|x-1| + \frac{3}{2}ln|x^2+4| + 2 \times \frac{1}{2}tan^{-1}\left(\frac{x}{2}\right) + C$$
$$= x + 2ln|x-1| + \frac{3}{2}ln|x^2+4| + tan^{-1}\left(\frac{x}{2}\right) + C$$

© HERIOT-WATT UNIVERSITY

One application exercise (page 255)

Q86:

$\int f(x)g'(x)\,dx = f(x)g(x) - \int f'(x)g(x)\,dx$

Let $f(x) = 3x + 1$ and $g'(x) = e^{2x}$ then $f'(x) = 3$ and $g(x) = \frac{1}{2}e^{2x}$

$$\begin{aligned}
\int (3x+1)e^{2x}\,dx &= (3x+1) \times \frac{1}{2}e^{2x} - \int 3 \times \frac{1}{2}e^{2x}\,dx \\
&= \frac{1}{2}(3x+1)e^{2x} - \frac{3}{2}\int e^{2x}\,dx \\
&= \frac{1}{2}(3x+1)e^{2x} - \frac{3}{2} \times \frac{1}{2}e^{2x} + C \\
&= \frac{1}{2}(3x+1)e^{2x} - \frac{3}{4}e^{2x} + C
\end{aligned}$$

Q87:

$\int f(x)g'(x)\,dx = f(x)g(x) - \int f'(x)g(x)\,dx$

Let $f(x) = 2x - 3$ and $g'(x) = \sin(2x)$ then $f'(x) = 2$ and $g(x) = -\frac{1}{2}\cos(2x)$

$$\begin{aligned}
\int (2x-3)\sin(2x)\,dx &= (2x-3)\left(-\frac{1}{2}\cos(2x)\right) - \int 2\left(-\frac{1}{2}\cos(2x)\right)dx \\
&= -\frac{1}{2}(2x-3)\cos(2x) + \int \cos(2x)\,dx \\
&= -\frac{1}{2}(2x-3)\cos(2x) + \frac{1}{2}\sin(2x) + C
\end{aligned}$$

Q88:

$\int f(x)g'(x)\,dx = f(x)g(x) - \int f'(x)g(x)\,dx$

Let $f(x) = 2\ln|x|$ and $g'(x) = x^{-4}$ then $f'(x) = 2x^{-1}$ and $g(x) = -\frac{1}{3}x^{-3}$

$$\begin{aligned}
\int_1^2 \frac{2\ln|x|}{x^4}\,dx &= \left[2\ln|x| \times \left(-\frac{1}{3}x^{-3}\right)\right]_1^2 - \int_1^2 2x^{-1} \times \left(-\frac{1}{3}x^{-3}\right)dx \\
&= \left[-\frac{2}{3x^3}\ln|x|\right]_1^2 + \frac{2}{3}\int_1^2 x^{-4}\,dx \\
&= -\frac{2}{3}\left[\frac{1}{x^3}\ln|x|\right]_1^2 + \frac{2}{3}\left[-\frac{1}{3}x^{-3}\right]_1^2 \\
&= -\frac{2}{3}\left[\frac{1}{x^3}\ln|x|\right]_1^2 - \frac{2}{9}\left[\frac{1}{x^3}\right]_1^2 \\
&= -\frac{2}{3}\left(\frac{1}{2^3}\ln|2| - \frac{1}{1^3}\ln|1|\right) - \frac{2}{9}\left(\frac{1}{2^3} - \frac{1}{1^3}\right) \\
&= -\frac{2}{3}\left(\frac{1}{8}\ln|2|\right) - \frac{2}{9}\left(\frac{1}{8} - 1\right) \\
&= -\frac{1}{12}\ln|2| + \frac{7}{36}
\end{aligned}$$

Repeated application exercise (page 257)

Q89:

$\int_a^b fg'dx = [fg]_a^b - \int_a^b f'g\,dx$

Let $f(x) = 3x^2$ and $g'(x) = e^{-x}$ then $f'(x) = 6x$ and $g(x) = -e^{-x}$

$$\int 3x^2 e^{-x}dx = 3x^2 \times (-e^{-x}) - \int 6x \times (-e^{-x})\,dx$$

$$= -3x^2 e^{-x} + 6\int xe^{-x}dx$$

Second application of integration by parts for $\int x\,e^{-x}dx$

Let $f(x) = x$ and $g'(x) = e^{-x}$ then $f'(x) = 1$ and $g(x) = -e^{-x}$

$$\int 3x^2 e^{-x}dx = -3x^2 e^{-x} + 6\left(x \times (-e^{-x}) - \int 1 \times (-e^{-x})\,dx\right)$$

$$= -3x^2 e^{-x} + 6\left(-xe^{-x} + \int e^{-x}dx\right)$$

$$= -3x^2 e^{-x} + 6\left(-xe^{-x} + e^{-x} + C\right)$$

$$= -3x^2 e^{-x} - 6xe^{-x} + 6e^{-x} + C$$

$$= -3e^{-x}(x^2 + 2x - 2) + C$$

Note: Since C is an arbitrary constant $6C$ is just C in the working.

Q90:

Step 1: The integrand is the product of two functions so we must use integration by parts.

Let
$f(x) = x^2$ then $f'(x) = 2x$
$g'(x) = \sin 3x$ then $g(x) = -\frac{1}{3}\cos 3x$

The formula for integration by parts $\int_a^b fg'dx = [fg]_a^b - \int_a^b f'g\,dx$ gives,

$$\int_0^{\frac{\pi}{6}} x^2 \sin 3x\,dx = \left[x^2 \times \left(-\frac{1}{3}\cos 3x\right)\right]_0^{\frac{\pi}{6}} - \int_0^{\frac{\pi}{6}} 2x \times \left(-\frac{1}{3}\cos 3x\right)dx$$

$$= \left[-\frac{1}{3}x^2 \cos 3x\right]_0^{\frac{\pi}{6}} + \frac{2}{3}\int_0^{\frac{\pi}{6}} x\cos 3x\,dx$$

At this stage we can evaluate the first term, $-\frac{1}{3}\left[x^2 \cos 3x\right]_0^{\frac{\pi}{6}}$

$$-\frac{1}{3}\left[x^2 \cos 3x\right]_0^{\frac{\pi}{6}} = -\frac{1}{3}\left(\frac{\pi^2}{6}\cos\frac{\pi}{2} - 0^2 \cos 0\right)$$

$$= -\frac{1}{3}(0 - 0)$$

$$= 0$$

The integration now becomes:

$\int_0^{\frac{\pi}{6}} x^2 \sin 3x\,dx = 0 + \frac{2}{3}\int_0^{\frac{\pi}{6}} x\cos 3x\,dx$ (1)

The second term is the product of two functions, $x\cos 3x$, so we need to apply integration by parts again.

Step 2: Second application of integration by parts for $\int_0^{\frac{\pi}{6}} x\cos(3x)dx$

Let
$f(x) = x$ then $f'(x) = 1$
$g'(x) = \cos 3x$ then $g(x) = \frac{1}{3}\sin 3x$

The formula for integration by parts $\int_a^b fg'dx = [fg]_a^b - \int_a^b f'gdx$ gives,

$$\int_0^{\frac{\pi}{6}} x\cos 3xdx = \left[x\left(\frac{1}{3}\sin 3x\right)\right]_0^{\frac{\pi}{6}} - \int_0^{\frac{\pi}{6}} 1 \times \left(\frac{1}{3}\sin 3x\right)dx$$

$$= \frac{1}{3}[x\sin 3x]_0^{\frac{\pi}{6}} - \frac{1}{3}\int_0^{\frac{\pi}{6}} \sin 3xdx$$

$$= \frac{1}{3}[x\sin 3x]_0^{\frac{\pi}{6}} - \frac{1}{3}\left[-\frac{1}{3}\cos 3x\right]_0^{\frac{\pi}{6}}$$

$$= \frac{1}{3}[x\sin 3x]_0^{\frac{\pi}{6}} + \frac{1}{9}[\cos 3x]_0^{\frac{\pi}{6}}$$

$$= \frac{1}{3}\left(\frac{\pi}{6}\sin\frac{\pi}{2} - 0\sin 0\right) + \frac{1}{9}\left(\cos\frac{\pi}{2} - \cos 0\right)$$

$$= \frac{1}{3}\left(\frac{\pi}{6} - 0\right) + \frac{1}{9}(0 - 1)$$

$$= \frac{\pi}{18} - \frac{1}{9}$$

$$= \frac{1}{18}(\pi - 2)$$

Step 3: Substitute the solution for $\int_0^{\frac{\pi}{6}} x\cos(3x)dx$ back into (1), the solution from step 1.

$$\int_0^{\frac{\pi}{6}} x^2\sin 3xdx = \frac{2}{3}\int_0^{\frac{\pi}{6}} x\cos 3xdx$$

$$= \frac{2}{3} \times \frac{1}{18}(\pi - 2)$$

$$= \frac{1}{27}(\pi - 2)$$

Getting back to the original function exercise (page 259)

Q91:
The integrand is the product of two functions so we must use integration by parts.
$\int fg'dx = fg - \int f'gdx$
Step 1: Apply integration by parts to $\int e^{3x}\cos 2xdx$ and substitute into $\int fg'dx = fg - \int f'gdx$
Let $f(x) = e^{3x}$ then $f'(x) = 3e^{3x}$
$g'(x) = \cos 2x$ then $g(x) = \frac{1}{2}\sin 2x$

Step 2: Substitute into $\int fg'dx = fg - \int f'g dx$

$$\int fg'dx = fg - \int f'g dx$$

$$\int e^{3x}\cos 2x dx = e^{3x} \times \frac{1}{2}\sin 2x - \int 3e^{3x} \times \frac{1}{2}\sin 2x dx$$

$$= \frac{1}{2}e^{3x}\sin 2x - \frac{3}{2}\int e^{3x}\sin 2x dx$$

The integral on the RHS, $\int e^{3x}\sin 2x dx$, is the product of two function so we must use integration by parts a second time.

Step 3: Apply integration by parts to $\int e^{3x}\sin 2x dx$ and substitute into $\int fg'dx = fg - \int f'g dx$
Let $f(x) = e^{3x}$ then $f'(x) = 3e^{3x}$
$g'(x) = \sin 2x$ then $g(x) = -\frac{1}{2}\cos 2x$

Step 4: Substitute into $\int fg'dx = fg - \int f'g dx$ and letting $I = \int e^{3x}\cos 2x dx$

$$\int fg'dx = fg - \int f'g dx$$

$$\int e^{3x}\sin 2x dx = e^{3x}\left(-\frac{1}{2}\cos 2x\right) - \int 3e^{3x}\left(-\frac{1}{2}\cos 2x\right)dx$$

$$= -\frac{1}{2}e^{3x}\cos 2x + \frac{3}{2}\int e^{3x}\cos 2x dx$$

$$= -\frac{1}{2}e^{3x}\cos 2x + \frac{3}{2}I$$

Step 5: Substitute the solution of $\int e^{3x}\sin 2x dx$ back into the first application of integration by parts in step 1 and letting $I = \int e^{3x}\cos 2x dx$

$$I = \frac{1}{2}e^{3x}\sin 2x - \frac{3}{2}\left(-\frac{1}{2}e^{3x}\cos 2x + \frac{3}{2}I\right)$$

$$I = \frac{1}{2}e^{3x}\sin 2x + \frac{3}{4}e^{3x}\cos 2x - \frac{9}{4}I$$

Step 6: Rearrange the equation for I.

$$I = \frac{1}{2}e^{3x}\sin 2x + \frac{3}{4}e^{3x}\cos 2x - \frac{9}{4}I$$

$$\frac{13}{4}I = \frac{1}{2}e^{3x}\sin 2x + \frac{3}{4}e^{3x}\cos 2x + C$$

$$I = \frac{4}{13}\left(\frac{1}{2}e^{3x}\sin 2x + \frac{3}{4}e^{3x}\cos 2x\right) + C$$

$$I = \frac{1}{13}\left(2e^{3x}\sin 2x + 3e^{3x}\cos 2x\right) + C$$

Therefore, $\int e^{3x}\cos 2x dx = \frac{1}{13}\left(2e^{3x}\sin 2x + 3e^{3x}\cos 2x\right) + C$

Q92:

The integrand is the product of two functions so we must use integration by parts.

$\int_a^b fg'dx = [fg]_a^b - \int_a^b f'g dx$

Step 1: Apply integration by parts to $\int_1^3 \frac{\ln|x|}{x}dx$ and substitute into $\int_a^b fg'dx = [fg]_a^b - \int_a^b f'g dx$
Let $f(x) = \ln|x|$ then $f'(x) = \frac{1}{x}$
$g'(x) = \frac{1}{x}$ then $g(x) = \ln|x|$

Step 2: Substitute into $\int_a^b fg'dx = [fg]_a^b - \int_a^b f'gdx$ and integrate.

$$\int_a^b fg'dx = [fg]_a^b - \int_a^b f'gdx$$

$$\int_1^3 \frac{\ln|x|}{x}dx = [\ln|x| \times \ln|x|]_1^3 - \int_1^3 \frac{1}{x} \times \ln|x|dx$$

$$= \left[\ln|x|^2\right]_1^3 - \int_1^3 \frac{\ln|x|}{x}dx$$

$$= \left(\ln|3|^2 - \ln|1|^2\right) - \int_1^3 \frac{\ln|x|}{x}dx$$

Recall that $\ln|1| = 0$

$$= \ln|3|^2 - \int_1^3 \frac{\ln|x|}{x}dx$$

Notice that $\int_1^3 \frac{\ln|x|}{x}dx$ appears on the LHS and the RHS.

Step 3: Let $I = \int_1^3 \frac{\ln|x|}{x}dx$ and rearrange for I.

$$\int_1^3 \frac{\ln|x|}{x}dx = \ln|3|^2 - \int_1^3 \frac{\ln|x|}{x}dx$$

$$I = \ln|3|^2 - I$$

$$2I = \ln|3|^2$$

$$I = \frac{1}{2}\ln|3|^2$$

Therefore, $\int_1^3 \frac{\ln|x|}{x}dx = \frac{1}{2}\ln|3|^2$

Introducing 1 as a 'dummy variable' exercise (page 260)

Q93:

Rewrite the integrand as $\ln|3x| \times 1$ then find the integral by using integration by parts as follows.

Let $f(x) = \ln|3x|$ then $f'(x) = \frac{1}{x}$
$g'(x) = 1$ then $g(x) = x$

The formula for integration by parts $\int fg' \, dx = fg - \int f'g \, dx$ then gives us,

$$\int \ln|3x| \times 1 \, dx = x\ln|3x| - \int \frac{1}{x}x \, dx$$

$$= x\ln|3x| - \int 1 \, dx$$

$$= x\ln|3x| - x + C$$

Q94:

Rewrite the integrand as $\tan^{-1}(2x) \times 1$ then find the integral by using integration by parts as follows.

Let $f(x) = \tan^{-1}(2x)$ then $f'(x) = \frac{2}{1+4x^2}$
$g'(x) = 1$ then $g(x) = x$

The formula for integration by parts $\int fg'\, dx = fg - \int f'g\, dx$ then gives us,

$$\int \tan^{-1}(2x) \times 1 dx = x\tan^{-1}(2x) - \int \frac{2x}{1+4x^2} dx$$
$$= x\tan^{-1}(2x) - 2\int \frac{x}{1+4x^2} dx$$

To integrate the integrand on the RHS recall the standard result $\int \frac{f'(x)}{f(x)} dx = \ln|f(x)| + C$

We need to make the numerator look like the derivative of the denominator i.e. $\frac{d}{dx}\{1+4x^2\} = 8x$

We therefore multiply the integrand by $\frac{8}{8}$.

$$\int \tan^{-1}(2x) \times 1 dx = x\tan^{-1}(2x) - \frac{2}{8}\int \frac{8x}{1+4x^2} dx$$
$$= x\tan^{-1}(2x) - \frac{2}{8}\ln|1+4x^2| + C$$
$$= x\tan^{-1}(2x) - \frac{1}{4}\ln|1+4x^2| + C$$

Q95:
Hints:

- Apply integration by parts twice.

Answer:

Rewrite the integrand as $[\ln|x|]^2 \times 1$ then find the integral by using integration by parts as follows.

Let $f(x) = [\ln|x|]^2$ then $f'(x) = \frac{2\ln|x|}{x}$
$g'(x) = 1$ then $g(x) = x$

The formula for integration by parts $\int fg'\, dx = fg - \int f'g\, dx$ then gives us,

$$\int [\ln|x|]^2 dx = [\ln|x|]^2 \times x - \int \frac{2\ln|x|}{x} \times x dx$$
$$= x[\ln|x|]^2 - 2\int \ln|x| dx$$

Second application of integration by parts on $\ln|x|$

Let $f(x) = \ln|x|$ then $f'(x) = \frac{1}{x}$
$g'(x) = 1$ then $g(x) = x$

© HERIOT-WATT UNIVERSITY

$$\int [\ln|x|]^2 \times 1 dx = [\ln|x|]^2 \times x - \int \frac{2\ln|x|}{x} \times x dx$$
$$= x[\ln|x|]^2 - 2\int \ln|x| dx$$
$$= x[\ln|x|]^2 - 2\int \ln|x| \times 1 dx$$
$$= x[\ln|x|]^2 - 2\left(\ln|x| \times x - \int \frac{1}{x} \times x dx\right)$$
$$= x[\ln|x|]^2 - 2\left(x\ln|x| - \int 1 dx\right)$$
$$= x[\ln|x|]^2 - 2(x\ln|x| - x) + C$$
$$= x\ln|x|^2 - 2x\ln|x| + 2x + C$$
$$= x\left([\ln|x|]^2 - 2\ln|x| + 2\right) + C$$
$$\int [\ln|x|]^2 dx = x\left([\ln|x|]^2 - 2\ln|x| + 2\right) + C$$

Challenge question 3 (page 260)

Expected answer

$\int \sin^{-1} x dx$

Rewrite the integrand as $\sin^{-1} x \times 1$, then find the integral by using integration by parts, $\int fg' dx = fg - \int f'g dx$

Let $f(x) = \sin^{-1} x$ then $f'(x) = \frac{2}{\sqrt{1-x^2}}$
$g'(x) = 1$ then $g(x) = x$

Using the formula for integration by parts then gives us:

$$\int fg' dx = fg - \int f'g dx$$
$$\int \sin^{-1} x dx = \sin^{-1} x \times x - \int \frac{1}{\sqrt{1-x^2}} \times x dx$$
$$= x\sin^{-1} x - \int \frac{x}{\sqrt{1-x^2}} dx \quad (1)$$

To work out $\int \frac{x}{\sqrt{1-x^2}} dx$ we use integration by substitution.

Let $u = 1 - x^2$, then $\frac{du}{dx} = -2x$ and $\frac{dx}{du} = -\frac{1}{2x}$

$$\int \frac{x}{\sqrt{1-x^2}}dx = \int \frac{x}{\sqrt{1-x^2}}\frac{dx}{du}du$$
$$= \int \frac{x}{\sqrt{u}}\left(-\frac{1}{2x}\right)du$$
$$= -\frac{1}{2}\int \frac{1}{\sqrt{u}}du$$
$$= -\frac{1}{2}\int u^{-\frac{1}{2}}du$$
$$= -\frac{1}{2} \times \frac{2}{1}u^{\frac{1}{2}} + C$$
$$= -u^{\frac{1}{2}} + C$$

Substituting $u = 1 - x^2$, give the solution: $\int \frac{x}{\sqrt{1-x^2}}dx = -\left(1-x^2\right)^{\frac{1}{2}} + C$

Substitute this back into (1)

$$\int \sin^{-1}x\,dx = x\sin^{-1}x - \int \frac{x}{\sqrt{1-x^2}}dx$$
$$= x\sin^{-1}x - \left(-\left(1-x^2\right)^{\frac{1}{2}} + C\right)$$
$$= x\sin^{-1}x + \sqrt{(1-x^2)} + C$$

Area between a curve and the x-axis exercise (page 266)

Q96:

To draw the curve we need to work out where it cuts the x-axis.
$x^2 - 3x + 2 = 0$
$(x-1)(x-2) = 0$
The curve cuts the x-axis at $x = 1$ and $x = 2$. It is a minimum since the constant in front of x^2 is positive.

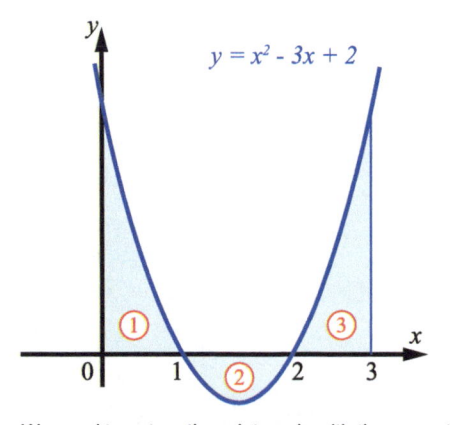

We need to set up three integrals with the correct limits to find the total area.

Area 1:

$$\int_0^1 x^2 - 3x + 2\,dx = \left[\frac{1}{3}x^3 - \frac{3}{2}x^2 + 2x\right]_0^1$$

$$= \left(\frac{1}{3}(1)^3 - \frac{3}{2}(1)^2 + 2(1)\right) - \left(\frac{1}{3}(0)^3 - \frac{3}{2}(0)^2 + 2(0)\right)$$

$$= \frac{1}{3} - \frac{3}{2} + 2$$

$$= \frac{5}{6} \text{ units}^2$$

Area 2:

$$\int_1^2 x^2 - 3x + 2\,dx = \left[\frac{1}{3}x^3 - \frac{3}{2}x^2 + 2x\right]_1^2$$

$$= \left(\frac{1}{3}(2)^3 - \frac{3}{2}(2)^2 + 2(2)\right) - \left(\frac{1}{3}(1)^3 - \frac{3}{2}(1)^2 + 2(1)\right)$$

$$= \left(\frac{8}{3} - 6 + 4\right) - \left(\frac{1}{3} - \frac{3}{2} + 2\right)$$

$$= \frac{2}{3} - \frac{5}{6}$$

$$= -\frac{1}{6}$$

$$= \frac{1}{6} \text{ units}^2$$

Area 3:

$$\int_2^3 x^2 - 3x + 2\,dx = \left[\frac{1}{3}x^3 - \frac{3}{2}x^2 + 2x\right]_2^3$$

$$= \left(\frac{1}{3}(3)^3 - \frac{3}{2}(3)^2 + 2(3)\right) - \left(\frac{1}{3}(2)^3 - \frac{3}{2}(2)^2 + 2(2)\right)$$

$$= \left(9 - \frac{27}{2} + 6\right) - \left(\frac{8}{3} - 6 + 4\right)$$

$$= \frac{3}{2} - \frac{2}{3}$$

$$= \frac{5}{6} \text{ units}^2$$

Total area = $1\frac{5}{6}$ square units

Q97:

A sketch of the curve shows that area is in three parts. Two parts are above the x-axis and have positive values for the definite integral and the other part is below the x-axis and has a negative value for the definite integral.

So the three areas needs to be calculated separately.

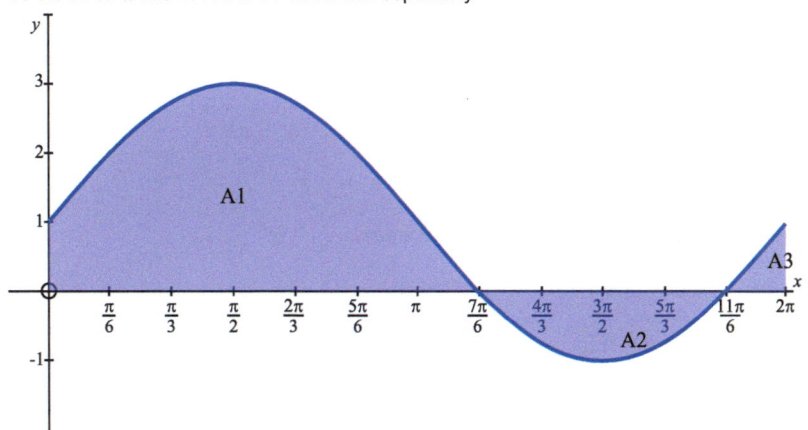

To work out where the curve cuts the x-axis we need to solve the equation:

$2 \sin x + 1 = 0$

$$\sin x = -\frac{1}{2}$$

$$x = \sin^{-1}\left(\frac{1}{2}\right)$$

$$x = \frac{\pi}{6}$$

$Q_3 : \pi + \frac{\pi}{6} = \frac{7\pi}{6}$

$Q_4 : 2\pi - \frac{\pi}{6} = \frac{11\pi}{6}$

The curve cuts the x-axis at $x = \frac{7\pi}{6}$ and $x = \frac{11\pi}{6}$

The shaded area above the x-axis for $x = 0$ and $x = \frac{7\pi}{6}$

$$\int_0^{\frac{7\pi}{6}} 2\sin x + 1\, dx = [-2\cos x + x]_0^{\frac{7\pi}{6}}$$

$$= \left(-2\cos\frac{7\pi}{6} + \frac{7\pi}{6}\right) - (-2\cos 0 + 0)$$

$$= \left(-2\left(-\cos\frac{\pi}{6}\right) + \frac{7\pi}{6}\right) - (-2(1))$$

$$= \left(-2\left(-\frac{\sqrt{3}}{2}\right) + \frac{7\pi}{6}\right) + 2$$

$$= \sqrt{3} + \frac{7\pi}{6} + 2$$

The shaded area below the x-axis for $x = \frac{7\pi}{6}$ to $x = \frac{11\pi}{6}$

$$\int_{\frac{7\pi}{6}}^{\frac{11\pi}{6}} 2\sin x + 1\, dx = [-2\cos x + x]_{\frac{7\pi}{6}}^{\frac{11\pi}{6}}$$

$$= \left(-2\cos\frac{11\pi}{6} + \frac{11\pi}{6}\right) - \left(-2\cos\frac{7\pi}{6} + \frac{7\pi}{6}\right)$$

$$= \left(-2\cos\frac{\pi}{6} + \frac{11\pi}{6}\right) - \left(-2\left(-\cos\frac{\pi}{6}\right) + \frac{7\pi}{6}\right)$$

$$= \left(-2\left(\frac{\sqrt{3}}{2}\right) + \frac{11\pi}{6}\right) - \left(-2\left(-\frac{\sqrt{3}}{2}\right) + \frac{7\pi}{6}\right)$$

$$= \left(-\sqrt{3} + \frac{11\pi}{6}\right) - \left(\sqrt{3} + \frac{7\pi}{6}\right)$$

$$= -2\sqrt{3} + \frac{4\pi}{6}$$

This shaded area is below the x-axis and so the actual area is $2\sqrt{3} - \frac{4\pi}{6}$

The shaded area above the x-axis for $x = \frac{11\pi}{6}$ to $x = 2\pi$

$$\int_{\frac{11\pi}{6}}^{2\pi} 2\sin x + 1\, dx = [-2\cos x + x]_{\frac{11\pi}{6}}^{2\pi}$$

$$= (-2\cos 2\pi + 2\pi) - \left(-2\cos\frac{11\pi}{6} + \frac{11\pi}{6}\right)$$

$$= (-2(1) + 2\pi) - \left(-2\cos\frac{\pi}{6} + \frac{11\pi}{6}\right)$$

$$= -2 + 2\pi - \left(-2\left(\frac{\sqrt{3}}{2}\right) + \frac{11\pi}{6}\right)$$

$$= -2 + 2\pi + \sqrt{3} - \frac{11\pi}{6}$$

$$= -2 + \sqrt{3} + \frac{\pi}{6}$$

Now we can find the total shaded area.

Total shaded area $= A_1 + A_2 + A_3$

$$= \left(\sqrt{3} + \frac{7\pi}{6} + 2\right) + \left(2\sqrt{3} - \frac{4\pi}{6}\right) + \left(-2 + \sqrt{3} + \frac{\pi}{6}\right)$$

$$= 4\sqrt{3} + \frac{4\pi}{6}$$

$$= 4\sqrt{3} + \frac{2\pi}{3}$$

Area between a curve and the y-axis exercise (page 269)

Q98:

Hints:

- $x = \sqrt{y-1}$

Steps:

- Sketch the graph of $y = x^2 + 1$.

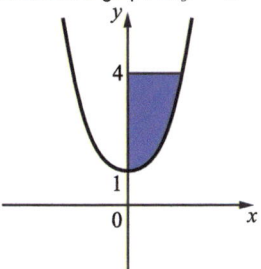

- Rearrange the equation of the curve in terms of y.
$$y = x^2 + 1$$
$$y - 1 = x^2$$
$$x = \sqrt{y-1}$$

- Integrate the rearranged equation.

Answer:

$$\int_1^4 x\,dy = \int_1^4 \sqrt{y-1}\,dy$$
$$= \int_1^4 (y-1)^{\frac{1}{2}}\,dy$$
$$= \left[\frac{2}{3}(y-1)^{\frac{3}{2}}\right]_1^4$$
$$= \left(\frac{2}{3}(4-1)^{\frac{3}{2}}\right) - \left(\frac{2}{3}(1-1)^{\frac{3}{2}}\right)$$
$$= \frac{2}{3}(3)^{\frac{3}{2}}$$
$$= \frac{2\sqrt{27}}{3} \text{ units}^2$$

Q99:
Steps:
- Sketch the graph of $y = \ln|x| + 5$.

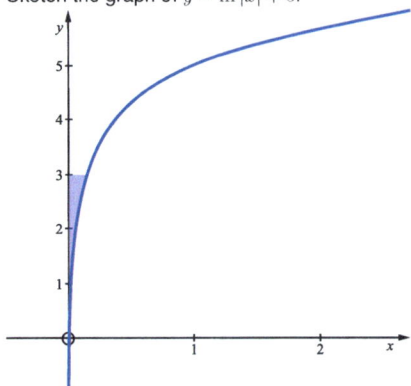

- Rearrange the equation of the curve in terms of y.
$x = e^{y-5}$
- Integrate the rearranged equation.

Answer:
$$\int_0^3 x\,dy = \int_0^3 \left(e^{y-5}\right) dy$$
$$= \left[e^{y-5}\right]_0^3$$
$$= \left(e^{3-5}\right) - \left(e^{0-5}\right)$$
$$= e^{-2} - e^{-5} \text{ units}^2$$

Q100:
Steps:
- Sketch the graph of $xy - y = 2$.

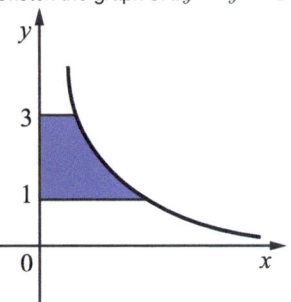

ANSWERS: UNIT 1 TOPIC 3

- Rearrange the equation of the curve in terms of y.

$$xy - y = 2$$
$$y(x - 1) = 2$$
$$x - 1 = \frac{2}{y}$$
$$x = \frac{2}{y} + 1$$

- Integrate the rearranged equation.

Answer:

$$\int_1^3 x\,dy = \int_1^3 \frac{2}{y} + 1\,dy$$
$$= [2\ln|y| + y]_1^3$$
$$= (2\ln|3| + 3) - (2\ln|1| + 1)$$
$$= 2\ln(3) + 2 \text{ units}^2$$

Volume of revolution about the x-axis exercise (page 272)

Q101:

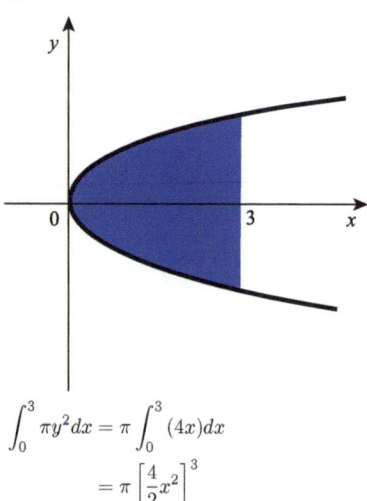

$$\int_0^3 \pi y^2 dx = \pi \int_0^3 (4x)dx$$
$$= \pi \left[\frac{4}{2}x^2\right]_0^3$$
$$= \pi \left(2(3)^2 - 2(0)^2\right)$$
$$= 18\pi \text{ units}^3$$

© HERIOT-WATT UNIVERSITY

Q102:

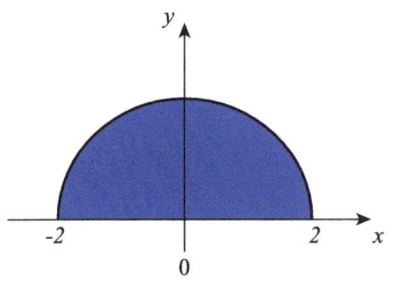

To calculate the volume of revolution about the x-axis we use $\int_a^b \pi y^2 dx$

This means that we must square y. i.e. work out y^2, then multiply it by π.

First we need to find y^2 by squaring the expression for y.

$y = \sqrt{4 - x^2}$
$y^2 = 4 - x^2$

Now we integrate and evaluate between $x = -2$ and $x = 2$.

Now substitute y^2 into $\int_a^b \pi y^2 dx$ where $y^2 = 4 - x^2$, $a = -2$ and $b = 2$.

Since π is a constant we can take it out as a factor to the front of the integral sign.

$$\int_{-2}^{2} \pi y^2 dx = \pi \int_{-2}^{2} 4 - x^2 dx$$

$$= \pi \left[4x - \frac{1}{3}x^3 \right]_{-2}^{2}$$

$$= \pi \left[\left(4(2) - \frac{1}{3}(2)^3 \right) - \left(4(-2) - \frac{1}{3}(-2)^3 \right) \right]$$

$$= \pi \left[\left(8 - \frac{8}{3} \right) - \left(-8 + \frac{8}{3} \right) \right]$$

$$= \pi \left(8 - \frac{8}{3} + 8 - \frac{8}{3} \right)$$

$$= \pi \left(8 + 8 - \frac{8}{3} - \frac{8}{3} \right)$$

$$= \pi \left(16 - \frac{16}{3} \right)$$

$$= \pi \left(\frac{48}{3} - \frac{16}{3} \right)$$

$$= \pi \left(\frac{32}{3} \right)$$

$$= 10\frac{2}{3} \pi \text{ units}^3$$

Q103:

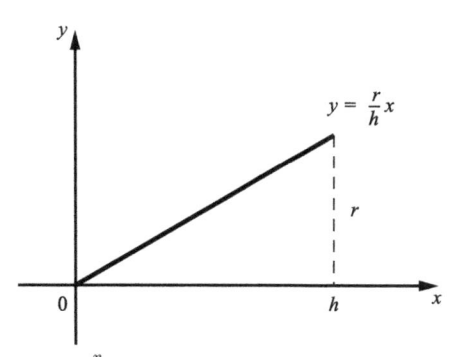

$y = \frac{r}{h}x$

$y^2 = \frac{r^2}{h^2}x^2$

$$\int_0^h \pi y^2 dx = \pi \int_0^h \frac{r^2}{h^2} x^2 dx$$

$$= \frac{\pi r^2}{h^2} \int_0^h x^2 dx$$

$$= \frac{\pi r^2}{h^2} \left[\frac{x^3}{3}\right]_0^h$$

$$= \frac{1}{3}\pi r^2 h \quad \text{units}^3$$

This is the formula for the volume of a cone with radius r and height h.

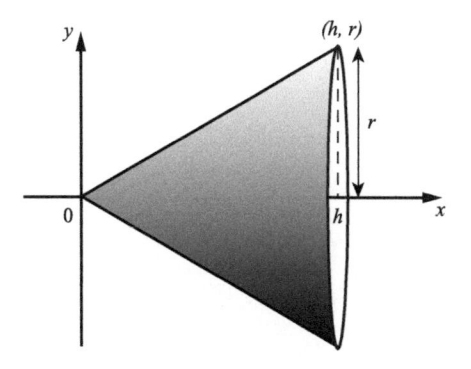

Volume of revolution about the y-axis exercise (page 275)

Q104:

Before integrating we must rearrange the equation:
$$y = 4x^2 - 1$$
$$y + 1 = 4x^2$$
$$\frac{1}{4}(y+1) = x^2$$
$$x^2 = \frac{1}{4}(y+1)$$

Using the formula:
$$V = \int_{-1}^{4} \pi x^2 dy$$
$$= \pi \int_{-1}^{4} \frac{1}{4}(y+1)dy$$
$$= \frac{\pi}{4}\left[\frac{1}{2}y^2 + y\right]_{-1}^{4}$$
$$= \frac{\pi}{4}\left(\left(\frac{1}{2}(4)^2 + 4\right) - \left(\frac{1}{2}(-1)^2 - 1\right)\right)$$
$$= \frac{\pi}{4}\left((8+4) - \left(\frac{1}{2} - 1\right)\right)$$
$$= 3\frac{1}{8}\pi \text{ units}^3$$

Q105:

Before integrating we must rearrange the equation:
$$xy^2 = 2$$
$$x = \frac{2}{y^2}$$
$$x^2 = \frac{4}{y^4}$$

Using the formula:
$$V = \int_{1}^{3} \pi x^2 dy$$
$$= \pi \int_{1}^{3} \frac{4}{y^4} dy$$
$$= 4\pi \int_{1}^{3} y^{-4} dy$$
$$= 4\pi \left[-\frac{1}{3}y^{-3}\right]_{1}^{3}$$
$$= 4\pi \left(\left(-\frac{1}{3}(3)^{-3}\right) - \left(-\frac{1}{3}(1)^{-3}\right)\right)$$
$$= 4\pi \left(\left(-\frac{1}{3} \times \frac{1}{27}\right) - \left(-\frac{1}{3} \times 1\right)\right)$$
$$= 1\frac{23}{81}\pi \text{ units}^3$$

ANSWERS: UNIT 1 TOPIC 3

Q106:
Before integrating we must rearrange the equation:
$y = \ln|x|$
$e^y = x$
$x^2 = e^{2y}$

Using the formula:
$$V = \int_1^4 \pi x^2 \, dy$$
$$= \pi \int_1^4 e^{2y} \, dy$$
$$= \pi \left[\frac{1}{2} e^{2y} \right]_1^4$$
$$= \pi \left(\frac{1}{2} e^8 - \frac{1}{2} e^2 \right)$$
$$= \frac{1}{2} \pi e^2 \left(e^4 - 1 \right) \text{ units}^3$$

End of topic 3 test (page 281)

Q107:
$$\int \frac{5}{\sqrt{1 - (2x)^2}} \, dx = 5 \int \frac{1}{\sqrt{1 - (2x)^2}} \, dx$$
$$= 5 \int \frac{1}{\sqrt{1 - 4x^2}} \, dx$$
$$= 5 \int \frac{1}{\sqrt{4 \left(\frac{1}{4} - x^2 \right)}} \, dx$$
$$= 5 \int \frac{1}{2\sqrt{\frac{1}{4} - x^2}} \, dx$$
$$= \frac{5}{2} \int \frac{1}{\sqrt{\left(\frac{1}{2} \right)^2 - x^2}} \, dx$$
$$= \frac{5}{2} \sin^{-1} \left(\frac{x}{\frac{1}{2}} \right) + C$$
$$= \frac{5}{2} \sin^{-1}(2x) + C$$

Q108:
$$\int 3\sec^2(5x) \, dx = 3 \int \sec^2(5x) \, dx$$
$$= 3 \times \frac{1}{5} \tan(5x) + C$$
$$= \frac{3}{5} \tan(5x) + C$$

© HERIOT-WATT UNIVERSITY

Q109:

$$\int 3e^{4x-1}dx = 3\int e^{4x-1}dx$$
$$= 3 \times \frac{1}{4}e^{4x-1} + C$$
$$= \frac{3}{4}e^{4x-1} + C$$

Q110:

$$\int_3^6 \frac{3}{2x-5}dx = 3\int_3^6 \frac{1}{2x-5}dx$$
$$= 3\left[\frac{1}{2}\ln|2x-5|\right]_3^6$$
$$= \frac{3}{2}[\ln|2x-5|]_3^6$$
$$= \frac{3}{2}(\ln|7| - \ln|1|)$$
$$= \frac{3}{2}(\ln|7|)$$
$$= \frac{3}{2}\ln|7|$$

Q111:

$$\int_0^{\frac{\pi}{3}} \frac{5}{1+(3x)^2}dx = 5\int_0^{\frac{\pi}{3}} \frac{1}{1+9x^2}dx$$
$$= 5\int_0^{\frac{\pi}{3}} \frac{1}{9\left(\frac{1}{9}+x^2\right)}dx$$
$$= \frac{5}{9}\int_0^{\frac{\pi}{3}} \frac{1}{\left(\frac{1}{3}\right)^2+x^2}dx$$
$$= \frac{5}{9}\left[\frac{1}{\frac{1}{3}}\tan^{-1}\left(\frac{x}{\frac{1}{3}}\right)\right]_0^{\frac{\pi}{3}}$$
$$= \frac{5}{9}\left[3\tan^{-1}(3x)\right]_0^{\frac{\pi}{3}}$$
$$= \frac{5}{3}\left[\tan^{-1}(3x)\right]_0^{\frac{\pi}{3}}$$
$$= \frac{5}{3}\left(\tan^{-1}(\pi) - \tan^{-1}(0)\right)$$
$$= \frac{5}{3}\tan^{-1}(\pi)$$
$$= 2 \cdot 104$$

ANSWERS: UNIT 1 TOPIC 3

Q112:

$\int 3x^2 (2x^3 + 5)^3 dx$

Let $u = 2x^3 + 5$, then $\frac{du}{dx} = 6x^2$ and $\frac{dx}{du} = \frac{1}{6x^2}$

$$\int 3x^2 (2x^3 + 5)^3 dx = \int 3x^2 (2x^3 + 5)^3 \frac{dx}{du} du$$
$$= \int 3x^2 u^3 \times \frac{1}{6x^2} du$$
$$= \frac{1}{2} \int u^3 du$$
$$= \frac{1}{2} \times \frac{u^4}{4} + C$$
$$= \frac{1}{8} u^4 + C$$
$$= \frac{1}{8} (2x^3 + 5)^4 + C$$

Q113:

$\int \cos^4 x \sin x dx$

Let $u = \cos(x)$, then $\frac{du}{dx} = -\sin x$ and $\frac{dx}{du} = -\frac{1}{\sin x}$

$$\int \cos^4 x \sin x dx = \int \cos^4 x \sin x \frac{dx}{du} du$$
$$= \int u^4 \times -\sin x \times -\frac{1}{\sin x} du$$
$$= \int u^4 du$$
$$= \frac{1}{5} u^5 + C$$
$$= \frac{1}{5} \cos^5 x + C$$

Q114:

$\int \frac{5x}{2x^2+1} dx$

Let $u = 2x^2 + 1$, then $\frac{du}{dx} = 4x$

$$\int \frac{5x}{2x^2 + 1} dx = \int \frac{5x}{2x^2 + 1} \frac{dx}{du} du$$
$$= \int \frac{5x}{u} \times \frac{1}{4x} du$$
$$= \frac{5}{4} \int \frac{1}{u} du$$
$$= \frac{5}{4} \ln |u| + C$$
$$= \frac{5}{4} \ln |2x^2 + 1| + C$$

© HERIOT-WATT UNIVERSITY

Q115:

$\int 2x^3 e^{4x^4+3} dx$

Let $u = 4x^4 + 3$, then $\frac{du}{dx} = 16x^3$ and $\frac{dx}{du} = \frac{1}{16x^3}$

$$\int 2x^3 e^{4x^4+3} dx = \int 2x^3 e^{4x^4+3} \frac{dx}{du} du$$
$$= \int 2x^3 e^u \times \frac{1}{16x^3} du$$
$$= \frac{1}{8} \int e^u du$$
$$= \frac{1}{8} e^u + C$$
$$= \frac{1}{8} e^{4x^4+3} + C$$

Q116:

$\int \tan x \, dx = \int \frac{\sin x}{\cos x} dx$

Let $u = \cos x$, then $\frac{du}{dx} = -\sin x$ and $\frac{dx}{du} = -\frac{1}{\sin x}$

$$\int \tan x \, dx = \int \frac{\sin x}{\cos x} \frac{dx}{du} du$$
$$= \int \frac{\sin x}{u} \times -\frac{1}{\sin x} du$$
$$= -\int \frac{1}{u} du$$
$$= -\ln|u| + C$$
$$= -\ln|\cos x| + C$$

Q117:

$\int_0^{\sqrt{2}} 4x\sqrt{1+3x^2} dx$

Let $u = 1 + 3x^2$, then $\frac{du}{dx} = 6x$ and $\frac{dx}{du} = \frac{1}{6x}$

When $x = 0$, then $u = 1 + 3(0)^2$, so $u = 5$
When $x = \sqrt{2}$, then $u = 1 + 3(\sqrt{2})^2$, so $u = 7$

$$\int_0^{\sqrt{2}} 4x\sqrt{1+3x^2} dx = \int_1^7 4x\sqrt{1+3x^2} \frac{dx}{du} du$$
$$= \int_1^7 4x \times \sqrt{u} \times \frac{1}{6x} du$$
$$= \frac{2}{3} \int_1^7 u^{\frac{1}{2}} du$$
$$= \frac{2}{3} \left[\frac{2}{3} u^{\frac{3}{2}} \right]_1^7$$
$$= \frac{4}{9} \left[u^{\frac{3}{2}} \right]_1^7$$
$$= \frac{4}{9} \left(\left(\sqrt{7}\right)^3 - (1)^3 \right)$$
$$\approx 7 \cdot 79$$

ANSWERS: UNIT 1 TOPIC 3

Q118:

$\int_0^{\frac{\pi}{2}} 2x^2 \sin(x^3+1)dx$
Let $u = x^3 + 1$ then $\frac{du}{dx} = 3x^2$ and $\frac{dx}{du} = \frac{1}{3x^2}$
When $x = 0$, then $u = 0^3 + 1$, so $u = 1$
When $x = \frac{\pi}{2}$, then $u = \left(\frac{\pi}{2}\right)^3 + 1$, so $u = \frac{\pi^3}{8} + 1$

$$\int_0^{\frac{\pi}{2}} 2x^2 \sin(x^3+1)dx = \int_1^{\frac{\pi^3}{8}+1} 2x^2 \sin(x^3+1) \frac{dx}{du} du$$

$$= \int_1^{\frac{\pi^3}{8}+1} 2x^2 \sin u \times \frac{1}{3x^2} du$$

$$= \frac{2}{3} \int_1^{\frac{\pi^3}{8}+1} \sin u \, du$$

$$= \frac{2}{3} [-\cos u]_1^{\frac{\pi^3}{8}+1}$$

$$= \frac{2}{3} \left(-\cos\left(\frac{\pi^3}{8}+1\right) + \cos(1)\right)$$

$$= 0 \cdot 252$$

Q119:

$\int_0^{\frac{\pi}{4}} \sec^2 x \tan^3 x \, dx$
Let $u = \tan x$, then $\frac{du}{dx} = \sec^2 x$ and $\frac{dx}{du} = \frac{1}{\sec^2 x}$
When $x = 0$, then $u = \tan 0$, so $u = 0$
When $x = \frac{\pi}{4}$, then $u = \tan \frac{\pi}{4}$, so $u = 1$

$$\int_0^{\frac{\pi}{4}} \sec^2 x \tan^3 x \, dx = \int_0^1 \sec^2 x \tan^3 x \frac{dx}{du} du$$

$$= \int_0^1 \sec^2 x \times \tan^3 x \times \frac{1}{\sec^2 x} du$$

$$= \int_0^1 u^3 du$$

$$= \left[\frac{u^4}{4}\right]_0^1$$

$$= \frac{1}{4} [u^4]_0^1$$

$$= \frac{1}{4}(1^4 - 0^4)$$

$$= \frac{1}{4}$$

© HERIOT-WATT UNIVERSITY

Q120:

$\int_{\frac{1}{5}}^{\frac{2}{5}} 3x(5x+2)^3 dx$

Let $u = 5x + 2$, then $\frac{du}{dx} = 5$ and $\frac{dx}{du} = \frac{1}{5}$
Also, since $u = 5x + 2$, then $x = \frac{1}{5}(u-2)$
When $x = \frac{1}{5}$, then $u = 5\left(\frac{1}{5}\right) + 2$, so $u = 3$
When $x = \frac{2}{5}$, then $u = 5\left(\frac{2}{5}\right) + 2$, so $u = 4$

$\int_{\frac{1}{5}}^{\frac{2}{5}} 3x(5x+2)^3 \, dx = \int_{3}^{4} 3x(5x+2)^3 \frac{dx}{du} \, du$

$= \int_{3}^{4} 3 \times \frac{1}{5}(u-2) \times u^3 \times \frac{1}{5} \, du$

$= \frac{3}{25} \int_{3}^{4} u^4 - 2u^3 \, du$

$= \frac{3}{25} \left[\frac{u^5}{5} - \frac{u^4}{2} \right]_{3}^{4}$

$= \frac{3}{25} \left(\left(\frac{4^5}{5} - \frac{4^4}{2} \right) - \left(\frac{3^5}{5} - \frac{3^4}{2} \right) \right)$

$= \frac{3}{25} \left(\left(\frac{1024}{5} - \frac{256}{2} \right) - \left(\frac{243}{5} - \frac{81}{2} \right) \right)$

$= \frac{3}{25} \left(\left(\frac{2048}{10} - \frac{1280}{10} \right) - \left(\frac{486}{10} - \frac{405}{10} \right) \right)$

$= \frac{3}{25} \left(\frac{768}{10} - \frac{81}{10} \right)$

$= \frac{3}{25} \left(\frac{687}{10} \right)$

$= \frac{2061}{250}$

Q121:

$\int_{2}^{\frac{15}{4}} \frac{3x}{\sqrt{4x+1}} dx$

Let $u = 4x + 1$, then $\frac{du}{dx} = 4$ and $\frac{dx}{du} = \frac{1}{4}$

Also, since $u = 4x + 1$, then $x = \frac{1}{4}(u-1)$

When $x = 2$, then $u = 4(2) + 1$, so $u = 9$
When $x = \frac{15}{4}$, then $u = 4\left(\frac{15}{4}\right) + 1$, so $u = 16$

$$\int_{2}^{\frac{15}{4}} \frac{3x}{\sqrt{4x+1}} dx = \int_{9}^{16} \frac{3x}{\sqrt{4x+1}} \frac{dx}{du} du$$

$$= \int_{9}^{16} 3 \times \frac{1}{4}(u-1) \times \frac{1}{\sqrt{u}} \times \frac{1}{4} du$$

$$= \frac{3}{16} \int_{9}^{16} u^{\frac{1}{2}} - u^{-\frac{1}{2}} du$$

$$= \frac{3}{16} \left[\frac{2}{3} u^{\frac{3}{2}} - 2u^{\frac{1}{2}}\right]_{9}^{16}$$

$$= \frac{6}{16} \left[\frac{1}{3} u^{\frac{3}{2}} - u^{\frac{1}{2}}\right]_{9}^{16}$$

$$= \frac{6}{48} \left[u^{\frac{3}{2}} - 3u^{\frac{1}{2}}\right]_{9}^{16}$$

$$= \frac{1}{8} \left(\left(\left(\sqrt{16}\right)^3 - 3\sqrt{16}\right) - \left(\left(\sqrt{9}\right)^3 - 3\sqrt{9}\right)\right)$$

$$= \frac{1}{8} ((64 - 12) - (27 - 9))$$

$$= \frac{1}{8}(34)$$

$$= \frac{17}{4}$$

$$= 4\frac{1}{4}$$

Q122:

$\int x^3 (3x^4 - 1)^2 dx$

Let $u = 3x^4 - 1$, then $\frac{du}{dx} = 12x^3$ and $\frac{dx}{du} = \frac{1}{12x^3}$

$$\int x^3 (3x^4 - 1)^2 dx = \int x^3 (3x^4 - 1)^2 \frac{dx}{du} du$$

$$= \int x^3 u^2 \times \frac{1}{12x^3} du$$

$$= \frac{1}{12} \int u^2 du$$

$$= \frac{1}{12} \times \frac{1}{3} u^3 + C$$

$$= \frac{1}{36} (3x^4 - 1)^3 + C$$

Q123:

$\int (2x+1)(3x+4)^2 dx$

Let $u = 3x + 4$, then $\frac{du}{dx} = 3$ and $\frac{dx}{du} = \frac{1}{3}$

Also, since $u = 3x + 4$, then $x = \frac{1}{3}(u-4)$

$$\int (2x+1)(3x+4)^2 dx = \int (2x+1)(3x+4)^2 \frac{dx}{du} du$$

$$= \int \left(2 \times \frac{1}{3}(u-4) + 1\right) u^2 \times \frac{1}{3} du$$

$$= \frac{1}{3} \int \left(\frac{2}{3}u - \frac{8}{3} + \frac{3}{3}\right) u^2 du$$

$$= \frac{1}{3} \int \left(\frac{2}{3}u - \frac{5}{3}\right) u^2 du$$

$$= \frac{1}{9} \int (2u-5) u^2 du$$

$$= \frac{1}{9} \int 2u^3 - 5u^2 du$$

$$= \frac{1}{9} \left(\frac{u^4}{2} - \frac{5u^3}{3}\right) + C$$

$$= \frac{1}{9} \left(\frac{3u^4}{6} - \frac{10u^3}{6}\right) + C$$

$$= \frac{1}{54} u^3 (3u - 10) + C$$

$$= \frac{1}{54} (3x+4)^3 (3(3x+4) - 10) + C$$

$$= \frac{1}{54} (3x+4)^3 (9x + 12 - 10) + C$$

$$= \frac{1}{54} (3x+4)^3 (9x - 2) + C$$

Q124:

$\int \frac{7x^2}{3x^3+1} dx$

Let $u = 3x^3 + 1$, then $\frac{du}{dx} = 9x^2$ and $\frac{dx}{du} = \frac{1}{9x^2}$

$$\int \frac{7x^2}{3x^3+1} dx = \int \frac{7x^2}{3x^3+1} \frac{dx}{du} du$$

$$= \int \frac{7x^2}{u} \times \frac{1}{9x^2} du$$

$$= \frac{7}{9} \int \frac{1}{u} du$$

$$= \frac{7}{9} \ln|u| + C$$

$$= \frac{7}{9} \ln|3x^3 + 1| + C$$

ANSWERS: UNIT 1 TOPIC 3

Q125:

$\int_0^{\frac{\pi}{2}} \sin\left(4x - \frac{\pi}{2}\right) dx$

Let $u = 4x - \frac{\pi}{2}$, then $\frac{du}{dx} = 4$ and $\frac{dx}{du} = \frac{1}{4}$

When $x = 0$, then $u = 4(0) - \frac{\pi}{2}$, so $u = -\frac{\pi}{2}$
When $x = \frac{\pi}{2}$, then $u = 4\left(\frac{\pi}{2}\right) - \frac{\pi}{2}$, so $u = \frac{3\pi}{2}$

$$\int_0^{\frac{\pi}{2}} \sin\left(4x - \frac{\pi}{2}\right) dx = \int_{-\frac{\pi}{2}}^{\frac{3\pi}{2}} \sin\left(4x - \frac{\pi}{2}\right) \frac{dx}{du} du$$

$$= \int_{-\frac{\pi}{2}}^{\frac{3\pi}{2}} \sin u \times \frac{1}{4} du$$

$$= \frac{1}{4} \int_{-\frac{\pi}{2}}^{\frac{3\pi}{2}} \sin u\, du$$

$$= \frac{1}{4} [-\cos u]_{-\frac{\pi}{2}}^{\frac{3\pi}{2}}$$

$$= -\frac{1}{4} [\cos u]_{-\frac{\pi}{2}}^{\frac{3\pi}{2}}$$

$$= -\frac{1}{4} \left(\cos \frac{3\pi}{2} - \cos\left(-\frac{\pi}{2}\right)\right)$$

$$= -\frac{1}{4}(0 - 0)$$

$$= 0$$

Q126:

$\int_{-1}^{1} 5x^2 e^{4x^3+1} dx$

Let $u = 4x^3 + 1$ then $\frac{du}{dx} = 12x^2$ and $\frac{dx}{du} = \frac{1}{12x^2}$

When $x = -1$, then $u = 4(-1)^3 + 1$, so $u = -3$
When $x = 1$, then $u = 4(1)^3 + 1$, so $u = 5$

$$\int_{-1}^{1} 5x^2 e^{4x^3+1} dx = \int_{-3}^{5} 5x^2 e^{4x^3+1} \frac{dx}{du} du$$

$$= \int_{-3}^{5} 5x^2 \times e^u \times \frac{1}{12x^2} du$$

$$= \frac{5}{12} \int_{-3}^{5} e^u du$$

$$= \frac{5}{12} [e^u]_{-3}^{5}$$

$$= \frac{5}{12} \left(e^5 - e^{-3}\right)$$

© HERIOT-WATT UNIVERSITY

Q127:
Use the standard result $\int \frac{1}{\sqrt{a^2-x^2}}dx = \sin^{-1}\left(\frac{x}{a}\right) + C$

$$\int \frac{2}{\sqrt{9-16x^2}}dx = 2\int \frac{1}{\sqrt{9-16x^2}}dx$$
$$= 2\int \frac{1}{\sqrt{16\left(\frac{9}{16}-x^2\right)}}dx$$
$$= 2\int \frac{1}{\sqrt{16}\sqrt{\left(\frac{3}{4}\right)^2-x^2}}dx$$
$$= \frac{2}{4}\int \frac{1}{\sqrt{\left(\frac{3}{4}\right)^2-x^2}}dx$$
$$= \frac{1}{2}\sin^{-1}\left(\frac{x}{\frac{3}{4}}\right) + C$$
$$= \frac{1}{2}\sin^{-1}\left(\frac{4x}{3}\right) + C$$

Q128:
Use the standard result $\int \frac{1}{a^2+x^2}dx = \frac{1}{a}\tan^{-1}\left(\frac{x}{a}\right) + C$

$$\int \frac{2}{9+16x^2}dx = 2\int \frac{1}{16\left(\frac{9}{16}-x^2\right)}dx$$
$$= \frac{1}{8}\int \frac{1}{\frac{9}{16}+x^2}dx$$
$$= \frac{1}{8}\int \frac{1}{\left(\frac{3}{4}\right)^2+x^2}dx$$
$$= \frac{1}{8}\times\frac{1}{\frac{3}{4}}\tan^{-1}\left(\frac{x}{\frac{3}{4}}\right) + C$$
$$= \frac{1}{8}\times\frac{4}{3}\tan^{-1}\left(\frac{4x}{3}\right) + C$$
$$= \frac{1}{6}\tan^{-1}\left(\frac{4x}{3}\right) + C$$

© HERIOT-WATT UNIVERSITY

Q129:
Use the standard result $\int \frac{1}{a^2+x^2}dx = \frac{1}{a}\tan^{-1}\left(\frac{x}{a}\right) + C$

$\int \frac{2}{x^2-6x+25}dx$

First we must complete the square on the denominator:

$x^2 - 6x + 25 = (x-3)^2 - 9 + 25$
$\qquad\qquad\qquad = (x-3)^2 + 16$

$\int \frac{2}{x^2-6x+25}dx = 2\int \frac{1}{(x-3)^2+16}dx$

Let $u = x + 3$, then $\frac{du}{dx} = 1$ and $\frac{dx}{du} = 1$

$2\int \frac{1}{(x-3)^2+16} dx = 2\int \frac{1}{(x-3)^2+16} \frac{dx}{du}du$

$\qquad\qquad\qquad = 2\int \frac{1}{u^2+16} \times 1\, du$

$\qquad\qquad\qquad = 2\int \frac{1}{u^2+4^2}\, du$

$\qquad\qquad\qquad = 2 \times \frac{1}{4}\tan^{-1}\left(\frac{u}{4}\right) + C$

$\qquad\qquad\qquad = \frac{1}{2}\tan^{-1}\left(\frac{x-3}{4}\right) + C$

Q130:
Use the standard result $\int \frac{1}{a^2+x^2}dx = \frac{1}{a}\tan^{-1}\left(\frac{x}{a}\right) + C$

$\int_0^2 \frac{2}{3x^2+4}dx = 2\int_0^2 \frac{1}{3x^2+4}dx$

$\qquad\qquad\qquad = 2\int_0^2 \frac{1}{3\left(x^2+\frac{4}{3}\right)}dx$

$\qquad\qquad\qquad = \frac{2}{3}\int_0^2 \frac{1}{x^2+\left(\frac{2}{\sqrt{3}}\right)^2}dx$

$\qquad\qquad\qquad = \frac{2}{3}\left[\frac{1}{\frac{2}{\sqrt{3}}}\tan^{-1}\left(\frac{x}{\frac{2}{\sqrt{3}}}\right)\right]_0^2$

$\qquad\qquad\qquad = \frac{\sqrt{3}}{3}\left[\tan^{-1}\left(\frac{\sqrt{3}x}{2}\right)\right]_0^2$

$\qquad\qquad\qquad = \frac{\sqrt{3}}{3}\left(\tan^{-1}\left(\sqrt{3}\right) - \tan^{-1}(0)\right)$

$\qquad\qquad\qquad = \frac{\sqrt{3}}{3}\left(\frac{\pi}{3} - 0\right)$

$\qquad\qquad\qquad = \frac{\sqrt{3}\pi}{9}$

© HERIOT-WATT UNIVERSITY

Q131:
Use the standard result $\int \frac{1}{\sqrt{a^2-x^2}}dx = \sin^{-1}\left(\frac{x}{a}\right) + C$

$$\int_0^{\frac{2}{\sqrt{3}}} \frac{1}{\sqrt{16-9x^2}}dx = \int_0^{\frac{2}{\sqrt{3}}} \frac{1}{\sqrt{9\left(\frac{16}{9}-x^2\right)}}dx$$

$$= \int_0^{\frac{2}{\sqrt{3}}} \frac{1}{\sqrt{9}\sqrt{\frac{16}{9}-x^2}}dx$$

$$= \frac{1}{3}\int_0^{\frac{2}{\sqrt{3}}} \frac{1}{\sqrt{\left(\frac{4}{3}\right)^2-x^2}}dx$$

$$= \frac{1}{3}\left[\sin^{-1}\left(\frac{x}{\frac{4}{3}}\right)\right]_0^{\frac{2}{\sqrt{3}}}$$

$$= \frac{1}{3}\left[\sin^{-1}\left(\frac{3x}{4}\right)\right]_0^{\frac{2}{\sqrt{3}}}$$

$$= \frac{1}{3}\left(\sin^{-1}\left(\frac{3}{2\sqrt{3}}\right) - \sin^{-1}(0)\right)$$

$$= \frac{1}{3}\left(\sin^{-1}\left(\frac{\sqrt{3}}{2}\right) - \sin^{-1}(0)\right)$$

$$= \frac{1}{3}\left(\frac{\pi}{3} - 0\right)$$

$$= \frac{\pi}{9}$$

Q132:
Use the standard result $\int \frac{1}{a^2+x^2}dx = \frac{1}{a}\tan^{-1}\left(\frac{x}{a}\right) + C$

$\int_{-2}^1 \frac{2}{x^2+4x+13} dx$

First we must complete the square on the denominator:

$x^2 + 4x + 13 = (x+2)^2 - 4 + 13$
$\qquad\qquad\qquad = (x+2)^2 + 9$

$\int_{-2}^1 \frac{2}{x^2+4x+13} dx = \int_{-2}^1 \frac{2}{(x+2)^2+9} dx$

Let $u = x + 2$, then $\frac{du}{dx} = 1$ and $\frac{dx}{du} = 1$

When $x = -2$, then $u = -2 + 2$, so $u = 0$
When $x = 1$, then $u = 1 + 2$, so $u = 3$

$$\int_{-2}^{1} \frac{2}{(x+2)^2+9}dx = \int_{0}^{3} \frac{2}{(x+2)^2+9}\frac{dx}{du}du$$
$$= \int_{0}^{3} \frac{2}{u^2+9} \times 1\, du$$
$$= \int_{0}^{3} \frac{1}{u^2+3^2} \times 1\, du$$
$$= \left[\frac{2}{3}\tan^{-1}\left(\frac{u}{3}\right)\right]_{0}^{3}$$
$$= \frac{2}{3}\left(\tan^{-1}(1) - \tan^{-1}(0)\right)$$
$$= \frac{2}{3}\left(\frac{\pi}{4} - 0\right)$$
$$= \frac{\pi}{6}$$

Q133:
Use the standard result $\int \frac{1}{f(x)}dx = \frac{1}{f'(x)}\ln|f(x)| + C$
Step 1: Set up the sum of partial fractions.
$\frac{-9}{(x+1)(x-2)} = \frac{A}{(x+1)} + \frac{B}{(x-2)}$
Step 2: Obtain a common denominator and equate the numerators.
$$\frac{-9}{(x+1)(x-2)} = \frac{A(x-2)}{(x+1)(x-2)} + \frac{B(x+1)}{(x+1)(x-2)}$$
$$-9 = A(x-2) + B(x+1)$$
Step 3: Select values of x to work out A and B.
Let $x = 2$ then,
$-9 = A(2-2) + B(2+1)$
$-9 = 3B$
$B = -3$
Let $x = -1$ and given $B = -3$ then,
$-9 = A(-1-2) - 3(-1+1)$
$-9 = A(-1-2) - 3(-1+1)$
$A = 3$
Step 4: Substitute $A = 3$ and $B = -3$ back into the sum of partial fractions.
$\frac{-9}{(x+1)(x-2)} = \frac{3}{(x+1)} - \frac{3}{(x-2)}$
Step 5: Integrate using standard results.
Use the standard result $\int \frac{1}{f(x)}dx = \frac{1}{f'(x)}\ln|f(x)| + C$
$$\int \frac{-9}{(x+1)(x-2)}dx = \int \frac{3}{x+1} - \frac{3}{x-2}dx$$
$$= 3\int \frac{1}{x+1}dx - 3\int \frac{1}{x-2}dx$$
$$= 3\ln|x+1| - 3\ln|x-2| + C$$

Q134:
Step 1: Factorise the denominator and set up the sum of partial fractions.

$$4x^2 + 4x - 3 = (2x - 1)(2x + 3)$$

$$\frac{8x - 8}{4x^2 + 4x - 3} = \frac{8x - 8}{(2x - 1)(2x + 3)}$$

$$\frac{8x - 8}{(2x - 1)(2x + 3)} = \frac{A}{(2x - 1)} + \frac{B}{(2x + 3)}$$

Step 2: Obtain a common denominator and equate the numerators.

$$\frac{8x - 8}{(2x - 1)(2x + 3)} = \frac{A(2x + 3)}{(2x - 1)(2x + 3)} + \frac{B(2x - 1)}{(2x - 1)(2x + 3)}$$

$$8x - 8 = A(2x + 3) + B(2x - 1)$$

Step 3: Select values of x to work out A and B.
Let $x = \frac{1}{2}$ then,

$$8\left(\frac{1}{2}\right) - 8 = A\left(2\left(\frac{1}{2}\right) + 3\right) + B\left(2\left(\frac{1}{2}\right) - 1\right)$$

$$-4 = 4A$$

$$A = -1$$

Let $x = 0$ and given $A = -1$ then,

$$8(0) - 8 = -1(2(0) + 3) + B(2(0) - 1)$$

$$-8 = -3 - B$$

$$-5 = -B$$

$$B = 5$$

Step 4: Substitute $A = -1$ and $B = 5$ back into the sum of partial fractions.

$$\frac{8x-8}{(2x-1)(2x+3)} = \frac{-1}{(2x-1)} + \frac{5}{(2x+3)}$$

Step 5: Integrate using standard results.

$$\int \frac{8x - 8}{(2x - 1)(2x + 3)} dx = \int \frac{-1}{2x - 1} + \frac{5}{2x + 3} dx$$

$$= -\frac{1}{2} \int \frac{2}{2x - 1} dx + \frac{5}{2} \int \frac{2}{2x + 3} dx$$

$$= -\frac{1}{2} \ln|2x - 1| + \frac{5}{2} \ln|2x + 3| + C$$

Q135:
Step 1: Set up the sum of partial fractions.

$$\frac{5x+7}{(x-1)(x+2)(x+3)} = \frac{A}{(x-1)} + \frac{B}{(x+2)} + \frac{C}{(x+3)}$$

Step 2: Obtain a common denominator and equate the numerators.

$$\frac{5x + 7}{(x - 1)(x + 2)(x + 3)} = \frac{A(x + 2)(x + 3)}{(x - 1)(x + 2)(x + 3)} + \frac{B(x - 1)(x + 3)}{(x - 1)(x + 2)(x + 3)} + \frac{C(x - 1)(x + 2)}{(x - 1)(x + 2)(x + 3)}$$

$$5x + 7 = A(x + 2)(x + 3) + B(x - 1)(x + 3) + C(x - 1)(x + 2)$$

ANSWERS: UNIT 1 TOPIC 3   547

Step 3: Select values of x to work out A, B and C.
Let $x = -2$ then,
$5(-2) + 7 = A(-2+2)(-2+3) + B(-2-1)(-2+3) + C(-2-1)(-2+2)$
$\quad -3 = -3B$
$\quad B = 1$
Let $x = -3$ and given $B = 1$ then,
$5(-3) + 7 = A(-3+2)(-3+3) + (-3-1)(-3+3) + C(-3-1)(-3+2)$
$\quad -8 = 4C$
$\quad C = -2$
Let $x = 1$ and given $B = 1$ and $C = -2$ then,
$5(1) + 7 = A(1+2)(1+3) + (1-1)(-3+3) - 2(1-1)(1+2)$
$\quad 12 = 12A$
$\quad A = 1$

Step 4: Substitute $A = 1$, $B = 1$ and $C = -2$ back into the sum of partial fractions.
$\frac{5x+7}{(x-1)(x+2)(x+3)} = \frac{1}{(x-1)} + \frac{1}{(x+2)} - \frac{2}{(x+3)}$

Step 5: Integrate using standard results.
$$\int \frac{5x+7}{(x-1)(x+2)(x+3)} dx = \int \frac{1}{x-1} + \frac{1}{x+2} - \frac{2}{x+3} dx$$
$$= \int \frac{1}{x-1} dx + \int \frac{1}{x+2} dx - 2 \int \frac{1}{x+3} dx$$
$$= \ln|x-1| + \ln|x+2| - 2\ln|x+3| + C$$

Q136:
Step 1: Factorise the denominator and set up the sum of partial fractions.
Using synthetic division try $x = 1$

1	1	-2	-1	2
	0	1	-1	-2
	1	-1	-2	0

Since the remainder is zero, then $x = 1$ is a root and so $(x - 1)$ is a factor.
$x^3 - 2x^2 - x + 2 = (x-1)(x^2 - x - 2)$
$\qquad\qquad\qquad\qquad = (x-1)(x+1)(x-2)$
$\frac{-6}{x^3 - 2x^2 - x + 2} = \frac{-6}{(x-1)(x+1)(x-2)}$
$\frac{-6}{(x-1)(x+1)(x-2)} = \frac{A}{(x-1)} + \frac{B}{(x+1)} + \frac{C}{(x-2)}$

Step 2: Obtain a common denominator and equate the numerators.
$\frac{-6}{(x-1)(x+1)(x-2)} = \frac{A(x+1)(x-2)}{(x-1)(x+1)(x-2)} + \frac{B(x-1)(x-2)}{(x-1)(x+1)(x-2)} + \frac{C(x-1)(x+1)}{(x-1)(x+1)(x-2)}$
$\qquad -6 = A(x+1)(x-2) + B(x-1)(x-2) + C(x-1)(x+1)$

© HERIOT-WATT UNIVERSITY

Step 3: Select values of x to work out A, B and C.
Let $x = -1$ then,
$-6 = A(-1+1)(-1-2) + B(-1-1)(-1-2) + C(-1-1)(-1+1)$
$-6 = 6B$
$B = -1$
Let $x = 2$ and given $B = -1$ then,
$-6 = A(2+1)(2-2) - (2-1)(2-2) + C(2-1)(2+1)$
$-6 = 3C$
$C = -2$
Let $x = 1$ and given $B = -1$ and $C = -2$ then,
$-6 = A(1+1)(1-2) - (1-1)(1-2) - 2(1-1)(1+1)$
$-6 = -2A$
$A = 3$
Step 4: Substitute $A = 3$, $B = -1$ and $C = -2$ back into the sum of partial fractions.
$\frac{-6}{(x-1)(x+1)(x-2)} = \frac{3}{(x-1)} - \frac{1}{(x+1)} - \frac{2}{(x-2)}$
Step 5: Integrate using standard results.
$$\int \frac{-6}{(x-1)(x+1)(x-2)} dx = \int \frac{3}{x-1} - \frac{1}{x+1} - \frac{2}{x-2} dx$$
$$= 3\int \frac{1}{x-1} dx - \int \frac{1}{x+1} dx - 2\int \frac{1}{x-2} dx$$
$$= 3\ln|x-1| - \ln|x+1| - 2\ln|x-2| + C$$

Q137:
Step 1: Set up the sum of partial fractions.
$\frac{-3x-5}{(x+2)^2} = \frac{A}{(x+2)} + \frac{B}{(x+2)^2}$
Step 2: Obtain a common denominator and equate the numerators.
$$\frac{-3x-5}{(x+2)^2} = \frac{A(x+2)}{(x+2)^2} + \frac{B}{(x+2)^2}$$
$-3x - 5 = A(x+2) + B$
Step 3: Select values of x to work out A and B.
Let $x = -2$ then,
$-3(-2) - 5 = A(-2+2) + B$
$\qquad 1 = B$
$\qquad B = 1$
Let $x = 0$ and given $B = 1$ then,
$-3(0) - 5 = A(0+2) + 1$
$\qquad -5 = 2A + 1$
$\qquad -6 = 2A$
$\qquad A = -3$
Step 4: Substitute $A = -3$ and $B = 1$ back into the sum of partial fractions.
$\frac{-3x-5}{(x+2)^2} = \frac{-3}{(x+2)} + \frac{1}{(x+2)^2}$

Step 5: Integrate using standard results.

$$\int \frac{-3x-5}{(x+2)^2}dx = \int \frac{-3}{x+2} + \frac{1}{(x+2)^2}dx$$

$$= -3\int \frac{1}{x+2}dx + \int (x+2)^{-2}dx$$

$$= -3\ln|x+2| - (x+2)^{-1} + C$$

$$= -3\ln|x+2| - \frac{1}{x+2} + C$$

Q138:

Step 1: Set up the sum of partial fractions.

$\frac{x^2-4x+5}{(x-1)^3} = \frac{A}{(x-1)} + \frac{B}{(x-1)^2} + \frac{C}{(x-1)^3}$

Step 2: Obtain a common denominator and equate the numerators.

$$\frac{x^2-4x+5}{(x-1)^3} = \frac{A(x-1)^2}{(x-1)^3} + \frac{B(x-1)}{(x-1)^3} + \frac{C}{(x-1)^3}$$

$$x^2 - 4x + 5 = A(x-1)^2 + B(x-1) + C$$

Step 3: Select values of x to work out A, B and C.

Let $x = 1$ then,

$$1^2 - 4(1) + 5 = A(1-1)^2 + B(1-1) + C$$
$$2 = C$$
$$C = 2$$

Let $x = 0$ and given $C = 2$ then,

$$0^2 - 4(0) + 5 = A(0-1)^2 + B(0-1) + 2$$
$$5 = A - B + 2$$
$$3 = A - B$$
$$A = B + 3$$

Let $x = 0$ and given $C = 2$ then,

$$(-1)^2 - 4(-1) + 5 = A(-1-1)^2 + B(-1-1) + 2$$
$$10 = 4A - 2B + 2$$
$$8 = 4A - 2B$$
$$4 = 2A - B$$

Substitute $A = B + 3$ and $4 = 2A - B$ then,

$$4 = 2(B+3) - B$$
$$4 = 2B + 6 - B$$
$$-2 = B$$
$$B = -2$$

Substitute $B = -2$ back into $A = B + 3$ then,

$$A = -2 + 3$$
$$A = 1$$

Step 4: Substitute $A = 1$, $B = -2$ and $C = 2$ back into the sum of partial fractions.

$\frac{x^2-4x+5}{(x-1)^3} = \frac{1}{(x-1)} - \frac{2}{(x-1)^2} + \frac{2}{(x-1)^3}$

Step 5: Integrate using standard results.

$$\int \frac{x^2 - 4x + 5}{(x-1)^3} dx = \int \frac{1}{x-1} - \frac{2}{(x-1)^2} + \frac{2}{(x-1)^3} dx$$

$$= \int \frac{1}{x-1} dx - 2\int \frac{1}{(x-1)^2} dx + 2\int \frac{1}{(x-1)^3} dx$$

$$= \int \frac{1}{x-1} dx - 2\int (x-1)^{-2} dx + 2\int (x-1)^{-3} dx$$

$$= \ln|x-1| + 2(x-1)^{-1} - \frac{2}{2}(x-1)^{-2} + C$$

$$= \ln|x-1| + \frac{2}{x-1} - \frac{1}{(x-1)^2} + C$$

Q139:

Step 1: Set up the sum of partial fractions.

$\frac{9x-9}{(2x+1)(x-2)^2} = \frac{A}{(2x+1)} + \frac{B}{(x-2)} + \frac{C}{(x-2)^2}$

Step 2: Obtain a common denominator and equate the numerators.

$$\frac{9x-9}{(2x+1)(x-2)^2} = \frac{A(x-2)^2}{(2x+1)(x-2)^2} + \frac{B(2x+1)(x-2)}{(2x+1)(x-2)^2} + \frac{C(2x+1)}{(2x+1)(x-2)^2}$$

$$9x - 9 = A(x-2)^2 + B(2x+1)(x-2) + C(2x+1)$$

Step 3: Select values of x to work out A, B and C.

Let $x = 2$ then,

$$9(2) - 9 = A(2-2)^2 + B(2(2)+1)(2-2) + C(2(2)+1)$$

$$9 = 5C$$

$$C = \frac{9}{5}$$

Let $x = -\frac{1}{2}$ and given $C = \frac{9}{5}$ then,

$$9\left(-\frac{1}{2}\right) - 9 = A\left(-\frac{1}{2} - 2\right)^2 + B\left(2\left(-\frac{1}{2}\right) + 1\right)\left(-\frac{1}{2} - 2\right) + \frac{9}{5}\left(2\left(-\frac{1}{2}\right) + 1\right)$$

$$-\frac{27}{2} = \frac{25}{4}A$$

$$A = -\frac{54}{25}$$

Let $x = 0$ and given $A = -\frac{54}{25}$ and $C = \frac{9}{5}$ then,

$$9(0) - 9 = -\frac{54}{25}(0-2)^2 + B(2(0)+1)(0-2) + \frac{9}{5}(2(0)+1)$$

$$-9 = -\frac{216}{25} - 2B + \frac{9}{5}$$

$$-9 = -\frac{171}{25} - 2B$$

$$-\frac{54}{25} = -2B$$

$$\frac{27}{25} = B$$

$$B = \frac{27}{25}$$

Step 4: Substitute $A = -\frac{54}{25}$, $B = \frac{27}{25}$ and $C = \frac{9}{5}$ back into the sum of partial fractions.
$\frac{9x-9}{(2x+1)(x-2)^2} = -\frac{54}{25(2x+1)} + \frac{27}{25(x-2)} + \frac{9}{5(x-2)^2}$

Step 5: Integrate using standard results.

$$\int \frac{9x-9}{(2x+1)(x-2)^2} dx = \int -\frac{54}{25(2x+1)} + \frac{27}{25(x-2)} + \frac{9}{5(x-2)^2} dx$$

$$= -\frac{54}{25} \int \frac{1}{2x+1} dx + \frac{27}{25} \int \frac{1}{x-2} dx + \frac{9}{5} \int (x-2)^{-2} dx$$

$$= -\frac{54}{25} \times \frac{1}{2} \int \frac{2}{2x+1} dx + \frac{27}{25} \int \frac{1}{x-2} dx + \frac{9}{5} \int (x-2)^{-2} dx$$

$$= -\frac{27}{25} \ln|2x+1| + \frac{27}{25} \ln|x-2| - \frac{9}{5}(x-2)^{-1} + C$$

$$= -\frac{27}{25} \ln|2x+1| + \frac{27}{25} \ln|x-2| - \frac{9}{5(x-2)} + C$$

$$= \frac{27}{25} \ln\left|\frac{x-2}{2x+1}\right| - \frac{9}{5(x-2)} + C$$

Q140:

Step 1: Factorise the denominator and set up the sum of partial fractions.
$\frac{1}{(x-1)(x^2+4)} = \frac{A}{(x-1)} + \frac{Bx+C}{(x^2+4)}$

Step 2: Obtain a common denominator and equate the numerators.

$$\frac{1}{(x-1)(x^2+4)} = \frac{A(x^2+4)}{(x-1)(x^2+4)} + \frac{(Bx+C)(x-1)}{(x-1)(x^2+4)}$$

$$1 = A(x^2+4) + (Bx+C)(x-1)$$

$$1 = A(x^2+4) + Bx(x-1) + C(x-1)$$

Step 3: Select values of x to work out A, B and C.

Let $x = 1$ then,

$1 = A(1^2+4) + B(1)(1-1) + C(1-1)$

$1 = 5A$

$A = \frac{1}{5}$

Let $x = 0$ and given $A = \frac{1}{5}$ then,

$1 = \frac{1}{5}(0^2+4) + B(0)(0-1) + C(0-1)$

$1 = \frac{4}{5} - C$

$\frac{1}{5} = -C$

$C = -\frac{1}{5}$

Let $x = -1$ and given $A = \frac{1}{5}$ and $C = -\frac{1}{5}$ then,

$$1 = \frac{1}{5}\left((-1)^2 + 4\right) + B(-1)(-1-1) - \frac{1}{5}(-1-1)$$

$$1 = 1 + 2B + \frac{2}{5}$$

$$-\frac{2}{5} = 2B$$

$$B = -\frac{1}{5}$$

Step 4: Substitute $A = \frac{1}{5}$, $B = -\frac{1}{5}$ and $C = -\frac{1}{5}$ back into the sum of partial fractions.

$\frac{1}{(x-1)(x^2+4)} = \frac{1}{5(x-1)} - \frac{x}{5(x^2+4)} - \frac{1}{5(x^2+4)}$

Step 5: Integrate using standard results.

$$\int \frac{1}{(x-1)(x^2+4)}dx = \int \frac{1}{5(x-1)} - \frac{x}{5(x^2+4)} - \frac{1}{5(x^2+4)}dx$$

$$= \frac{1}{5}\int \frac{1}{x-1}dx - \frac{1}{5} \times \frac{1}{2}\int \frac{2x}{x^2+4}dx - \frac{1}{5}\int \frac{1}{x^2+4}dx$$

$$= \frac{1}{5}\ln|x-1| - \frac{1}{10}\ln|x^2+4| - \frac{1}{5} \times \frac{1}{2}\tan^{-1}\left(\frac{x}{2}\right) + C$$

$$= \frac{1}{5}\ln|x-1| - \frac{1}{10}\ln|x^2+4| - \frac{1}{10}\tan^{-1}\left(\frac{x}{2}\right) + C$$

Q141:

Step 1: Perform long division and set up the sum of partial fractions.

```
                    x  -  1
         ┌─────────────────────
x² + 3x + 2│ x³ + 2x² + 3x + 1
           x³ + 3x² + 2x      ↓
          ─────────────
              - x²  +  x  + 1
              - x²  - 3x  - 2
             ─────────────
                    4x  + 3
```

$\frac{x^3+2x^2+3x+1}{x^2+3x+2} = x - 1 + \frac{4x+3}{x^2+3x+2}$

Factorise $x^2 + 3x + 2$

$x^2 + 3x + 2 = (x+1)(x+2)$

$\frac{x^3+2x^2+3x+1}{x^2+3x+2} = x - 1 + \frac{4x+3}{(x+1)(x+2)}$

Set up the form of the sum of partial fractions.

$\frac{4x+3}{(x+1)(x+2)} = \frac{A}{(x+1)} + \frac{B}{(x+2)}$

Step 2: Obtain a common denominator and equate the numerators.

$$\frac{4x+3}{(x+1)(x+2)} = \frac{A(x+2)}{(x+1)(x+2)} + \frac{B(x+1)}{(x+1)(x+2)}$$

$$4x + 3 = A(x+2) + B(x+1)$$

Step 3: Select values of x to work out A and B.
Let $x = -2$ then,
$$4(-2) + 3 = A(-2+2) + B(-2+1)$$
$$-5 = -B$$
$$B = 5$$
Let $x = -1$ and given $B = 5$ then,
$$4(-1) + 3 = A(-1+2) + (1)(-1+1)$$
$$-1 = A$$
$$A = -1$$
Step 4: Substitute $A = -1$ and $B = 5$ back into the sum of partial fractions and the original expression.
$$\frac{2x+3}{(x+1)(x+2)} = -\frac{1}{(x+1)} + \frac{5}{(x+2)}$$
Therefore,
$$\frac{x^3 + 2x^2 + 3x + 1}{x^2 + 3x + 2} = x - 1 + \frac{4x+3}{(x+1)(x+2)}$$
$$= x - 1 - \frac{1}{x+1} + \frac{5}{x+2}$$
Step 5: Integrate using standard results.
$$\int \frac{x^3 + 2x^2 + 3x + 1}{x^2 + 3x + 2} dx = \int x - 1 - \frac{1}{x+1} + \frac{5}{x+2} dx$$
$$= \int x\,dx - \int 1\,dx - \int \frac{1}{x+1}dx + \int \frac{5}{x+2}dx$$
$$= \frac{x^2}{2} - x - \ln|x+1| + 5\ln|x+2| + C$$

Q142:
Step 1: Set up the sum of partial fractions.
$$\frac{5x-2}{(x-1)(x+2)} = \frac{A}{(x-1)} + \frac{B}{(x+2)}$$
Step 2: Obtain a common denominator and equate the numerators.
$$\frac{5x-2}{(x-1)(x+2)} = \frac{A(x+2)}{(x-1)(x+2)} + \frac{B(x-1)}{(x-1)(x+2)}$$
$$5x - 2 = A(x+2) + B(x-1)$$
Step 3: Select values of x to work out A and B.
Let $x = -2$ then,
$$5(-2) - 2 = A(-2+2) + B(-2-1)$$
$$-12 = -3B$$
$$B = 4$$
Let $x = 1$ and given $B = 4$ then,
$$5(1) - 2 = A(1+2) + 4(1-1)$$
$$3 = 3A$$
$$A = 1$$

Step 4: Substitute $A = 1$ and $B = 4$ back into the sum of partial fractions.
$\frac{5x-2}{(x-1)(x+2)} = \frac{1}{(x-1)} + \frac{4}{(x+2)}$

Step 5: Integrate using standard results and evaluate for the limits.
$$\int_2^3 \frac{5x-2}{(x-1)(x+2)} dx = \int_2^3 \frac{1}{x-1} + \frac{4}{x+2} dx$$
$$= \int_2^3 \frac{1}{x-1} dx + 4\int_2^3 \frac{1}{x+2} dx$$
$$= [\ln|x-1|]_2^3 + 4[\ln|x+2|]_2^3$$
$$= (\ln|2| - \ln|1|) + 4(\ln|5| - \ln|4|)$$
$$= \ln|2| + 4\ln|5| - 4\ln|4|$$
$$= 1 \cdot 5857$$

Q143:

Step 1: Factorise the denominator and set up the sum of partial fractions.
$2x^2 + 7x + 3 = (x+3)(2x+1)$
$\frac{5}{2x^2 + 7x + 3} = \frac{5}{(x+3)(2x+1)}$

Step 2: Obtain a common denominator and equate the numerators.
$\frac{5}{(x+3)(2x+1)} = \frac{A}{(x+3)} + \frac{B}{(2x+1)}$
$5 = A(2x+1) + B(x+3)$

Step 3: Select values of x to work out A and B.
Let $x = -3$ then,
$5 = A(2(-3)+1) + B(-3+3)$
$5 = -5A$
$A = -1$
Let $x = 0$ and given $A = -1$ then,
$5 = -1(2(0)+1) + B(0+3)$
$5 = -1 + 3B$
$6 = 3B$
$B = 2$

Step 4: Substitute $A = -1$ and $B = 2$ back into the sum of partial fractions.
$\frac{5}{(x+3)(2x+1)} = \frac{-1}{(x+3)} + \frac{2}{(2x+1)}$

ANSWERS: UNIT 1 TOPIC 3

Step 5: Integrate using standard results and evaluate for the limits.

$$\int_0^3 \frac{5}{(x+3)(2x+1)} dx = \int_0^3 \frac{-1}{x+3} + \frac{2}{2x+1} dx$$

$$= -\int_0^3 \frac{1}{x+3} dx + \int_0^3 \frac{2}{2x+1} dx$$

$$= -[\ln|x+3|]_0^3 + [\ln|2x+1|]_0^3$$

$$= -(\ln|6| - \ln|3|) + (\ln|7| - \ln|1|)$$

$$= -\ln|6| + \ln|3| + \ln|7|$$

$$= \ln\left|\frac{3 \times 7}{6}\right|$$

$$= \ln\left|\frac{21}{6}\right|$$

$$= 1 \cdot 253$$

Q144:
Step 1: Set up the sum of partial fractions.

$$\frac{2}{(x-1)(x-2)(x-3)} = \frac{A}{(x-1)} + \frac{B}{(x-2)} + \frac{C}{(x-3)}$$

Step 2: Obtain a common denominator and equate the numerators.

$$\frac{2}{(x-1)(x-2)(x-3)} = \frac{A(x-2)(x-3)}{(x-1)(x-2)(x-3)} + \frac{B(x-1)(x-3)}{(x-1)(x-2)(x-3)} + \frac{C(x-1)(x-2)}{(x-1)(x-2)(x-3)}$$

$$2 = A(x-2)(x-3) + B(x-1)(x-3) + C(x-1)(x-2)$$

Step 3: Select values of x to work out A, B and C.
Let $x = 2$ then,
$2 = A(2-2)(2-3) + B(2-1)(2-3) + C(2-1)(2-2)$
$2 = -B$
$B = -2$
Let $x = 3$ and given $B = -2$ then,
$2 = A(3-2)(3-3) - 2(3-1)(3-3) + C(3-1)(3-2)$
$2 = 2C$
$C = 1$
Let $x = 1$ and given $B = -2$ and $C = 1$ then,
$2 = A(1-2)(1-3) - 2(1-1)(1-3) + (1-1)(1-2)$
$2 = 2A$
$A = 1$

Step 4: Substitute $A = 1$, $B = -2$ and $C = 1$ back into the sum of partial fractions.

$$\frac{2}{(x-1)(x-2)(x-3)} = \frac{1}{(x-1)} - \frac{2}{(x-2)} + \frac{1}{(x-3)}$$

© HERIOT-WATT UNIVERSITY

Step 5: Integrate using standard results and evaluate for the limits.

$$\int_4^5 \frac{2}{(x-1)(x-2)(x-3)} dx = \int_4^5 \frac{1}{x-1} - \frac{2}{x-2} + \frac{1}{x-3} dx$$

$$= \int_4^5 \frac{1}{x-1} dx - 2\int_4^5 \frac{1}{x-2} dx + \int_4^5 \frac{1}{x-3} dx$$

$$= [\ln|x-1|]_4^5 - 2[\ln|x-2|]_4^5 + [\ln|x-3|]_4^5$$

$$= (\ln|4| - \ln|3|) - 2(\ln|3| - \ln|2|) + (\ln|2| - \ln|1|)$$

$$= \ln|2^2| - \ln|3| - 2\ln|3| + 2\ln|2| + \ln|2|$$

$$= 2\ln|2| + 3\ln|2| - 3\ln|3|$$

$$= 5\ln|2| - 3\ln|3|$$

$$= 0 \cdot 1699$$

Q145:

Step 1: Set up the sum of partial fractions.

Using synthetic division try $x = 1$

```
1 | 2  -12   22  -12
  | 0    2  -10   12
  |_____
    2  -10   12 |  0
```

Since the remainder is zero, then $x = 1$ is a root and so $(x - 1)$ is a factor.

$$2x^3 - 12x^2 + 22x - 12 = (x-1)(2x^2 - 10x + 12)$$
$$= (x-1)(2x-4)(x-3)$$

$$\frac{11x^2 - 46x + 47}{2x^3 - 12x^2 + 22x - 12} = \frac{11x^2 - 46x + 47}{(x-1)(2x-4)(x-3)}$$

$$\frac{11x^2 - 46x + 47}{(x-1)(2x-4)(x-3)} = \frac{A}{(x-1)} + \frac{B}{(2x-4)} + \frac{C}{(x-3)}$$

Step 2: Obtain a common denominator and equate the numerators.

$$\frac{11x^2 - 46x + 47}{(x-1)(2x-4)(x-3)} = \frac{A(2x-4)(x-3)}{(x-1)(2x-4)(x-3)}$$

$$+ \frac{B(x-1)(x-3)}{(x-1)(2x-4)(x-3)} + \frac{C(x-1)(2x-4)}{(x-1)(2x-4)(x-3)}$$

$$11x^2 - 46x + 47 = A(2x-4)(x-3) + B(x-1)(x-3) + C(x-1)(2x-4)$$

Step 3: Select values of x to work out A, B and C.

Let $x = 2$ then,

$$11(2)^2 - 46(2) + 47 = A(2(2) - 4)(2-3) + B(2-1)(2-3) + C(2-1)(2(2)-4)$$

$$-1 = -B$$

$$B = 1$$

Let $x = 3$ and given $B = 1$ then,

$$11(3)^2 - 46(3) + 47 = A(2(3) - 4)(3-3) + (1)(3-1)(3-3) + C(3-1)(2(3)-4)$$

$$8 = 4C$$

$$C = 2$$

ANSWERS: UNIT 1 TOPIC 3

Let $x = 1$ and given $B = 1$ and $C = 2$ then,
$$11(1)^2 - 46(1) + 47 = A(2(1) - 4)(1 - 3) + (1)(1 - 1)(1 - 3) + 2(1 - 1)(2(1) - 4)$$
$$12 = 4A$$
$$A = 3$$

Step 4: Substitute $A = 3$, $B = 1$ and $C = 2$ back into the sum of partial fractions.
$$\frac{11x^2 - 46x + 47}{(x-1)(2x-4)(x-3)} = \frac{3}{(x-1)} + \frac{1}{(2x-4)} + \frac{2}{(x-3)}$$

Step 5: Integrate using standard results and evaluate for the limits.

$$\int_4^5 \frac{11x^2 - 46x + 47}{(x-1)(2x-4)(x-3)} dx = \int_4^5 \frac{3}{x-1} + \frac{1}{2x-4} + \frac{2}{x-3} dx$$
$$= 3\int_4^5 \frac{1}{x-1} dx + \int_4^5 \frac{1}{2x-4} dx + 2\int_4^5 \frac{1}{x-3} dx$$
$$= 3[\ln|x-1|]_4^5 + \frac{1}{2}[\ln|2x-4|]_4^5 + 2[\ln|x-3|]_4^5$$
$$= 3(\ln|4| - \ln|3|) + \frac{1}{2}(\ln|6| - \ln|4|) + 2(\ln|2| - \ln|1|)$$
$$= 3\ln|4| - 3\ln|3| + \frac{1}{2}\ln|6| - \frac{1}{2}\ln|4| + 2\ln|2|$$
$$= 2 \cdot 4521$$

Q146:

Step 1: Factorise the denominator and set up the sum of partial fractions.
$$x^2 - 2x + 1 = (x-1)(x-1)$$
$$= (x-1)^2$$
$$\frac{3x-2}{x^2 - 2x + 1} = \frac{3x-2}{(x-1)^2}$$
$$\frac{3x-2}{(x-1)^2} = \frac{A}{(x-1)} + \frac{B}{(x-1)^2}$$

Step 2: Obtain a common denominator and equate the numerators.
$$\frac{3x-2}{(x-1)^2} = \frac{A(x-1)}{(x-1)^2} + \frac{B}{(x-1)^2}$$
$$3x - 2 = A(x-1) + B$$

Step 3: Select values of x to work out A and B.
Let $x = 1$ then,
$$3(1) - 2 = A(1-1) + B$$
$$1 = B$$
$$B = 1$$
Let $x = 0$ and given $B = 1$ then,
$$3(0) - 2 = A(0-1) + 1$$
$$-2 = -A + 1$$
$$-3 = -A$$
$$A = 3$$

Step 4: Substitute $A = 3$ and $B = 1$ back into the sum of partial fractions.
$$\frac{3x-2}{(x-1)^2} = \frac{3}{(x-1)} + \frac{1}{(x-1)^2}$$

© HERIOT-WATT UNIVERSITY

Step 5: Integrate using standard results and evaluate for the limits.

$$\int_3^5 \frac{3x-2}{(x-1)^2} dx = \int_3^5 \frac{3}{x-1} + \frac{1}{(x-1)^2} dx$$

$$= 3 \int_3^5 \frac{1}{x-1} dx + \int_3^5 \frac{1}{(x-1)^2} dx$$

$$= 3 \int_3^5 \frac{1}{x-1} dx + \int_3^5 (x-1)^{-2} dx$$

$$= 3 \left[\ln|x-1| \right]_3^5 + \left[-(x-1)^{-1} \right]_3^5$$

$$= 3 \left[\ln|x-1| \right]_3^5 - \left[\frac{1}{x-1} \right]_3^5$$

$$= 3 \left(\ln|5-1| - \ln|3-1| \right) - \left(\frac{1}{5-1} - \frac{1}{3-1} \right)$$

$$= 3 \left(\ln|4| - \ln|2| \right) - \left(\frac{1}{4} - \frac{1}{2} \right)$$

$$= 3 \ln|2| + \frac{1}{4}$$

Q147:

Step 1: Factorise the denominator and set up the sum of partial fractions.

Using synthetic division try $x = 2$

```
-2 | 1   -6   12   -8
   | 0    2   -8    8
   -------------------
     1   -4    4  | 0
```

Since the remainder is zero, then $x = 2$ is a root and so $(x - 2)$ is a factor.

$$x^3 - 6x^2 + 12x - 8 = (x-2)(x^2 - 4x + 4)$$
$$= (x-2)(x-2)(x-2)$$
$$= (x-2)^3$$

$$\frac{3x^2 - 13x + 18}{x^3 - 6x^2 + 12x - 8} = \frac{3x^2 - 13x + 18}{(x-2)^3}$$

$$\frac{3x^2 - 13x + 18}{(x-2)^3} = \frac{A}{(x-2)} + \frac{B}{(x-2)^2} + \frac{C}{(x-2)^3}$$

Step 2: Obtain a common denominator and equate the numerators.

$$\frac{3x^2 - 13x + 18}{(x-2)^3} = \frac{A(x-2)^2}{(x-2)^3} + \frac{B(x-2)}{(x-2)^3} + \frac{C}{(x-2)^3}$$

$$3x^2 - 13x + 18 = A(x-2)^2 + B(x-2) + C$$

Step 3: Select values of x to work out A, B and C.

Let $x = 2$ then,

$$3(2)^2 - 13(2) + 18 = A(2-2)^2 + B(2-2) + C$$
$$4 = C$$
$$C = 4$$

Let $x = 0$ and given $C = 4$ then,
$3(0)^2 - 13(0) + 18 = A(0-2)^2 + B(0-2) + 4$
$$18 = 4A - 2B + 4$$
$$14 = 4A - 2B$$
$$7 = 2A - B$$
$$B = 2A - 7$$
Let $x = 1$ and given $C = 4$ then,
$3(1)^2 - 13(1) + 18 = A(1-2)^2 + B(1-2) + 4$
$$8 = A - B + 4$$
$$4 = A - B$$
Substitute $B = 2A - 7$ into $4 = A - B$
$$4 = A - (2A - 7)$$
$$4 = A - 2A + 7$$
$$-3 = -A$$
$$A = 3$$
Substitute $A = 3$ into $B = 2A - 7$
$$B = 2(3) - 7$$
$$B = -1$$

Step 4: Substitute $A = 3$, $B = -1$ and $C = 4$ back into the sum of partial fractions.
$$\frac{3x^2 - 13x + 18}{(x-2)^3} = \frac{3}{(x-2)} - \frac{1}{(x-2)^2} + \frac{4}{(x-2)^3}$$

Step 5: Integrate using standard results.

$$\int_4^5 \frac{3x^2 - 13x + 18}{(x-2)^3} dx = \int_4^5 \frac{3}{(x-2)} - \frac{1}{(x-2)^2} + \frac{4}{(x-2)^3} dx$$

$$= 3\int_4^5 \frac{1}{x-2} dx - \int_4^5 \frac{1}{(x-2)^2} dx + 4\int_4^5 \frac{1}{(x-2)^3} dx$$

$$= 3\int_4^5 \frac{1}{x-2} dx - \int_4^5 (x-2)^{-2} dx + 4\int_4^5 (x-2)^{-3} dx$$

$$= 3\left[\ln|x-2|\right]_4^5 - \left[-(x-2)^{-1}\right]_4^5 + 4\left[-\frac{1}{2}(x-2)^{-2}\right]_4^5$$

$$= 3\left[\ln|x-2|\right]_4^5 + \left[\frac{1}{x-2}\right]_4^5 - 2\left[\frac{1}{(x-2)^2}\right]_4^5$$

$$= 3(\ln|3| - \ln|2|) + \left(\frac{1}{5-2} - \frac{1}{4-2}\right) - 2\left(\frac{1}{(5-2)^2} - \frac{1}{(4-2)^2}\right)$$

$$= 3\ln\left|\frac{3}{2}\right| + \left(\frac{1}{3} - \frac{1}{2}\right) - 2\left(\frac{1}{9} - \frac{1}{4}\right)$$

$$= 3\ln\left|\frac{3}{2}\right| - \frac{1}{6} - 2\left(-\frac{5}{36}\right)$$

$$= \cdots$$

© HERIOT-WATT UNIVERSITY

$$\int_4^5 \frac{3x^2 - 13x + 18}{(x-2)^3} dx = \cdots$$

$$= 3\ln\left|\frac{3}{2}\right| - \frac{1}{6} + \frac{5}{18}$$

$$= 3\ln\left|\frac{3}{2}\right| + \frac{2}{18}$$

$$= 3\ln\left|\frac{3}{2}\right| + \frac{1}{9}$$

Q148:
Step 1: Set up the sum of partial fractions.
$$\frac{4x^2 - 22x - 8}{(2x+3)(2x-1)^2} = \frac{A}{(2x+3)} + \frac{B}{(2x-1)} + \frac{C}{(2x-1)^2}$$
Step 2: Obtain a common denominator and equate the numerators.
$$\frac{4x^2 - 22x - 8}{(2x+3)(2x-1)^2} = \frac{A(2x-1)^2}{(2x+3)(2x-1)^2} + \frac{B(2x+3)(2x-1)}{(2x+3)(2x-1)^2} + \frac{C(2x+3)}{(2x+3)(2x-1)^2}$$
$$4x^2 - 22x - 8 = A(2x-1)^2 + B(2x+3)(2x-1) + C(2x+3)$$
Step 3: Select values of x to work out A, B and C.
Let $x = \frac{1}{2}$ then,
$$4\left(\frac{1}{2}\right)^2 - 22\left(\frac{1}{2}\right) - 8 = A\left(2\left(\frac{1}{2}\right) - 1\right)^2 + B\left(2\left(\frac{1}{2}\right) + 3\right)\left(2\left(\frac{1}{2}\right) - 1\right) + C\left(2\left(\frac{1}{2}\right) + 3\right)$$
$$-18 = 4C$$
$$C = -\frac{18}{4}$$
$$C = -\frac{9}{2}$$
Let $x = -\frac{3}{2}$ and given $C = -\frac{9}{2}$ then,
$$4\left(-\frac{3}{2}\right)^2 - 22\left(-\frac{3}{2}\right) - 8 = A\left(2\left(-\frac{3}{2}\right) - 1\right)^2 + B\left(2\left(-\frac{3}{2}\right) + 3\right)\left(2\left(-\frac{3}{2}\right) - 1\right) - \frac{9}{2}\left(2\left(-\frac{3}{2}\right) + 3\right)$$
$$34 = 16A$$
$$A = \frac{34}{16}$$
$$A = \frac{17}{8}$$
Let $x = 0$ and given $A = \frac{17}{8}$ and $C = -\frac{9}{2}$ then,
$$4(0)^2 - 22(0) - 8 = \frac{17}{8}(2(0) - 1)^2 + B(2(0) + 3)(2(0) - 1) - \frac{9}{2}(2(0) + 3)$$
$$-8 = \frac{17}{8} - 3B - \frac{27}{2}$$
$$-8 = -\frac{91}{8} - 3B$$
$$\frac{27}{8} = -3B$$
$$B = -\frac{9}{8}$$

Step 4: Substitute $A = \frac{17}{8}$, $B = -\frac{9}{8}$ and $C = -\frac{9}{2}$ back into the sum of partial fractions.

$$\frac{4x^2-22x-8}{(2x+3)(2x-1)^2} = \frac{17}{8(2x+3)} - \frac{9}{8(2x-1)} - \frac{9}{2(2x-1)^2}$$

Step 5: Integrate using standard results and evaluate for the limits.

$$\int_1^2 \frac{4x^2 - 22x - 8}{(2x+3)(2x-1)^2} dx = \int_1^2 \frac{17}{8(2x+3)} - \frac{9}{8(2x-1)} - \frac{9}{2(2x-1)^2} dx$$

$$= \frac{17}{8}\int_1^2 \frac{1}{2x+3} dx - \frac{9}{8}\int_1^2 \frac{1}{2x-1} dx - \frac{9}{2}\int_1^2 \frac{1}{(2x-1)^2} dx$$

$$= \frac{17}{16}\int_1^2 \frac{2}{2x+3} dx - \frac{9}{16}\int_1^2 \frac{2}{2x-1} dx - \frac{9}{2}\int_1^2 (2x-1)^{-2} dx$$

$$= \frac{17}{16}\left[\ln|2x+3|\right]_1^2 - \frac{9}{16}\left[\ln|2x-1|\right]_1^2 - \frac{9}{2}\left[-\frac{1}{1} \times \frac{1}{2}(2x-1)^{-1}\right]_1^2$$

$$= \frac{17}{16}\left[\ln|2x+3|\right]_1^2 - \frac{9}{16}\left[\ln|2x-1|\right]_1^2 + \frac{9}{4}\left[(2x-1)^{-1}\right]_1^2$$

$$= \frac{17}{16}\left[\ln|2x+3|\right]_1^2 - \frac{9}{16}\left[\ln|2x-1|\right]_1^2 + \frac{9}{4}\left[\frac{1}{2x-1}\right]_1^2$$

$$= \frac{17}{16}(\ln|7| - \ln|5|) - \frac{9}{16}(\ln|3| - \ln|1|) + \frac{9}{4}\left(\frac{1}{3} - 1\right)$$

$$= \frac{17}{16}\ln\left|\frac{7}{5}\right| - \frac{9}{16}\ln|3| - \frac{3}{2}$$

$$= -1 \cdot 7605$$

Q149:

Step 1: Factorise the denominator and set up the sum of partial fractions.

Using synthetic division try $x = 3$

```
3 |  1    1   -8  -12
  |  0    3   12   12
  |_____
     1    4    4 |  0
```

Since the remainder is zero, then $x = 3$ is a root and so $(x - 3)$ is a factor.

$$x^3 + x^2 - 8x - 12 = (x - 3)(x^2 + 4x + 4)$$
$$= (x - 3)(x + 2)(x + 2)$$
$$= (x - 3)(x + 2)^2$$

$$\frac{25}{x^3 + x^2 - 8x - 12} = \frac{25}{(x - 3)(x + 2)^2}$$

$$\frac{25}{(x - 3)(x + 2)^2} = \frac{A}{(x - 3)} + \frac{B}{(x + 2)} + \frac{C}{(x + 2)^2}$$

Step 2: Obtain a common denominator and equate the numerators.

$$\frac{25}{(x - 3)(x + 2)^2} = \frac{A(x + 2)^2}{(x - 3)(x + 2)^2} + \frac{B(x - 3)(x + 2)}{(x - 3)(x + 2)^2} + \frac{C(x - 3)}{(x - 3)(x + 2)^2}$$

$$25 = A(x + 2)^2 + B(x - 3)(x + 2) + C(x - 3)$$

Step 3: Select values of x to work out A, B and C.
Let $x = -2$ then,
$25 = A(-2+2)^2 + B(-2-3)(-2+2) + C(-2-3)$
$25 = -5C$
$C = -5$
Let $x = 3$ and given $C = -5$ then,
$25 = A(3+2)^2 + B(3-3)(3+2) - 5(3-3)$
$25 = 25A$
$A = 1$
Let $x = 0$ and given $A = 1$ and $C = -5$ then,
$25 = (0+2)^2 + B(0-3)(0+2) - 5(0-3)$
$25 = 4 - 6B + 15$
$6 = -6B$
$B = -1$

Step 4: Substitute $A = 1$, $B = -1$ and $C = -5$ back into the sum of partial fractions.
$\frac{25}{(x-3)(x+2)^2} = \frac{1}{(x-3)} - \frac{1}{(x+2)} - \frac{5}{(x+2)^2}$

Step 5: Integrate using standard results.

$$\int_4^5 \frac{25}{(x-3)(x+2)^2}dx = \int_4^5 \frac{1}{x-3} - \frac{1}{x+2} - \frac{5}{(x+2)^2}dx$$

$$= \int_4^5 \frac{1}{x-3}dx - \int_4^5 \frac{1}{x+2}dx - 5\int_4^5 \frac{1}{(x+2)^2}dx$$

$$= \int_4^5 \frac{1}{x-3}dx - \int_4^5 \frac{1}{x+2}dx - 5\int_4^5 (x+2)^{-2}dx$$

$$= [\ln|x-3|]_4^5 - [\ln|x+2|]_4^5 - 5\left[-(x+2)^{-1}\right]_4^5$$

$$= [\ln|x-3|]_4^5 - [\ln|x+2|]_4^5 + 5\left[\frac{1}{x+2}\right]_4^5$$

$$= (\ln|2| - \ln|1|) - (\ln|7| - \ln|6|) + 5\left(\frac{1}{7} - \frac{1}{6}\right)$$

$$= \ln|2| - \ln|7| + \ln|6| + 5\left(-\frac{1}{42}\right)$$

$$= \ln\left|\frac{12}{7}\right| - \frac{5}{42}$$

ANSWERS: UNIT 1 TOPIC 3

Q150:

Step 1: Factorise the denominator and set up the sum of partial fractions.
Using synthetic division try $x = 1$

$$\begin{array}{c|cccc} 1 & 2 & -7 & 8 & -3 \\ & 0 & 2 & -5 & 3 \\ \hline & 2 & -5 & 3 & 0 \end{array}$$

Since the remainder is zero, then $x = 1$ is a root and so $(x - 1)$ is a factor.

$$2x^3 - 7x^2 + 8x - 3 = (x - 1)(2x^2 - 5x + 3)$$
$$= (x - 1)(2x - 3)(x - 1)$$
$$= (2x - 3)(x - 1)^2$$

$$\frac{2x^2 + 2x + 7}{2x^3 - 7x^2 + 8x - 3} = \frac{2x^2 + 2x + 7}{(2x - 3)(x - 1)^2} = \frac{A}{(2x - 3)} + \frac{B}{(x - 1)} + \frac{C}{(x - 1)^2}$$

Step 2: Obtain a common denominator and equate the numerators.

$$\frac{2x^2 + 2x + 7}{(2x - 3)(x - 1)^2} = \frac{A(x - 1)^2}{(2x - 3)(x - 1)^2} + \frac{B(x - 1)(2x - 3)}{(2x - 3)(x - 1)^2} + \frac{C(2x - 3)}{(2x - 3)(x - 1)^2}$$

$$2x^2 + 2x + 7 = A(x - 1)^2 + B(x - 1)(2x - 3) + C(2x - 3)$$

Step 3: Select values of x to work out A, B and C.

Let $x = 1$ then,

$$2(1)^2 + 2(1) + 7 = A(1 - 1)^2 + B(1 - 1)(2(1) - 3) + C(2(1) - 3)$$
$$11 = -C$$
$$C = -11$$

Let $x = \frac{3}{2}$ and given $C = -11$ then,

$$2\left(\frac{3}{2}\right)^2 + 2\left(\frac{3}{2}\right) + 7 = A\left(\frac{3}{2} - 1\right)^2 + B\left(\frac{3}{2} - 1\right)\left(2\left(\frac{3}{2}\right) - 3\right) - 11\left(2\left(\frac{3}{2}\right) - 3\right)$$

$$\frac{29}{2} = \frac{1}{4}A$$

$$A = 58$$

Let $x = 0$ and given $A = 58$ and $C = -11$ then,

$$2(0)^2 + 2(0) + 7 = 58(0 - 1)^2 + B(0 - 1)(2(0) - 3) - 11(2(0) - 3)$$
$$7 = 58 + 3B + 33$$
$$7 = 91 + 3B$$
$$-84 = 3B$$
$$B = -28$$

Step 4: Substitute $A = 58$, $B = -28$ and $C = -11$ back into the sum of partial fractions.

$$\frac{2x^2 + 2x + 7}{(2x - 3)(x - 1)^2} = \frac{58}{(2x - 3)} - \frac{28}{(x - 1)} - \frac{11}{(x - 1)^2}$$

© HERIOT-WATT UNIVERSITY

Step 5: Integrate using standard results.

$$\int_2^4 \frac{2x^2+2x+7}{2x^3-7x^2+8x-3}\,dx = \int_2^4 \frac{58}{2x-3} - \frac{28}{x-1} - \frac{11}{(x-1)^2}\,dx$$

$$= 58\int_2^4 \frac{1}{2x-3}\,dx - 28\int_2^4 \frac{1}{x-1}\,dx - 11\int_2^4 \frac{1}{(x-1)^2}\,dx$$

$$= 58 \times \frac{1}{2}\int_2^4 \frac{2}{2x-3}\,dx - 28\int_2^4 \frac{1}{x-1}\,dx - 11\int_2^4 (x-1)^{-2}\,dx$$

$$= 29\left[\ln|2x-3|\right]_2^4 - 28\left[\ln|x-1|\right]_2^4 - 11\left[-(x-1)^{-1}\right]_2^4$$

$$= 29\left[\ln|2x-3|\right]_2^4 - 28\left[\ln|x-1|\right]_2^4 + 11\left[\frac{1}{x-1}\right]_2^4$$

$$= 29\left(\ln|5| - \ln|1|\right) - 28\left(\ln|3| - \ln|1|\right) + 11\left(\frac{1}{3} - 1\right)$$

$$= 29\ln|5| - 28\ln|3| + \ln|6| + 11\left(-\frac{2}{3}\right)$$

$$= 29\ln|5| - 28\ln|3| - \frac{22}{3}$$

$$= 8 \cdot 579$$

Q151:

Step 1: Check for factorisation and set up the sum of partial fractions.

For $x^2 + 3$,

$a = 1, b = 0, c = 3$

$b^2 - 4ac = 0^2 - 4(1)(3)$

$\quad\quad\quad = 0^2 - 12$

$\quad\quad\quad = -12$

$\quad\quad\quad < 0$

$\frac{2x^2-x+5}{(x+1)(x^2+3)} = \frac{A}{(x+1)} + \frac{Bx+C}{(x^2+3)}$

Since $b^2 - 4ac < 0$, then $x^2 + 3$ has no real roots.

Step 2: Obtain a common denominator and equate the numerators.

$$\frac{2x^2-x+5}{(x+1)(x^2+3)} = \frac{A(x^2+3)}{(x+1)(x^2+3)} + \frac{(Bx+C)(x+1)}{(x+1)(x^2+3)}$$

$$2x^2 - x + 5 = A(x^2+3) + (Bx+C)(x+1)$$

$$2x^2 - x + 5 = A(x^2+3) + Bx(x+1) + C(x+1)$$

Step 3: Select values of x to work out A, B and C.

Let $x = -1$ then,

$$2(-1)^2 - (-1) + 5 = A\left((-1)^2 + 3\right) + B(-1)(-1+1) + C(-1+1)$$

$$8 = 4A$$

$$2 = A$$

$$A = 2$$

Let $x = 0$ and given $A = 2$ then,

© HERIOT-WATT UNIVERSITY

ANSWERS: UNIT 1 TOPIC 3 565

$$2(0)^2 - 0 + 5 = 2\left(0^2 + 3\right) + B(0)(0+1) + C(0+1)$$
$$5 = 2(3) + C$$
$$5 = 6 + C$$
$$-1 = C$$
$$C = -1$$

Let $x = 1$ and given $A = 2$ and $C = -1$ then,
$$2(1)^2 - 1 + 5 = 2\left(1^2 + 3\right) + B(1)(1+1) + (-1)(1+1)$$
$$6 = 2(4) + 2B - 2$$
$$6 = 8 + 2B - 2$$
$$6 = 6 + 2B$$
$$0 = 2B$$
$$B = 0$$

Step 4: Substitute $A = 2$, $B = 0$ and $C = -1$ back into the sum of partial fractions.

$$\frac{2x^2 - x + 5}{(x+1)(x^2+3)} = \frac{2}{(x+1)} + \frac{0x - 1}{(x^2+3)}$$

$$\frac{2x^2 - x + 5}{(x+1)(x^2+3)} = \frac{2}{(x+1)} - \frac{1}{(x^2+3)}$$

Step 5: Integrate using standard results.

$$\int_1^{\sqrt{3}} \frac{2x^2 - x + 5}{(x+1)(x^2+3)} dx = \int_1^{\sqrt{3}} \frac{2}{x+1} - \frac{1}{x^2+3} dx$$

$$= 2\int_1^{\sqrt{3}} \frac{1}{x+1} dx - \int_1^{\sqrt{3}} \frac{1}{x^2+3} dx$$

$$= 2\int_1^{\sqrt{3}} \frac{1}{x+1} dx - \int_1^{\sqrt{3}} \frac{1}{x^2 + (\sqrt{3})^2} dx$$

$$= 2[\ln|x+1|]_1^{\sqrt{3}} - \frac{1}{\sqrt{3}} \left[\tan^{-1}\left(\frac{x}{\sqrt{3}}\right)\right]_1^{\sqrt{3}}$$

$$= 2[\ln|x+1|]_1^{\sqrt{3}} - \frac{1}{\sqrt{3}} \left[\tan^{-1}\left(\frac{x}{\sqrt{3}}\right)\right]_1^{\sqrt{3}}$$

$$= 2\left(\ln\left|\sqrt{3}+1\right| - \ln|1+1|\right) - \frac{1}{\sqrt{3}}\left(\tan^{-1}(1) - \tan^{-1}\left(\frac{1}{\sqrt{3}}\right)\right)$$

$$= 2\left(\ln\left|\sqrt{3}+1\right| - \ln|2|\right) - \frac{1}{\sqrt{3}}\left(\frac{\pi}{4} - \frac{\pi}{6}\right)$$

$$= 2\ln\left|\frac{\sqrt{3}+1}{2}\right| - \frac{\sqrt{3}\pi}{36}$$

$$= \ln\left|\frac{(\sqrt{3}+1)^2}{4}\right| - \frac{\sqrt{3}\pi}{36}$$

© HERIOT-WATT UNIVERSITY

Q152:
$\int (3x+4) e^x dx$
The integrand is the product of two functions so we must use integration by parts.
$\int fg' dx = fg - \int f'g dx$
Step 1:
Let $f(x) = 3x + 4$ then $f'(x) = 3$
$g'(x) = e^x$ then g(x) = e^x
Substitute into $\int fg' dx = fg - \int f'g dx$ and integrate:
$$\int (3x+4) e^x dx = (3x+4) e^x - \int 3e^x dx$$
$$= \frac{1}{3}(3x+4) e^x - 3e^x + C$$

Q153:
$\int (4x^2 + 1) \cos 3x dx$
The integrand is the product of two functions so we must use integration by parts.
$\int fg' dx = fg - \int f'g dx$
Step 1:
Let $f(x) = 4x^2 + 1$ then $f'(x) = 8x$
$g'(x) = \cos 3x$ then $g(x) = \frac{1}{3} \sin 3x$
Substitute into $\int fg' dx = fg - \int f'g dx$ and integrate:
$$\int (4x^2 + 1) \cos 3x dx = (4x^2 + 1) \times \frac{1}{3} \sin 3x - \int 8x \times \frac{1}{3} \sin 3x dx$$
$$= \frac{1}{3}(4x^2 + 1) \sin 3x - \frac{8}{3} \int x \sin 3x dx \quad (1)$$
The integral on the RHS, $\int x \sin 3x dx$, is still the product of two functions and so we must use integration by parts a second time.
Step 2:
Let $f(x) = x$ then $f'(x) = 1$
$g'(x) = \cos 3x$ then $g(x) = \frac{1}{3} \sin 3x$
Substitute into $\int fg' dx = fg - \int f'g dx$ and integrate:
$$\int x \sin 3x dx = x \times -\frac{1}{3} \cos 3x - \int 1 \times -\frac{1}{3} \cos 3x dx$$
$$= -\frac{1}{3} x \cos 3x + \frac{1}{3} \int \cos 3x dx$$
$$= -\frac{1}{3} x \cos 3x + \frac{1}{3} \times \frac{1}{3} \sin 3x + C$$
$$= -\frac{1}{3} x \cos 3x + \frac{1}{9} \sin 3x + C$$

ANSWERS: UNIT 1 TOPIC 3

Step 3: Substitute the solution to $\int x \sin 3x dx$ back into (1)

$$\int (4x^2 + 1) \cos 3x dx = \frac{1}{3}(4x^2 + 1) \sin 3x - \frac{8}{3}\int x \sin 3x dx$$

$$= \frac{1}{3}(4x^2 + 1) \sin 3x - \frac{8}{3}\left(-\frac{1}{3}x \cos 3x + \frac{1}{9}\sin 3x + C\right)$$

$$= \frac{1}{3}(4x^2 + 1) \sin 3x + \frac{8}{9}x \cos 3x - \frac{8}{27}\sin 3x + C$$

$$= \frac{9}{27}(4x^2 + 1) \sin 3x + \frac{8}{9}x \cos 3x - \frac{8}{27}\sin 3x + C$$

$$= \frac{1}{27}(36x^2 + 9 - 8) \sin 3x + \frac{8}{9}x \cos 3x + C$$

$$= \frac{1}{27}(36x^2 + 1) \sin 3x + \frac{8}{9}x \cos 3x + C$$

Note that C is an arbitrary constant and so expansion of the brackets gives $-\frac{8}{3}C$ which we can replace with $+C$.

Q154:

$\int 2 \sin 3x \cos 2x dx$

The integrand is the product of two functions so we must use integration by parts.

$\int fg' dx = fg - \int f'g dx$

Step 1:

Let $f(x) = 2 \sin 3x$ then $f'(x) = 6 \cos 3x$

$g'(x) = \cos 2x$ then $g(x) = \frac{1}{2}\sin 2x$

Substitute into $\int fg' dx = fg - \int f'g dx$ and integrate.

$$\int fg' dx = fg - \int f'g dx$$

$$\int 2 \sin 3x \cos 2x dx = 2 \sin 3x \times \frac{1}{2}\sin 2x - \int 6 \cos 3x \times \frac{1}{2}\sin 2x dx$$

$$= \sin 3x \sin 2x - 3 \int \cos 3x \sin 2x dx$$

The integral on the RHS, $\int \cos 3x \sin 2x dx$, is still the product of two functions and so we must use integration by parts a second time.

Step 2:

Let $f(x) = \cos 3x$ then $f'(x) = -3 \sin 3x$

$g'(x) = \sin 2x$ then $g(x) = -\frac{1}{2}\cos 2x$

Substitute into $\int fg' dx = fg - \int f'g dx$ and integrate.

$$\int fg' dx = fg - \int f'g dx$$

$$\int \cos 3x \sin 2x dx = \cos 3x \times -\frac{1}{2}\cos 2x - \int -3 \sin 3x \times -\frac{1}{2}\cos 2x dx$$

$$= -\frac{1}{2}\cos 3x \cos 2x - \frac{3}{2}\int \sin 3x \cos 2x dx$$

© HERIOT-WATT UNIVERSITY

Step 3: Substitute the solution to $\int \cos 3x \sin 2x \, dx$ back into (1)

$$\int 2\sin 3x \cos 2x \, dx = \sin 3x \sin 2x - 3\int \cos 3x \sin 2x \, dx$$

$$\int 2\sin 3x \cos 2x \, dx = \sin 3x \sin 2x - 3\left(-\frac{1}{2}\cos 3x \cos 2x - \frac{3}{2}\int \sin 3x \cos 2x \, dx\right)$$

$$2\int \sin 3x \cos 2x \, dx = \sin 3x \sin 2x + \frac{3}{2}\cos 3x \cos 2x + \frac{9}{2}\int \sin 3x \cos 2x \, dx$$

Step 4: Let $I = \int \sin 3x \cos 2x \, dx$ then rearrange for I.

$$2I = \sin 3x \sin 2x + \frac{3}{2}\cos 3x \cos 2x + \frac{9}{2}I + C$$

$$-\frac{5}{2}I = \sin 3x \sin 2x + \frac{3}{2}\cos 3x \cos 2x + C$$

$$I = -\frac{2}{5}\left(\sin 3x \sin 2x + \frac{3}{2}\cos 3x \cos 2x\right) + C$$

$$2I = -\frac{4}{5}\left(\sin 3x \sin 2x + \frac{3}{2}\cos 3x \cos 2x\right) + C$$

$$2I = -\frac{4}{5}\sin 3x \sin 2x - \frac{6}{5}\cos 3x \cos 2x + C$$

Therefore, $\int 2\sin 3x \cos 2x \, dx = -\frac{4}{5}\sin 3x \sin 2x - \frac{6}{5}\cos 3x \cos 2x + C$

Q155:

$\int e^x \cos x \, dx$

The integrand is the product of two functions so we must use integration by parts.

$\int fg' \, dx = fg - \int f'g \, dx$

Step 1:

Let $f(x) = e^x$ then $f'(x) = e^x$

$g'(x) = \cos x$ then $g(x) = \sin x$

Substitute into $\int fg' \, dx = fg - \int f'g \, dx$ and integrate.

$$\int fg' \, dx = fg - \int f'g \, dx$$

$$\int e^x \cos x \, dx = e^x \times \sin x - \int e^x \times \sin x \, dx$$

$$= e^x \sin x - \int e^x \sin x \, dx$$

The integral on the RHS, $\int e^x \sin x \, dx$, is still the product of two functions and so we must use integration by parts a second time.

Step 2:

Let $f(x) = e^x$ then $f'(x) = e^x$

$g'(x) = \sin x$ then $g(x) = -\cos x$

Substitute into $\int fg' \, dx = fg - \int f'g \, dx$ and integrate.

$$\int fg' \, dx = fg - \int f'g \, dx$$

$$\int e^x \sin x \, dx = e^x \times -\cos x - \int e^x \times -\cos x \, dx$$

$$= -e^x \cos x + \int e^x \cos x \, dx$$

ANSWERS: UNIT 1 TOPIC 3

Step 3: Substitute the solution to $\int e^x \sin x\, dx$ back into (1)

$$\int e^x \cos x\, dx = e^x \sin x - \int e^x \sin x\, dx$$

$$\int e^x \cos x\, dx = e^x \sin x - \left(-e^x \cos x + \int e^x \cos x\, dx\right)$$

$$\int e^x \cos x\, dx = e^x \sin x + e^x \cos x - \int e^x \cos x\, dx$$

Step 4: Let $I = \int e^x \cos x\, dx$ then rearrange for I.

$I = e^x \sin x + e^x \cos x - I + C$

$2I = e^x \sin x + e^x \cos x + C$

$I = \dfrac{1}{2}\left(e^x \sin x + e^x \cos x\right) + C$

Therefore, $\int e^x \cos x\, dx = \tfrac{1}{2} e^x \left(\sin x + \cos x\right) + C$

Q156:

$\int [\ln |x|]^2 dx$

We must multiply the integrand by 1, then use integration by parts to work out this integration.

$\int [\ln |x|]^2 \times 1\, dx$

Step 1:

Let $f(x) = [\ln |x|]^2$ then $f'(x) = \dfrac{2}{x} \ln |x|$

$g'(x) = 1$ then $g(x) = x$

Step 2: Substitute into

$$\int f g'\, dx = fg - \int f' g\, dx$$

and integrate.

$$\int [\ln |x|]^2 dx = [\ln |x|]^2 \times x - \int \dfrac{2}{x} \ln |x| \times x\, dx$$

$$= x[\ln |x|]^2 - 2 \int \ln |x|\, dx \quad (1)$$

Note that $\int \ln |x|\, dx$ cannot be integrated directly. We must multiply the integrand by 1, then apply integration by parts a second time to work out this integration.

Step 3: Apply integration by parts to $\int \ln |x| \times 1\, dx$.

Let $f(x) = \ln |x|$ then $f'(x) = \dfrac{1}{x}$

$g'(x) = 1$ then $g(x) = x$

Substitute into

$$\int f g'\, dx = fg - \int f' g\, dx$$

and integrate.

$$\int \ln |x|\, dx = \ln |x| \times x - \int \dfrac{1}{x} \times x\, dx$$

$$= x \ln |x| - \int 1\, dx$$

$$= x \ln |x| - x + C$$

Step 4: Substitute the solution to $\int \ln |x| dx$ back into (1).

$$\int [\ln |x|]^2 dx = x[\ln |x|]^2 - 2\int \ln |x| dx$$
$$= x[\ln |x|]^2 - 2(x \ln |x| - x + C)$$
$$= x[\ln |x|]^2 - 2x \ln |x| + 2x + C$$
$$= x\left([\ln |x|]^2 - 2 \ln |x| + 2\right) + C$$

Q157:

$\int \cos^{-1} 2x dx$

We must multiply the integrand by 1, then use integration by parts to work out this integration.

$\int \cos^{-1} 2x \times 1 dx$

Step 1:

Let $f(x) = \cos^{-1} 2x$ then $f'(x) = -\frac{2}{\sqrt{1-4x^2}}$

$g'(x) = 1$ then $g(x) = x$

Step 2: Substitute into $\int fg' dx = fg - \int f'g dx$ and integrate.

$$\int \cos^{-1} 2x dx = \cos^{-1} 2x \times x - \int \frac{-2}{\sqrt{1-4x^2}} \times x dx$$
$$= x \cos^{-1} 2x + 2 \int \frac{x}{\sqrt{1-4x^2}} dx$$

Step 3: Integrate $\int \frac{x}{\sqrt{1-4x^2}} dx$ using integration by substitution.

Let $u = 1 - 4x^2$, then $\frac{du}{dx} = -8x$ and $\frac{dx}{du} = -\frac{1}{8x}$

$$\int \frac{x}{\sqrt{1-4x^2}} dx = \int \frac{x}{\sqrt{1-4x^2}} \frac{dx}{du} du$$
$$= \int \frac{x}{\sqrt{u}} \times \left(-\frac{1}{8x}\right) du$$
$$= -\frac{1}{8} \int u^{-\frac{1}{2}} du$$
$$= -\frac{1}{8} \times 2u^{\frac{1}{2}} + C$$
$$= -\frac{1}{4} u^{\frac{1}{2}} + C$$
$$= -\frac{1}{4}(1 - 4x^2)^{\frac{1}{2}} + C$$
$$= -\frac{1}{4}\sqrt{1 - 4x^2} + C$$

Step 4: Substitute $-\frac{1}{4}\sqrt{1-4x^2}$ back into the solution to step 2.

$$\int \cos^{-1} 2x dx = x \cos^{-1} 2x + 2 \int \frac{x}{\sqrt{1-4x^2}} dx$$
$$= x \cos^{-1} 2x + 2\left(-\frac{1}{4}\sqrt{1-4x^2}\right) + C$$
$$= x \cos^{-1} 2x - \frac{1}{2}\sqrt{1-4x^2} + C$$

Q158:

$\int_{\pi/6}^{\pi/2} 5x \cos 2x \, dx$

The integrand is the product of two functions so we must use integration by parts.

$\int_a^b fg' dx = [fg]_a^b - \int_a^b f'g \, dx$

Step 1:

Let $f(x) = 5x$ then $f'(x) = 5$

$g'(x) = \cos 2x$ then $g(x) = \frac{1}{2}\sin 2x$

Step 2: Substitute into $\int_a^b fg' dx = [fg]_a^b - \int_a^b f'g \, dx$ and integrate.

$$\int_a^b fg' \, dx = [fg]_a^b - \int_a^b f'g \, dx$$

$$\int_{\pi/6}^{\pi/2} 5x \cos 2x \, dx = \left[5x \times \frac{1}{2}\sin 2x\right]_{\pi/6}^{\pi/2} - \int_{\pi/6}^{\pi/2} 5 \times \frac{1}{2}\sin 2x \, dx$$

$$= \frac{5}{2}[x \sin 2x]_{\pi/6}^{\pi/2} - \frac{5}{2}\int_{\pi/6}^{\pi/2} \sin 2x \, dx$$

$$= \frac{5}{2}[x \sin 2x]_{\pi/6}^{\pi/2} - \frac{5}{2}\left[-\frac{1}{2}\cos 2x\right]_{\pi/6}^{\pi/2}$$

$$= \frac{5}{2}\left(\frac{\pi}{2}\sin \pi - \frac{\pi}{6}\sin \frac{\pi}{3}\right) + \frac{5}{4}\left(\cos \pi - \cos \frac{\pi}{3}\right)$$

$$= \frac{5}{2}\left(0 - \frac{\pi}{6} \cdot \frac{\sqrt{3}}{2}\right) + \frac{5}{4}\left(-1 - \frac{1}{2}\right)$$

$$= \frac{5}{2}\left(-\frac{\sqrt{3}\pi}{12}\right) + \frac{5}{4}\left(-\frac{3}{2}\right)$$

$$= -\frac{5\sqrt{3}\pi}{24} - \frac{15}{8}$$

$$= -\frac{1}{24}\left(5\sqrt{3}\pi + 45\right)$$

$$= -\frac{5}{24}\left(\sqrt{3}\pi + 9\right)$$

Q159:

$\int_0^2 x^2 e^{3x} dx$

The integrand is the product of two functions so we must use integration by parts.

$\int_a^b fg' dx = [fg]_a^b - \int_a^b f'g \, dx$

Step 1:

Let $f(x) = x^2$ then $f'(x) = 2x$

$g'(x) = e^{3x}$ then $g(x) = \frac{1}{3}e^{3x}$

Step 2: Substitute into $\int_a^b fg'dx = [fg]_a^b - \int_a^b f'g dx$ and integrate.

$$\int_a^b fg'dx = fg - \int_a^b f'g dx$$

$$\int_0^2 x^2 e^{3x} dx = \left[\frac{1}{3}x^2 e^{3x}\right]_0^2 - \int_0^2 2x \times \frac{1}{3}e^{3x} dx$$

$$= \frac{1}{3}\left[x^2 e^{3x}\right]_0^2 - \frac{2}{3}\int_0^2 xe^{3x} dx$$

$$= \frac{1}{3}\left(2^2 e^6 - 0^2 e^0\right) - \frac{2}{3}\int_0^2 xe^{3x} dx$$

$$= \frac{4}{3}e^6 - \frac{2}{3}\int_0^2 xe^{3x} dx \ \text{(1)}$$

The integral on the RHS, $\int_0^2 xe^{3x}dx$, is still the product of two functions and so we must use integration by parts a second time.

Step 3:
Let $f(x) = x$ then $f'(x) = 1$
$g'(x) = e^{3x}$ then $g(x) = \frac{1}{3}e^{3x}$
Substitute into $\int_a^b fg'dx = [fg]_a^b - \int_a^b f'g dx$ and integrate.

$$\int_a^b fg'dx = [fg]_a^b - \int_a^b f'g dx$$

$$\int_0^2 xe^{3x} dx = \left[\frac{1}{3}xe^{3x}\right]_0^2 - \int_0^2 \frac{1}{3}e^{3x} dx$$

$$= \frac{1}{3}\left[xe^{3x}\right]_0^2 - \frac{1}{3}\int_0^2 e^{3x} dx$$

$$= \frac{1}{3}\left[xe^{3x}\right]_0^2 - \frac{1}{9}\left[e^{3x}\right]_0^2$$

$$= \frac{1}{3}\left(2e^6 - 0e^0\right) - \frac{1}{9}\left(e^6 - e^0\right)$$

$$= \frac{2}{3}e^6 - \frac{1}{9}e^6 + \frac{1}{9}$$

$$= \frac{6}{9}e^6 - \frac{1}{9}e^6 + \frac{1}{9}$$

$$= \frac{5}{9}e^6 + \frac{1}{9}$$

Step 4: Substitute the solution to $\int_0^2 xe^{3x}dx$ back into (1)

$$\int_0^2 x^2 e^{3x} dx = \frac{4}{3}e^6 - \frac{2}{3}\int_0^2 xe^{3x} dx$$

$$= \frac{4}{3}e^6 - \frac{2}{3}\left(\frac{5}{9}e^6 + \frac{1}{9}\right)$$

$$= \frac{4}{3}e^6 - \frac{10}{27}e^6 - \frac{2}{27}$$

$$= \frac{36}{27}e^6 - \frac{10}{27}e^6 - \frac{2}{27}$$

$$= \frac{26}{27}e^6 - \frac{2}{27}$$

$$= \frac{2}{27}\left(13e^6 - 1\right)$$

Q160:

$\int_0^\pi 5x^2 \sin 2x \, dx$

The integrand is the product of two functions so we must use integration by parts.

$\int_a^b fg' dx = [fg]_a^b - \int_a^b f'g \, dx$

Step 1:
Let $f(x) = 5x^2$ then $f'(x) = 10x$
$g'(x) = \sin 2x$ then $g(x) = -\frac{1}{2}\cos 2x$

Step 2: Substitute into $\int_a^b fg' dx = [fg]_a^b - \int_a^b f'g \, dx$ and integrate.

$$\int_a^b fg' dx = [fg]_a^b - \int_a^b f'g \, dx$$

$$\int_0^\pi 5x^2 \sin 2x \, dx = \left[5x^2 \times -\frac{1}{2}\cos 2x\right]_0^\pi - \int_0^\pi 10x \times -\frac{1}{2}\cos 2x \, dx$$

$$= -\frac{5}{2}\left[x^2 \cos 2x\right]_0^\pi + 5\int_0^\pi x\cos 2x \, dx$$

$$= -\frac{5}{2}\left(\pi^2 \cos 2\pi - 0^2 \cos 0\right) + 5\int_0^\pi x\cos 2x \, dx$$

$$= -\frac{5}{2}\pi^2 + 5\int_0^\pi x\cos 2x \, dx \quad (1)$$

The integral on the RHS, $\int_0^\pi x\cos 2x \, dx$, is still the product of two functions and so we must use integration by parts a second time.

Step 3:
Let $f(x) = x$ then $f'(x) = 1$
$g'(x) = \cos 2x$ then $g(x) = \frac{1}{2}\sin 2x$

Substitute into $\int_a^b fg' dx = [fg]_a^b - \int_a^b f'g \, dx$ and integrate

$$\int_a^b fg' dx = [fg]_a^b - \int_a^b f'g \, dx$$

$$\int_0^\pi x\cos 2x \, dx = \left[x \times \frac{1}{2}\sin 2x\right]_0^\pi - \int_0^\pi \frac{1}{2}\sin 2x \, dx$$

$$= \frac{1}{2}[x\sin 2x]_0^\pi - \frac{1}{2}\int_0^\pi \sin 2x \, dx$$

$$= \frac{1}{2}[x\sin 2x]_0^\pi - \frac{1}{2}\left[-\frac{1}{2}\cos 2x\right]_0^\pi$$

$$= \frac{1}{2}[x\sin 2x]_0^\pi + \frac{1}{4}[\cos 2x]_0^\pi$$

$$= \frac{1}{2}(\pi \sin 2\pi - 0\sin 0) + \frac{1}{4}(\cos 2\pi - \cos 0)$$

$$= \frac{1}{4}(1 - 1)$$

$$= 0$$

Step 4: Substitute the solution to $\int_0^\pi x\cos 2x\,dx$ back into (1)

$$\int_0^\pi 5x^2\sin 2x\,dx = -\frac{5}{2}\pi^2 + 5\int_0^\pi x\cos 2x\,dx$$
$$= -\frac{5}{2}\pi^2 + 5\times 0$$
$$= -\frac{5}{2}\pi^2$$

Q161:

$\int_1^3 \ln|4x|\,dx$

We must multiply the integrand by 1, then use integration by parts to work out this integration.

$\int_1^3 \ln|4x|\times 1\,dx$

Step 1:

Let $f(x) = \ln|4x|$ then $f'(x) = \frac{1}{x}$

$g'(x) = 1$ then $g(x) = x$

Step 2: Substitute into $\int_a^b fg'\,dx = [fg]_a^b - \int_a^b f'g\,dx$ and integrate

$$\int_1^3 \ln|4x|\,dx = [\ln|4x|\times x]_1^3 - \int_1^3 \frac{1}{x}\times x\,dx$$
$$= [x\ln|4x|]_1^3 - \int_1^3 1\,dx$$
$$= [x\ln|4x|]_1^3 - [x]_1^3$$
$$= (3\ln|12| - \ln|4|) - (3-1)$$
$$= 3\ln|12| - \ln|4| - 2$$
$$\approx 4\cdot 0684$$

Q162:

Step 1: Sketch the graph of the curve $y = (x-2)^3$, for y between $y = 2$ and $y = 8$.

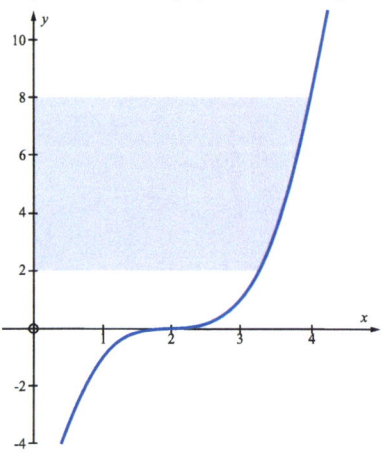

ANSWERS: UNIT 1 TOPIC 3

Step 2: Find the area, but first write the function in terms of y.
Given $y = (x-2)^3$, then $x = \sqrt[3]{y} + 2$.
Now integrate $x = \sqrt[3]{y} + 2$ with respect to y for $y = 2$ to $y = 8$.

$$\begin{aligned}
\text{Area} &= \int_2^8 \sqrt[3]{y} + 2 \, dy \\
&= \int_2^8 y^{\frac{1}{3}} + 2 \, dy \\
&= \left[\frac{3}{4} y^{\frac{4}{3}} + 2y\right]_2^8 \\
&= \left[\frac{3}{4} (\sqrt[3]{y})^4 + 2y\right]_2^8 \\
&= \left(\frac{3}{4} (\sqrt[3]{8})^4 + 2 \times 8\right) - \left(\frac{3}{4} (\sqrt[3]{2})^4 + 2 \times 2\right) \\
&= \left(\frac{3}{4} (16) + 16\right) - \left(\frac{3}{4} (\sqrt[3]{2})^4 + 4\right) \\
&= 28 - \left(\frac{3}{4} (\sqrt[3]{2})^4 + 4\right) \\
&= 22 \cdot 110 \text{ units}^2
\end{aligned}$$

Q163:
Step 1: Sketch the graph of the curve $y = \frac{1}{x^4}$, for y between $y = \frac{1}{16}$ and $y = 1$.

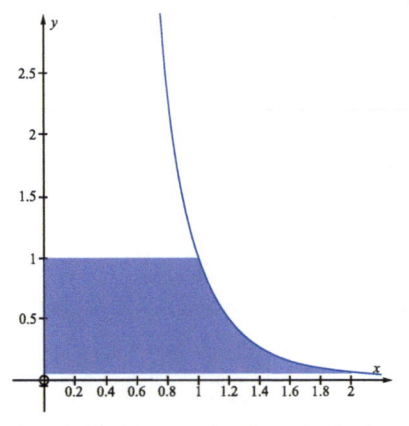

Step 2: Find the area, but first write the function in terms of y.
Given $y = \frac{1}{x^4}$, then $x = \frac{1}{y^{\frac{1}{4}}}$ i.e. $x = y^{-\frac{1}{4}}$.

Now integrate $x = y^{-\frac{1}{4}}$ with respect to y for $y = \frac{1}{16}$ to $y = 1$.

© HERIOT-WATT UNIVERSITY

Area $= \int_{\frac{1}{16}}^{1} y^{-\frac{1}{4}} dy$

$= \left[\frac{4}{3} y^{\frac{3}{4}}\right]_{\frac{1}{16}}^{1}$

$= \frac{4}{3} \left[(\sqrt[4]{y})^3\right]_{\frac{1}{16}}^{1}$

$= \frac{4}{3} \left((\sqrt[4]{1})^3 - \left(\sqrt[4]{\frac{1}{16}}\right)^3\right)$

$= \frac{4}{3} \left(1 - \frac{1}{8}\right)$

$= \frac{4}{3} \left(\frac{7}{8}\right)$

$= \frac{7}{6}$

$= 1 \cdot 167$ units^2

Q164:
Step 1: Sketch the graph of the curve $y^2 = 16 - x$, for y between $y = 4$ and $y = -4$.

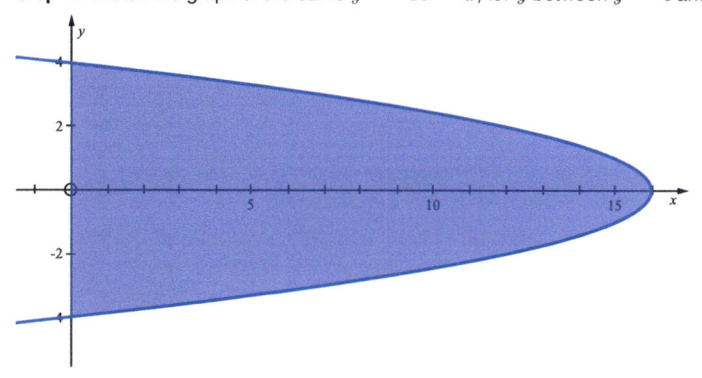

Step 2: Find the area, but first write the function in terms of y.
Given $y^2 = 16 - x$, then $x = 16 - y^2$.
Now integrate with respect to y for $y = 4$ and $y = -4$.

Area $= \int_{-4}^{4} 16 - y^2 \, dy$

$= \left[16y - \frac{1}{3} y^3\right]_{-4}^{4}$

$= \left(16(4) - \frac{1}{3}(4)^3\right) - \left(16(-4) - \frac{1}{3}(-4)^3\right)$

$= \left(64 - \frac{64}{3}\right) - \left(-64 + \frac{64}{3}\right)$

$= 85 \cdot 333$ units^2

ANSWERS: UNIT 1 TOPIC 3

Q165:

The volume of revolution is given by $\int_a^b \pi y^2 dx$.

Step 1: Work out y^2

$$y^2 = \left(2x - \frac{3}{x}\right)^2$$

$$= 4x^2 - 12 + \frac{9}{x^2}$$

Step 2: Substitute y^2 into $\int_a^b \pi y^2 dx$ for $x = 1$ and $x = 2$ and integrate.

$$\text{Volume} = \pi \int_1^2 4x^2 - 12 + \frac{9}{x^2} dx$$

$$= \pi \int_1^2 4x^2 - 12 + 9x^{-2} dx$$

$$= \pi \left[\frac{4}{3}x^3 - 12x - 9x^{-1}\right]_1^2$$

$$= \pi \left[\frac{4}{3}x^3 - 12x - \frac{9}{x}\right]_1^2$$

$$= \pi \left\{\left(\frac{4}{3} \times 2^3 - 12 \times 2 - \frac{9}{2}\right) - \left(\frac{4}{3} \times 1^3 - 12 \times 1 - \frac{9}{1}\right)\right\}$$

$$= \frac{11}{6}\pi \text{ units}^3$$

Q166:

The volume of revolution is given by $\int_a^b \pi y^2 dx$.

Step 1: Work out the limits of integration in the first quadrant.

The curve is bound by the y-axis i.e. when $x = 0$, and the x-axis when $y = 0$.

When $y = 0$, $0^2 = 6 - x^2$, then $x^2 = 6$, so $x = \sqrt{6}$

Step 2: Substitute y^2 into $\int_a^b \pi y^2 dx$ for $x = 0$ and $x = \sqrt{6}$ and integrate.

$$\text{Volume} = \pi \int_0^{\sqrt{6}} 6 - x^2 dx$$

$$= \pi \left[6x - \frac{x^3}{3}\right]_0^{\sqrt{6}}$$

$$= \pi \left\{\left(6 \times \sqrt{6} - \frac{\sqrt{6}^3}{3}\right) - \left(6 \times 0 - \frac{0^3}{3}\right)\right\}$$

$$= \pi \left(6\sqrt{6} - 2\sqrt{6}\right)$$

$$= 4\sqrt{6}\pi \text{ units}^3$$

Q167:

The volume of revolution is given by $\int_a^b \pi y^2 dx$.

Step 1: Work out y^2

$$y^2 = (\cos \theta)^2$$

$$= \cos^2 \theta$$

© HERIOT-WATT UNIVERSITY

Step 2: Substitute y^2 into $\int_a^b \pi y^2 dx$ for $x = 0$ and $x = \frac{\pi}{4}$ and integrate.

Volume $= \pi \int_0^{\frac{\pi}{4}} \cos^2\theta d\theta$

$= \pi \int_0^{\frac{\pi}{4}} \frac{1}{2}(\cos 2\theta + 1) d\theta$

$= \frac{\pi}{2} \left[\frac{1}{2} \sin 2\theta + \theta \right]_0^{\frac{\pi}{4}}$

$= \frac{\pi}{2} \left\{ \left(\frac{1}{2} \sin \frac{\pi}{2} + \frac{\pi}{4} \right) - \left(\frac{1}{2} \sin 0 + 0 \right) \right\}$

$= \frac{\pi}{2} \left(\frac{1}{2} + \frac{\pi}{4} \right)$

$= \frac{\pi}{4} + \frac{\pi^2}{8}$ units^3

Q168:

The volume of revolution is given by $\int_a^b \pi y^2 dx$.

Step 1: Work out y^2

$(x-3)^2 + y^2 = 2^2$

$y^2 = 4 - (x-3)^2$

$= 4 - (x^2 - 6x + 9)$

$= 4 - x^2 + 6x - 9$

$= -x^2 + 6x - 5$

Step 2: Work out the limits of integration.

The circle has a radius of 2 centred at (3,0).

The limits go from $x = 1$ to $x = 5$.

Step 3: Substitute y^2 into $\int_a^b \pi y^2 dx$ for $x = 1$ and $x = 5$ and integrate.

Volume $= \pi \int_1^5 -x^2 + 6x - 5 \, dx$

$= \pi \left[-\frac{x^3}{3} + 3x^2 - 5x \right]_1^5$

$= \pi \left(\left(-\frac{5^3}{3} + 3 \times 5^2 - 5 \times 5 \right) - \left(-\frac{1^3}{3} + 3 \times 1^2 - 5 \times 1 \right) \right)$

$= \pi \left(\left(-\frac{125}{3} + 75 - 25 \right) - \left(-\frac{1}{3} + 3 - 5 \right) \right)$

$= \pi \left(-\frac{125}{3} + 50 + \frac{1}{3} + 2 \right)$

$= \frac{32}{3}\pi$ units^3

Q169:

The volume of revolution is given by $\int_a^b \pi x^2 dy$.

Step 1: Substitute x^2 into $\int_a^b \pi x^2 dy$ for $y = -1$ and $y = 2$ and integrate.

$$\text{Volume} = \pi \int_{-1}^{2} 3y \, dy$$

$$= \pi \left[\frac{3}{2} y^2 \right]_{-1}^{2}$$

$$= \frac{3\pi}{2} \left[y^2 \right]_{-1}^{2}$$

$$= \frac{3\pi}{2} \left(2^2 - (-1)^2 \right)$$

$$= \frac{3\pi}{2} (4 - 1)$$

$$= \frac{9\pi}{2} \text{ units}^3$$

Q170:

The volume of revolution is given by $\int_a^b \pi x^2 dy$.

Step 1: Work out x^2

$y = 5 - x^2$

$x^2 = 5 - y$

Step 2: Substitute x^2 into $\int_a^b \pi x^2 dy$ for $y = -5$ and $y = 5$ and integrate.

$$\text{Volume} = \pi \int_{-5}^{5} 5 - y \, dy$$

$$= \pi \left[5y - \frac{y^2}{2} \right]_{-5}^{5}$$

$$= \pi \left\{ \left(5(5) - \frac{5^2}{2} \right) - \left(5(-5) - \frac{(-5)^2}{2} \right) \right\}$$

$$= \pi \left\{ \left(25 - \frac{25}{2} \right) - \left(-25 - \frac{25}{2} \right) \right\}$$

$$= \pi \left(25 - \frac{25}{2} + 25 + \frac{25}{2} \right)$$

$$= 50\pi \text{ units}^3$$

Q171:

The volume of revolution is given by $\int_a^b \pi x^2 dy$.

Step 1: Work out x^2

$3y = \ln |x|$

$e^{3y} = x$

$x = e^{3y}$

$x^2 = \left(e^{3y} \right)^2$

$x^2 = e^{6y}$

Step 2: Substitute x^2 into $\int_a^b \pi x^2 dy$ for $y = 2$ and $y = 4$ and integrate.

Volume $= \pi \int_2^4 e^{6y} \, dy$

$= \pi \left[\dfrac{1}{6} e^{6y} \right]_2^4$

$= \dfrac{\pi}{6} \left[e^{6y} \right]_2^4$

$= \dfrac{\pi}{6} \left(e^{6(4)} - e^{6(2)} \right)$

$= \dfrac{\pi}{6} \left(e^{24} - e^{12} \right)$

$= \dfrac{\pi}{6} e^{12} \left(e^{12} - 1 \right)$

$= 13869586696$ units^3

Q172:

The volume of revolution is given by $\int_a^b \pi x^2 dy$.

Step 1: Work out x^2

$xy^2 = 3$

$x = \dfrac{3}{y^2}$

$x^2 = \left(\dfrac{3}{y^2} \right)^2$

$x^2 = \dfrac{9}{y^4}$

Step 2: Substitute x^2 into $\int_a^b \pi x^2 dy$ for $y = 1$ and $y = 3$ and integrate.

Volume $= \pi \int_1^3 \dfrac{9}{y^4} dy$

$= 9\pi \int_1^3 y^{-4} dy$

$= 9\pi \left[-\dfrac{1}{3} y^{-3} \right]_1^3$

$= -3\pi \left[\dfrac{1}{y^3} \right]_1^3$

$= -3\pi \left(\dfrac{1}{3^3} - \dfrac{1}{1^3} \right)$

$= -3\pi \left(\dfrac{1}{3^3} - \dfrac{1}{1^3} \right)$

$= -3\pi \left(\dfrac{1}{27} - 1 \right)$

$= -3\pi \left(-\dfrac{26}{27} \right)$

$= \dfrac{78}{27} \pi$

$= \dfrac{26}{9} \pi$ units^3

Q173:

The volume of revolution is given by $\int_a^b \pi\theta^2 dy$.

Step 1: Work out θ^2

$y = \sin\theta^2$

$\theta^2 = \sin^{-1} y$

Step 2: Substitute θ^2 into $\int_a^b \pi\theta^2 dy$ for $y = 0$ and $y = \frac{1}{2}$ and integrate.

$$\text{Volume} = \pi \int_0^{\frac{1}{2}} \sin^{-1} y \, dy$$

$$= \pi \left[\frac{1}{\sqrt{1-y^2}}\right]_0^{\frac{1}{2}}$$

$$= \pi \left(\frac{1}{\sqrt{1-\left(\frac{1}{2}\right)^2}} - \frac{1}{\sqrt{1-0^2}}\right)$$

$$= \pi \left(\frac{1}{\sqrt{1-\frac{1}{4}}} - \frac{1}{\sqrt{1}}\right)$$

$$= \pi \left(\frac{1}{\sqrt{\frac{3}{4}}} - 1\right)$$

$$= \pi \left(\frac{2}{\sqrt{3}} - 1\right) \text{ units}^3$$

Topic 4: Differential equations

Laws of exponentials exercise (page 295)

Q1: $\frac{x^6 \times x^2}{x^3} = x^{6+2-3} = x^5$

Q2:
$$2x^{\frac{1}{4}} \times 5x^{\frac{1}{4}} = 2 \times 5 \times x^{\frac{1}{4}} \times x^{\frac{1}{4}}$$
$$= 2 \times 5 \times x^{\frac{1}{4} + \frac{1}{4}}$$
$$= 10 \times x^{\frac{2}{4}}$$
$$= 10x^{\frac{1}{2}}$$

Q3: $(3y^4)^3 = 3^3 \times y^{4 \times 3} = 27y^{12}$

Q4: $(5m^{\frac{3}{2}})^2 = 5^2 \times m^{\frac{3}{2} \times 2} = 25m^{\frac{6}{2}} = 25m^3$

Q5: $(3m^{-4})^2 = 3^2 \times m^{(-4) \times 2} = 9m^{-8} = \frac{9}{m^8}$

Q6: $y^{(-10) + \frac{3}{2} - \frac{1}{2}} = y^{(-10)+1} = y^{-9} = \frac{1}{y^9}$

Q7: $16^{\frac{1}{4}} = \sqrt[4]{16} = 2$ because $2^4 = 16$

Q8: $27^{\frac{2}{3}} = \sqrt[3]{27^2} = 9$ because $\sqrt[3]{27} = 3$ and $3^2 = 9$

Q9: $4^{\frac{3}{2}} = \sqrt{4^3} = 8$ because $\sqrt{4} = 2$ and $2^3 = 8$

Laws of logarithms exercise (page 299)

Q10: 2

Q11: 10

Q12: 1

Q13: $\log_{10} 8$

Q14: 1000

Q15: 6

Q16: 6

ANSWERS: UNIT 1 TOPIC 4	583

Differentiating composite functions exercise (page 301)

Q17:

Step 1: Turn the square root into a power. $h(x) = (x^2 + 6x)^{\frac{1}{2}}$

Step 2: Bring down the power. $\frac{1}{2}$

Step 3: Write down the bracket. $\frac{1}{2}(x^2 + 6x)$

Step 4: Reduce the power by 1. $\frac{1}{2}(x^2 + 6x)^{-\frac{1}{2}}$

Step 5: Differentiate the bracket. $\frac{1}{2}(x^2 + 6x)^{-\frac{1}{2}} \times (2x + 6)$

Step 6: Simplify the answer. $\frac{dy}{dx} = \frac{2x + 6}{2\sqrt{x^2 + 6x}}$

$= \frac{2(x + 3)}{2\sqrt{x^2 + 6x}}$

$= \frac{x + 3}{\sqrt{x^2 + 6x}}$

Q18:

Step 1: Remember: $\cos^3 x = (\cos x)^3$

Step 2: Bring down the power. 3

Step 3: Write down the bracket. $3(\cos x)$

Step 4: Reduce the power by 1. $3(\cos x)^2$

Step 5: Differentiate the bracket. $3(\cos x)^2 \times (-\sin x)$

Step 6: Simplify the answer. $f'(x) = -3\cos^2 x \sin x$

First order linear differential equations introduction exercise (page 306)

Q19: b) Second order linear

Q20: c) Neither

Q21: a) First order linear

Q22: a) First order linear

Q23: c) Neither

Q24: b) Second order linear

© HERIOT-WATT UNIVERSITY

Q25:

a)

Integrate to find $g(x)$:

$$\int 3x^2 + 2\, dx = \frac{3x^3}{3} + 2x + C$$
$$= x^3 + 2x + C$$

so, $g(x) = x^3 + 2x$

b)

The curve passes through the point $(2, 11)$ so when $x = 2$, then $f(x) = 11$:

$$f(x) = x^3 + 2x + C$$
$$11 = 2^3 + 2(2) + C$$
$$11 = 12 + C$$
$$C = -1$$

therefore, $f(x) = x^3 + 2x - 1$

Q26:

a)

Integrate to find $h(t)$:

$$h(t) = \int 13 - 10t\, dt$$
$$= 13t - \frac{10t^2}{2} + C$$
$$= 13t - 5t^2 + C$$

When $t = 0$, then $h(t) = 6$:

$$h(t) = 13t - 5t^2 + C$$
$$6 = 13(0) - 5(0)^2 + C$$
$$C = 6$$

so, $h(t) = 6 + 13t - 5t^2$

b)

When $t = 1 \cdot 5$, evaluate $h(t)$:

$h(1 \cdot 5) = 6 + 13(1 \cdot 5) - 5(1 \cdot 5)^2$
$h(1 \cdot 5) = 14 \cdot 25$ m

ANSWERS: UNIT 1 TOPIC 4

Q27:
a)
Hints:

- $B(t) = \int 12e^{0 \cdot 6t}\, dt$

Answer:
Integrate to find $B(t)$:

$$B(t) = \int 12e^{0 \cdot 6t}\, dt$$
$$= \frac{12}{0 \cdot 6} e^{0 \cdot 6t} + C$$
$$= 20e^{0 \cdot 6t} + C$$

When $t = 0$, $B(0) = 20$:
$$B(0) = 20e^{0 \cdot 6(0)} + C$$
$$20 = 20 + C$$
$$C = 0$$

therefore, $B(t) = 20e^{0 \cdot 6t}$

b)
When $t = 10$, evaluate $B(t)$:
$B(10) = 20e^{0 \cdot 6(10)}$
$B(10) = 8068 \cdot 575 \ldots$
$B(10) = 8068$ bacteria

The context of this problem requires that we round down to the nearest whole bacteria.

First order linear differential equations with separable variables extra help (page 316)

Q28: b) $x\, dy = y\, dx$

Q29: a) $\frac{dy}{y} = \frac{dx}{x}$

Q30: c) $\ln y = \ln x + C$

Q31: c) $\ln y = \ln(Ax) \Rightarrow y = Ax$

Q32: c) $x\, dy = dx$

Q33: a) $dy = \frac{dx}{x}$

Q34: a) $y = \ln(x) + C$

Q35: b) $y = \ln(Ax)$

First order linear differential equations with separable variables exercise (page 318)

Q36:

$\frac{dy}{dx} = \sqrt{xy}$ can be solved by the separable variable method:

$$\frac{dy}{dx} = \sqrt{xy}$$
$$\frac{dy}{dx} = \sqrt{x}\sqrt{y}$$
$$\frac{dy}{\sqrt{y}} = \sqrt{x}\, dx$$

Q37: $x\frac{dy}{dx} - \frac{y}{x} = 3x$ cannot be solved by the separable variable method.

Q38:

$\frac{dy}{dx} = 1 - x + y - xy$ can be solved by the separable variable method.

$$\frac{dy}{dx} = 1 - x + y - xy$$
$$\frac{dy}{dx} = (1-x)(1+y)$$
$$\frac{dy}{1+y} = (1-x)\, dx$$

Q39: $\frac{dy}{dx} = x^2 + xy - 3$ cannot be solved by the separable variable method.

Q40: $x\frac{dy}{dx} = \frac{dy}{dx} + y^2 + xy$ cannot be solved by the separable variable method.

Q41:

$x\frac{dy}{dx} = \frac{dy}{dx} + y^2$ can be solved by the separable variable method.

$$x\frac{dy}{dx} = \frac{dy}{dx} + y^2$$
$$x\frac{dy}{dx} - \frac{dy}{dx} = y^2$$
$$(x-1)\frac{dy}{dx} = y^2$$
$$\frac{dy}{dx} = \frac{y^2}{x-1}$$
$$\frac{dy}{y^2} = \frac{dx}{x-1}$$

ANSWERS: UNIT 1 TOPIC 4

Q42:

a)

Separate variables and integrate to find the general solution in logarithmic form:

$$\frac{dy}{dx} = \frac{y+2}{x-2}$$

$$\int \frac{1}{y+2} \, dy = \int \frac{1}{x-2} \, dx$$

$$\ln(y+2) = \ln(x-2) + C$$

b)

Using $y = 2$ when $x = 3$:

$$\ln(2+2) = \ln(3-2) + C$$

$$\ln(4) = \ln(1) + C$$

Recall that $\ln(1) = 0$

$C = \ln(4)$

c)

$\ln(y+2) = \ln(x-2) + \ln(4)$
$\ln(y+2) = \ln(4(x-2))$

Take the exponential of both sides:

$e^{\ln(y+2)} = e^{\ln(4(x-2))}$

$$y + 2 = 4x - 8$$

$$y = 4x - 10$$

Q43:

a)

Separate variables and integrate:

$$\int \frac{dh}{h^2} = \int -\frac{3}{4} \, dt$$

$$\int h^{-2} \, dh = \int -\frac{3}{4} \, dt$$

$$-\frac{1}{h} = -\frac{3}{4}t + C$$

Use the initial conditions to evaluate C, i.e. when $t = 0$, $h = 1$

Therefore,

$$-\frac{1}{1} = -\frac{3}{4}(0) + C$$

$$C = -1$$

$$-\frac{1}{h} = -\frac{3}{4}t - 1$$

$$-\frac{1}{h} = -\left(\frac{3t+4}{4}\right)$$

$$h = \frac{4}{3t+4}$$

© HERIOT-WATT UNIVERSITY

b)
The depth of water in the tank, h, is given in cm but needed in metres.
$$h = 10 \, cm = 0 \cdot 1 \, m$$
$$0 \cdot 1 = \frac{4}{3t+4}$$
$$0 \cdot 3t + 0 \cdot 4 = 4$$
$$0 \cdot 3t = 3 \cdot 6$$
$$t = 12 \, \text{min}$$

Q44:

a)

Factorising gives: $\frac{dy}{dx} = (1+y)(1-x)$

b)

Separate variables and integrate:
$$\int \frac{dy}{1+y} = \int (1-x) \, dx$$
$$\ln(1+y) = x - \frac{x^2}{2} + C$$

c)

Take the exponential of both sides:
$$e^{\ln(1+y)} = e^{\left(x - \frac{x^2}{2} + C\right)}$$
$$1 + y = e^{\left(x - \frac{x^2}{2}\right)} e^C$$

Remember that $e^{x+y} = e^x e^y$ and in this case e^C is a constant so $e^C = A$

$$1 + y = Ae^{\left(x - \frac{x^2}{2}\right)}$$
$$y = Ae^{\left(x - \frac{x^2}{2}\right)} - 1$$

Q45:

a)

Separate the variables and write in the form:
$$\frac{dy}{dx} = f(x) g(y)$$
$$(1+x) \frac{dy}{dx} - y(y+1) = 0$$
$$\frac{dy}{dx} = \frac{y(y+1)}{1+x}$$
$$\int \frac{1}{y(y+1)} \, dy = \int \frac{1}{1+x} \, dx$$

ANSWERS: UNIT 1 TOPIC 4

b)
Write the LHS as a sum of partial fractions:
$$\frac{1}{y(y+1)} = \frac{A}{y} + \frac{B}{y+1}$$
$$1 \equiv A(y+1) + By$$
Let $y = 0:\ 1 = A \Rightarrow A = 1$
Let $y = -1:\ 1 = -B \Rightarrow B = -1$

c)
Write the LHS in partial fractions and integrate:
$$\int \frac{1}{y}\,dy - \int \frac{1}{y+1}\,dy = \int \frac{1}{1+x}\,dx$$
$$\ln(y) - \ln(y+1) = \ln(1+x) + C$$
Using the initial conditions, $y = 1$ when $x = 1$:
$$\ln(1) - \ln(2) = \ln(2) + C$$
$$C = -2\ln(2)$$
$$C = \ln(2^{-2})$$
$$C = \ln\left(\frac{1}{4}\right)$$
$$C = \ln(1) - \ln(4)$$
$$C = -\ln(4)$$

d)
The function y:
$$\ln y - \ln(y+1) = \ln(1+x) - \ln(4)$$
$$\ln\left(\frac{y}{y+1}\right) = \ln\left(\frac{1+x}{4}\right)$$
Take the exponential of both sides:
$$e^{\ln\left(\frac{y}{y+1}\right)} = e^{\ln\left(\frac{1+x}{4}\right)}$$
$$\frac{y}{y+1} = \frac{1+x}{4}$$
$$\frac{y+1}{y} = \frac{4}{1+x}$$
$$1 + \frac{1}{y} = \frac{4}{1+x}$$
$$\frac{1}{y} = \frac{4}{1+x} - 1$$
$$\frac{1}{y} = \frac{4-(1+x)}{1+x}$$
$$\frac{1}{y} = \frac{3-x}{1+x}$$
$$y = \frac{1+x}{3-x}$$

© HERIOT-WATT UNIVERSITY

Growth and decay: Half-life of a radioactive element (page 325)

Q46:

a) The differential equation for N is $\frac{dN}{dt} = -kN$

b) Separate the variables and integrate:

$$\frac{dN}{dt} = -kN$$

$$\int \frac{dN}{N} = \int -k\, dt$$

$$\ln(N) = -kt + C$$

$$N = e^{-kt+C}$$

$$N = Ae^{-kt} \text{ since } e^C = A \text{ is a constant}$$

When $t = 0$, $N = 1000$

$$1000 = Ae^{-k(0)}$$

$$A = 1000$$

so $N = 1000e^{-kt}$

Q47: 500 nuclei

Q48: 40 seconds

Growth and decay exercise (page 326)

Q49:

a)

The differential equation for n is $\frac{dn}{dt} = kn$

b)

Separate the variables and integrate:

$$\frac{dn}{dt} = kn$$

$$\int \frac{dn}{dt} = \int k\, dt$$

$$\ln(n) = kt + C$$

$$n = e^{kt+C}$$

$$n = Ae^{kt} \text{ since } e^C = A \text{ is a constant}$$

when $t = 0$, $n = 3$

$$3 = Ae^{k(0)}$$

$$A = 3$$

So $n(t) = 3e^{kt}$

ANSWERS: UNIT 1 TOPIC 4

c)
When $t = 2$, then $n(2) = 60$:
so $n(t) = 3e^{kt}$
$$60 = 3e^{k(2)}$$
$$20 = e^{k(2)}$$
$$\ln(20) = 2k$$
$$k = \frac{1}{2}\ln(20)$$
$$k = 1 \cdot 498 \text{ (3 d.p.)}$$

d)
$n(3) = 3e^{1 \cdot 498(3)} = 268$ people

Q50:

a)
The differential equation for P is $\frac{dP}{dt} = kP$

b)
Separate the variables and integrate:
$$\frac{dP}{dt} = kP$$
$$\int \frac{dP}{dt} = \int k\, dt$$
$$\ln(P) = kt + C$$
$$P = e^{kt+C}$$
$P = Ae^{kt}$ since $e^C = A$ is a constant
when $t = 0$, $P = 10$
$$10 = Ae^{k(0)}$$
$$A = 10$$
so $P(t) = 10e^{kt}$

c)
t is in hours so when $t = 0 \cdot 5$, then $P(0 \cdot 5) = 20$
so $P(t) = 10e^{kt}$
$$20 = 10e^{k(0 \cdot 5)}$$
$$2 = e^{k(0 \cdot 5)}$$
$$\ln(2) = 0 \cdot 5k$$
$$k = 2\ln(2)$$
$$k = 1 \cdot 386 \text{ (3 d.p.)}$$

d)
$P(4) = 10e^{1 \cdot 386(4)} = 2557\text{g}$

© HERIOT-WATT UNIVERSITY

Further applications of differential equations exercise (page 330)

Q51:

Set up the differential equation.

$\frac{dT}{dt} = -k(T - T_S)$

Separate the variables by arranging the T variables on the LHS and everything else on the RHS and integrate:

$\int \frac{dT}{(T-T_S)} = \int -k \, dt$

Integrate and rearrange for T.

Recall that T_S is a constant, not a variable. When integrating we therefore treat T_S as a number:

$\ln|T - T_S| = -kt + C$

$e^{\ln|T-T_S|} = e^{-kt+C}$

$T - T_S = e^{-kt} e^C$

$T - T_S = Ae^{-kt}$ where $A = e^C$

$T = Ae^{-kt} + T_S$

The general solution is therefore: $T_t = Ae^{-kt} + T_S$

Calculate the value of A.

To obtain A we will substitute for when $t = 0$.

$T_0 = Ae^{-k(0)} + T_S$

$T_0 = A + T_S$

$A = T_0 - T_S$

Therefore our solution becomes:

$T_t = (T_0 - T_S)e^{-kt} + T_S$

Evaluate k to obtain the particular solution.

Use the conditions given in the question to work out the value of k.

From the question, we know that the surrounding temperature, T_S, is 16°C.

Also, when $t = 5$, $T_5 = 40 \Rightarrow 40 = (T_0 - 16)e^{-5k} + 16 \Rightarrow 24 = (T_0 - 16)e^{-5k}$ (*)

Also, when $t = 20$, $T_{20} = 20 \Rightarrow 20 = (T_0 - 16)e^{-20k} + 16 \Rightarrow 4 = (T_0 - 16)e^{-20k}$ (+)

Looking at the LHS of the two equations above, (*) and (+), we can see that if we multiply (+) by 6 we get 24, the same value as the LHS of (*). This means that we can equate the equations as follows (*) = 6 × (+).

$(T_0 - 16)e^{-5k} = 6(T_0 - 16)e^{-20k}$

Dividing both sides by $(T_0 - 16)$: $e^{-5k} = 6e^{-20k}$

Dividing both sides by e^{-20k} and using the rule $\frac{a^m}{a^n} = a^{m-n}$ gives:

$\frac{e^{-5k}}{e^{-20k}} = 6$

$e^{15k} = 6$

… ANSWERS: UNIT 1 TOPIC 4

Taking the logarithm of both sides gives:

$\ln \left| e^{15k} \right| = \ln |6|$

$15k = \ln |6|$

$k = \dfrac{\ln |6|}{15}$

$k = 0 \cdot 119$

The particular solution is therefore $T_t = (T_0 - 16)e^{-0 \cdot 119 t} + 16$

Work out the initial temperature of the egg, T_0.

Use one of the conditions given in the question, say at time 5 minutes, $t = 5$, the temperature of the egg is 40°C, i.e. $T_5 = 40$:

$T_5 = (T_0 - 16) \, e^{-0 \cdot 119(5)} + 16$

$40 = (T_0 - 16) \, e^{-0 \cdot 119(5)} + 16$

$24 = (T_0 - 16) \, e^{-0 \cdot 119(5)}$

$\dfrac{24}{e^{-0 \cdot 119(5)}} = (T_0 - 16)$

$T_0 = \dfrac{24}{e^{-0 \cdot 119(5)}} + 16$

$T_0 = 59 \cdot 5127 \ldots$

$T_0 = 60^\circ C$

Q52:

a)

Using Newton's law of cooling, T is the temperature of the body at time t, t is the time, in hours, since the body's temperature was first taken, T_S is the temperature of the room and k is the constant of proportionality.

$\dfrac{dT}{dt} = -k(T - T_s)$

$\dfrac{dT}{dt} = -k(T - 32 \cdot 2)$

b)

Separate variables and integrate:

$\dfrac{dT}{dt} = -k(T - 32 \cdot 2)$

$\displaystyle\int \dfrac{1}{T - 32 \cdot 2} \, dT = \int -k \, dt$

$\ln |T - 32 \cdot 2| = -kt + C$

$e^{\ln|T - 32 \cdot 2|} = e^{-kt + C}$

$T - 32 \cdot 2 = Ae^{-kt}$ since $e^C = A$ is a constant

$T = 32 \cdot 2 + Ae^{-kt}$

c)

When $t = 0, T = 34 \cdot 8^\circ C$:

$T = 32 \cdot 2 + Ae^{-kt}$

$34 \cdot 8 = 32 \cdot 2 + Ae^{-k(0)}$

$A = 2 \cdot 6$

© HERIOT-WATT UNIVERSITY

When $t = 1, T = 34 \cdot 1°C$:
$$T = 32 \cdot 2 + 2 \cdot 6e^{-kt}$$
$$34 \cdot 1 = 32 \cdot 2 + 2 \cdot 6e^{-k(1)}$$
$$e^{-k} = \frac{1 \cdot 9}{2 \cdot 6}$$
$$-k = \ln\left(\frac{1 \cdot 9}{2 \cdot 6}\right)$$
$$k = -\ln\left(\frac{1 \cdot 9}{2 \cdot 6}\right)$$
$$k = 0 \cdot 314 \text{ (3 d.p.)}$$

d)
Calculate t when $T = 37°C$
$$T = 32 \cdot 2 + 2 \cdot 6e^{-0 \cdot 314 t}$$
$$37 = 32 \cdot 2 + 2 \cdot 6e^{-0 \cdot 314 t}$$
$$e^{-0 \cdot 314 t} = \frac{4 \cdot 8}{2 \cdot 6}$$
$$-0 \cdot 314 t = \ln\left(\frac{4 \cdot 8}{2 \cdot 6}\right)$$
$$t = -\frac{1}{0 \cdot 134} \ln\left(\frac{4 \cdot 8}{2 \cdot 6}\right)$$
$$t = -1 \cdot 9525 \text{ hours}$$
$$t = -1 \text{ hour } 57 \text{ min}$$
02:30 - 01:57 (i.e. 1 hour 57 minutes prior to 2:30am) = 00:33

Method of integrating factor: Identifying $P(x)$ and $f(x)$ exercise (page 332)

Q53: $P(x) = -2$
$f(x) = e^{3x}$

Q54: Rearranging into standard form:
$$\frac{dy}{dx} = \frac{xy + 4}{x^2}$$
$$x^2 \frac{dy}{dx} = xy + 4$$
$$x^2 \frac{dy}{dx} - xy = 4$$
$$\frac{dy}{dx} - \frac{y}{x} = \frac{4}{x^2}$$
$$P(x) = -\frac{1}{x}$$
$$f(x) = \frac{4}{x^2}$$

ANSWERS: UNIT 1 TOPIC 4

Q55: Rearranging into standard form:

$$x\frac{dy}{dx} = \frac{\cos x}{x} - 2y$$

$$x\frac{dy}{dx} + 2y = \frac{\cos x}{x}$$

$$\frac{dy}{dx} + \frac{2y}{x} = \frac{\cos x}{x^2}$$

$$P(x) = \frac{2}{x}$$

$$f(x) = \frac{\cos x}{x^2}$$

Q56: Rearranging into standard form:

$$\frac{dy}{dx} + \frac{1}{2}e^{-x} = 4y$$

$$\frac{dy}{dx} - 4y = -\frac{1}{2}e^{-x}$$

$$P(x) = -4$$

$$f(x) = -\frac{1}{2}e^{-x}$$

Q57: Rearranging into standard form:

$$2\frac{dy}{dx} - 2y = e^{\frac{x}{2}}$$

$$\frac{dy}{dx} - y = \frac{1}{2}e^{\frac{x}{2}}$$

$$P(x) = -1$$

$$f(x) = \frac{1}{2}e^{\frac{x}{2}}$$

Q58: $P(x) = 3x^2$

$f(x) = 6x^2$

Method of integrating factor extra help: Differential equations (page 336)

Q59: a) $P(x) = 2$

Q60: a) $\int 2\, dx = 2x$

Q61: c) e^{2x}

Q62: a) $e^{2x}y = \int e^x\, dx$

Q63: b) $e^{2x}y = e^x + C$

Q64: a) $y = e^{-x} + Ce^{-2x}$

Q65: b) $P(x) = \tan x$

Q66: a) $\ln(\sec x)$

Q67: a) $\sec x$

© HERIOT-WATT UNIVERSITY

Q68: c) $\sec x \bullet y = 2 \int \sin x \, dx$

Q69: b) $\sec x \bullet y = -2\cos x + C$

Q70: b) $y = -2\cos^2 x + C \cos x$

Q71: c) $C = 3$

Method of integrating factor exercise (page 339)

Q72:

a)
Integrating factor:
$$I(x) = e^{\int P(x) \, dx}$$
$$= e^{\int \frac{2}{x} \, dx}$$
$$= e^{2\ln(x)}$$
$$= e^{\ln(x^2)}$$
$$= x^2$$

b)
Using the integrating factor:
$$I(x) y = \int I(x) f(x) \, dx$$
$$x^2 y = \int x^3 \, dx$$
$$x^2 y = \frac{x^4}{4} + C$$
$$y = \frac{x^2}{4} + \frac{C}{x^2}$$
$$y = \frac{1}{4}x^2 + Cx^{-2}$$
$$g(x) = \frac{1}{4}x^2$$
$$h(x) = x^{-2}$$

c)
General solution:
$y = \frac{1}{4}x^2 + Cx^{-2}$

Q73:

a)
Integrating factor:
$$I(x) = e^{\int P(x) \, dx}$$
$$= e^{\int -2x \, dx}$$
$$= e^{-x^2}$$

ANSWERS: UNIT 1 TOPIC 4

b)
Using the integrating factor:

$$I(x)y = \int I(x)f(x)\,dx$$

$$e^{-x^2}y = \int 3xe^{-x^2}\,dx$$

Using the substitution $u = x^2$, then $\frac{du}{dx} = 2x$ so that $dx = \frac{du}{2x}$

$$e^{-u}y = \int 3xe^{-u}\frac{du}{2x}$$

$$e^{-u}y = \frac{3}{2}\int e^{-u}\,du$$

$$e^{-u}y = -\frac{3}{2}e^{-u} + C$$

$$y = -\frac{3}{2} + \frac{C}{e^{-u}}$$

$$y = \cdots$$

$$y = \cdots$$

$$y = -\frac{3}{2} + Ce^u$$

$$y = -\frac{3}{2} + Ce^{x^2}$$

$$g(x) = -\frac{3}{2}$$

$$h(x) = e^{x^2}$$

c)
General solution:

$y = -\frac{3}{2} + Ce^{x^2}$

Q74:

a)
Rearrange $x^2 \frac{dy}{dx} - x^3 + xy = 0$ into standard form $\frac{dy}{dx} + P(x)y = f(x)$ and so the integrating factor is:

$$\frac{dy}{dx} + \frac{y}{x} = x$$

$$I(x) = e^{\int P(x)\,dx}$$

$$= e^{\int \frac{1}{x}\,dx}$$

$$= e^{\ln(x)}$$

$$= x$$

© HERIOT-WATT UNIVERSITY

b)
Using the integrating factor:

$$I(x)y = \int I(x)f(x)\,dx$$
$$xy = \int x^2\,dx$$
$$xy = \frac{x^3}{3} + C$$
$$y = \frac{x^2}{3} + \frac{C}{x}$$
$$y = \frac{x^2}{3} + Cx^{-1}$$
$$g(x) = \frac{x^2}{3}$$
$$h(x) = x^{-1}$$

c)
Let $x = 1$, $y = 1$
$$1 = \frac{1}{3} + C(1)^{-1}$$
$$C = \frac{2}{3}$$

Particular solution:
$$y = \tfrac{1}{3}x^2 + \tfrac{2}{3}x^{-1}$$

Q75:

a)
Integrating factor:
$$I(x) = e^{\int P(x)\,dx}$$
$$= e^{\int -3\,dx}$$
$$= e^{-3x}$$

b)
Using the integrating factor:
$$I(x)y = \int I(x)f(x)\,dx$$
$$e^{-3x}y = \int xe^{-3x}\,dx$$

Use integration by parts on the RHS $\int fg' = fg - \int f'g$

$f(x) = x$

$f'(x) = 1$

$g'(x) = e^{-3x}$

$g(x) = -\frac{1}{3}e^{-3x}$

So $e^{-3x}y = -\frac{1}{3}xe^{-3x} - \int -\frac{1}{3}e^{-3x}\,dx$

$e^{-3x}y = -\frac{1}{3}xe^{-3x} + \frac{1}{3}\left(-\frac{1}{3}e^{-3x}\right) + C$

$e^{-3x}y = -\frac{1}{3}xe^{-3x} - \frac{1}{9}e^{-3x} + C$

$y = -\frac{1}{3}x - \frac{1}{9} + Ce^{3x}$

$g(x) = -\frac{1}{3}x - \frac{1}{9}$

$h(x) = e^{3x}$

c)

Let $x = 0$, $y = 1$

$1 = -\frac{1}{3}(0) - \frac{1}{9} + Ce^{3(0)}$

$1 = -\frac{1}{9} + C$

$C = \frac{10}{9}$

Particular solution:

$y = -\frac{1}{3}x - \frac{1}{9} + \frac{10}{9}e^{3x}$

Save the fish (page 343)

Expected answer

Let S = total mass of salt in the tank and let w = flow of water in litres/minute. Then the differential equation is:

$\frac{dS}{dt} = -\frac{w}{5000}S$

In standard linear form this becomes: $\frac{dS}{dt} + \frac{w}{5000}S = 0$, with $P(t) = \frac{w}{5000}$ and $f(t) = 0$

This gives the differential equation: $\frac{d}{dt}\left(e^{-\frac{wt}{5000}}S\right) = 0$

This can be integrated and solved for S to give $S = Ce^{-\frac{wt}{5000}}$ where C is an arbitrary constant.

There is initially $S(0) = \frac{20 \times 5000}{1000} = 100$ kg of salt in the tank, therefore $C = 100$.

There has to be $S = \frac{12 \times 5000}{1000} = 60\ kg$ when the fish is put in for it to survive so the equation is $60 = 100e^{-\frac{wt}{5000}}$.

The formula for the correct answer is $w = -\frac{5000\ln(0\cdot 6)}{t}$

Therefore, for $t = 30$, $w = 85$ litres/minute.

© HERIOT-WATT UNIVERSITY

Integrating factor applications exercise (page 343)

Q76:

a) Differential equation:
$\frac{dW}{dt} - kW = 0$

b) Integrating factor:
$I(t) = e^{\int P(t)\,dt}$
$= e^{\int -k\,dt}$
$= e^{-kt}$

c) Using the integrating factor:
$$I(t)y = \int I(t)f(t)\,dt$$
$$e^{-kt}W = \int e^{-kt}(0)\,dt$$
$$e^{-kt}W = \int 0\,dx$$
$$e^{-kt}W = C$$
$$W = \frac{C}{e^{-kt}}$$
$$W = Ce^{kt}$$
$g(t) = 0$
$h(t) = e^{kt}$

d) When $t = 0, W = 2 \cdot 5$
$W = Ce^{kt}$
$2 \cdot 5 = Ce^{k(0)}$
$C = 2 \cdot 5$

When $t = 14, W = 3 \cdot 7$
$W = 2 \cdot 5 e^{kt}$
$3 \cdot 7 = 2 \cdot 5 e^{k(14)}$
$\frac{3 \cdot 7}{2 \cdot 5} = e^{14k}$
$\ln\left(\frac{3 \cdot 7}{2 \cdot 5}\right) = \ln\left(e^{14k}\right)$
$14k = \ln\left(\frac{3 \cdot 7}{2 \cdot 5}\right)$
$k = \frac{1}{14}\ln\left(\frac{3 \cdot 7}{2 \cdot 5}\right)$
$k = 0 \cdot 028$

e) Particular solution:
$W = 2 \cdot 5 e^{0 \cdot 028 t}$
Substitute for $t = 5$ days:
$W = 2 \cdot 5 e^{0 \cdot 028(5)}$
$= 2 \cdot 88$ kg

Q77:

a) Rate of change in mass $\frac{dM}{dt}$
Rate at which solution enters = 5×3 = 15 kg/min
Rate at which solution leaves = $\frac{3}{300} M = 0 \cdot 01$ kg/min
Rate of change of mass of solution $\frac{dM}{dt} = 15 - 0 \cdot 01 M$
Therefore, in standard from $\frac{dM}{dt} + 0 \cdot 01 M = 15$

b) Integrating factor:
$$I(t) = e^{\int P(t)\, dt}$$
$$= e^{\int 0 \cdot 01\, dt}$$
$$= e^{0 \cdot 01 t}$$

c) Using the integrating factor:
$$I(t) y = \int I(t) f(t)\, dt$$
$$e^{0 \cdot 01 t} M = \int 15 e^{0 \cdot 01 t}\, dt$$
$$e^{0 \cdot 01 t} M = \frac{15}{0 \cdot 01} e^{0 \cdot 01 t} + C$$
$$M = 1500 + \frac{C}{e^{0 \cdot 01 t}}$$
$$M = 1500 + C e^{-0 \cdot 01 t}$$
$$g(t) = 1500$$
$$h(t) = e^{-0 \cdot 01 t}$$

d) Let $t = 0$, $M = 100$
$$M = 1500 + C e^{-0 \cdot 01 t}$$
$$100 = 1500 + C e^{-0 \cdot 01 (0)}$$
$$100 = 1500 + C$$
$$C = -1400$$

e) $M = 1500 - 1400 e^{-0 \cdot 01 t}$
t is measured in minutes so $t = 60$
$$M = 1500 - 1400 e^{-0 \cdot 01 (60)}$$
$$= 1500 - 1400 e^{-0 \cdot 6}$$
$$= 731 \cdot 7 \text{ kg}$$

f) Given $M = 1000$
$$1000 = 1500 - 1400 e^{-0 \cdot 01 t}$$
$$-500 = -1400 e^{-0 \cdot 01 t}$$
$$\frac{-500}{-1400} = e^{-0 \cdot 01 t}$$
$$-0 \cdot 01 t = \ln\left(\frac{500}{1400}\right)$$
$$t = -\frac{1}{0 \cdot 01} \ln\left(\frac{500}{1400}\right)$$
$$t = 103 \text{ min}$$

g) As $t \to \infty$, $M = 1500 - 1400 e^{-0 \cdot 01} \to 1500 kg$ since $e^{-0 \cdot 01} \to 0$

Real and distinct roots exercise (page 349)

Q78:

a) The auxiliary equation for $\frac{d^2y}{dx^2} + \frac{dy}{dx} - 2y = 0$ is $m^2 + m - 2 = 0$
b) Factorise this to give $(m + 2)(m - 1) = 0 \Rightarrow (m + 2) = 0$ or $(m - 1) = 0$
Therefore, the roots of the auxiliary equation are $m_1 = -2$ and $m_2 = 1$
c) Hence, the general solution of the differential equation is $y(x) = Ae^{-2x} + Be^{x}$

Q79:

The auxiliary equation for $\frac{d^2y}{dx^2} + 2\frac{dy}{dx} - 15y = 0$ is $m^2 + 2m - 15 = 0$
Factorise this to give $(m - 3)(m + 5) = 0 \Rightarrow (m - 3) = 0$ or $(m + 5) = 0$
Therefore, the roots of the auxiliary equation are $m_1 = 3$ or $m_2 = -5$
Hence, the general solution of the differential equation is $y = Ae^{3x} + Be^{-5x}$

Q80:

The auxiliary equation for $2\frac{d^2y}{dx^2} - \frac{dy}{dx} - 6y = 0$ is $2m^2 - m - 6 = 0$
Factorise this to give $(2m + 3)(m - 2) = 0 \Rightarrow (2m + 3) = 0$ or $(m - 2) = 0$
Therefore, the roots of the auxiliary equation are $m_1 = -\frac{3}{2}$ or $m_2 = 2$
Hence, the general solution of the differential equation is $y = Ae^{-\frac{3}{2}x} + Be^{2x}$

Q81:

The auxiliary equation for $6\frac{d^2y}{dx^2} - \frac{dy}{dx} - 2y = 0$ is $6m^2 - m - 2 = 0$
Factorise this to give $(2m + 1)(3m - 2) = 0 \Rightarrow (2m + 1) = 0$ or $(3m - 2) = 0$
Therefore, the roots of the auxiliary equation are $m_1 = -\frac{1}{2}$ or $m_2 = \frac{2}{3}$
Hence, the general solution of the differential equation is $y = Ae^{-\frac{1}{2}x} + Be^{\frac{2}{3}x}$

Equal roots exercise (page 351)

Q82:

a) The auxiliary equation for $\frac{d^2y}{dx^2} + 4\frac{dy}{dx} + 4y = 0$ is $m^2 + 4m + 4 = 0$
b) Factorise this to give $(m + 2)(m + 2) = 0 \Rightarrow (m + 2) = 0$ (repeated root)
Therefore, the repeated root of the auxiliary equation is $m = -2$
c) Hence, the equation has independent solutions $y_1 = e^{-2x}$ and $y_2 = xe^{-2x}$ and therefore has general solution $y(x) = Ae^{-2x} + Bxe^{-2x}$
In factorised form, the general solution is: $y = (A + Bx)e^{-2x}$.

Q83:

The auxiliary equation for $\frac{d^2y}{dx^2} - 6\frac{dy}{dx} + 9y = 0$ is $m^2 - 6m + 9 = 0$
Factorise this to give $(m - 3)(m - 3) = 0 \Rightarrow (m - 3) = 0$ (repeated root)
Therefore, the repeated root of the auxiliary equation is $m = 3$

Hence, the equation has independent solutions $y_1 = e^{3x}$ and $y_2 = xe^{3x}$ and therefore has general solution $y = Ae^{3x} + Bxe^{3x}$
In factorised form, the general solution is: $y = (A + Bx)e^{3x}$.

Q84:

The auxiliary equation for $9\frac{d^2y}{dx^2} - 6\frac{dy}{dx} + y = 0$ is $9m^2 - 6m + 1 = 0$
Factorise this to give $(3m - 1)(3m - 1) = 0 \Rightarrow (3m - 1) = 0$ (repeated root)
Therefore, the repeated root of the auxiliary equation is $m = \frac{1}{3}$

Hence, the equation has independent solutions $y_1 = e^{\frac{1}{3}x}$ and $y_2 = xe^{\frac{1}{3}x}$ and therefore has general solution $y = Ae^{\frac{1}{3}x} + Bxe^{\frac{1}{3}x}$
In factorised form the general solution is: $y = (A + Bx)e^{\frac{1}{3}x}$.

Q85:

The auxiliary equation for $4\frac{d^2y}{dx^2} + 12\frac{dy}{dx} + 9y = 0$ is $4m^2 + 12m + 9 = 0$
Factorise this to give $(2m + 3)(2m + 3) = 0 \Rightarrow (2m + 3) = 0$ (repeated root)
Therefore, the repeated root of the auxiliary equation is $m = -\frac{3}{2}$

Hence, the equation has independent solutions $y_1 = e^{-\frac{3}{2}x}$ and $y_2 = xe^{-\frac{3}{2}x}$ and therefore has general solution $y = Ae^{-\frac{3}{2}x} + Bxe^{-\frac{3}{2}x}$
In factorised form the general solution is: $y = (A + Bx)e^{-\frac{3}{2}x}$

Complex roots exercise (page 355)

Q86:

a) The auxiliary equation for $\frac{d^2y}{dx^2} - 6\frac{dy}{dx} + 13y = 0$ is $m^2 - 6m + 13 = 0$
b) This equation has roots given by the following:

$$m = \frac{6 \pm \sqrt{(-6)^2 - 4(1)(13)}}{2(1)}$$

$$m = \frac{6 \pm \sqrt{-16}}{2}$$

$$m = \frac{6 \pm 4i}{2}$$

$m = 3 \pm 2i$
$m_1 = 3 + 2i$ or $m_2 = 3 - 2i$
Therefore, $p = 3$ and $q = 2$ and the differential equation has independent solutions:
$y_1 = e^{3x} \cos 2x$ and $y_2 = e^{3x} \sin 2x$

c) Hence, the general solution is $y(x) = e^{3x}(A\cos(2x) + B\sin(2x))$

Q87:

The auxiliary equation for $\frac{d^2y}{dx^2} + 4\frac{dy}{dx} + 5y = 0$ is $m^2 + 4m + 5 = 0$

This equation has roots given by the following:

$m = \dfrac{-4 \pm \sqrt{4^2 - 4(1)(5)}}{2(1)}$

$m = \dfrac{-4 \pm \sqrt{-4}}{2}$

$m = \dfrac{-4 \pm 2i}{2}$

$m = -2 \pm i$

$m_1 = -2 + i$ or $m_2 = -2 - i$

Therefore, $p = -2$ and $q = 1$ and the differential equation has independent solutions:

$y_1 = e^{-2x}\cos x$ and $y_2 = e^{-2x}\sin x$

Hence, the general solution is $y(x) = e^{-2x}(A\cos x + B\sin x)$

Q88:

The auxiliary equation for $\frac{d^2y}{dx^2} + \frac{dy}{dx} + y = 0$ is $m^2 + m + 1 = 0$

This equation has roots given by the following:

$m = \dfrac{-1 \pm \sqrt{1^2 - 4(1)(1)}}{2(1)}$

$m = \dfrac{-1 \pm \sqrt{-3}}{2}$

$m = \dfrac{-1 \pm i\sqrt{3}}{2}$

$m = -\dfrac{1}{2} \pm i\dfrac{\sqrt{3}}{2}$

$m_1 = -\frac{1}{2} + i\frac{\sqrt{3}}{2}$ or $m_2 = -\frac{1}{2} - i\frac{\sqrt{3}}{2}$

Therefore, $p = -\frac{1}{2}$ and $q = \frac{\sqrt{3}}{2}$ and the differential equation has independent solutions:

$y_1 = e^{-\frac{1}{2}x}\cos\left(\frac{\sqrt{3}}{2}x\right)$ and $y_2 = e^{-\frac{1}{2}x}\sin\left(\frac{\sqrt{3}}{2}x\right)$

Hence, the general solution is $y(x) = e^{-\frac{1}{2}x}\left(A\cos\left(\frac{\sqrt{3}}{2}x\right) + B\sin\left(\frac{\sqrt{3}}{2}x\right)\right)$

Q89:

The auxiliary equation for $\frac{d^2y}{dx^2} + 6y = 0$ is $m^2 + 0m + 6 = 0$

This equation has roots given by the following:

$m^2 = -6$

$m = \pm\sqrt{-6}$

$m = \pm i\sqrt{6}$

$m_1 = i\sqrt{6}$ or $m_2 = -i\sqrt{6}$

Therefore, $p = 0$ and $q = \sqrt{6}$ and the differential equation has independent solutions:
$y_1 = \cos(\sqrt{6}x)$ and $y_2 = \sin(\sqrt{6}x)$
Hence, the general solution is $y(x) = A\cos(\sqrt{6}x) + B\sin(\sqrt{6}x)$

Initial value problems exercise (page 359)

Q90:

a) General solution:
$$\frac{d^2y}{dx^2} = -(10)^2 x$$
so, $\frac{d^2y}{dx^2} + 100x = 0$
$$m^2 + 100 = 0$$
$$m = \pm\sqrt{-100}$$
$$m = \pm 10i$$
So $x(t) = A\cos(10t) + B\sin(10t)$

b) When $t = 0$, $x(0) = 5$ so $5 = A\cos(10(0)) + B\sin(10(0)) \Rightarrow A = 5$

c) Arbitrary constant:
$x(t) = 5\cos(10t) + B\sin(10t)$
$$\frac{dx}{dt} = -50\sin(10t) + 10B\cos(10t)$$
When $t = 0$, $\frac{dx}{dt} = -10$
$-10 = -50\sin(10(0)) + 10B\cos(10(0))$
$-10 = 10B$
$B = -1$

d) $x(t) = 5\cos(10t) - \sin(10t)$

Q91:

After obtaining the general solution and applying the initial conditions, use simultaneous equations to work out the arbitrary constants.

Auxiliary equation: $m^2 - 5m + 6 = 0$
Factorising gives: $(m - 2)(m - 3) = 0$
$(m - 2) = 0$ or $(m - 3) = 0$
$m = 2$ or $m = 3$
General solution: $y(x) = Ae^{2x} + Be^{3x}$
Derivative of the general solution: $y'(x) = 2Ae^{2x} + 3Be^{3x}$
Initial condition $y(0) = 1$:
$1 = Ae^{2(0)} + Be^{3(0)}$
$1 = A + B$ **(Equation 1)**

Initial condition $y'(0) = 1$:
$1 = 2Ae^{2(0)} + 3Be^{3(0)}$
$1 = 2A + 3B$ (Equation 2)
Simultaneous equations: -2 x Equation 1 $\Rightarrow -2 = -2A - 2B$ (Equation 3)
Equation 3 + Equation 2: $-1 = B \Rightarrow B = -1$
Substitute $B = -1$ into Equation 1: $1 = A + (-1) \Rightarrow A = 2$
Particular solution: $y(x) = 2e^{2x} - e^{3x}$

Q92:

Set up and solve the auxiliary equation, then write down the general solution which takes one of the three forms:

1. $x(t) = Ae^{m_1 t} + Be^{m_2 t}$ (two real distinct roots)
2. $x(t) = (A + Bt)e^{mt}$ (equal or repeated roots)
3. $x(t) = e^{px}(A\cos(qt) + B\sin(qt))$ (complex roots)

Apply the initial conditions at $t = 0$, $x(0) = 0$ and $\frac{dx}{dt} = -0 \cdot 6$
Use simultaneous equations to solve for the arbitrary constants A and B.
Auxiliary equation: $m^2 + 20m + 64 = 0$
$(m + 4)(m + 16) = 0$
$(m + 4) = 0$ or $(m + 16) = 0$
$m = -4$ or $m = -16$
General Solution: $x(t) = Ae^{-4t} + Be^{-16t}$
Derivative of the general solution: $x'(t) = -4Ae^{-4t} - 16Be^{-16t}$
Initial condition $x(0) = 0 \cdot 2$:
$0 \cdot 2 = Ae^{-4(0)} + Be^{-16(0)}$
$0 \cdot 2 = A + B$ (Equation 1)
Initial condition $x'(0) = -0 \cdot 6$:
$0 \cdot 6 = -4Ae^{-4(0)} - 16Be^{-16(0)}$
$0 \cdot 6 = -4A - 16B$ (Equation 2)
Using simultaneous equations:
(Equation 1) × 4: $0 \cdot 8 = 4A + 4B$ (Equation 3)
(Equation 2): $0 \cdot 6 = -4A - 16B$
(Eqn 3 + Eqn 2): $1 \cdot 4 = -12B$
$B = 0 \cdot 12$
Substitute for $B = 0 \cdot 1$ into Equation 1:
$0 \cdot 2 = A + 0 \cdot 14$
$A = 0 \cdot 08$
The particular solution is therefore: $x(t) = 0 \cdot 08e^{-4t} + 0 \cdot 12e^{-16t}$

Non-homogeneous second order linear differential equations exercise (page 362)

Q93:
Auxiliary equation: $2m^2 - m - 3 = 0$
$(2m - 3)(m + 1) = 0$
$2m - 3 = 0$ or $m + 1 = 0$
$m = \frac{3}{2}$ or $m = -1$
Complementary function: $y_c = Ae^{\frac{3}{2}x} + Be^{-x}$
Particular integral: $y_p = e^{2x}$, $\frac{dy_p}{dx} = 2e^{2x}$ and $\frac{d^2y_p}{dx^2} = 4e^{2x}$
$2(4e^{2x}) - (2e^{2x}) - 3(e^{2x}) = 8e^{2x} - 2e^{2x} - 3e^{2x} = 3e^{2x}$
General solution: $y(x) = e^{2x} + Ae^{\frac{3}{2}x} + Be^{-x}$

Q94:
Auxiliary equation: $2m^2 - 4m + 4 = 0$
$(m - 2)(m - 2) = 0$
$m - 2 = 0$ (repeated)
$m = 2$
Complementary function: $y_c = (A + Bx)e^{2x}$
Particular integral: $y_p = e^{-3x}$, $\frac{dy_p}{dx} = -3e^{-3x}$ and $\frac{d^2y_p}{dx^2} = 9e^{-3x}$
$(9e^{-3x}) - 4(-3e^{-3x}) + 4(e^{-3x}) = 9e^{-3x} + 12e^{-3x} + 4e^{-3x} = 25e^{-3x}$
General solution: $y(x) = e^{-3x} + (A + Bx)e^{2x}$

Q95:
Auxiliary equation: $m^2 + 4m + 5 = 0$
$m = \dfrac{-4 \pm \sqrt{16 - 20}}{2}$
$m = \dfrac{-4 \pm \sqrt{-4}}{2}$
$m = -2 \pm i$
$p = -2$ and $q = 1$
Complementary function: $y_c = e^{-2x}(A \cos x + B \sin x)$
Particular integral: $y_p = e^{5x}$, $\frac{dy_p}{dx} = 5e^{5x}$ and $\frac{d^2y_p}{dx^2} = 25e^{5x}$
$(25e^{5x}) + 4(5e^{5x}) + 5(e^{5x}) = 25e^{5x} + 20e^{5x} + 5e^{5x} = 50e^{5x}$
General solution: $y(x) = e^{5x} + e^{-2x}(A \cos x + B \sin x)$

Q96:
Auxiliary equation: $m^2 + m - 12 = 0$
$(m - 3)(m + 4) = 0$
$m - 3 = 0$ or $m + 4 = 0$
$m = 3$ or $m = -4$
Complementary function: $y_c = Ae^{3x} + Be^{-4x}$

Particular integral: $Y_p = e^{-x}$, $\frac{dy_p}{dx} = -e^{-x}$ and $\frac{d^2y_p}{dx^2} = e^{-x}$
$(e^{-x}) + (-e^{-x}) - 12(e^{-x}) = e^{-x} - e^{-x} - 12e^{-x} = -12e^{-x}$
General solution: $y_C = e^{-x} + Ae^{3x} + Be^{-4x}$

Particular integrals of exponential functions (page 368)

Q97:
The auxiliary equation is given by:
$$m^2 - m - 6 = 0$$
$$(m+2)(m-3) = 0$$
$(m+2) = 0$ or $(m-3) = 0$
$m = -2$ or $m = 3$
The complementary function therefore takes the form $y_C = Ae^{-2x} + Be^{3x}$.
Neither root coincides with $f(x) = 4e^{2x}$ and so the particular integral takes the form $y_P = Ce^{2x}$.

Q98:
The auxiliary equation is given by:
$$m^2 + m - 6 = 0$$
$$(m+3)(m-2) = 0$$
$(m+3) = 0$ or $(m-2) = 0$
$m = -3$ or $m = 2$
The complementary function therefore takes the form $y_C = Ae^{-3x} + Be^{2x}$.
One root, $m = 2$, coincides with $f(x) = 10e^{2x}$ and so the particular integral takes the form $y_P = Cxe^{2x}$.

Q99:
The auxiliary equation is given by:
$$m^2 + 8m + 16 = 0$$
$$(m+4)(m+4) = 0$$
$(m+4) = 0$ or $(m+4) = 0$
$m = -4$ or $m = -4$ (repeated roots)
The complementary function therefore takes the form $y_C = (A + Bx)e^{-4x}$.
Two roots, $m = -4$, coincide with $f(x) = 6e^{-4x}$ and so the particular integral takes the form $y_P = Cx^2e^{-4x}$.

Q100:
The auxiliary equation is given by:
$$2m^2 - m - 1 = 0$$
$$(2m+1)(m-1) = 0$$
$(2m+1) = 0$ or $(m-1) = 0$
$m = -\frac{1}{2}$ or $m = 1$

ANSWERS: UNIT 1 TOPIC 4   609

The complementary function therefore takes the form $y_c = Ae^{-\frac{1}{2}x} + Be^x$.
One root, $m = 1$, coincides with $f(x) = 9e^x$ and so the particular integral takes the form $y_P = Cxe^x$.

Q101:
The auxiliary equation is given by:
$$m^2 - 6m + 9 = 0$$
$$(m-3)(m-3) = 0$$
$(m-3) = 0$ or $(m-3) = 0$
$m = 3$ or $m = 3$ (repeated roots)

The complementary function therefore takes the form $y_C = (A + Bx)e^{3x}$.
Two roots, $m = 3$, coincide with $f(x) = 8e^{3x}$ and so the particular integral takes the form $y_P = Cx^2e^{3x}$.

Sum of terms (page 371)

Q102:
Auxiliary equation:
$m^2 - m - 2 = 0$
$(m-2)(m+1) = 0$
$m - 2 = 0$ or $m + 1 = 0$
$m = 2$ or $m = -1$

The root $m = 2$ coincides with $3e^{2x}$ so
$y_{p1} = Axe^{2x}$ and
$y_{p2} = Bx^2 + Cx + D$ so
$y_p = Axe^{2x} + Bx^2 + Cx + D$

Q103:
Auxiliary equation:
$m^2 - 4m + 4 = 0$
$(m-2)(m-2) = 0$
$m - 2 = 0$ (repeated)
$m = 2$

The root $m = 2$ coincides with $4e^{2x}$ and is repeated so
$y_{p1} = Ax^2e^{2x}$ and
$y_{p2} = Bx + C$ so
$y_p = Ax^2e^{2x} + Bx + C$

Q104:
Auxiliary equation:
$m^2 + m - 6 = 0$
$(m+3)(m-2) = 0$
$m + 3 = 0$ or $m - 2 = 0$
$m = -3$ or $m = 2$

© HERIOT-WATT UNIVERSITY

The root $m = -3$ coincides with $4e^{-3x}$ so
$y_{p1} = Axe^{-3x}$ and
$y_{p2} = C \sin x + D \cos x$ so
$y_p = Axe^{-3x} + C \sin x + D \cos x$

Q105:

Auxiliary equation:
$m^2 - 5m + 4 = 0$
$(m - 1)(m - 4) = 0$
$m - 1 = 0$ or $m - 4 = 0$
$m = 1$ or $m = 4$

No coincidental or repeated roots so
$y_{p1} = Ae^{-3x}$ and
$y_{p2} = C \sin(2x) + D \cos(2x)$ so
$y_p = Ae^{-3x} + C \sin(2x) + D \cos(2x)$

Finding particular integrals exercise (page 372)

Q106:

Start working out the complementary function, y_c: $\frac{d^2y}{dx^2} - 2\frac{dy}{dx} - 3y = 0$

Auxiliary equation:
$m^2 - 2m - 3 = 0$
$(m + 1)(m - 3) = 0$
$m + 1 = 0$ or $m - 3 = 0$
$m = -1$ or $m = 3$

Complimentary function: $y_c = Ae^{-x} + Be^{3x}$

Particular integral, y_p, will have the general form $y_p = C \cos x + D \sin x$

Differentiate and substitute into the LHS of the differential equation:

$y_p = C \cos x + D \sin x$
$\frac{dy_p}{dx} = -C \sin x + D \cos x$
$\frac{d^2y}{dx^2} = -C \cos x - D \sin x$

Therefore:

$$\frac{d^2y}{dx^2} - 2\frac{dy}{dx} - 3y = -20 \cos x$$
$$(-C \cos x - D \sin x) - 2(-C \sin x + D \cos x) - 3(C \cos x + D \sin x) = -20 \cos x$$
$$(-C - 2D - 3C) \cos x + (-D + 2C - 3D) \sin x = -20 \cos x$$
$$(-4C - 2D) \cos x + (2C - 4D) \sin x = -20 \cos x$$

Equate the coefficients of $\cos x$ and $\sin x$:

$\cos x$: $-4C - 2D = -20$ (Equation 1)
$\sin x$: $2C - 4D = 0$ (Equation 2)

ANSWERS: UNIT 1 TOPIC 4   611

2 × Equation 2: $4C - 8D = 0$ (Equation 3)
Equation 1 + Equation 3: $-10D = -20 \Rightarrow D = 2$
Substituting $D = 2$ into Equation 2 gives:
$2C - 4(2) = 0$
$\quad 2C = 8$
$\quad\ \ C = 4$
Therefore, the general solution to the differential equation is: $y = 4\cos x + 2\sin x + Ae^{-x} + Be^{3x}$

Q107:

Start working out the complementary function, y_c: $\frac{d^2y}{dx^2} + 4y = 0$
Auxiliary equation:
$m^2 + 4 = 0$
$m^2 = -4$
$m = \pm\sqrt{-4}$
$m = \pm 2i$
Complimentary function: $y_c = A\sin(2x) + B\cos(2x)$
Particular integral, y_p, will have the general form $y_p = Cx + D$
Differentiate and substitute into the LHS of the differential equation: $y_p = Cx + D$, $\frac{dy_p}{dx} = C$, $\frac{d^2y}{dx^2} = 0$

$$\frac{d^2y}{dx^2} + 4y = 3x$$
$$0 + 4(Cx + D) = 3x$$
$$4Cx + 4D = 3x$$

Equate coefficients of x^0 and x:
$x^0 : 4D = 0 \Rightarrow D = 0$
$x : 4C = 3 \Rightarrow C = \frac{3}{4}$
Therefore, the general solution to the differential equation is: $y = \frac{3}{4}x + A\sin(2x) + B\cos(2x)$
Using the initial condition $y = 2$ when $x = 0$:
$$2 = \frac{3}{4}(0) + A\sin(2(0)) + B\cos(2(0))$$
$2 = B$
$B = 2$
Using the initial condition $y = \frac{\pi}{4}$ when $x = \frac{\pi}{4}$:
$$\frac{\pi}{4} = \frac{3}{4}\left(\frac{\pi}{4}\right) + A\sin\left(2\left(\frac{\pi}{4}\right)\right) + B\cos\left(2\left(\frac{\pi}{4}\right)\right)$$
$$\frac{\pi}{4} = \frac{3\pi}{16} + A\sin\left(\frac{\pi}{2}\right) + B\cos\left(\frac{\pi}{2}\right)$$
$$A = \frac{\pi}{16}$$
Therefore, the particular solution is: $y = \frac{3}{4}x + \frac{\pi}{16}\sin(2x) + 2\cos(2x)$

© HERIOT-WATT UNIVERSITY

End of topic 4 test (page 385)

Q108:

$$\frac{1}{y} dy = x^2 dx$$
$$\int \frac{1}{y} dy = \int x^2 dx$$
$$\ln y = \frac{x^3}{3} + C$$
$$y = e^{\frac{x^3}{3}+C}$$
$$y = Ae^{\frac{x^3}{3}} \text{ where } A = e^C$$

Q109:

$$\frac{dy}{2y} = \frac{dx}{2x+1}$$
$$\int \frac{1}{2y} dy = \int \frac{1}{2x+1} dx$$
$$\frac{1}{2} \ln y = \frac{1}{2} \ln(2x+1) + C$$
$$\ln y = \ln(2x+1) + C$$
$$y = e^{\ln(2x+1)+C}$$
$$y = e^{\ln(2x+1)} e^C$$
$$y = A(2x+1) \text{ where } A = e^C$$

Q110:

$$y \, dy = (x+3) \, dx$$
$$\int y \, dy = \int x + 3 \, dx$$
$$\frac{y^2}{2} = \frac{x^2}{2} + 3x + C$$
$$y^2 = x^2 + 6x + C$$

Q111:

$$\frac{dy}{\sqrt{y}} = \sqrt{x} \, dx$$
$$\int \frac{1}{y^{\frac{1}{2}}} dy = \int x^{\frac{1}{2}} dx$$
$$2y^{\frac{1}{2}} = \frac{2}{3} x^{\frac{3}{2}} + C$$
$$\sqrt{y} = \frac{1}{3} x^{\frac{3}{2}} + C$$
$$y = \left(\frac{1}{3} x^{\frac{3}{2}} + C \right)^2$$

ANSWERS: UNIT 1 TOPIC 4

Q112:

$$\frac{dy}{dx} = (1-2x)(y+1)$$

$$\frac{dy}{y+1} = (1-2x)\ dx$$

$$\int \frac{1}{y+1}\ dy = \int 1 - 2x\ dx$$

$$\ln(y+1) = x - x^2 + C$$

$$y + 1 = e^{x-x^2+C}$$

$$y = e^{x-x^2} e^C - 1$$

$$y = Ae^{x-x^2} - 1 \text{ where } A = e^C$$

Q113:

$$\frac{dy}{dx}(1-x) = y^2$$

$$\frac{dy}{y^2} = \frac{1}{1-x}\ dx$$

$$\int \frac{1}{y^2}\ dy = \int \frac{1}{1-x}\ dx$$

$$-\frac{1}{y} = -\ln(1-x) + C$$

$$y = \frac{1}{\ln(1-x) + C}$$

Q114:

$$\frac{dy}{y} = \cos x\ dx$$

$$\int \frac{1}{y}\ dy = \int \cos x\ dx$$

$$\ln y = \sin x + C$$

$$y = e^{\sin x + C}$$

$$y = e^{\sin x} e^C$$

$$y = Ae^{\sin x} \text{ where } A = e^C$$

Given that $y(0) = 1$ then $1 = Ae^0 \Rightarrow A = 1$

$y = e^{\sin x}$

Q115:

$$\frac{dy}{\sqrt{1-y^2}} = dx$$

$$\int \frac{1}{\sqrt{1-y^2}}\ dy = \int 1\ dx$$

$$\sin^{-1} y = x + C$$

$$y = \sin(x + C)$$

© HERIOT-WATT UNIVERSITY

Given that $y = 0$ when $x = \frac{\pi}{6}$ then:

$$0 = \sin\left(\frac{\pi}{6} + C\right)$$
$$\frac{\pi}{6} + C = \sin^{-1}(0)$$
$$\frac{\pi}{6} + C = 0$$
$$C = -\frac{\pi}{6}$$

Therefore, $y = \sin\left(x - \frac{\pi}{6}\right)$

Q116:

$$2(x-1)\,dy = 2y(y+1)\,dx$$
$$\frac{dy}{y(y+1)} = \frac{dx}{x-1}$$
$$\int \frac{1}{y(y+1)}\,dy = \int \frac{1}{x-1}\,dx$$
$$\int \frac{1}{y} - \frac{1}{y+1}\,dy = \ln(x-1) + C$$
$$\ln y - \ln(y+1) = \ln(x-1) + C$$
$$\ln\left(\frac{y}{y+1}\right) = \ln(x-1) + C$$
$$e^{\ln\left(\frac{y}{y+1}\right)} = e^{\ln(x-1)+C}$$
$$\ldots$$
$$\ldots$$
$$\frac{y}{y+1} = A(x-1) \text{ where } A = e^C$$
$$\frac{y+1}{y} = \frac{A}{(x-1)}$$
$$\frac{1}{y} = \frac{A}{(x-1)} - 1$$
$$\frac{1}{y} = \frac{A}{(x-1)} - \frac{(x-1)}{(x-1)}$$
$$\frac{1}{y} = \frac{A - (x-1)}{x-1}$$
$$y = \frac{x-1}{A+1-x}$$

Given that $y(2) = 1$ then $1 = \frac{1}{A-1} \Rightarrow A = 2$

$y = \frac{x-1}{3-x}$

ANSWERS: UNIT 1 TOPIC 4

Q117:
Differential equation:
$$\frac{dm}{dt} = -km$$
$$\frac{dm}{m} = -k\,dt$$
$$\int \frac{1}{m}\,dm = -\int k\,dt$$
$$\ln m = -kt + C$$
$$m = e^{-kt+C}$$
$$m = Ae^{-kt} \text{ where } A = e^C$$
Given that $m = m_0$ when $t = 0$ then $m_0 = Ae^0 \Rightarrow A = m_0$
$m_t = m_0 e^{-kt}$
Given that $m_{10} = \frac{1}{2}m_0$ when $t = 10$ then $\frac{1}{2}m_0 = m_0 e^{-10k} \Rightarrow k = 0 \cdot 069$
$m_t = m_0 e^{-0 \cdot 069 t}$

Q118:
Differential equation:
$$\frac{dy}{dt} = -ky$$
$$\frac{dy}{y} = -k\,dt$$
$$\int \frac{1}{y}\,dy = -\int k\,dt$$
$$\ln y = -kt + C$$
$$y = e^{-kt+C}$$
$$y = Ae^{-kt} \text{ where } A = e^C$$
Given that $y_0 = 70$ when $t = 0$ then $70 = Ae^0 \Rightarrow A = 70$
$y_t = 70e^{-kt}$
Given that $y_{40} = 60$ when $t = 40$ then $60 = 70e^{-40k} \Rightarrow k = 0 \cdot 004$
$y_t = 70e^{-0 \cdot 004 t}$

Q119:
a) Differential equation:
$$\frac{dP}{dt} = kP$$
$$\frac{dP}{P} = k\,dt$$
$$\int \frac{1}{P}\,dP = \int k\,dt$$
$$\ln P = kt + C$$
$$P = e^{kt+C}$$
$$P = e^{kt}e^C$$
$$P = Ae^{kt} \text{ where } A = e^C$$

© HERIOT-WATT UNIVERSITY

Given that $P_0 = 0 \cdot 1$ when $t = 0$ then $0 \cdot 1 = Ae^0 \Rightarrow A = 0 \cdot 1$
$P_t = 0 \cdot 1 e^{-kt}$

b) Given that $P_{10} = 0 \cdot 4$ when $t = 10$ then $0 \cdot 4 = 0 \cdot 1 e^{10k} \Rightarrow k = 0 \cdot 139$
$P_t = 0 \cdot 1 e^{0 \cdot 139 t}$
When $t = 14$, $P_{14} = 0 \cdot 1 e^{0 \cdot 139(14)} = 0 \cdot 7\% \Rightarrow 700$ people are infected after 14 days so 300 more people will have contracted the virus after 14 days than after 10 days.

Q120:

a) Separating variables and integrating:
$$\frac{dh}{\sqrt{h}} = -2\, dt$$
$$\int h^{-\frac{1}{2}}\, dh = -\int 2\, dt$$
$$2\sqrt{h} = -2t + C$$
$$h = (C - t)^2$$

b) Given that $h = 100$ when $t = 0$ then $100 = C^2 \Rightarrow C = 10$
$h = (10 - t)^2$

c) When $h = 25$, $25 = (10 - t)^2 \Rightarrow t = 5$ minutes

Q121:

$$\frac{dT}{dt} = -k(T - T_S)$$
$$\frac{dT}{(T - T_S)} = -k\, dt$$
$$\int \frac{1}{(T - T_S)}\, dT = -\int k\, dt$$
$$\ln(T - T_S) = -kt + C$$
$$T - T_S = e^{-kt+C}$$
$$T - T_S = e^{-kt} e^C$$
$$T - T_S = Ae^{-kt} \text{ where } A = e^C$$

Given that $T_S = 17$ and $T_0 = 80$ then $80 - 17 = Ae^0 \Rightarrow A = 63$
$T - 17 = 63 e^{-kt}$
Given that $T_{20} = 60$ then $60 - 17 = 63 e^{-20k} \Rightarrow k = 0 \cdot 019$
$T - 17 = 63 e^{-0 \cdot 019 t}$
When $T = 40$ then $40 - 17 = 63 e^{-0 \cdot 019 t} \Rightarrow t = 3$ so it takes 33 minutes longer for the soup to cool from 60°C to 40°C.

Q122:

$P(x) = 5$ and $f(x) = e^{3x}$
$I(x) = e^{\int P(x)\, dx}$
$I(x) = e^{\int 5\, dx}$
$I(x) = e^{5x}$

ANSWERS: UNIT 1 TOPIC 4

Hence:
$$\frac{d}{dx}[I(x)y] = I(x)f(x)$$
$$\frac{d}{dx}\left[e^{5x}y\right] = e^{5x}e^{3x}$$
$$e^{5x}y = \int e^{8x}\,dx$$
$$y = \frac{e^{8x}}{8e^{5x}} + \frac{C}{e^{5x}}$$
$$y = \frac{1}{8}e^{3x} + Ce^{-5x}$$

Q123:
$P(x) = \frac{2}{x}$ and $f(x) = 6x^3$
$I(x) = e^{\int P(x)\,dx}$
$I(x) = e^{\int \frac{2}{x}\,dx}$
$I(x) = e^{2\ln x}$
$I(x) = x^2$

Hence:
$$\frac{d}{dx}[I(x)y] = I(x)f(x)$$
$$\frac{d}{dx}\left[x^2 y\right] = 6x^5$$
$$x^2 y = \int 6x^5\,dx$$
$$x^2 y = x^6 + C$$
$$y = x^4 + Cx^{-2}$$

Q124:
$P(x) = 5$ and $f(x) = e^{-x}$
$I(x) = e^{\int P(x)\,dx}$
$I(x) = e^{\int 5\,dx}$
$I(x) = e^{5x}$

Hence:
$$\frac{d}{dx}[I(x)y] = I(x)f(x)$$
$$\frac{d}{dx}\left[e^{5x}y\right] = e^{5x}e^{-x}$$
$$e^{5x}y = \int e^{4x}\,dx$$
$$y = \frac{e^{4x}}{4e^{5x}} + \frac{C}{e^{5x}}$$
$$y = \frac{1}{4}e^{-x} + Ce^{-5x}$$

© HERIOT-WATT UNIVERSITY

Q125:

$\frac{dy}{dx} + 2xy = e^{-x^2}\cos x$ where $P(x) = 2x$ and $f(x) = e^{-x^2}\cos x$

$\quad I(x) = e^{\int P(x)\,dx}$

$\quad I(x) = e^{\int 2x\,dx}$

$\quad I(x) = e^{x^2}$

Hence:

$\frac{d}{dx}[I(x)y] = I(x)f(x)$

$\quad \frac{d}{dx}\left[e^{x^2}y\right] = e^{x^2}e^{-x^2}\cos x$

$\quad\quad e^{x^2}y = \int \cos x\,dx$

$\quad\quad e^{x^2}y = \sin x + C$

$\quad\quad y = e^{-x^2}(\sin x + C)$

Q126:

$\frac{dy}{dx} + \frac{5y}{x} = 3x$ where $P(x) = \frac{5}{x}$ and $f(x) = 3x$

$\quad I(x) = e^{\int P(x)\,dx}$

$\quad I(x) = e^{\int \frac{5}{x}\,dx}$

$\quad I(x) = e^{5\ln x}$

$\quad I(x) = x^5$

Hence:

$\frac{d}{dx}[I(x)y] = I(x)f(x)$

$\quad \frac{d}{dx}\left[x^5 y\right] = 3x^6$

$\quad\quad x^5 y = \int 3x^6\,dx$

$\quad\quad x^5 y = \frac{3}{7}x^7 + C$

$\quad\quad y = \frac{3}{7}x^2 + Cx^{-5}$

Q127:

$\frac{dy}{dx} + \frac{y}{x} = \sin x$ where $P(x) = \frac{1}{x}$ and $f(x) = \sin x$

$\quad I(x) = e^{\int P(x)\,dx}$

$\quad I(x) = e^{\int \frac{1}{x}\,dx}$

$\quad I(x) = e^{\ln x}$

$\quad I(x) = x$

Hence:
$$\frac{d}{dx}\left[I\left(x\right)y\right] = I\left(x\right)f\left(x\right)$$
$$\frac{d}{dx}\left[xy\right] = x\sin x$$
$$xy = \int x\sin x\, dx$$
$$xy = \sin x - x\cos x + C$$
$$y = \frac{\sin x}{x} - \cos x + Cx^{-1}$$

Q128:
$P(x) = 1$ and $f(x) = \sin x$
$I\left(x\right) = e^{\int P(x)\, dx}$
$I\left(x\right) = e^{\int 1\, dx}$
$I\left(x\right) = e^{x}$
Hence:
$$\frac{d}{dx}\left[I\left(x\right)y\right] = I\left(x\right)f\left(x\right)$$
$$\frac{d}{dx}\left[e^{x}y\right] = e^{x}\sin x$$
$$e^{x}y = \int e^{x}\sin x\, dx$$

Use integration by parts on the RHS $\int fg' = fg - \int f'g$:
Taking the RHS only let:
$f = e^{x} \Rightarrow f' = e^{x}$
$g' = \sin x \Rightarrow g = -\cos x$
$$\int e^{x}\sin x\, dx = -e^{x}\cos x - \int -e^{x}\cos x\, dx$$
$$= -e^{x}\cos x + \int e^{x}\cos x\, dx$$

Apply integration by parts again on $\int e^{x}\cos x\, dx$
$f = e^{x} \Rightarrow f' = e^{x}$
$g' = \cos x \Rightarrow g = \sin x$
We therefore obtain:
$$\int e^{x}\sin x\, dx = -e^{x}\cos x + e^{x}\sin x - \int e^{x}\sin x\, dx$$
$$2\int e^{x}\sin x\, dx = -e^{x}\cos x + e^{x}\sin x + C$$
$$\int e^{x}\sin x\, dx = \frac{1}{2}e^{x}\left(\sin x - \cos x\right) + C$$
$$e^{x}y = \frac{1}{2}e^{x}\left(\sin x - \cos x\right) + C$$
$$y = \frac{1}{2}\left(\sin x - \cos x\right) + Ce^{-x}$$

© HERIOT-WATT UNIVERSITY

Q129:

$\dfrac{dy}{dx} + \dfrac{2y}{x} = \dfrac{\cos x}{x^2}$ where $P(x) = \dfrac{2}{x}$ and $f(x) = \dfrac{\cos x}{x^2}$

$I(x) = e^{\int P(x)\, dx}$

$I(x) = e^{\int \frac{2}{x}\, dx}$

$I(x) = e^{2\ln x}$

$I(x) = x^2$

Hence:

$\dfrac{d}{dx}[I(x)y] = I(x)f(x)$

$\dfrac{d}{dx}[x^2 y] = \dfrac{x^2 \cos x}{x^2}$

$x^2 y = \displaystyle\int \cos x\, dx$

$x^2 y = \sin x + C$

$y = \dfrac{\sin x + C}{x^2}$

Given that $y(\pi) = 0$, $0 = \dfrac{\sin(\pi) + C}{\pi^2} \Rightarrow C = 0$

$y = \dfrac{\sin x}{x^2}$

Q130:

$\dfrac{dy}{dx} + \dfrac{3y}{2} = \dfrac{e^{-5x}}{4}$ where $P(x) = \dfrac{3}{2}$ and $f(x) = \dfrac{e^{-5x}}{4}$

$I(x) = e^{\int P(x)\, dx}$

$I(x) = e^{\int \frac{3}{2}\, dx}$

$I(x) = e^{\frac{3}{2}x}$

Hence:

$\dfrac{d}{dx}[I(x)y] = I(x)f(x)$

$\dfrac{d}{dx}\left[e^{\frac{3}{2}x} y\right] = \dfrac{e^{\frac{3}{2}x} e^{-5x}}{4}$

$e^{\frac{3}{2}x} y = \displaystyle\int \dfrac{e^{-\frac{7}{2}x}}{4}\, dx$

$e^{\frac{3}{2}x} y = -\dfrac{1}{14} e^{-\frac{7}{2}x} + C$

$y = -\dfrac{1}{14} e^{-5x} + C e^{-\frac{3}{2}x}$

Given that $y(0) = 0$: $0 = -\dfrac{1}{14}e^0 + Ce^0 \Rightarrow C = \dfrac{1}{14}$

$y = -\dfrac{1}{14}\left(e^{-5x} - e^{-\frac{3}{2}x}\right)$

ANSWERS: UNIT 1 TOPIC 4

Q131:

$\dfrac{dy}{dx} + \dfrac{2y}{x} = 6x^3$ where $P(x) = \dfrac{2}{x}$ and $f(x) = 6x^3$

$\quad I(x) = e^{\int P(x)\, dx}$

$\quad I(x) = e^{\int \frac{2}{x}\, dx}$

$\quad I(x) = e^{2\ln x}$

$\quad I(x) = x^2$

Hence:

$\quad \dfrac{d}{dx}[I(x)\, y] = I(x)\, f(x)$

$\quad \dfrac{d}{dx}[x^2 y] = 6x^5$

$\quad x^2 y = \displaystyle\int 6x^5\, dx$

$\quad x^2 y = x^6 + C$

$\quad y = x^4 + C x^{-2}$

Given that $y(1) = 5$ then $5 = 1 + C \Rightarrow C = 4$

$y = x^4 + 4x^{-2}$

Q132:

$P(x) = \tan x$ and $f(x) = \cos^2 x$

$I(x) = e^{\int P(x)\, dx}$

$I(x) = e^{\int \tan x\, dx}$

$I(x) = e^{-\ln(\cos x)}$

$I(x) = \dfrac{1}{\cos x}$

Hence:

$\quad \dfrac{d}{dx}[I(x)\, y] = I(x)\, f(x)$

$\quad \dfrac{d}{dx}\left[\dfrac{1}{\cos x} y\right] = \dfrac{\cos^2 x}{\cos x}$

$\quad \dfrac{1}{\cos x} y = \displaystyle\int \cos x\, dx$

$\quad \dfrac{1}{\cos x} y = \sin x + C$

$\quad y = \sin x \cos x + C \cos x$

© HERIOT-WATT UNIVERSITY

Given that $x = \frac{\pi}{4}$ when $y = \frac{3}{2}$ then:

$$\frac{3}{2} = sin\left(\frac{\pi}{4}\right) cos\left(\frac{\pi}{4}\right) + C cos\left(\frac{\pi}{4}\right)$$

$$\frac{3}{2} = \frac{1}{\sqrt{2}}\frac{1}{\sqrt{2}} + C\frac{1}{\sqrt{2}}$$

$$\frac{3}{2} = \frac{1}{2} + \frac{1}{\sqrt{2}}C$$

$$1 = \frac{1}{\sqrt{2}}C$$

$$C = \sqrt{2}$$

Therefore: $y = \cos x \left(\sin x + \sqrt{2}\right)$

Q133:

$$\frac{dB}{dt} = kB$$

$$\frac{dB}{B} - kB = 0$$

$P(t) = -k$ and $f(t) = 0$

$I(t) = e^{\int P(t)\, dt}$

$I(t) = e^{\int -k\, dt}$

$I(t) = e^{-kt}$

Hence:

$$\frac{d}{dt}[I(t)B] = I(t)f(t)$$

$$\frac{d}{dt}\left[e^{-kt}B\right] = 0$$

$$e^{-kt}B = \int 0\, dt$$

$$e^{-kt}B = C$$

$$B = Ce^{kt}$$

Given that $B = 150$ when $t = 0$ then $150 = Ce^0 \Rightarrow C = 150$

$B = 150e^{kt}$

Q134:

$\frac{dG}{dt} + 0 \cdot 005 G = \frac{1}{25}$

$P(t) = 0 \cdot 005$ and $f(t) = \frac{1}{25}$

$I(t) = e^{\int P(t)\, dt}$

$I(t) = e^{\int 0 \cdot 005\, dt}$

$I(t) = e^{0 \cdot 005 t}$

ANSWERS: UNIT 1 TOPIC 4

Hence:
$$\frac{d}{dt}\left[I\left(t\right)G\right] = I\left(t\right)f\left(t\right)$$
$$\frac{d}{dt}\left[e^{0.005t}G\right] = \frac{e^{0.005t}}{25}$$
$$e^{0.005t}G = \int \frac{e^{0.005t}}{25}\,dt$$
$$e^{0.005t}G = 8e^{0.005t} + C$$
$$G = 8 + Ce^{-0.005t}$$

Given that $G = 110$ when $t = 0$ then $110 = 8 + Ce^0 \Rightarrow C = 102$
$$G = 8 + 102e^{-0.005t}$$

Q135:
$\frac{dI}{dt} + \frac{2}{5}I = \frac{V}{10}$
$P(t) = \frac{2}{5}$ and $f(t) = \frac{V}{10}$
$I(t) = e^{\int P(t)\,dt}$
$I(t) = e^{\int \frac{2}{5}\,dt}$
$I(t) = e^{\frac{2}{5}t}$

Hence:
$$\frac{d}{dt}\left[I\left(t\right)I\right] = I\left(t\right)f\left(t\right)$$
$$\frac{d}{dt}\left[e^{\frac{2}{5}t}I\right] = e^{\frac{2}{5}t}\frac{V}{10}$$
$$e^{\frac{2}{5}t}I = \int \frac{Ve^{\frac{2}{5}t}}{10}\,dt$$
$$e^{\frac{2}{5}t}I = \frac{Ve^{\frac{2}{5}t}}{4} + C$$
$$I = \frac{V}{4} + Ce^{-\frac{2}{5}t}$$

Given that $I = 0$ when $t = 0$ then $0 = \frac{V}{4} + Ce^0 \Rightarrow C = -\frac{V}{4}$
$$I = \frac{V}{4}\left(1 - e^{-\frac{2}{5}t}\right)$$

Q136:
$\frac{dM}{dt} + 0 \cdot 008M = 0$
$P(t) = 0 \cdot 008$ and $f(t) = 0$
$I(t) = e^{\int P(t)\,dt}$
$I(t) = e^{\int 0 \cdot 008\,dt}$
$I(t) = e^{0 \cdot 008t}$

Hence:
$$\frac{d}{dt}[I(t)M] = I(t)f(t)$$
$$\frac{d}{dt}[e^{0.008t}M] = 0$$
$$e^{0.008t}M = \int 0\, dt$$
$$M = Ce^{-0.008t}$$
Given that $M = 50 \times 5000 = 250000$ when $t = 0$ then $250000 = Ce^0 \Rightarrow C = 250000$
$M = 250000e^{-0.008t}$
When $M = 16 \times 5000 = 80000$, then $80000 = 250000e^{-0.008t} \Rightarrow t = 142$ minutes

Q137:
$\frac{dM}{dt} + 0.016M = 20$
$P(t) = 0.016$ and $f(t) = 20$
$I(t) = e^{\int P(t)\, dt}$
$I(t) = e^{\int 0.016\, dt}$
$I(t) = e^{0.016t}$
Hence:
$$\frac{d}{dt}[I(t)M] = I(t)f(t)$$
$$\frac{d}{dt}[e^{0.016t}M] = e^{0.016t} \times 20$$
$$e^{0.016t}M = \int 20e^{0.016t}\, dt$$
$$e^{0.016t}M = 1250e^{0.016t} + C$$
$$M = 1250 + Ce^{-0.016t}$$
Given that $M = 30$ when $t = 0$ then $30 = 1250 + Ce^0 \Rightarrow C = -1220$
$M = 1250 - 1220e^{-0.016t}$
When $t = 40$ then $M = 1250 - 1220e^{-0.016 \times 40} = 607$ kg so the concentration is 1·21 kg/m³

Q138:
Auxiliary equation: $m^2 - 6m + 8 = 0$
Factorise: $(m-2)(m-4) = 0 \Rightarrow m - 2 = 0$ or $m - 4 = 0$
Roots: $m_1 = 2$ or $m_2 = 4$
General solution: $y(x) = Ae^{2x} + Be^{4x}$
Given that $y = 9$ when $x = 0$ then $9 = A + B$ (Equation 1)
$y'(x) = 2Ae^{2x} + 4Be^{4x}$
Given that $\frac{dy}{dx} = 20$ when $x = 0$ then $20 = 2A + 4B$ (Equation 2)
Using simultaneous equations:
-2 × Equation 1 + Equation 2:
$-18 + 20 = -2A + 2A - 2B + 4B \Rightarrow B = 1 \Rightarrow A = 8$
Particular solution: $y(x) = 8e^{2x} + e^{4x}$

ANSWERS: UNIT 1 TOPIC 4 625

Q139:
Auxiliary equation: $m^2 + 4m + 8 = 0$
$$m = \frac{-4 \pm \sqrt{4^2 - 4(1)(8)}}{2(1)}$$
$$m = -2 \pm \frac{\sqrt{-16}}{2}$$
$$m = -2 \pm \sqrt{-4}$$
$$m = -2 \pm 2i$$
Roots: $m_1 = -2 + 2i$ or $m_2 = -2 - 2i$
$p = -2$ and $q = 2$
General solution: $y(x) = e^{-2x}(A\cos(2x) + B\sin(2x))$
Given that $y = 2$ when $x = 0$ then $2 = e^0(A\cos 0 + B\sin 0) \Rightarrow A = 2$
Differentiate the general solution using the product rule, $\frac{dy}{dx} = u'v + uv'$:
Let $u = e^{-2x} \Rightarrow u' = -2e^{-2x}$
Let $v = A\cos(2x) + B\sin(2x) \Rightarrow v' = -2A\sin(2x) + 2B\cos(2x)$
$y'(x) = -2e^{-2x}(A\cos(2x) + B\sin(2x)) + e^{-2x}(-2A\sin(2x) + 2B\cos(2x))$
$= -2Ae^{-2x}\cos(2x) - 2Be^{-2x}\sin(2x) - 2Ae^{-2x}\sin(2x) + 2Be^{-2}x\cos(2x)$
$y'(x) = -2Ae^{-2x}(cos(2x) + sin(2x)) + 2Be^{-2x}(cos(2x) - sin(2x))$
Given that $\frac{dy}{dx} = 10$ when $x = 0$ and $A = 2$ then:
$10 = -2(2)e^{-2(0)}(cos(2(0)) + sin(2(0))) + 2Be^{-2(0)}(cos(2(0)) - sin(2(0)))$
$10 = -4 + 2B$
$B = 7$
Particular solution: $y(x) = e^{-2x}(2\cos(2x) + 7\sin(2x))$

Q140:
Auxiliary equation: $m^2 - 4m + 4 = 0$
Factorise: $(m - 2)(m - 2) = 0 \Rightarrow m - 2 = 0$ (repeated)
Roots: $m_1 = 2$ or $m_2 = 2$
General solution: $y(x) = (A + Bx)e^{2x}$
Given that $y = 7$ when $x = 0$ then $7 = (A + B(0))e^0 \Rightarrow A = 7$
$y'(x) = 2Ae^{2x} + Be^{2x}(2x + 1)$
Given that $\frac{dy}{dx} = 5$ when $x = 0$ then $5 = 14e^0 + Be^0 \Rightarrow B = -9$
Particular solution: $y(x) = (7 - 9x)e^{2x}$

Q141:
Auxiliary equation: $m^2 - 5m + 6 = 0$
Factorise: $(m - 2)(m - 3) = 0 \Rightarrow m - 2 = 0$ or $m - 3 = 0$
Roots: $m_1 = 2$ or $m_2 = 3$
General solution: $y(x) = Ae^{2x} + Be^{3x}$
Given that $y = 1$ when $x = 0$ then $1 = A + B$ (Equation 1)
$y'(x) = 2Ae^{2x} + 3Be^{3x}$

© HERIOT-WATT UNIVERSITY

Given that $y'(0) = 1$ when $x = 0$ then $1 = 2A + 3B$ (Equation 2)
Using simultaneous equations:
-3 × Equation 1 + Equation 2
$-3 + 1 = -3A + 2A - 3B + 3B \Rightarrow A = 2 \Rightarrow B = -1$
Particular solution: $y(x) = 2e^{2x} - e^{3x}$

Q142:

Auxiliary equation: $m^2 + 2m + 1 = 0$
Factorise: $(m + 1)(m + 1) = 0 \Rightarrow m + 1 = 0$ (repeated)
Roots: $m_1 = -1$ or $m_2 = -1$
General solution: $y(x) = (A + Bx)e^{-x}$
Given that $y = 1$ when $x = 0$ then $1 = (A + B(0))e^0 \Rightarrow A = 1$
Differentiate the general solution using the product rule, $\frac{dy}{dx} = u'v + uv'$:
Let $u = A + Bx \Rightarrow u' = B$
Let $v = e^{-x} \Rightarrow v' = -e^{-x}$
$y'(x) = Be^{-x} + (A + Bx)(-e^{-x})$
$= Be^{-x} - Ae^{-x} - Bxe^{-x}$
$y'(x) = -Ae^{-x} - Be^{-x}(x - 1)$
Given that $y'(0) = 3$ when $x = 0$ then $3 = -e^0 + Be^0 \Rightarrow B = 4$
Particular solution: $y(x) = (1 + 4x)e^{-x}$

Q143:

Auxiliary equation: $m^2 - 2m + 2 = 0$

$m = \dfrac{2 \pm \sqrt{(-2)^2 - 4(1)(2)}}{2(1)}$

$m = 1 \pm \dfrac{\sqrt{-4}}{2}$

$m = 1 \pm i$

Roots: $m_1 = 1 + i$ or $m_2 = 1 - i$
$p = 1$ and $q = 1$
General solution: $y(x) = e^x(A\cos x + B\sin x)$
Given that $y(0) = 3$ when $x = 0$ then $3 = e^0(A\cos 0 + B\sin 0) \Rightarrow A = 3$
Differentiate the general solution using the product rule, $\frac{dy}{dx} = u'v + uv'$:
Let $u = e^x \Rightarrow u' = e^x$
Let $v = A\cos x + B\sin x \Rightarrow v' = -A\sin x + B\cos x$
$y'(x) = e^x(A\cos(x) + B\sin(x)) + e^x(-A\sin(x) + B\cos(x))$
$= Ae^x\cos(x) + Be^x\sin(x) - Ae^x\sin(x) + Be^x\cos(x)$
$y'(x) = Ae^x(\cos x - \sin x) + Be^x(\cos x + \sin x)$
Given that $y'(0) = 2$ when $x = 0$ then $2 = 3e^0(\cos 0 - \sin 0) + Be^0(\cos 0 + \sin 0) \Rightarrow B = -1$
Particular solution: $y(x) = e^x(3\cos x - \sin x)$

Q144:

Auxiliary equation: $m^2 + 2m + 1 = 0$
Factorise: $(m + 1)(m + 1) = 0 \Rightarrow m + 1 = 0$ (repeated)
Roots: $m_1 = -1$ or $m_2 = -1$
General solution: $y(x) = (A + Bx)e^{-x}$
Given that $y = 1$ when $x = 2$ then $1 = (A + 2B)e^{-2} = Ae^{-2} + 2Be^{-2}$ (Equation 1)
Differentiate the general solution using the product rule, $\frac{dy}{dx} = u'v + uv'$:
Let $u = A + Bx \Rightarrow u' = B$
Let $v = e^{-x} \Rightarrow v' = -e^{-x}$
$y'(x) = Be^{-x} + (A + Bx)(-e^{-x})$
$\quad = Be^{-x} - Ae^{-x} - Bxe^{-x}$
$y'(x) = -Ae^{-x} - Be^{-x}(x - 1)$
Given that $y'(2) = 2$ when $x = 2$ then $2 = -Ae^{-2} - Be^{-2}$ (Equation 2)
Using simultaneous equations:
Equation 1 + Equation 2:
$1 + 2 = Ae^{-2} - Ae^{-2} + 2Be^{-2} - Be^{-2} \Rightarrow B = 3e^2 \Rightarrow A = -5e^2$
Particular solution: $y(x) = (-5e^2 + 3e^2 x)e^{-x} = (3x - 5)e^{2-x}$

Q145:

Auxiliary equation: $9m^2 - 72m + 288 = 0$

$m = \dfrac{72 \pm \sqrt{(-72)^2 - 4(9)(288)}}{2(9)}$

$m = 4 \pm \dfrac{\sqrt{-5184}}{18}$

$m = 4 \pm \dfrac{\sqrt{-(72^2)}}{18}$

$m = 4 \pm 4i$

Roots: $m_1 = 4 + 4i$ or $m_2 = 4 - 4i$
$p = 4$ and $q = 4$
General solution: $y(x) = e^{4x}(A\cos(4x) + B\sin(4x))$
Given that $y = 3$ when $x = 0$ then $3 = e^0(A\cos 0 + B\sin 0) \Rightarrow A = 3$
Differentiate the general solution using the product rule, $\frac{dy}{dx} = u'v + uv'$:
Let $u = e^{4x} \Rightarrow u' = 4e^{4x}$
let $v = A\cos(4x) + B\sin(4x) \Rightarrow v' = -4A\sin(4x) + 4B\cos(4x)$
$y'(x) = 4e^{4x}(A\cos(4x) + B\sin(4x)) + e^{4x}(-4A\sin(4x) + 4B\cos(4x))$
$\quad = 4Ae^{4x}\cos(4x) + 4Be^{4x}\sin(4x) - 4Ae^{4x}\sin(4x) + 4Be^{4x}\cos(4x)$
$y'(x) = 4Ae^{4x}(\cos(4x) - \sin(4x)) + 4Be^{4x}(\sin(4x) + \cos(4x))$
Given that $\frac{dy}{dx} = 20$ when $x = 0$ then $20 = 12e^0(\cos 0 - \sin 0) + 4Be^0(\sin 0 + \cos 0) \Rightarrow B = 2$
Particular solution: $y(x) = e^{4x}(3\cos(4x) + 2\sin(4x))$

© HERIOT-WATT UNIVERSITY

Q146:

Auxiliary equation: $m^2 + 1 = 0$

$m = \sqrt{-1}$

$m = \pm i$

Roots: $m_1 = i$ or $m_2 = -i$

$p = 0$ and $q = 1$

General solution: $x(t) = e^{0t}(A \cos t + B \sin t) = A \cos t + B \sin t$

Given that $x(t) = 10$ when $t = 0$ then $10 = A \cos 0 + B \sin 0 \Rightarrow A = 10$

$x'(t) = -A \sin t + B \cos t$

Given that $\frac{dx}{dt} = -14$ when $t = 0$ then $-14 = -10 \sin 0 + B \cos 0 \Rightarrow B = -14$

Particular solution: $x(t) = 10 \cos t - 14 \sin t$

When $t = 4$ then $x(4) = 10 \cos 4 - 14 \sin 4 = 4 \cdot 1$ cm

Q147:

Auxiliary equation: $m^2 + 2m + 37 = 0$

$m = \dfrac{-2 \pm \sqrt{2^2 - 4(1)(37)}}{2(1)}$

$m = -1 \pm \dfrac{\sqrt{-144}}{2}$

$m = -1 \pm \dfrac{\sqrt{-(12^2)}}{2}$

$m = -1 \pm 6i$

Roots: $m_1 = -1 + 6i$ or $m_2 = -1 - 6i$

$p = -1$ and $q = 6$

General solution: $x(t) = e^{-t}(A \cos(6t) + B \sin(6t))$

Given that $x(t) = 55$ when $t = 0$ then $55 = e^0(A \cos 0 + B \sin 0) \Rightarrow A = 55$

$y'(x) = -6Ae^{-t}\sin(6t) - Ae^{-t}\cos(6t)) + 6Be^{-t}\cos(6t) - Be^{-t} \sin(6t))$

Given that $\frac{dx}{dt} = -8$ when $t = 0$ then $-8 = -330e^0 \sin 0 - 55e^0 \cos 0 + 6Be^0 \cos 0 - Be^0 \sin 0$
$\Rightarrow B = \frac{47}{6}$

Particular solution: $x(t) = e^{-t}\left(55 \cos(6t) + \frac{47}{6} \sin(6t)\right)$

When $t = 2$ then $x(2) = e^{-2}\left(55 \cos(12) + \frac{47}{6} \sin(12)\right)$ = 5·712 cm

Q148:

Auxiliary equation: $m^2 + 1000m + 200000 = 0$

Use the quadratic formula to solve for m:

$m = \dfrac{-1000 \pm \sqrt{1000^2 - 4(1)(200000)}}{2(1)}$

$m = \dfrac{-1000 \pm \sqrt{200000}}{2}$

$m = \dfrac{-1000 \pm 200\sqrt{5}}{2}$

$m = -500 \pm 100\sqrt{5}$

Roots: $m_1 = -500 + 100\sqrt{5}$ or $m_2 = -500 - 100\sqrt{5}$
General solution: $i(t) = Ae^{(-500 + 100\sqrt{5})t} + Be^{(-500 - 100\sqrt{5})t}$

Q149:

Auxiliary equation: $0 \cdot 1m^2 + 0 \cdot 2m + 1 = 0$

$$m = \frac{-0 \cdot 2 \pm \sqrt{(0 \cdot 2)^2 - 4(0 \cdot 1)(1)}}{2(0 \cdot 1)}$$

$$m = -1 \pm \frac{\sqrt{-0 \cdot 36}}{0 \cdot 2}$$

$m = -1 \pm 3i$

Roots: $m_1 = -1 + 0 \cdot 3i$ or $m_2 = -1 - 0 \cdot 3i$

$p = -1$ and $q = 3$

General solution: $\theta(t) = e^{-t}(A\cos(3t) + B\sin(3t))$

Given that $\theta(0) = 0$ when $t = 0$ then $0 = e^0(A\cos 0 + B\sin 0) \Rightarrow A = 0$

Differentiate the general solution using the product rule, $\frac{dy}{dx} = u'v + uv'$:

Let $u = e^{-t} \Rightarrow u' = -e^{-t}$

Let $v = A\cos(3t) + B\sin(3t) \Rightarrow v' = -3A\sin(3t) + 3B\cos(3t)$

$\theta'(x) = -e^{-t}(A\cos(3t) + B\sin(3t)) + e^{-t}(-3A\sin(3t) + 3B\cos(3t))$

$= -Ae^{-t}\cos(3t) - Be^{-t}\sin(3t) - 3Ae^{-t}\sin(3t) + 3Be^{-t}\cos(3t)$

$= -Ae^{-t}(\cos(3t) + \sin(3t)) + Be^{-t}(3\cos(3t) - \sin(3t))$

Given that $\theta'(0) = 0 \cdot 3$ when $t = 0$ and $A = 0$:

$0 \cdot 3 = -0e^{-0}(\cos(0) + \sin(0)) + Be^{-0}(3\cos(0) - \sin(0))$

$0 \cdot 3 = 3B$

$B = 0 \cdot 1$

$\theta'(t) = -Ae^{-t}(3\cos(3t) + \sin(3t)) + Be^{-t}(3\cos(3t) - \sin(3t))$

Given that $\theta'(0) = 0 \cdot 3$ when $t = 0$ then $0 \cdot 3 = Be^0(3\cos 0 - \sin 0) \Rightarrow B = 0 \cdot 1$

Particular solution: $\theta(t) = e^{-t}(0 \cdot 1 \sin(3t))$

Q150:

Auxiliary equation: $m^2 + 4m + 3 = 0$

Factorise: $(m+1)(m+3) = 0 \Rightarrow m + 1 = 0$ or $m + 3 = 0$

Roots: $m_1 = -1$ or $m_2 = -3$

General solution: $x(t) = Ae^{-t} + Be^{-3t}$

Given that $x = 10$ when $t = 0$ then $10 = A + B$ (Equation 1)

$x'(t) = -Ae^{-t} - 3Be^{-3t}$

Given that $x'(0) = 0$ when $t = 0$ then $0 = -A - 3B$ (Equation 2)

Using simultaneous equations:

Equation 1 + Equation 2:

$10 + 0 = A - A + B - 3B \Rightarrow B = -5 \Rightarrow A = 15$

Particular solution: $x(t) = 15e^{-t} - 5e^{-3t}$ (formula for displacement)

$x'(t) = -15e^{-t} + 15e^{-3t}$ (formula for velocity)

© HERIOT-WATT UNIVERSITY

Q151:

Auxiliary equation: $m^2 - 2m - 24 = 0$
Factorise: $(m + 4)(m - 6) = 0 \Rightarrow m + 4 = 0$ or $m - 6 = 0$
Roots: $m = -4$ or $m = 6$
Complementary function: $y_c = Ae^{-4x} + Be^{6x}$
Let $y_p = ke^{-3x}$ so $\frac{dy_p}{dx} = -3ke^{-3x}$ and $\frac{d^2y_p}{dx^2} = 9ke^{-3x}$
Hence, y_p is a particular integral provided that:
$$9ke^{-3x} - 2(-3ke^{-3x}) - 24ke^{-3x} = e^{-3x}$$
$$9ke^{-3x} + 6ke^{-3x} - 24ke^{-3x} = e^{-3x}$$
$$-9ke^{-3x} = e^{-3x}$$
$$k = -\frac{1}{9}$$

Particular solution: $y_p = -\frac{1}{9}e^{-3x}$
General solution: $y = -\frac{1}{9}e^{-3x} + Ae^{-4x} + Be^{6x}$

Q152:

Auxiliary equation: $m^2 + 2m + 17 = 0$
$$m = \frac{-2 \pm \sqrt{2^2 - 4(1)(17)}}{2(1)}$$
$$m = -1 \pm \frac{\sqrt{-64}}{2}$$
$$m = -1 \pm 4i$$
$m = -1 + 4i$ or $m = -1 - 4i$
$p = -1$ and $q = 4$
Complementary function: $y_c = e^{-x}(A\cos(4x) + B\sin(4x))$
Let $y_p = p\sin x + q\cos x$ so $\frac{dy_p}{dx} = p\cos x - q\sin x$ and $\frac{d^2y_p}{dx^2} = -p\sin x - q\cos x$
Hence, y_p is a particular integral provided that:
$$-p\sin x - q\cos x + 2(p\cos x - q\sin x) + 17(p\sin x + q\cos x) = -520\cos x$$
$$(16p - 2q)\sin x + (2p + 16q)\cos x = -520\cos x$$
Equating coefficients of $\cos x$ and $\sin x$:
$\cos x: 2p + 16q = -520$ (Equation 1)
$\sin x: 16p - 2q = 0$ (Equation 2)
Using simultaneous equations:
Equation 1 + 8 × Equation 2:
$2p + 128P + 16q - 16q = -520 \Rightarrow p = -4 \Rightarrow q = -32$
Particular solution: $y_p = -4\sin x - 32\cos x$
General solution: $y = -4\sin x - 32\cos x + e^{-x}(A\cos(4x) + B\sin(4x))$

Q153:
Auxiliary equation: $m^2 + 12m + 36 = 0$
Factorise: $(m + 6)(m + 6) = 0 \Rightarrow m + 6 = 0$ (repeated)
Roots: $m = -6$
Complementary function: $y_c = (A + Bx)e^{-6x}$
Let $y_p = ax^2 + bx + c$ so $\frac{dy_p}{dx} = 2ax + b$ and $\frac{d^2y_p}{dx^2} = 2a$
Hence, y_p is a particular integral provided that:
$$2a + 12(2ax + b) + 36(ax^2 + bx + c) = 108x^2$$
$$2a + 24ax + 12b + 36ax^2 + 36bx + 36c = 108x^2$$
$$36ax^2 + (24a + 36b)x + (2a + 12b + 36c) = 108x^2$$
Equating coefficients of x^2, x and x^0:
$x^2: 36a = 108 \Rightarrow a = 3$
$x: 24a + 36b = 0 \Rightarrow b = -2$
$x^0: 2a + 12b + 36c = 0 \Rightarrow c = \frac{1}{2}$
Particular integral: $y_p = 3x^2 - 2x + \frac{1}{2}$
General solution: $y = 3x^2 - 2x + \frac{1}{2} + Ae^{-6x} + Bxe^{-6x}$

Q154:
Auxiliary equation: $m^2 - 4m - 5 = 0$
Factorise: $(m - 5)(m + 1) = 0 \Rightarrow m - 5 = 0$ or $m + 1 = 0$
Roots: $m = 5$ or $m = -1$
Complementary function: $y_c = Ae^{5x} + Be^{-x}$
Let $y_p = kxe^{5x}$ so $\frac{dy_p}{dx} = ke^{5x}(5x + 1)$ and $\frac{d^2y_p}{dx^2} = 5ke^{5x}(5x + 1) + 5ke^{5x}$
Hence, y_p is a particular integral provided that:
$$\left(5ke^{5x}(5x+1) + 5ke^{5x}\right) - 4\left(ke^{5x}(5x+1)\right) - 5kxe^{5x} = 12e^{5x}$$
$$25kxe^{5x} + 10ke^{5x} - 20kxe^{5x} - 4ke^{5x} - 5kxe^{5x} = 12e^{5x}$$
$$6ke^{5x} = 12e^{5x}$$
$$k = 2$$
Particular solution: $y_p = 2xe^{5x}$
General solution: $y = 2xe^{5x} + Ae^{5x} + Be^{-x}$

Q155:
Auxiliary equation: $m^2 + 5m + 4 = 0$
Factorise: $(m + 1)(m + 4) = 0 \Rightarrow m + 1 = 0$ or $m + 4 = 0$
Roots: $m = -1$ or $m = -4$
Complementary function: $y_c = Ae^{-x} + Be^{-4x}$
Let $y_p = p\sin x + q\cos x$ so $\frac{dy_p}{dx} = p\cos x - q\sin x$ and $\frac{d^2y_p}{dx^2} = -p\sin x - q\cos x$

© HERIOT-WATT UNIVERSITY

Hence, y_p is a particular integral provided that:
$$-p\sin x - q\cos x + 5(p\cos x - q\sin x) + 4(p\sin x + q\cos x) = 34\sin x$$
$$(3p - 5q)\sin x + (5p + 3q)\cos x = 34\sin x$$
Equating coefficients for sin x and cos x:
sin x : $3p - 5q = 34$ (Equation 1)
cos x : $5p + 3q = 0$ (Equation 2)
Using simultaneous equations (3 × Equation 1 + 5 × Equation 2):
$9p + 25p - 15q + 15q = 102 \Rightarrow p = 3 \Rightarrow q = -5$
Particular solution: $y_p = 3\sin x - 5\cos x$
General solution: $y = 3\sin x - 5\cos x + Ae^{-x} + Be^{-4x}$

Q156:

Auxiliary equation: $m^2 - 10m + 25 = 0$
Factorise: $(m - 5)(m - 5) = 0 \Rightarrow m - 5 = 0$ (repeated)
Roots: $m = 5$
Complementary function: $y_c = (A + Bx)e^{5x}$
Let $y_p = kx^2e^{5x}$ so $\frac{dy_p}{dx} = kxe^{5x}(5x + 2)$ and $\frac{d^2y_p}{dx^2} = ke^{5x}(25x^2 + 20x + 2)$
Hence, y_p is a particular integral provided that:
$$ke^{5x}(25x^2 + 20x + 2) - 10(5kx^2e^{5x} + 2kxe^{5x}) + 25kx^2e^{5x} = e^{5x}$$
$$25kx^2e^{5x} + 20kxe^{5x} + 2ke^{5x} - 50kx^2e^{5x} - 20kxe^{5x} + 25kx^2e^{5x} = e^{5x}$$
$$2ke^{5x} = e^{5x}$$
$$k = \frac{1}{2}$$
Particular solution: $y_p = \frac{1}{2}x^2e^{5x}$
General solution: $y = \frac{1}{2}x^2e^{5x} + Ae^{5x} + Bxe^{5x}$

Q157:

The differential equation is given by $0 \cdot 005\frac{d^2q}{dt^2} + 6\frac{dq}{dt} + 1000q = 100$
First find the auxiliary equation and solve for the homogeneous solution.
Auxiliary equation: $0 \cdot 005m^2 + 6m + 1000 = 0$
Solve for m using the quadratic formula, where $a = 0 \cdot 005$, $b = 6$ and $c = 1000$.
$$m = \frac{-6 \pm \sqrt{6^2 - 4(0 \cdot 005)(1000)}}{2(0 \cdot 005)}$$
$$= \frac{-6 \pm \sqrt{16}}{0 \cdot 01}$$
$$= \frac{-6 \pm 4}{0 \cdot 01}$$
$$= -600 \pm 400$$

ANSWERS: UNIT 1 TOPIC 4

Roots: $m_1 = -1000$ or $m_2 = -200$
The complementary function is therefore $y_C = Ae^{-1000t} + Be^{-200t}$
The particular integral takes the form of a constant since the RHS of the differential equation is a constant.
so, $y_P = a$
$$\frac{dy_P}{dx} = 0$$
$$\frac{d^2y_P}{dx^2} = 0$$
Substituting into the differential equation gives:
$$0 \cdot 005\,(0) + 6\,(0) + 1000a = 100$$
$$1000a = 100$$
$$a = \frac{1}{10}$$
The general solution is therefore:
$$q(t) = y_C + y_P$$
$$q(t) = Ae^{-1000t} + Be^{-200t} + \frac{1}{10}$$

Q158:
Auxiliary equation: $m^2 + 6m + 5 = 0$
Factorise: $(m+1)(m+5) = 0 \Rightarrow m+1 = 0$ or $m+5 = 0$
Roots: $m = -1$ or $m = -5$
Complementary function: $y_c = Ae^{-x} + Be^{-5x}$
$f(x)$ is of the form xe^{-x} which coincides with the root $m_1 = -1$ from the auxiliary equation.
Therefore, let $y_p = kxe^{-x}$ so $\frac{dy_p}{dx} = -ke^{-x}(x-1)$ and $\frac{d^2y_p}{dx^2} = ke^{-x}(x-2)$
Hence, y_p is a particular integral provided that:
$$ke^{-x}(x-2) + 6\left(-ke^{-x}(x-1)\right) + 5\left(kxe^{-x}\right) = 4e^{-x}$$
$$kxe^{-x} - 2ke^{-x} - 6kxe^{-x} + 6ke^{-x} + 5kxe^{-x} = 4e^{-x}$$
$$4ke^{-x} = 4e^{-x}$$
$$k = 1$$
Particular solution: $y_p = xe^{-x}$
General solution: $y = xe^{-x} + Ae^{-x} + Be^{-5x}$
Given that $y(x) = 0$ when $x = 0$ then:
$$0 = (0)e^0 + Ae^0 + Be^0$$
$$0 = A + B \text{ (Equation 1)}$$
We need the derivative before using the second set of initial conditions:
$$y'(x) = e^{-x} - xe^{-x} - Ae^{-x} - 5Be^{-5x}$$
Given that $y'(x) = 0$ when $x = 0$ then:
$$0 = e^0 - (0)e^{-0} - Ae^{-0} - 5Be^{-5(0)}$$
$$-1 = -A - 5B \text{ (Equation 2)}$$

© HERIOT-WATT UNIVERSITY

Using simultaneous equations (Equation 1 + Equation 2:
$-1 = A - A + B - 5B \Rightarrow B = \frac{1}{4} \Rightarrow A = -\frac{1}{4}$
Solution: $y(x) = xe^{-x} - \frac{1}{4}e^{-x} + \frac{1}{4}e^{-5x}$

Q159:

Auxiliary equation: $2m^2 + m - 1 = 0$
Factorise: $(2m - 1)(m + 1) = 0 \Rightarrow 2m - 1 = 0$ or $m + 1 = 0$
Roots: $m = \frac{1}{2}$ or $m = -1$
Complementary function: $y_c = Ae^{\frac{1}{2}x} + Be^{-x}$
Let $y_p = ax + b$ so $\frac{dy_p}{dx} = a$ and $\frac{d^2y_p}{dx^2} = 0$
Hence, y_p is a particular integral provided that:
$0 + a - ax - b = x$
$-ax + (a - b) = x$
Equating coefficients for x and x^0:
$x: \; -a = 1 \Rightarrow a = -1$
$x^0: \; a - b = 0 \Rightarrow b = -1$
Particular integral: $y_p = -x - 1$
General solution: $y = -x - 1 + Ae^{\frac{1}{2}x} + Be^{-x}$
Given that $y(x) = 0$ when $x = 0$ then:
$0 = -0 - 1 + A^0 + B^0$
$A + B = 1$ (Equation 1)
Given that $y'(x) = 1$ when $x = 0$ then:
$1 = -1 + \frac{1}{2}Ae^0 - Be^0$
$\frac{1}{2}A - B = 2$ (Equation 2)
Using simultaneous equations (Equation 1 + Equation 2):
$A + \frac{1}{2}A + B - B = 1 + 2 \Rightarrow A = 2 \Rightarrow B = -1$
Solution: $y(x) = -x - 1 + 2e^{\frac{1}{2}x} - e^{-x}$

Q160:

Auxiliary equation: $m^2 - m - 6 = 0$
Factorise: $(m + 2)(m - 3) = 0 \Rightarrow m + 2 = 0$ or $m - 3 = 0$
Roots: $m = -2$ or $m = 3$
Complementary function: $y_c = Ae^{-2x} + Be^{3x}$
Let $y_p = ax^2 + bx + c$ so $\frac{dy_p}{dx} = 2ax + b$ and $\frac{d^2y_p}{dx^2} = 2a$
Hence, y_p is a particular integral provided that:
$2a - (2ax + b) - 6(ax^2 + bx + c) = 5 + 2x - 12x^2$
$2a - 2ax - b - 6ax^2 - 6bx - 6c = 5 + 2x - 12x^2$
$(2a - b - 6c) + (-2a - 6b)x - 6ax^2 = 5 + 2x - 12x^2$

ANSWERS: UNIT 1 TOPIC 4

Equating coefficients for x^2, x and x^0:
$x^2: -6a = -12 \Rightarrow a = 2$
$x: -2a - 6b = 2 \Rightarrow b = -1$
$x^0: 2a - b - 6c = 5 \Rightarrow c = 0$
Particular integral: $y_p = 2x^2 - x$
General solution: $y = 2x^2 - x + Ae^{-2x} + Be^{3x}$
Derivative of general solution: $y = 4x - 1 - 2Ae^{-2x} + 3Be^{3x}$
Given that $y(x) = 12$ when $x = 0$ then:
$12 = 0 - 0 + A^0 + B^0$
$A + B = 12$ (Equation 1)
Given that $y'(x) = 10$ when $x = 0$ then:
$10 = 0 - 1 - 2Ae^0 + 3Be^0$
$3B - 2A = 11$ (Equation 2)
Using simultaneous equations (2 × Equation 1 + Equation 2):
$2A - 2A + 2B + 3B = 24 + 11 \Rightarrow B = 7 \Rightarrow A = 5$
Solution: $y(x) = 2x^2 - x + 5e^{-2x} + 7e^{3x}$

© HERIOT-WATT UNIVERSITY